Remote Sensing, Big Data, and GeoAI

Earth Observation Series

Remote Sensing, Big Data, and GeoAI

Exploring Applications with Geospatial Insights

Edited by

Erin Bunting

Department of Geography, Environment, and Spatial Sciences, Michigan State University, East Lansing, MI, United States

Jane Southworth

Department of Geography, University of Florida, Gainesville, FL, United States

Cerian Gibbes

Department of Geography & Environmental Studies, University of Colorado Colorado Springs, Colorado Springs, CO, United States

Hannah Herrero

Department of Geography & Sustainability, University of Tennessee, Knoxville, TN, United States

Series Editor

George P. Petropoulos

ELSEVIER

Elsevier
Radarweg 29, PO Box 211, 1000 AE Amsterdam, Netherlands
125 London Wall, London EC2Y 5AS, United Kingdom
50 Hampshire Street, 5th Floor, Cambridge, MA 02139, United States

ISBN: 978-0-443-26741-3

For information on all Elsevier publications visit our website at
https://www.elsevier.com/books-and-journals

Publisher: Peter Llewellyn
Acquisitions Editor: Menka Mehta
Editorial Project Manager: Ashavari Bhattacharya
Production Project Manager: Punithavathy Govindaradjane
Cover Designer: S. Bakyalakshmi

Typeset by TNQ Tech

Dedication

Erin Bunting

This book began in the summer of 2023, when a remarkable group of women came together for fieldwork and realized through surveys, shared meals, safaris, and long meetings filled with laughter how we could make a meaningful contribution to the field. To my fellow editors—Jane Southworth, who first shaped my path in her undergraduate classroom; Cerian Gibbes; and Hannah Herrero—it has been a true privilege to collaborate with you on this project. I am deeply grateful to my parents, Clyde and Jeanne Bunting, and to my friends for their steady support and encouragement. Lastly, to my sweet pup, Phoebe, who I lost in the fall of 2025, thank you for being a steadfast companion through the long hours of work from graduate school to my years as a professor.

Jane Southworth

I am deeply grateful to my husband, Michael Kohlhaas, and to our sons, Romer and Parker, for their constant love and for sharing in many geospatial fieldwork adventures. I also thank my colleagues and students, past and present, including my three fellow editors, all former PhD students of mine at the University of Florida, whose collaboration and leadership continue to shape the future of our field. I dedicate this book to my brother, William "Bill" Southworth, whose passing on December 10, 2025, reminds me that love and the people we share life with are what truly endure.

Cerian Gibbes

My deepest gratitude goes to my siblings and mother who have supported me throughout my professional journey and life's adventures. A special thanks to all the extraordinary women, including my mother, sister, swim-sisters, Wednesday-lime, and coeditors, who walked beside me throughout my cancer treatment. You carried hope for me when I struggled to hold onto it. You refused to let me face fear alone and this made all the difference. I dedicate this work to my best friend and husband, Phillip, who shares every joy and stands with me in the darkest hours. You celebrate victories with genuine delight and when life feels heavy, you steady me. Your love is my safe harbor, your partnership, my greatest gift. This is for you.

Hannah Herrero

To the landscapes and communities that inspired this work—may our responsible use of GeoAI help us to better understand and care for the world we study and depend upon.

Contents

Contributors

Kathleen A. Alexander Department of Fish and Wildlife Conservation, Virginia Tech, Blacksburg, VA, United States; Chobe Research Institute, CARACAL, Kasane, Botswana

Ali R. Alruzuq Department of Geography and Geographic Information Systems, Imam University, Riyadh, Saudi Arabia

Assaf Anyamba Geospatial Science and Human Security Division, Oak Ridge National Laboratory, Oak Ridge, TN, United States

Kevin D. Ash Department of Geography, University of Florida, Gainesville, FL, United States

Mia M. Bennett Department of Geography, University of Washington, Seattle, WA, United States

Jason K. Blackburn Spatial Epidemiology & Ecology Research Laboratory, Department of Geography, University of Florida, Gainesville, FL, United States; Emerging Pathogens Institute, University of Florida, Gainesville, FL, United States

Rachel Bump Department of Geography & Environmental Studies, University of Colorado Colorado Springs, Colorado Springs, CO, United States

Erin Bunting Department of Geography, Environment, and Spatial Sciences, Michigan State University, East Lansing, MI, United States

Torit Chakraborty Department of Geography, University of Florida, Gainesville, FL, United States

Anusha Chaudhary Department of Geography, University of Florida, Gainesville, FL, United States

Zachary Davison Department of Geography & Environmental Studies, University of Colorado Colorado Springs, Colorado Springs, CO, United States

Thi Hong Diep Dao Department of Geography & Environmental Studies, University of Colorado Colorado Springs, Colorado Springs, CO, United States

Cerian Gibbes Department of Geography & Environmental Studies, University of Colorado Colorado Springs, Colorado Springs, CO, United States

Ted L. Hadfield Spatial Epidemiology & Ecology Research Laboratory, Department of Geography, University of Florida, Gainesville, FL, United States; Emerging Pathogens Institute, University of Florida, Gainesville, FL, United States

Gabriela Hamerlinck Department of Geography, University of Florida, Gainesville, FL, United States

Md Muyeed Hasan Department of Geography, University of Florida, Gainesville, FL, United States

Yaqian He Department of Geosciences, University of Arkansas, Fayetteville, AR, United States; Department of Geography, Indiana University Bloomington, Bloomington, IN, United States

Hannah Herrero Department of Geography & Sustainability, University of Tennessee, Knoxville, TN, United States

Xiao Huang Department of Environmental Sciences, Emory University, Atlanta, GA, United States

Stephanie A. Insalaco-Wyner Environmental Studies Program, Southwestern University, Georgetown, TX, United States

Treenate Jiranantasak Spatial Epidemiology & Ecology Research Laboratory, Department of Geography, University of Florida, Gainesville, FL, United States; Emerging Pathogens Institute, University of Florida, Gainesville, FL, United States

David Keellings Department of Geography, University of Florida, Gainesville, FL, United States

Russanne Low Institute for Global Environmental Strategies, Arlington, VA, United States

Matthew J. McCarthy Geospatial Science and Human Security Division, Oak Ridge National Laboratory, Oak Ridge, TN, United States

Morgan C. Metrailer Spatial Epidemiology & Ecology Research Laboratory, Department of Geography, University of Florida, Gainesville, FL, United States; Emerging Pathogens Institute, University of Florida, Gainesville, FL, United States

Kati Migliaccio Department of Geography, University of Florida, Gainesville, FL, United States

Jessica Mitchell Spatial Analysis Lab, University of Montana, Missoula, MT, United States

Carly S. Muir School of Life Sciences & Sustainability, Virginia Commonwealth University, Richmond, VA, United States

Peder Nelson College of Earth, Ocean, and Atmospheric Sciences, Oregon State University, Corvallis, OR, United States

Michael H. Norris Emerging Pathogens Institute, University of Florida, Gainesville, FL, United States; Pathogen Analysis and Translational Health Group, School of Life Sciences, University of Hawai'i at Mānoa, Honolulu, HI, United States

Bryn Nurczyk Spatial Epidemiology & Ecology Research Laboratory, Department of Geography, University of Florida, Gainesville, FL, United States; Emerging Pathogens Institute, University of Florida, Gainesville, FL, United States

Kyle Owens Department of Geography & Environmental Studies, University of Colorado Colorado Springs, Colorado Springs, CO, United States

Brad Peter Department of Geosciences, University of Arkansas, Fayetteville, AR, United States

Evan Pickard Department of Geography & Environmental Studies, University of Colorado Colorado Springs, Colorado Springs, CO, United States

Mashoukur Rahaman Department of Geography, University of Florida, Gainesville, FL, United States

Mostafa Rezaali Department of Geography, University of Florida, Gainesville, FL, United States

Mohammad Safaei Department of Geography, University of Florida, Gainesville, FL, United States

Katherine A. Serafin Department of Geography, University of Florida, Gainesville, FL, United States

Audrey Culver Smith Department of Geography, University of Florida, Gainesville, FL, United States

Jane Southworth Department of Geography, University of Florida, Gainesville, FL, United States

Bewuket B. Tefera Department of Geography, University of Florida, Gainesville, FL, United States

Vasily Tolmanov Department of Geography, Environment, and Spatial Sciences, Michigan State University, East Lansing, MI, United States

Heidi Tubbs Geospatial Science and Human Security Division, Oak Ridge National Laboratory, Oak Ridge, TN, United States

Faith Vire Department of Geography, University of Central Arkansas, Conway, AR, United States

Brandon Vogt Department of Geography & Environmental Studies, University of Colorado Colorado Springs, Colorado Springs, CO, United States

Suiyuan Wang Department of Geography, State University of New York at Buffalo, Buffalo, NY, United States

Dan Wanyama Department of Geography, Sustainability, Community, and Urban Studies, University of Connecticut, Storrs, CT, United States

Phillipe A. Wernette Department of Geography, Environment, and Spatial Sciences, Michigan State University, East Lansing, MI, United States

Di Yang Department of Geography, University of Florida, Gainesville, FL, United States

Fuzhen Yin Department of Geography, State University of New York at Buffalo, Buffalo, NY, United States

About the editors

Erin Bunting is an Assistant Professor at Michigan State University (MSU) in the Department of Geography, Environment, and Spatial Sciences. Her research is highly applied and centers on the development and use of geospatial technologies, with particular expertise in time-series remote sensing, climatology, geospatial modeling, and GeoAI, to address challenges related to environmental management and climate variability. Her work spans various spatial scales and includes unoccupied aerial vehicles (UAVs) to satellite data. A defining feature of her work is the integration of remote sensing into community-collaborative and community science projects, emphasizing co-produced knowledge and real-world impact. Her research experience includes extensive work across Africa, the Southeastern and Southwestern United States, and Michigan, with more limited research conducted in South America. Previously, Erin served as Director of Remote Sensing and GIS Research and Outreach Services (RS&GIS) at MSU. In this role, she led interdisciplinary research and applied collaborations, delivering innovative remote sensing and GIS-based solutions in partnership with academic, governmental, and community stakeholders.

Jane Southworth is a Professor of Geography at the University of Florida and an internationally recognized scholar in land change science, remote sensing, and GeoAI. Her work sits at the intersection of environmental science, geography, and artificial intelligence, with a strong emphasis on interdisciplinary, collaborative research teams that integrate social, physical, and data sciences to understand complex human–environment systems. Her research focuses on time-series analysis of Earth observation data, machine learning and GeoAI approaches, vegetation dynamics, and land-use and land-cover change. She is particularly known for linking remotely sensed change to climatic drivers, addressing issues of scale and scaling, and examining interactions between people, protected areas, and conservation landscapes under conditions of climate variability and change. Since joining UF in 2002, Southworth has participated in more than $18 million in funded interdisciplinary research, published two books and over 120 peer-reviewed journal articles, and advised 35 graduate students to completion, with eight additional students currently under her supervision. Mentorship and interdisciplinary training are central to her academic identity. Southworth has also played a major leadership role in shaping AI education and research at scale. She served as Chair of UF's Department of Geography for 9 years (2016–2025) and co-led two Provost-appointed, university-wide initiatives: AI Across the Curriculum, embedding AI literacy across undergraduate education, and the UF AI BlueSky

Visioning Taskforce, which developed a forward-looking strategy for AI across research, teaching, and engagement. She currently serves as Specialty Chief Editor for *Frontiers in Remote Sensing: Remote Sensing Time Series Analysis* and is a 2025 American Association of Geographers Fellow, a lifetime honor recognizing sustained contributions to the discipline. Across research, leadership, and mentorship, Southworth is known for building bridges between disciplines, technologies, and people to advance GeoAI science and education.

Cerian Gibbes is a Professor of Geography & Environmental Studies at the University of Colorado, Colorado Springs (UCCS). Her work examines the social and ecological impacts of landscape change with a particular focus on understanding land change in spaces where agriculture and conservation intermingle. Dr. Gibbes' research uses multiscalar and mixed-methods approaches that integrate remote sensing and geospatial technologies. Her work is primarily situated in southern Africa, Latin America, and the Caribbean. She is a University of Colorado Presidential Teaching Scholar (PTSP). The PTSP honors faculty throughout the University of Colorado system who excel in both scholarship and teaching. Dr. Gibbes' contributions extend beyond academia. She utilizes a highly collaborative approach in which she actively engages with local communities, government agencies, and nonprofit organizations to foster sustainable practices in environmental management.

Hannah Herrero is an Associate Professor and the Director of Sustainability in the Department of Geography & Sustainability at the University of Tennessee, Knoxville. She is the founder and principal investigator of the Drylands And Remote sensing for Conservation Laboratory (DARC Lab) at UTK.

Dr. Herrero's research examines human-environment interactions within the field of land change science using remote sensing and GeoAI technologies. Working on interdisciplinary teams, she specializes in vegetation dynamics of drylands, land-use/land-cover change, people-park relationships, vegetation-herbivore interactions, environmental health, and seagrass mapping. Using mixed methods that integrate remote sensing with stakeholder surveys, Dr. Herrero investigates environmental change and its impacts on people, with particular attention to scale across coupled human-natural systems. Her research focuses on the conservation of terrestrial and marine protected areas and savanna ecosystems, with an emphasis on Southern Africa and Florida.

Dr. Herrero is also the co-founder and co-director of the Lagoon Watermen Alliance (LWA), a Florida-based nonprofit organization that translates scientific research into community action around the Indian River Lagoon through stakeholder engagement, workshops, and advocacy initiatives.

Foreword

Remote sensing has undergone significant transformation over the past decades, driven by major advances in data availability, sensor technology, computational infrastructure, and, more recently, by the integration of artificial intelligence into geospatial analysis workflows. One of the most notable developments has been the rapid expansion of open-access satellite data. Programs such as the Landsat program and the European Union's Copernicus program have made high-quality Earth observation data freely available to users worldwide, fundamentally changing how researchers, practitioners, and decision-makers access and use geospatial information.

The launches of Landsat 8 in 2013 and Landsat 9 in 2021 have ensured the continuity of one of the longest-running Earth observation records, providing consistent, medium-resolution imagery for monitoring land surface change since Landsat 5 was launched in 1984. Similarly, the Copernicus program has introduced a new generation of sensors, including Sentinel-1, launched in 2014, and Sentinel-2, launched in 2015. Together, these missions have greatly enhanced our ability to observe the Earth with improved spatial, spectral, and temporal resolution.

These advances have led to an unprecedented increase in data volume. Today, remote sensing archives contain petabytes of imagery, creating both opportunities and challenges for data storage, processing, and analysis. In parallel, the rise of commercial satellite providers, such as Maxar and Planet Labs, has further expanded the availability of high-resolution imagery, offering near-daily global coverage and enabling new applications across science, industry, and the government.

Equally important has been the development of cloud-based geospatial platforms that allow users to analyze large datasets without the need for local storage or high-performance computing infrastructure. The launch of Google Earth Engine in 2010 marked a turning point in how remote sensing data are accessed and processed, providing a scalable environment for planetary-scale analysis. More recently, the Microsoft Planetary Computer, introduced in 2021, has further advanced this paradigm by integrating open data catalogs with cloud-native tools for analysis and machine learning.

This book comes at a particularly important moment. GeoAI is emerging as a powerful framework that brings together remote sensing, machine learning, and big data analytics. Rather than treating these areas as separate domains, the chapters in this volume show how they can be integrated to address spatial and temporal problems with greater depth, speed, and precision. The book also makes clear that these advances build on a long history of innovation in Earth observation, data infrastructure, and geospatial analysis.

One of the strengths of this volume is its balance between conceptual foundations and practical applications. The early chapters introduce the evolution of GeoAI and the technological developments that have made it possible. They then move into the critical steps of data acquisition, preprocessing, management, and integration, which remain at the heart of reliable remote sensing workflows. From there, the book examines a broad range of analytical approaches, including time-series analysis, vegetation monitoring, spatial statistics, machine learning, and deep learning for land cover and land use mapping.

This book also demonstrates the real-world value of these methods through a wide range of applications. The readers will encounter examples related to climate and human health, coastal systems, conservation, disaster management, and medical geography. These case studies show that GeoAI is not simply about improving computational efficiency; it is about strengthening our ability to understand complex environmental and social processes and support more informed decision-making.

Another important contribution of this book is its attention to the broader context in which GeoAI is being developed and applied. Ethical and legal issues, including privacy, fairness, and governance, are increasingly central to geospatial practice. This volume also highlights the growing role of citizen science, emerging technologies, and educational innovation in shaping the future of the field. As remote sensing and AI become more closely intertwined, there is a clear need to prepare students and professionals who can work across disciplinary boundaries while remaining attentive to both technical rigor and social responsibility.

Taken together, the chapters in this book offer a timely and valuable perspective on the current state and future direction of GeoAI in remote sensing. They reflect a field that is advancing rapidly, but also one that is becoming more mature in its methods, applications, and responsibilities. For students, researchers, and practitioners alike, this volume provides both a solid foundation and a forward-looking guide.

I am pleased to introduce this book to the readers. It offers a valuable resource for understanding how artificial intelligence is transforming remote sensing and geospatial data science. I hope that it will inspire readers to think critically, explore new ideas, and collaborate across disciplines as we continue working to better observe, understand, and manage our changing planet.

Qiusheng Wu, PhD
University of Tennessee, Knoxville

The Evolution of Big Data and GeoAI Within the Field of Remote Sensing

1

Erin Bunting[1], *Jane Southworth*[2], *Hannah Herrero*[3], *Cerian Gibbes*[4], *Stephanie A. Insalaco-Wyner*[5] *and Mashoukur Rahaman*[2]

[1]Department of Geography, Environment, and Spatial Sciences, Michigan State University, East Lansing, MI, United States; [2]Department of Geography, University of Florida, Gainesville, FL, United States; [3]Department of Geography & Sustainability, University of Tennessee, Knoxville, TN, United States; [4]Department of Geography & Environmental Studies, University of Colorado Colorado Springs, Colorado Springs, CO, United States; [5]Environmental Studies Program, Southwestern University, Georgetown, TX, United States

1. Introduction: From remote sensing to GeoAI systems

GeoAI represents the intersection of geospatial data and artificial intelligence; however, its current state cannot be understood without recognizing the foundational role of remote sensing within that intersection. The contemporary GeoAI landscape reflects more than the application of machine learning to spatial datasets; it embodies the convergence of large-scale Earth observation systems, expanding computational infrastructures, and advanced analytical models capable of learning from spatial and temporal complexity (Fig. 1.1). With the proliferation of high-resolution satellite imagery, aerial surveys, drone platforms, and continuously georeferenced data streams, artificial intelligence techniques, particularly machine learning and deep learning, are being deployed to extract patterns and insights from vast and complex spatial archives. GeoAI now underpins advances in urban planning, environmental monitoring, disaster response, precision agriculture, and resource management, offering enhanced capabilities for automating tasks, generating predictions, and informing spatial decision-making processes.

Remote sensing provides the structural backbone of this transformation. Advances in sensor technology, increased spatial and temporal resolution, and the emergence of dense time-series data have reshaped how environmental and social processes are observed. Simultaneously, cloud computing environments, distributed storage systems, and scalable analytical frameworks have enabled the processing of petabyte-scale geospatial archives. The integration of spatial data analysis, machine learning, and geospatial technologies has produced sophisticated algorithmic systems capable of interpreting heterogeneous and multi-sensor datasets. This integration reflects not simply technical progress, but a reconfiguration of the geospatial analytical ecosystem itself.

The institutionalization of GeoAI within academic literature mirrors this structural shift. An analysis of Web of Science citation data for peer-reviewed articles containing

Remote Sensing, Big Data, and GeoAI. https://doi.org/10.1016/B978-0-443-26741-3.01001-1

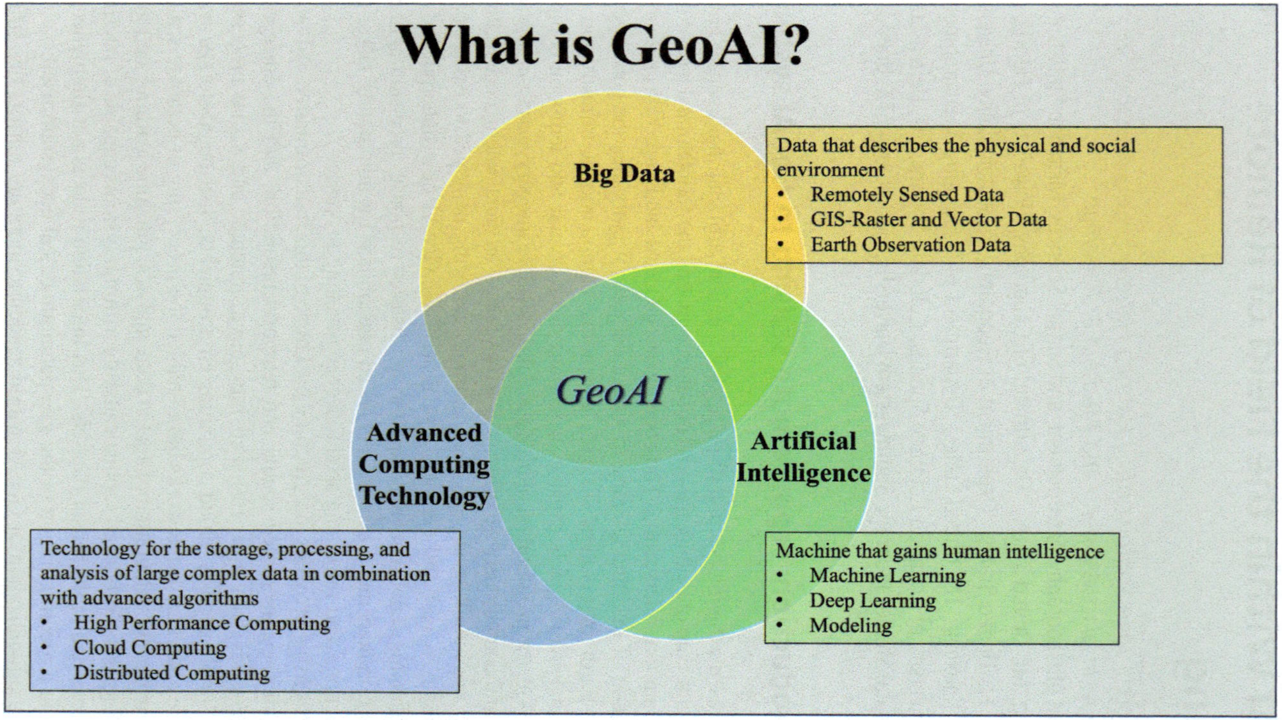

Figure 1.1 Defining GeoAI within the broader geographical context including how it fits in relation to big data, computing technology, and artificial intelligence.

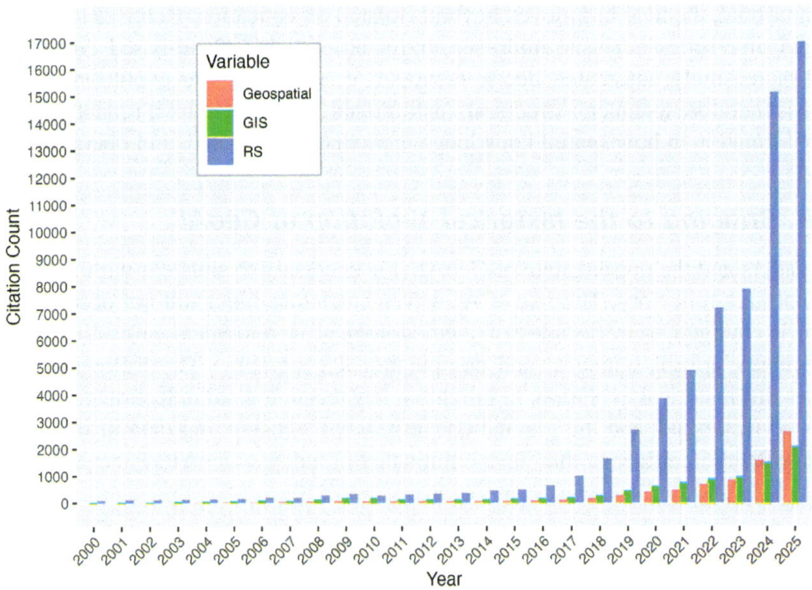

Figure 1.2 Citation trends related to GeoAI as it pertains to geospatial applications, particularly remote sensing and GIS.

the terms GeoAI, artificial intelligence, machine learning, neural network, or deep learning between 2000 and 2023 demonstrates a dramatic acceleration in scholarly output. Remote sensing shows a particularly pronounced inflection beginning in the mid-2010s, peaking in 2022 with 7886 citations. GIScience exhibits a similar upward trajectory beginning around 2019, though at a more moderate scale. These patterns, illustrated in Fig. 1.2, do more than reflect growing interest; they signal a paradigmatic transition in how geospatial inquiry is conceptualized and conducted. Remote sensing, long central to environmental observation, has become deeply intertwined with AI-driven modeling frameworks, while GIScience has increasingly incorporated predictive, automated, and data-intensive methodologies. The citation surge therefore represents not simply adoption, but disciplinary restructuring toward GeoAI systems that integrate sensing, computation, and inference at scale.

GeoAI is now firmly embedded within the academic and applied geospatial landscape. Its applications range from predicting spatial patterns and modeling environmental change to optimizing resource allocation and supporting rapid disaster response. Ongoing developments in satellite constellations, sensor fusion, and AI-driven analytical tools continue to expand our capacity to derive actionable insights from geospatial data. As such, GeoAI has emerged as a pivotal force in addressing complex global challenges at the intersection of geography, computation, and decision-making. At the same time, the rapid evolution of the field requires continuous reassessment of methodological assumptions, computational infrastructures, and epistemological implications.

This chapter positions remote sensing not merely as a data source, but as the structural backbone of contemporary GeoAI systems. As the subsequent chapters demonstrate, the field has moved beyond isolated classification exercises toward integrated time-series analysis, multi-sensor fusion, scalable cloud-based computation, and increasingly automated decision-support ecosystems. The evolution described here is therefore technical, structural, institutional, and epistemic; it reflects a broader transformation in how spatial knowledge is produced, validated, and operationalized.

1.1 Bridging to the historical evolution of GeoAI

To understand the present state of GeoAI, it is necessary to situate it within its longer disciplinary trajectory. The rise of AI-driven spatial analytics did not occur in isolation; it emerged from decades of methodological development in remote sensing, GIScience, spatial statistics, and computational geography. The following section traces this historical evolution, highlighting key technological inflection points, shifts in analytical scale, and the gradual convergence of sensing, big data infrastructures, and artificial intelligence that have culminated in today's GeoAI systems.

1.2 Purpose and structure of this volume

This volume is structured to bridge foundational theory, methodological evolution, and applied practice. The chapters progress from historical and computational foundations to methodological inflection points in remote sensing and GeoAI, and then into domain-specific applications including environmental monitoring, climate analytics, conservation, agriculture, disaster response, urban systems, and governance. Each chapter integrates conceptual framing with empirical demonstration, ensuring that theoretical developments are consistently grounded in real-world problem contexts. The volume is written for researchers, advanced students, practitioners, and policy-oriented analysts seeking to understand both the technical and institutional dimensions of GeoAI.

A defining feature of this volume is the inclusion of two to three structured case studies within each chapter. These case studies are written in plain language and are designed to demonstrate how the methodological and conceptual material discussed in each chapter manifests in real-world research and decision contexts. Rather than functioning as technical appendices, they serve as interpretive bridges between theory and application. Each case study foregrounds a concrete research problem, the data and methods employed, and the broader implications for environmental management, urban systems, conservation, agriculture, or governance. In doing so, they reinforce the book's core objective: to move beyond abstract discussion of GeoAI and big data toward tangible examples of how these systems operate in practice.

2. Historical foundations: Sensor, archive, and computational expansion

The emergence of GeoAI did not occur in isolation; it is the product of intertwined developments in sensing technologies, data accumulation, and computational capacity

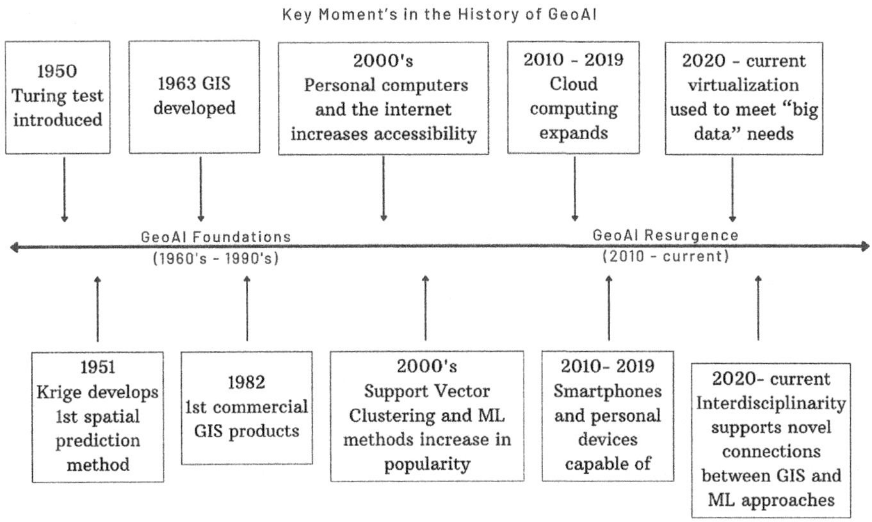

Figure 1.3 Key highlights in GeoAI history in terms of original establishment and more recent resurgence.

(Fig. 1.3). The convergence of computing enhancements, statistical innovation, and increasing availability of spatial data began to reshape geographic research in the mid-twentieth century. Advancements linked to decision theory, robustness in statistical methods, and operating systems capable of managing large resources led to a sustained intertwinement of computer science, statistics, and geography. This relationship has waxed and waned in prominence, at times celebrated and at other times subject to critique, particularly regarding its theoretical grounding and epistemological implications (Barnes, 2004; Dobson, 1983; Harvey, 1972; Smith, 1984).

Yet despite debates, three parallel trajectories steadily expanded the analytical horizon of geography: the proliferation of sensors, the growth of planetary-scale data archives, and the transformation of computational infrastructure. Together, these trajectories created the conditions under which GeoAI would move from experimental technique to structural paradigm.

2.1 Sensor proliferation and spectral innovation

The development of earth observing satellite systems in the 1960s and 1970s marked a foundational shift in geographic inquiry. Early multispectral systems enabled systematic observation of land surface processes; subsequent spectral expansion, radar technologies, and LiDAR systems extended observational capacity beyond visible wavelengths and into structural and temporal dimensions of the landscape. Remote sensing solidified quantitative analysis as central to the study of Earth's processes, embedding geographic inquiry within technologically mediated observation.

During the 1980s and 1990s, the increasing accessibility of personal computers combined with advancements in sampling methods and spatial analysis to reinforce what Openshaw (1994, p. 500) described as "a totally computer-dependent paradigm in geography." Geographic information systems became fundamental to spatial analysis, while computational techniques increasingly mediated interpretation of remotely sensed imagery (Fig. 1.3).

At the same time, developments in machine learning and neural networks were progressing within computer science. Early applications of machine learning for classifying satellite imagery and detecting spatial patterns represent some of the first integrations of geographic data and artificial intelligence. Although these methods were limited by computational constraints, they foreshadowed the analytic transformations that would later define GeoAI.

2.2 Archive growth and the emergence of big data

If sensor proliferation expanded observational capacity, archive growth expanded temporal and spatial scope. Continuity in satellite missions, including Landsat and later the Sentinel constellation, combined with open data policies to generate unprecedented archives of consistent, long-term earth observation data. By the early 2000s, global time series datasets from instruments such as MODIS and AVHRR were widely available at monthly and even finer temporal resolutions.

However, analytical capability initially lagged behind data availability. Long-term time series datasets were accessible across large regions, yet statistical methods capable of disentangling seasonal dynamics, drivers of change, and cross-scale interactions were still developing. As a result, early analyses often provided only partial understandings of complex environmental systems.

This imbalance between data abundance and analytical capacity is a recurring theme in GeoAI's history. While spatially explicit big data was available, tools for extracting insight from it were limited. Over time, advances in machine learning and deep learning began to address these limitations, enabling the modeling of nonlinear interactions and high-dimensional relationships driving landscape patterns, environmental health, and land use change.

Archive growth also reshaped institutional practices. Open data initiatives and the normalization of large-scale geospatial repositories embedded remote sensing within global research infrastructures, transforming it from specialized dataset to shared analytical backbone.

2.3 Computational transformation

Parallel to sensor and archive expansion, computational capacity underwent profound transformation (Fig. 1.3). Early desktop GIS environments enabled spatial modeling and visualization but were constrained by local processing power and storage. As datasets expanded in scale and complexity, distributed computing architectures and high-performance computing environments became increasingly important.

The release of cloud-based platforms such as Google Earth Engine in 2008 marked a turning point, enabling planetary-scale environmental data analysis without local data management burdens. Cloud platforms, combined with advances in GPU acceleration and scalable machine learning frameworks, shifted geospatial analysis from locally constrained workflows to globally distributed infrastructures.

These computational transformations also reframed theoretical debates. Earlier concerns regarding the theoretical foundations of computation in geography, including discussions of logic and set theory (Couclelis, 1986), were complemented by renewed attention to algorithmic transparency, data ethics, and interdisciplinary collaboration. Workshops and symposia focused on GeoAI began emerging within professional communities (AAG Annual Meeting, 2018; Lavallin & Downs, 2021; Mao et al., 2017), signaling institutional recognition of the field's maturation. Publication trends reflect this shift, with substantial increases in GeoAI-related scholarship over the past decade.

The excitement surrounding GeoAI stems from its capacity to extend artificial intelligence into explicitly spatial contexts. AI, often defined as the use of machine simulations to replicate or approximate aspects of human intelligence (Janowicz et al., 2020), has influenced numerous domains beyond geography (Chui et al., 2018). Its integration with remote sensing and geospatial infrastructures expands geography's long-standing concern with human-environment interactions into algorithmically mediated systems of observation and inference.

These parallel evolutions in sensor design, data availability, and computational infrastructure created the conditions under which GeoAI could emerge not as an incremental methodological refinement, but as a systemic transformation in how geographic knowledge is produced, scaled, and operationalized (Fig. 1.4).

3. Methodological inflection points in remote sensing

The evolution of GeoAI cannot be understood without examining the methodological transitions within remote sensing itself. While big data and cloud infrastructure reshaped scale and access, the deeper transformation occurred in how remotely sensed data were analyzed. The field has progressed through identifiable inflection phases, each marked by shifts in analytical philosophy, computational capacity, and epistemological assumptions about how patterns should be derived from spatial data (Table 1.1).

3.1 Classical remote sensing

Classical remote sensing was grounded in per-scene analysis, spectral separability, and analyst-guided classification. Supervised and unsupervised classification approaches relied on training data selected by experts, with statistical decision rules applied to multispectral imagery. Spectral indices such as NDVI became foundational tools for monitoring vegetation dynamics and environmental change. These approaches were computationally tractable within desktop environments and were well suited to discrete imagery products (Table 1.1).

During this period, geographic analysis remained closely tied to statistical robustness and theoretical debates about quantitative geography, as discussed by Barnes (2004), Dobson (1983), Harvey (1972) and Couclelis (1986) emphasized that much of the computational logic underlying these approaches was rooted in set theory and formal logic, even if practitioners did not always explicitly engage with those theoretical foundations. The result was a tightly coupled relationship between GIS, quantitative methods, and early remote sensing classification.

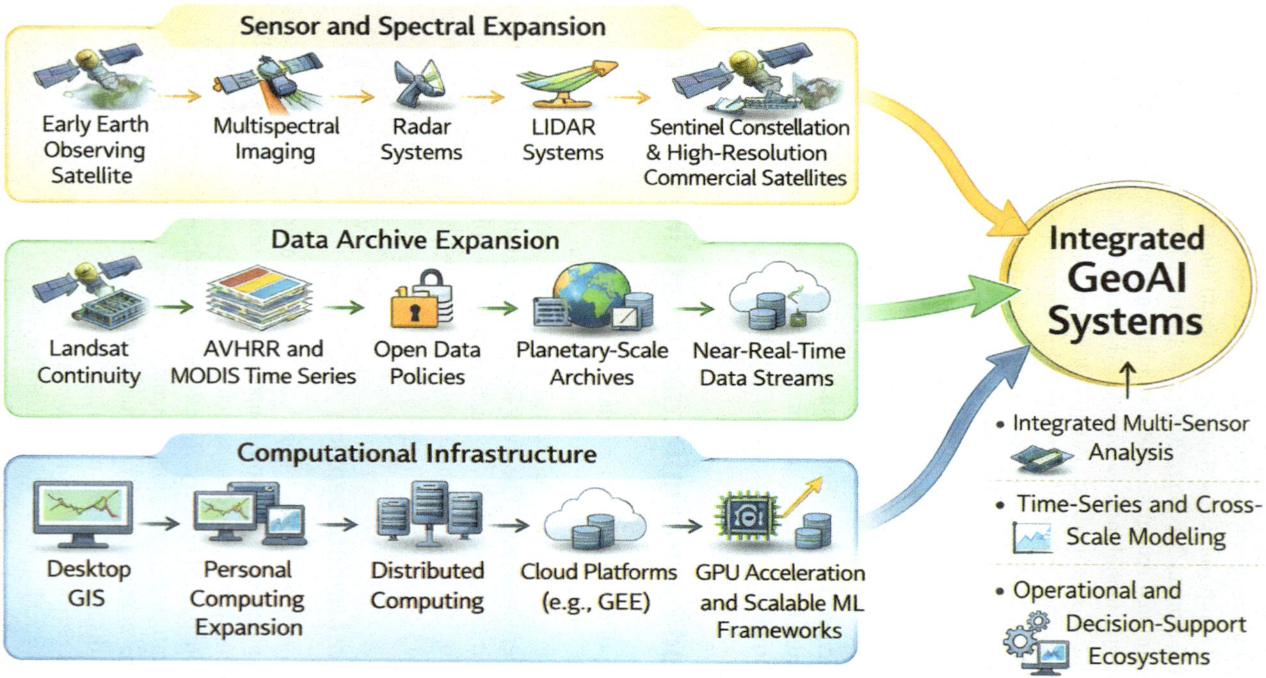

Figure 1.4 Parallel historical trajectories in sensor innovation, archive expansion, and computational infrastructure that collectively enabled the emergence of GeoAI as a systems-level transformation rather than a discrete methodological shift.

Table 1.1 Book chapters by methodological inflection phase.

Methodological phase	Defining characteristics	Primary book chapters	Secondary alignment
3.1 Classical remote sensing	Spectral indices; per-scene classification; physically based interpretation; statistical modeling; single-sensor emphasis	Chapter 2—Foundations of remote sensing and earth observation	Chapter 4—(Index theory roots) Chapter 5—(As baseline comparison)
3.2 Statistical machine learning	Feature engineering; RF, SVM, ensemble classifiers; structured evaluation; CPU-based workflows	Chapter 5—ML versus DL in land use and land cover classification	Chapter 4—(Temporal ML integration) Chapter 8—(Random forest in citizen science integration)
3.3 Deep learning and representation learning	CNNs; segmentation networks; RNNs; abstraction of hierarchical features; GPU/cloud acceleration	Chapter 5—Deep learning architectures in classification	Chapter 8—(Deep convolutional models in citizen-science fusion) Chapter 3—(AI-driven preprocessing automation)
3.4 Time-series and big data integration	Dense multi-temporal stacks; NDVI trajectories; spatial autocorrelation; cloud-based analytics; cross-sensor harmonization	Chapter 4—Big data analytics for geospatial applications	Chapter 3—(Cloud-native infrastructure) Chapter 2—(Archive expansion foundations)
3.5 Toward foundation and integrated socio-technical models	Multi-modal fusion; participatory sensing; governance integration; infrastructure awareness; transfer learning	Chapter 14—Citizen science and hybrid GeoAI systems	Chapter 13—(Policy-integrated GeoAI systems) Chapter 15—(AI literacy as infrastructure)

However, analyses were often scene-based, temporally discrete, and limited in their ability to model complex nonlinear relationships across large archives.

3.2 Statistical machine learning

The emergence of statistical machine learning marked the first major methodological expansion beyond classical remote sensing workflows. Algorithms such as random forests, support vector machines, K-nearest neighbors, and ensemble methods enabled more flexible classification across high-dimensional feature spaces (Table 1.1). These approaches reduced reliance on strict parametric assumptions and allowed more robust handling of mixed pixels, heterogeneous landscapes, and nonlinear relationships.

Machine learning demonstrated strong performance in supervised classification of optical satellite imagery, including medium and low-resolution data, as shown by Maxwell et al. (2018) and Aziz et al. (2024). Feature engineering became central; analysts constructed texture metrics, topographic variables, and multi-temporal composites to improve model performance. Incorporating semantic segmentation label uncertainties further improved classification accuracy, as discussed by Alemohammad (2023).

Despite these advances, machine learning remained partially analyst dependent. Features were still designed by experts, temporal aggregation was often predefined, and many workflows remained constrained by desktop computational limits. At the same time, large remote sensing archives such as MODIS and AVHRR provided extensive time series datasets, yet statistical tools lagged the data availability. As noted earlier, long-term monthly datasets were widely accessible, but analytical approaches capable of disentangling drivers of seasonal and interannual variability were still developing. This phase represented a clear methodological expansion, but not yet a structural transformation.

3.3 Deep learning and representation learning

The deep learning era marked a shift from feature engineering to feature learning. Convolutional neural networks, recurrent neural networks, and temporal architectures enabled hierarchical abstraction directly from raw or minimally processed imagery. Rather than specifying vegetation indices or handcrafted metrics, models learned internal representations optimized for prediction (Table 1.1).

Deep learning significantly enhanced capabilities for semantic segmentation, object detection, and spatio-temporal modeling. Temporal architectures improved analysis of land use change, environmental degradation, and dynamic processes. Neural networks trained with advanced word embeddings also demonstrated value in integrating social media streams into disaster response analytics, as discussed by Wang et al. (2023). Similarly, GeoAI methods have been used to predict pollutant concentrations and environmental performance metrics through artificial neural networks and hybrid approaches, as described by Yadav et al. (2024); Chen et al. (2025); Zhang et al. (2022).

Deep learning expanded beyond pixel-level classification toward dynamic modeling of processes. It supported predictive analytics in climate modeling, agriculture, public health, and urban systems. However, this phase also introduced new dependencies on GPU acceleration, large, labeled training datasets, and cloud infrastructure. The interpretability of models became more complex, and the distance between analyst and algorithm widened.

3.4 Toward foundation and integrated models

The most recent phase extends beyond individual task-specific models toward integrated, multi-modal, and cross-sensor architectures. Transfer learning enables models trained in one region or sensor domain to adapt to others. Multi-sensor integration combines optical, radar, LiDAR, and IoT data streams. Cloud-based platforms such as Google Earth Engine enable planetary-scale model deployment and iterative refinement (Table 1.1). GeoAI is increasingly embedded within decision-support ecosystems. Real-time monitoring systems ingest satellite data, IoT streams, and geospatial web applications to inform disaster management, urban mobility planning, air quality forecasting, and agricultural decision-making. Integration across domains has become a defining characteristic; environmental data are linked with public health metrics, urban infrastructure models, and socioeconomic indicators. For example, GeoAI has been used to map vector-born disease risk using satellite remote sensing, as shown by Abbasi (2025), and to monitor air quality through IoT integration, as described by Chinnaswamy and Walton (2022). This stage reflects movement toward scalable, transferable, and interoperable systems. It also introduces deeper governance questions, including data quality, interpretability, environmental impacts of high-performance computation, and ethical deployment, as emphasized by Janowicz et al. (2020) and Arashpour et al. (2023).

The transition from handcrafted features to representation learning marked a shift from analyst-guided classification toward data-driven abstraction. However, as later chapters demonstrate, automation introduces new tensions around interpretability, governance, and sustainability. Taken together, the chapters in this volume trace a clear methodological progression rather than presenting isolated case studies (Table 1.1). Early chapters establish the sensor, archive, and computational foundations upon which contemporary GeoAI systems depend. Subsequent contributions demonstrate the shift from classical spectral interpretation to statistical machine learning and deep representation learning, while later chapters move beyond algorithmic performance toward integrated, participatory, and governance-aware systems. The organization of the book therefore mirrors the maturation of the field itself; remote sensing evolves from image classification toward multi-sensor, time-series, and socio-technical infrastructures that increasingly support decision-making. This structural alignment underscores a central argument of the volume: GeoAI represents a reorganization of the geospatial analytical ecosystem itself—reshaping how knowledge is generated, scaled, validated, and operationalized.

4. Structural transformations that define the GeoAI era

The preceding sections traced the historical and methodological evolution of remote sensing and GeoAI. When examined across the chapters in this volume, a deeper pattern emerges. The development of GeoAI is not simply an algorithmic story; it is a structural transformation unfolding across four interrelated dimensions: data complexity, analytical ambition, computational infrastructure, and epistemic orientation. These shifts mark the transition from image-based mapping toward integrated, predictive, and decision-embedded geospatial systems.

4.1 Data complexity shift

Early remote sensing workflows were constrained by single-sensor imagery and per-scene analysis. Landsat scenes, AVHRR composites, or early radar acquisitions were often analyzed independently, with limited integration across sensors or time. Across this volume, the chapters demonstrate the acceleration of this shift. Multi-sensor integration now includes optical, SAR, LiDAR, UAV, and crowdsourced data. Dense time-series stacks enable trajectory modeling rather than static comparison. Cross-scale integration links pixel-level observations to regional processes and socio-ecological indicators. Data abundance is no longer exceptional; it is the baseline condition. This transformation can be summarized in Fig. 1.5 below, and it is important to note that the challenge has now shifted from acquisition to harmonization, fusion, and interpretability.

4.2 Analytical shift

Parallel to expanding data complexity is a redefinition of analytical purpose. Classical remote sensing emphasized descriptive mapping and classification accuracy. Analytical outputs were categorical and scene bound. The chapters in this volume collectively illustrate a different trajectory. Time-series modeling supports dynamic monitoring. Machine learning and deep architectures enable predictive modeling of hazards, environmental stress, and urban exposure. Hybrid systems embed outputs into operational planning and resource allocation. The analytical shift follows a recognizable progression as shown in Fig. 1.6. Analysis is no longer an end in itself; it is increasingly embedded in allocation, adaptation, and anticipatory systems.

4.3 Infrastructure shift

These analytical and data transformations are inseparable from infrastructural change. Desktop GIS environments once defined the limits of geospatial computation.

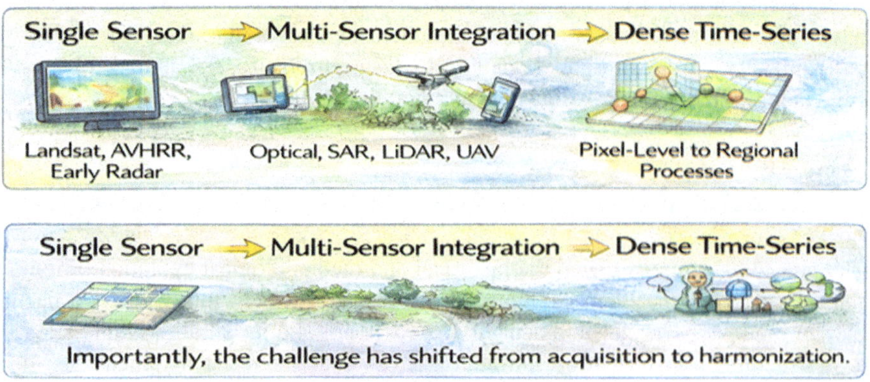

Figure 1.5 The data complexity shift in GeoAI: From single-sensor to cross-scale integration.

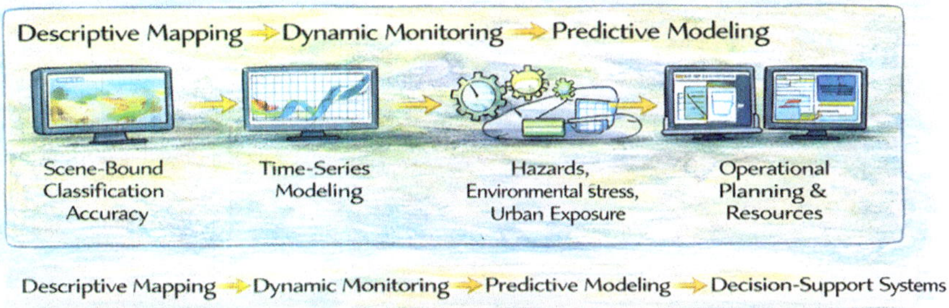

Figure 1.6 The analytical shift in GeoAI: from descriptive mapping to decision-support systems.

Distributed computing introduced parallel processing. Cloud-native platforms and centralized computational ecosystems now enable planetary-scale analysis. Cloud platforms democratize access to archives while simultaneously centralizing computational authority. Infrastructure shapes not only performance but also equity and dependency. Computational ecosystems have become as structurally significant as sensors themselves. The infrastructure trajectory can be seen in Fig. 1.7.

4.4 *Epistemic shift*

The most consequential transformation is epistemic. The object of inquiry has shifted. Earlier paradigms centered on pixel-level classification and land cover categories. Contemporary GeoAI increasingly models landscape processes, resilience trajectories, vulnerability patterns, and socio-ecological dynamics. Chapters focusing on conservation systems, urban heat exposure, compound hazards, and disaster analytics all demonstrate this shift from static mapping toward integrated systems thinking. The epistemic progression is clear from Fig. 1.8. Also, to clarify these transformations, Table 1.2 summarizes the progression across early, transitional, and contemporary phases of GeoAI development.

Figure 1.7 The infrastructure shift in GeoAI: From desktop GIS to centralized compute ecosystems.

Figure 1.8 The epistemic shift in GeoAI: from pixel-level classifications to integrated systems thinking.

Table 1.2 highlights that the evolution of GeoAI is cumulative rather than abrupt. Each phase retains elements of prior methods while expanding analytical ambition and institutional relevance. The field does not abandon classification; it embeds it within larger, predictive systems. Table 1.3 illustrates how the chapters in this volume align with these transformations.

This table demonstrates that the chapters are not isolated case studies; they collectively map the structural maturation of the field. Foundational chapters establish archive and computational shifts. Middle chapters operationalize multi-sensor and time-series integration. Application chapters embed GeoAI within resilience, health, climate, and governance systems. Remote sensing is no longer confined to mapping land cover; it now operates within predictive, policy-relevant, and interdisciplinary decision frameworks. The GeoAI era is defined not by any single algorithm, but by the convergence of data complexity, analytical ambition, computational infrastructure, and epistemic integration (Fig. 1.9). These structural shifts also foreshadow the tensions explored later in this book. As GeoAI becomes more predictive, centralized, and decision-proximate, questions of governance, interpretability, sustainability, and equity become inseparable from methodological innovation.

Table 1.2 Structural transformations across the GeoAI era.

Dimension	Early phase	Transitional phase	GeoAI era phase
Data structure	Single-sensor scenes	Multi-sensor integration	Dense time-series, cross-scale fusion
Analytical focus	Classification	Monitoring and change detection	Predictive and decision-support systems
Infrastructure	Desktop GIS	Distributed computing	Cloud-native, centralized computing ecosystems
Epistemic orientation	Pixel-level mapping	Landscape process modeling	Integrated socio-ecological system analysis
Policy proximity	Low	Moderate	Embedded in governance and resilience frameworks

Table 1.3 Chapter alignment with structural transformations.

Chapter	Data complexity	Analytical shift	Infrastructure shift	Epistemic shift
2	Archive expansion	Foundational analytics	Open data shift	Infrastructure framing
3	Cloud-native ingestion	Pipeline automation	Distributed computing	Platform ecosystem
4	Dense time-series	Resilience modeling	Scalable workflows	Landscape persistence
6	Multi-sensor fusion	Feature and decision fusion	Cloud analytics	Integrated sensing
8	Cross-scale vulnerability	Exposure modeling	Platform-dependent	Socio-environment integration
9	Hybrid physics-AI data	Surrogate modeling	HPC-enabled	Hazard systems
11	Multi-stream disaster data	Anticipatory analytics	Operational cloud systems	Governance embedding

Fig. 1.9 synthesizes these structural transformations across four dimensions: data complexity, analytical orientation, computational infrastructure, and epistemic framing. It situates chapters along these trajectories, illustrating how foundational contributions, methodological consolidation, and applied systems collectively define the GeoAI era. The figure reinforces a central argument of this volume: GeoAI is shaped not by any single algorithmic breakthrough, but by the convergence of data richness, computational scale, analytical ambition, and interdisciplinary embedding.

5. Cross-cutting tensions in remote sensing, big data, and GeoAI

The preceding sections have traced the historical, methodological, and infrastructural transformations that define the GeoAI era. Across chapters, however, a set of recurring tensions emerges (Table 1.4). These tensions are not problems to be solved once and for all; they are structural conditions of contemporary geospatial science. They surface differently in land cover classification, coastal modeling, conservation analytics, urban heat mapping, disaster response, geovisualization, and cloud-based preprocessing, yet their underlying logic remains consistent. This section introduces five cross-cutting tensions that serve as analytical lenses for the remainder of the

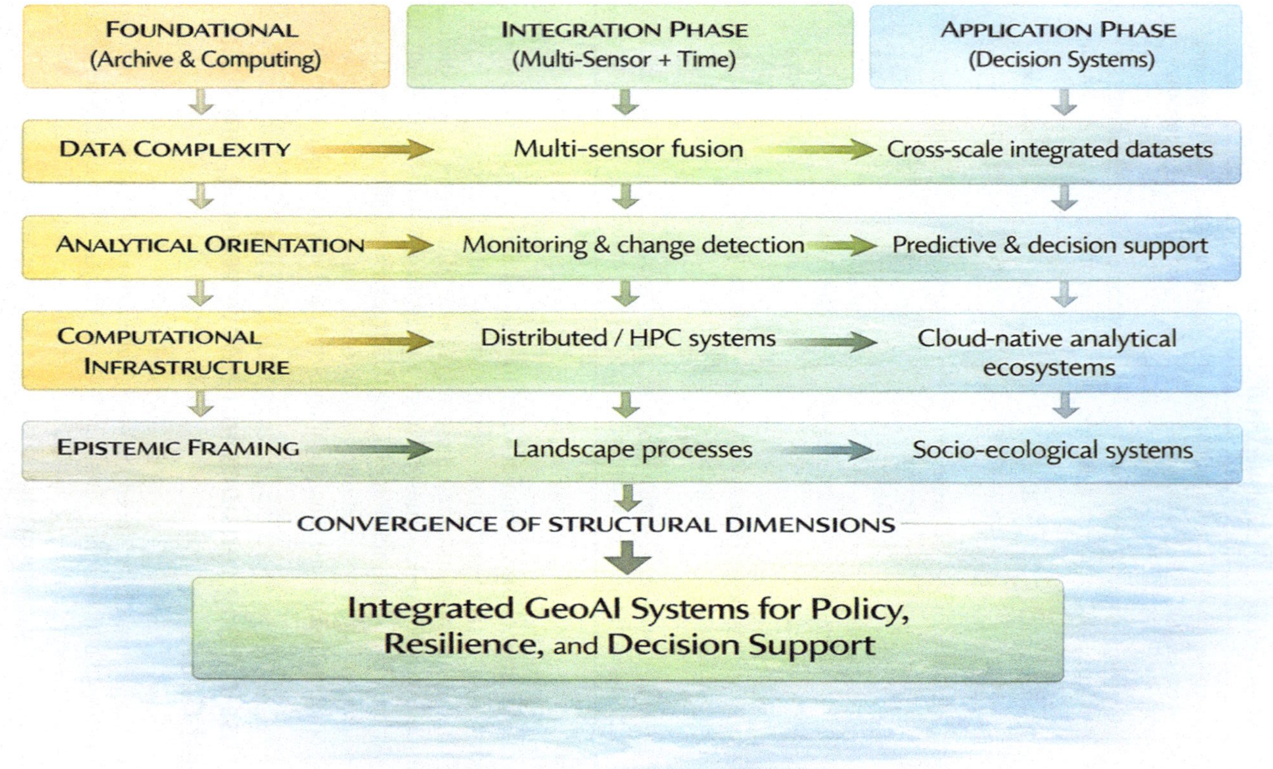

Figure 1.9 Structural maturation of GeoAI across four interrelated dimensions: data complexity, analytical orientation, computational infrastructure, and epistemic framing.

Table 1.4 Cross-cutting tensions matrix highlighting which tensions appear in which book chapter, with closed versus open circles indicating • = Core engagement ○ = Secondary engagement.

Chapter	Resolution versus sustainability	Automation versus explainability	Data abundance versus training scarcity	Cloud centralization versus local control	Innovation versus governance
1—Integrative GeoAI framing	○	•	•	•	•
2—Foundations of remote sensing	○	○	•	•	•
3—Big data & preprocessing	•	•	•	•	•
4—Time-series & spatial analytics	•	•	•	○	○
5—ML versus DL in LULC classification	•	•	•	•	•
6—Multi-sensor data fusion	•	•	•	•	•
7—Geovisualization & interpretation	○	•	○	•	•
8—Urban heat & public health	•	•	○	○	•
9—Coastal hazard modeling	•	•	○	○	•
10—Conservation & SDG monitoring	•	•	○	•	•

Continued

Table 1.4 Cross-cutting tensions matrix highlighting which tensions appear in which book chapter, with closed versus open circles indicating ● = Core engagement ○ = Secondary engagement.—cont'd

Chapter	Resolution versus sustainability	Automation versus explainability	Data abundance versus training scarcity	Cloud centralization versus local control	Innovation versus governance
11—Disaster management systems	●	●	○	●	●
12—Environmental health & exposure modeling	●	●	○	○	●
13—Climate and atmospheric prediction systems	●	●	●	●	●
14—Socio-ecological integration & policy systems	●	●	○	●	●
15—Integrated GeoAI decision frameworks	●	●	●	●	●

volume. They help frame why certain methodological choices recur, why infrastructure matters as much as algorithms, and why equity and governance cannot be separated from technical design.

5.1 Resolution versus sustainability

Across chapters, higher spatial, spectral, and temporal resolution is consistently associated with improved analytic precision. Dense NDVI time series, multi-sensor fusion, 2 m wetland mosaics, UAV-scale coastal mapping, and neighborhood-scale heat exposure models all illustrate the analytical power of increased resolution. Yet resolution is not free. Higher resolution increases storage demands, computational intensity, GPU requirements, cloud processing costs, and energy consumption. Deep neural networks trained on high-resolution imagery amplify these costs further. Even continuous monitoring pipelines, such as conservation systems or disaster dashboards, carry environmental and infrastructural footprints. This tension appears repeatedly:

- In land classification, marginal accuracy gains from deep learning must be weighed against computational burden.
- In coastal modeling, event-scale resolution increases processing complexity.
- In conservation monitoring, continental-scale mosaics require high-performance computing environments.
- In urban heat analytics, finer spatial grids demand denser data integration and validation.

Resolution therefore becomes not only a technical decision, but a sustainability decision. The field increasingly faces a design question: when is higher resolution substantively necessary, and when does it represent diminishing returns?

5.2 Automation versus explainability

GeoAI systems automate classification, shoreline detection, poacher identification, vulnerability scoring, and disaster damage extraction. Automation enhances scalability, reduces response time, and enables planetary-scale inference. However, increasing automation often reduces interpretability. Vegetation index trajectories remain interpretable; convolutional neural networks trained on hyperspectral imagery do not. Bayesian fusion models provide uncertainty estimates; some deep learning pipelines operate as opaque feature extractors. Vulnerability indices for urban heat can encode assumptions that are difficult to interrogate once embedded in machine learning models. This tension is not simply technical. It directly affects trust, governance, and policy uptake. In conservation enforcement, disaster response, and public health planning, black-box outputs may undermine legitimacy. Several chapters implicitly resolve this tension by combining automated pattern extraction with transparent validation, uncertainty reporting, and human oversight. The volume collectively suggests that automation without interpretability risks fragility, while interpretability without scalability limits impact. Mature GeoAI systems must navigate both.

5.3 Data abundance versus training scarcity

Remote sensing now operates in a context of petabyte-scale archives. Landsat continuity, Sentinel constellations, SAR, LiDAR, UAV systems, IoT streams, and multi-decadal time series represent unprecedented abundance. Yet labeled training data remain scarce, unevenly distributed, and often geographically biased. Coastal hazard models struggle with extreme event imbalance. Deep learning land classification requires curated ground truth. Urban vulnerability mapping depends on incomplete socio-demographic data. Fusion models require temporally aligned training pairs that are not always available. This paradox, abundance of imagery, scarcity of reliable labels, recurs across domains. Chapters address it through.

- Semi-supervised learning,
- Transfer learning and pretrained architectures,
- Feature engineering in ML approaches,
- Hybrid physics-AI modeling,
- Integration of ancillary datasets to compensate for sparse labels.

The tension reveals a deeper epistemic issue. Data volume does not automatically translate into knowledge quality. Training data governance, validation rigor, and representativeness become central constraints in the GeoAI era.

5.4 Cloud centralization versus local control

Cloud platforms such as Google Earth Engine, Azure, and distributed dashboards enable planetary-scale analytics, multi-user collaboration, and scalable processing. They reduce hardware barriers and facilitate global monitoring. At the same time, they introduce structural dependencies. GPU-intensive workflows privilege institutions with cloud credits or infrastructure partnerships. Platform governance, pricing structures, data retention policies, and computational access become external constraints on scientific practice. This tension appears across.

- Big data preprocessing pipelines,
- Conservation monitoring infrastructures,
- ML/DL classification workflows,
- Disaster response systems,
- Geovisualization dashboards.

Cloud centralization expands capability but consolidates control. It democratizes analysis in principle while concentrating infrastructure in practice. For regions in the Global South or under-resourced institutions, infrastructure inequity may replace data cost as the primary barrier. The field increasingly faces questions of computational sovereignty, federated models, open infrastructure alternatives, and energy-aware design.

5.5 Innovation versus governance

GeoAI innovation is accelerating. Deep architectures, hybrid fusion systems, foundation-style models, immersive visualization environments, and predictive hazard analytics expand methodological capability at unprecedented speed. Governance frameworks, however, evolve more slowly. Privacy concerns in IoT integration, surveillance implications in conservation monitoring, zoning consequences in flood prediction, equity issues in heat vulnerability mapping, and algorithmic bias in training datasets require regulatory and ethical reflection. Across chapters, governance appears in different forms:

- Open data policy as structural democratization,
- FAIR principles in geovisualization,
- Environmental justice in urban heat analytics,
- SDG alignment in conservation monitoring,
- Liability concerns in coastal hazard prediction.

Innovation without governance risks unintended consequences. Governance without innovation risks stagnation. The GeoAI era requires both to co-evolve.

These five sets of tensions (Fig. 1.10) recur throughout this volume and serve as analytical lenses through which subsequent application chapters can be interpreted. They are not obstacles to be eliminated, but structural dynamics to be navigated.

1. Resolution must be balanced with sustainability.
2. Automation must be paired with explainability.
3. Data abundance must be tempered by training rigor.
4. Cloud scalability must be weighed against infrastructural equity.
5. Innovation must evolve alongside governance.

Together, these tensions define the conditions under which remote sensing, big data, and GeoAI mature from promising technical advances into accountable, policy-relevant, and socially embedded systems.

The Tensions Matrix (Table 1.4) reveals a clear structural progression across the volume. Early chapters establish the infrastructural and methodological foundations of GeoAI, engaging strongly with data abundance, cloud centralization, and governance as structural conditions that shape the field. As the book moves into methodological inflection points, tensions around automation and explainability become more pronounced, reflecting the shift from supervised classification toward deep learning and integrated architectures. In the applications cluster, resolution pressures intensify, as high-frequency, fine-scale monitoring intersects with sustainability, computational cost, and environmental impact. Governance emerges not as a peripheral concern but as a recurring theme across domains including health, climate, conservation, and disaster response. Notably, training scarcity remains persistent even within data-rich environments, underscoring that petabyte archives do not automatically translate into labeled or equitable datasets. The matrix also demonstrates that no chapter is isolated within a single tension; rather, each contributes to a layered system of methodological, infrastructural, and ethical interdependencies. Taken together, the

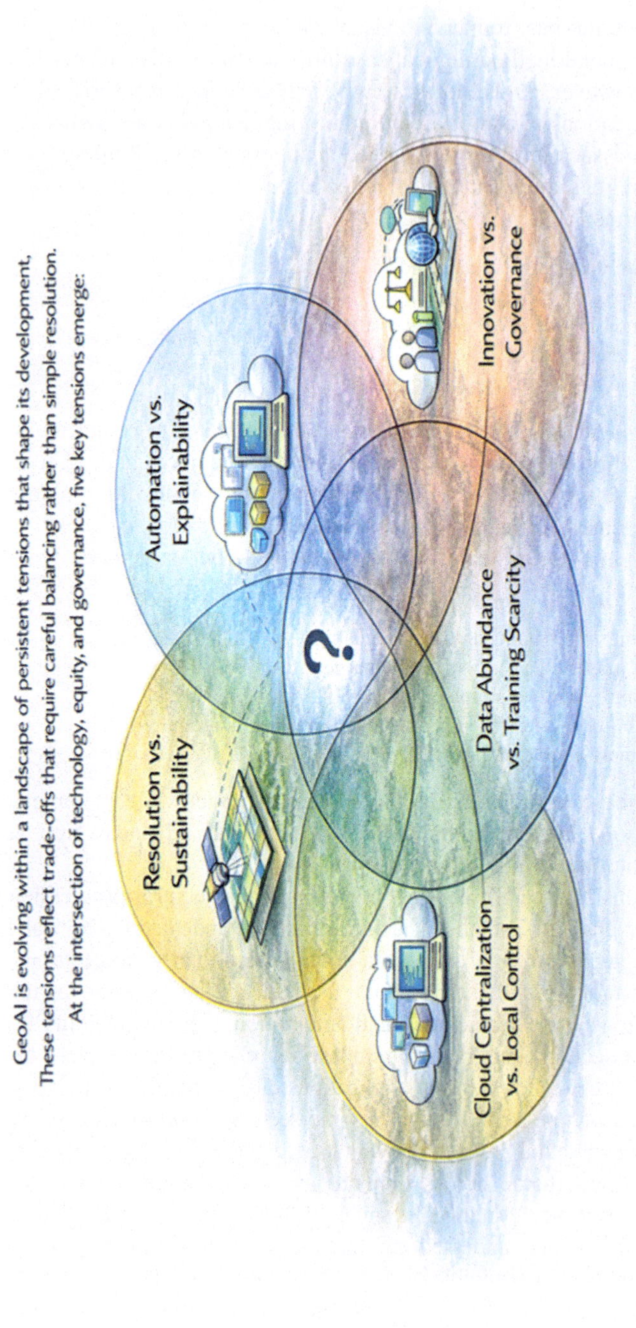

Figure 1.10 Cross-cutting tensions in remote sensing and GeoAI.

volume presents GeoAI not as a linear technological evolution, but as a structurally entangled field defined by recurring trade-offs that shape both its capabilities and its constraints. The tensions introduced here are diagnostic rather than prescriptive. They serve as organizing lenses for interpreting the methodological and application chapters that follow. Chapter 16 revisits these tensions not merely as analytical categories, but as interacting structural forces that shape the future trajectory of GeoAI.

6. Maturation of the field: From innovation to integration

The evolution of GeoAI and big data within remote sensing can be understood not simply as a sequence of technological milestones, but as a gradual maturation of the field across multiple decades. While Chapter 2 provides a more detailed historical account, it is useful here to position the present moment within a broader arc of disciplinary transformation. Across this arc, the emphasis has shifted from sensor development to data accumulation, to algorithmic expansion, and now toward systemic integration, governance, and institutional adaptation.

1990s: Sensor Physics and Quantitative Foundations: In the 1990s, the dominant intellectual energy within remote sensing centered on sensor physics, spectral theory, and quantitative spatial analysis. The development and refinement of satellite systems in the 1960s and 1970s had already established remote sensing as a foundational component of geographic inquiry. By the 1980s and 1990s, the emergence of personal computing and expanding sampling methodologies further solidified what Openshaw (1994) described as a "totally computer-dependent paradigm in geography". During this period, theoretical debates about computation and quantitative methods were prominent. Scholars such as Barnes (2004), Dobson (1983), and Harvey (1972) interrogated the epistemological implications of computational geography, while Couclelis (1986) emphasized the mathematical and logical foundations underpinning spatial analysis. The dominant methodological orientation focused on per-scene classification, spectral indices, and physically grounded interpretations of reflectance patterns. Machine learning techniques existed, but they remained peripheral rather than central to mainstream remote sensing practice.

2000s: Archive Expansion and Big Data Emergence: The 2000s marked a structural shift driven by archive expansion. Long-term datasets from sensors such as MODIS and AVHRR became widely available, creating unprecedented opportunities for temporal analysis. However, analytical capacity initially lagged behind data availability. As noted in earlier sections, time-series datasets at monthly resolution were accessible across large regions, yet the statistical tools required to fully disentangle seasonal dynamics and drivers of change were still emerging. This period also saw the introduction of cloud-based platforms such as Google Earth Engine in 2008, which fundamentally altered the infrastructure of environmental data analysis. Rather than downloading scenes individually and processing them locally, researchers could increasingly access planetary-scale archives through centralized compute environments. The concept of "big data" became inseparable from remote sensing practice, not only because of data volume, but because of velocity and variety. Yet, even during this expansion, the dominant workflow remained largely analyst-driven. Features were engineered manually, classifications were parameterized by domain expertise, and statistical modeling retained strong ties to traditional inferential frameworks.

2010s: Machine Learning Integration: The 2010s represent the period in which machine learning became structurally embedded within geospatial research. The rapid uptick in GeoAI-related citations in remote sensing and GIS, particularly after the mid-2010s, reflects this integration. Professional workshops and symposia dedicated to GeoAI at major conferences in 2016 and 2018 signaled institutional recognition of this shift (Lavallin & Downs, 2021; Mao et al., 2017). Random forests, support vector machines, ensemble approaches, and other statistical learning techniques became widely adopted for classification and prediction tasks. Importantly, these approaches allowed researchers to manage higher-dimensional feature spaces and more complex relationships than traditional parametric models. GeoAI moved from being a peripheral methodological curiosity to a central analytical framework. The integration of machine learning did not replace earlier quantitative traditions; rather, it extended them. Spatial analysis, GIS workflows, and remote sensing interpretation were increasingly augmented by algorithmic approaches capable of handling nonlinear interactions and large predictor sets. The field began to transition from static mapping toward dynamic monitoring and predictive modeling.

2020s: Deep Learning Dominance, Time-Series Integration, and Cloud Scaling: In the 2020s, deep learning architectures have become increasingly prominent. Convolutional neural networks, deep neural networks, hybrid machine learning and deep learning approaches are now widely used for segmentation, classification, and predictive analysis, as illustrated in the case studies included in this chapter (e.g., Insalaco et al., 2024; Rahaman et al., 2025). These approaches move beyond handcrafted features toward representation learning, where hierarchical abstractions are learned directly from imagery. Simultaneously, time-series analysis has matured. Rather than treating satellite scenes as independent observations, researchers increasingly analyze dense temporal stacks, allowing for trajectory analysis, anomaly detection, and seasonal modeling at scale. Cloud platforms, high performance computing, and GPU acceleration have made these approaches operationally feasible. The citation patterns discussed earlier in this chapter (Fig. 1.2) underscore the magnitude of this transition. Remote sensing publications referencing GeoAI-related terms have increased dramatically since the mid-2010s, with 2022 marking a notable peak. This is not merely a growth in publication volume; it reflects a reorientation of methodological norms within the discipline.

Present Phase: Governance, Equity, Education, and Sustainability: The contemporary phase is characterized less by the invention of new sensors or algorithms and more by system-level integration, ethical reflection, and institutional adaptation. As discussed in earlier sections, the integration of IoT systems, cloud platforms, and AI-driven analytics has created highly interconnected geospatial ecosystems. At the same time, concerns regarding data quality, interpretability, legal compliance, environmental impacts, and equitable outcomes have become increasingly prominent (Arashpour, 2023; Radanliev, 2025). GeoAI is no longer simply about improving classification accuracy. It now intersects with governance frameworks, educational curricula, interdisciplinary collaboration, and sustainability debates. The field is grappling with questions of transparency, bias, energy consumption, and the environmental costs of large-scale computation. These issues signal a maturation of the discipline, in which technical innovation is increasingly evaluated alongside social responsibility. This contemporary moment therefore represents a transition from innovation to integration. Remote sensing, big data, and GeoAI are embedded within policy-relevant decision frameworks, cross-sector collaborations, and educational infrastructures. The remainder of this volume reflects this maturation, moving beyond isolated methodological advances to examine integrated systems, applications, and forward-looking challenges.

6.1 Remote sensing as foundational infrastructure

Despite the prominence of artificial intelligence terminology, remote sensing remains the physical and epistemic foundation of GeoAI. Contemporary advances in machine learning and deep learning extend, rather than replace, the scientific grounding of Earth observation. Sensor physics, radiometric calibration, spectral integrity, spatial resolution, and temporal consistency continue to shape the reliability of derived products and the validity of downstream decisions. As multiple chapters in this volume demonstrate, model performance is inseparable from data provenance and sensor characteristics. Multi-sensor fusion, dense time-series integration, and cross-scale modeling all depend on careful attention to acquisition parameters and measurement uncertainty. In this sense, GeoAI represents an expansion of remote sensing capability, not a departure from it. Maintaining this grounding is essential for preserving disciplinary coherence as computational abstraction increases.

6.2 Education and institutional adaptation

The maturation of GeoAI is not solely technical; it is institutional and pedagogical. As analytical systems become more automated and computationally intensive, demands for AI literacy, interdisciplinary fluency, and ethical awareness grow correspondingly. Workforce development now requires familiarity with cloud platforms, data engineering pipelines, and hybrid modeling architectures, alongside traditional geospatial analysis. At the same time, disparities in access to computational infrastructure and training risk reinforcing existing inequalities across institutions and regions. Several chapters in this volume highlight the need for curriculum reform, collaborative research structures, and governance frameworks that align technological innovation with societal responsibility. The long-term stability of GeoAI will depend not only on algorithmic advancement, but on the capacity of institutions to adapt responsibly and inclusively.

7. Roadmap to the volume: Themes across chapters

The chapters that follow explore how the structural transformations outlined above manifest across domains that range from foundational sensor systems to climate analytics, agriculture, conservation, disaster response, socio-ecological integration, and governance. While the application contexts differ, the methodological trajectories are strikingly consistent. Across the volume, we observe recurrent shifts toward multi-sensor fusion, dense time-series modeling, cloud-based computation, hybrid machine learning architectures, and increasing proximity to operational decision systems. Table 1.5 synthesizes the orientation of Chapters 1 through 15 across five structural dimensions: application domain, methodological inflection, automation level, computational scale, and decision proximity. Rather than serving as a simple catalog of topics, the table reveals the intellectual architecture of the book.

Table 1.5 demonstrates that the progression across chapters is cumulative rather than linear; methodological consolidation and governance integration intensify together rather than sequentially. Several structural patterns emerge. First, the early chapters establish the epistemic and infrastructural substrate of GeoAI. Chapters 1 through 3 frame the conceptual terrain, document the historical arc of sensor proliferation and archive expansion, and describe the computational ecosystems that make planetary-scale analysis possible. These chapters operate primarily at the level of infrastructural and theoretical development, with limited automation emphasis and indirect decision proximity. Second, Chapters 4 through 7 represent methodological consolidation. Here the field transitions from per-scene classification to dense time-series analytics, multi-sensor integration, ensemble machine learning, and AI-assisted visualization systems. Automation increases but remains embedded within interpretive frameworks. Computational scales expand toward cloud environments, yet the focus remains on refining analytical capability rather than building fully operational systems. Third, Chapters 8 through 13 mark the clearest movement into applied GeoAI systems. Urban heat analytics, coastal hazard modeling, conservation monitoring, disaster management, climate-health modeling, and agricultural intelligence demonstrate how methodological inflection points translate into predictive, policy-adjacent, and in some cases operational systems. Automation intensifies, computation demands escalate, and decision proximity moves from analytical support toward institutional embedding. Finally, Chapters 14 and 15 signal maturation. Rather than accelerating computational scale further, these chapters foreground socio-ecological integration, governance alignment, education, and institutional adaptation. Automation recedes in prominence relative to interpretability, interdisciplinarity, and regulatory reflection. In structural terms, the field shifts from innovation-centric expansion to integration-centric consolidation.

Across Table 1.5, a progression becomes visible. Computational intensity peaks in the middle-to-late application chapters, particularly where dense time-series modeling and predictive analytics dominate. Yet the final chapters demonstrate that maturity is not defined by algorithmic novelty alone. It is defined by the ability to embed GeoAI within broader social, environmental, and institutional systems. The extended table

Table 1.5 Roadmap of the thematic and methodological orientation across book chapters.

Chapter	Application domain	Primary methodological inflection	Automation level	Computational scale	Decision proximity
1	Structural framing of GeoAI	Classification → integrated GeoAI systems	Conceptual	N/A	Foundational framing
2	Foundations of earth observation	Data scarcity → archive abundance; open access Landsat	Low automation	Desktop → early cloud	Indirect; epistemic
3	Big data acquisition & preprocessing	Desktop workflows → cloud-native analysis-ready ecosystems	Moderate automation in preprocessing	Cloud/HPC dependent	Foundational; infrastructural
4	Time-series analytics & vegetation dynamics	Single-date mapping → dense trajectory modeling	Moderate; interpretable ML integration	Cloud-scale time series	Applied environmental intelligence
5	Land use and land cover classification	Algorithm competition → decision-tree method governance	Moderate to high; conditional	CPU ML → GPU/ cloud DL	Applied; near-operational
6	Multi-sensor data fusion	Single-sensor analysis → scale-aware spatiotemporal integration	Semi-automated	Desktop → Cloud/ GPU	Methodological; applied
7	Geovisualization & interpretive systems	Static maps → interactive, AI-integrated visualization	Human-guided; AI-supported	Desktop + cloud	Decision-support interface
8	Urban heat & climate-health analytics	Descriptive UHI → predictive vulnerability modeling	Moderate ML automation	Desktop → moderate cloud	Policy-relevant; urban planning

Continued

Table 1.5 Roadmap of the thematic and methodological orientation across book chapters.—cont'd

Chapter	Application domain	Primary methodological inflection	Automation level	Computational scale	Decision proximity
9	Coastal hazard modeling	Physics-dominant → hybrid ML surrogate systems	Semi-automated	Moderate to high HPC	Emerging operational
10	Conservation & SDG monitoring	Static land cover → continuous AI-enabled monitoring	Moderate to high	Cloud + HPC	Operational conservation intelligence
11	Disaster management systems	Post-event mapping → anticipatory AI-enabled systems	Increasing automation	Cloud + GPU-intensive	Operational; institutionalized
12	Medical geography & environmental health	Observational datasets → predictive, multivariate climate-health integration	Moderate; model-assisted	Cloud + large archive processing	Policy-adjacent; public health relevance
13	Agricultural and food system intelligence	Field-scale mapping → multi-sensor crop and yield prediction systems	Moderate to high; hybrid ML/DL	Cloud-dependent; time-series intensive	Operational agricultural planning
14	Socio-ecological system integration	Remote sensing outputs → integrated socio-environmental modeling frameworks	Semi-automated; strongly human-guided	Cross-platform; moderate cloud	Strategic decision-support; planning scale
15	Governance, equity, and system integration	Technical innovation → institutional embedding and governance alignment	Low automation emphasis; reflective	Platform-dependent but not compute-intensive	Policy-integrated; institutional and educational

therefore reveals that the volume is organized not merely by domain but by structural evolution. It traces a trajectory from sensor and archive foundations, through methodological inflection, into predictive and operational systems, and ultimately toward governance-aware, interdisciplinary integration. This progression reflects the broader maturation of the field itself.

8. Conclusion: From remote sensing to integrated GeoAI systems

The trajectory outlined in this chapter reflects more than technological acceleration. It reveals a structural transformation in how remote sensing, big data, and artificial intelligence are conceptualized and deployed. What began as sensor-driven measurement and per-scene classification has evolved into dense time-series modeling, multi-sensor integration, cloud-based infrastructure, and predictive decision frameworks. Yet the contemporary phase of GeoAI is characterized less by the invention of new algorithms than by system-level integration, interdisciplinary convergence, and institutional adaptation.

Remote sensing remains foundational, but it now operates within computational ecosystems that link physical measurement to predictive analytics and policy-relevant decision-making. As automation increases and analytical abstraction deepens, new tensions emerge around sustainability, interpretability, governance, and equitable access. These tensions do not signal instability; rather, they indicate maturation. A field becomes mature not when innovation slows, but when reflection, integration, and responsibility accompany expansion. The chapters that follow illustrate how these structural shifts manifest across environmental monitoring, climate analytics, conservation, agriculture, disaster response, urban systems, and governance. Together they demonstrate that the future of remote sensing and GeoAI will not be defined solely by higher resolution or deeper networks, but by how effectively these tools are integrated into transparent, equitable, and sustainable decision frameworks.

Case studies as structural anchors

The structural trajectory outlined above is not presented as abstraction alone. Consistent with the architecture of this volume, each chapter integrates two to three case studies that demonstrate how conceptual frameworks and methodological inflection points translate into applied research and decision contexts (Table 1.6). These case studies are written deliberately in accessible language and are designed to illustrate the operational implications of GeoAI systems rather than simply their technical specifications. They highlight concrete research problems, data integration strategies, analytical workflows, and institutional or policy relevance.

Chapter 1 models this structure through two case studies that examine GeoAI applications in distinct environmental contexts: first, the quantification of environmental change within a lagoon ecosystem in Florida; and second, the analysis of air pollution and vegetation dynamics within an urban landscape in Bangladesh. Together, these examples illustrate how remote sensing archives, machine learning workflows, and evolving computational infrastructures support the detection, modeling, and interpretation of landscape change across ecological and urban systems. In developing these case studies, complex methodological details have been translated into clear, accessible prose so that readers from multiple disciplinary backgrounds can engage with the material. At the same time, the scientific rigor of the original research is preserved. Each case study provides full citations to the underlying peer-reviewed publications where available, allowing readers to trace the analytical methods, datasets, and technical contributions in greater depth where appropriate. This dual structure, accessible synthesis paired with direct linkage to the scholarly literature, reflects the broader pedagogical approach of the book.

Table 1.6 Titles of Every Case Studies Found in Each Book Chapter in this Volume.

Chapter number	**Case study titles**
1	#1: Tracking seagrass decline in mosquito lagoon, Florida: A deep learning approach
	#2: Understanding urban air pollution and vegetation change using GeoAI in Dhaka, Bangladesh
2	#1: Traditional pixel-based classification
	#2: Object oriented classification
3	#1: Spatiotemporal trends in heat waves across the United States and their impact on environmental health
	#2: Agricultural expansion and natural vegetation decline in Eastern Africa
4	#1: Moving beyond Land cover classes: Time series analysis of vegetation dynamics in Southern African Savannas
	#2: Understanding what drives vegetation change in Southern Africa using combined time-series analysis and machine learning
	#3: Assessing vegetation directional persistence in Southern African national parks (2000–2016)
5	#1: Balancing precision and practicality: classifying almond crops from space
	#2: Blending technology and ecology: Monitoring land-cover transformations in Addo Elephant national Park

Table 1.6 Titles of Every Case Studies Found in Each Book Chapter in this Volume.—cont'd

Chapter number	Case study titles
6	#1: A successful integration project in land cover and land use classification (Land cover mapping via crowdsourced multi-directional views: The more directional views, the better,
	#2: A failed integration project in agriculture
7	#1: Visualizing changes in the surface area of water and ice in lakes and glaciers through image vectorization
	#2: Remote sensing applications for annual population estimation at the fine spatial scale: A case study of Colorado Springs
8	#1: Spatiotemporal evolution of heat wave severity and coverage across the United States
	#2: An ensemble deep learning approach to spatiotemporal tropospheric ozone forecasting: A case study of Tehran, Iran
9	#1: Benthic habitat mapping
	#2: Extracting shorelines globally from satellite data
	#3: Modeling thc influence of compounding coastal and river processes on extreme along-river flooding
10	#1: Microsoft AI for conservation
	#2: Identifying Sahel vegetation drivers
11	#1: Detecting hurricane-related damage using Sentinel-2 imagery in Valdosta, Georgia
	#2: Mapping tornado damage with Sentinel-2 imagery in the Southeastern U.S.
12	#1: Time specific ecological niche modeling reveals environmental reservoirs for pathogen persistence: A case study with Bacillus anthracis in Chobe national Park, Bostwana
	#2: Building ecological niche models of specific Bacillus anthracis lineages in West Texas to improve risk predictions
13	#1: Commercial satellite imagery and privacy: What the public says
	#2: The EU AI Act and remote sensing, GeoAI
14	#1: OpenStreetMap—Land use crowdsourced mapping
	#2: Land cover mapping via crowdsourced multi-directional views
15	#1: AI-focused ICURE classes
	#2: GeoGator summer undergraduate research experience

Case Study 1: Tracking seagrass decline in Mosquito Lagoon, Florida: A deep learning approach

Insalaco, S. A., Herrero, H. V., Limber, R., Oliver, C., & Wolfson, W. B. (2024). Monitoring an Ecosystem in Crisis: Measuring Seagrass Meadow Loss Using Deep Learning in Mosquito Lagoon, Florida. Photogrammetric Engineering & Remote Sensing, 90(6), 363-370.

This study quantified the loss of seagrass meadows in Mosquito Lagoon, Florida, using cutting-edge GeoAI technology, specifically Deep Neural Networks (DNNs). This approach is comprehensive, blending satellite image analysis, on-the-ground fieldwork, and advanced deep learning techniques. The focus is on understanding the changes in seagrass coverage over 2 decades, from 2000 to 2020.

Seagrass is a key component of marine ecosystems, offering habitats and nutrients for numerous wildlife (Hughes et al., 2009; Orth et al., 2006; Reynolds, 2018; St. Johns River Water Management District, 2020). Moreover, seagrass meadows are important in carbon sequestration, surpassing even terrestrial forests in their ability to capture carbon (Fourqurean et al., 2012). In Mosquito Lagoon, seagrass is not just important for the local marine life; it is an indicator of the health of the entire ecosystem (St. Johns River Water Management District, 2020). Since the 2010s, this area has seen a significant decline in seagrass, signaling an overall ecosystem crisis. This decline is due to various factors, ranging from environmental changes like temperature fluctuations and disease outbreaks to human-induced pressures such as boating, overfishing, and pollution, though one of the most important factors is that of pollution/nutrient loading (Reynolds, 2018; Valentine & Duffy, 2006). The aim of this research is twofold: to document the extent of seagrass loss in Mosquito Lagoon (Case Study Figure 1.1) and to evaluate the effectiveness of DNNs in predicting future seagrass coverage. This case study not only sheds light on the challenges facing this critical ecosystem but also showcases the potential of cutting-edge AI techniques in environmental monitoring and prediction.

To quantify seagrass decline in Mosquito Lagoon, this study used satellite imagery, specifically Landsat 5 and 8, capturing snapshots of the lagoon from 2000 to 2020. Images from the boreal summer months were used, the season when seagrass is at peak biomass, as a benchmark of the greatest potential seagrass meadow extent and density. This method is not only cost-effective but also allows for tracking changes across vast areas and over many years, using available software and machine learning techniques to sift through the data. To train the deep learning model, field data was collected during the summers of 2020, 2021, and 2022 to confirm where seagrass was present. This in situ data collection helped in the creation of accurate training samples for the model (Case Study Figure 1.1).

This study used a Deep Neural Network (DNN) modeling approach. DNN is a network of digital neurons, with each layer processing information received from the previous one (Arora et al., 2018). This type of model is well-suited for tasks where training data is limited, which was the case here. While Convolutional Neural Networks (CNNs) are often the go-to for image classifications, they need a higher volume of data to learn effectively (Rathi et al., 2017, pp. 1–6; Tan et al., 2020). The DNN, on

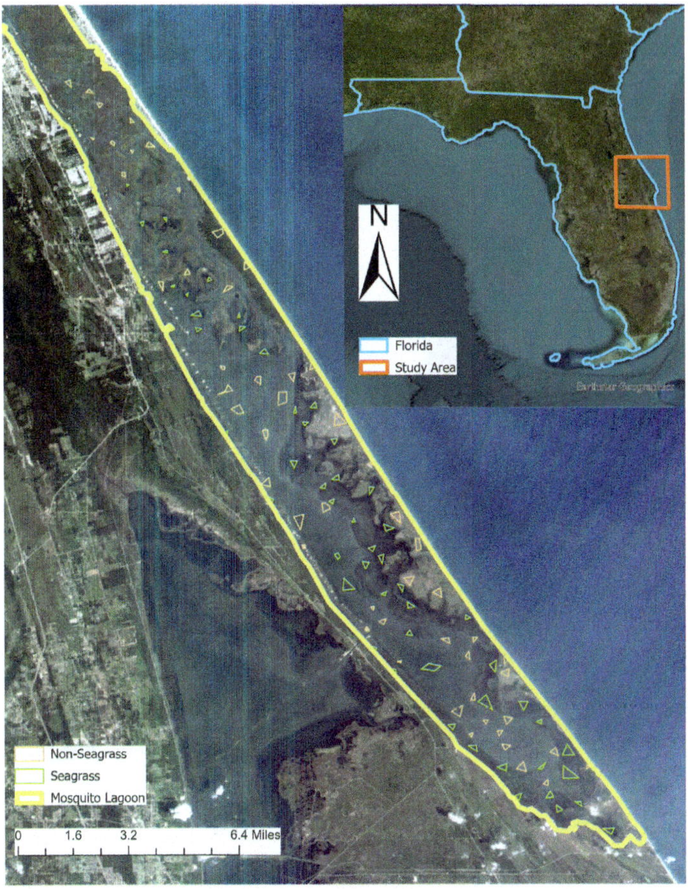

Case Study Figure 1.1 Map of study area and training samples for Landsat imagery of 2000.

the other hand, excels in analyzing individual pixels, appropriate for mapping seagrass on a granular level, especially given that some years had a very limited extent/density of seagrass (Wang et al., 2020, pp. 3570–3575; Weidmann et al., 2019). For each selected year, a separate model was built and then measured its accuracy, precision, recall, and F1-score. A confusion matrix and a change trajectory analysis were used to track the seagrass loss over the years.

The results demonstrate a severe decline in seagrass coverage in Mosquito Lagoon. In 2007, about 16.6% of ML was covered with seagrass. By 2020, that number decreased to less than 1% (Case Study Figure 1.2). The most severe loss occurred between 2013 and 2020, aligning with an intense algal bloom after anomalous freeze events and excessive nutrient pollution (Lopez et al., 2021; Phlips et al., 2015). The pixel-by-pixel analysis confirmed this trend, showing a consistent decrease in seagrass coverage over the years. Almost all the changes observed were from seagrass areas turning into non-seagrass, underlining the severity of the loss.

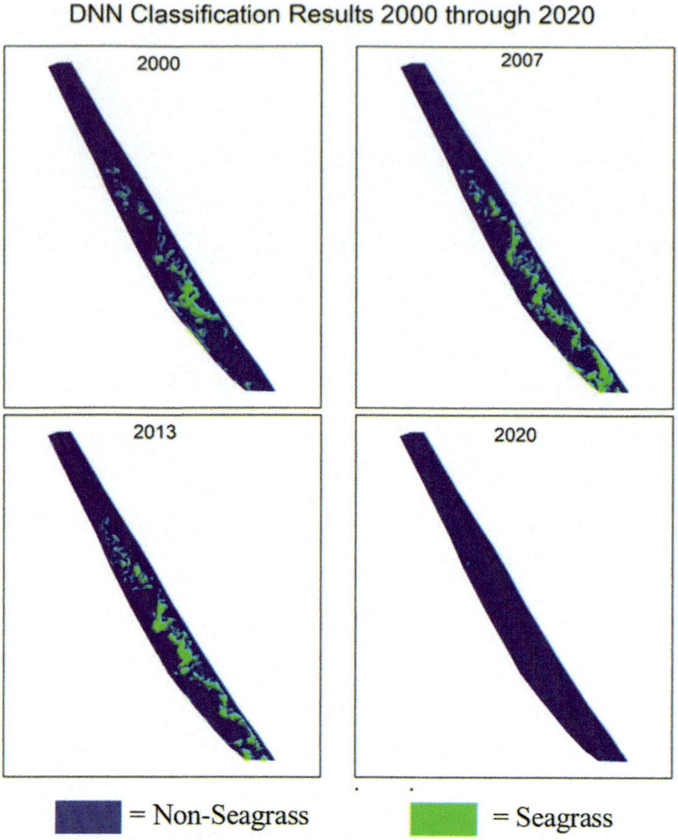

Year	2000	2007	2013	2020
Non-Seagrass	189,913 (89.9%)	176,175 (83.4%)	179,244 (84.8%)	211,178 (99.9%)
Seagrass	21,384 (10.1%)	35,122 (16.6%)	32,053 (15.2%)	119 (0.1%)

Case Study Figure 1.2 Classification results and pixel counts.

The evaluation metrics used, like precision, accuracy, recall, and the F1-Score, underline the DNN model's effectiveness in classifying seagrass. Precision and accuracy were notably high, with precision above 0.786 and accuracy over 86.9% in all years. Over the years, precision improved significantly, achieving a perfect score in 2020. This suggests that the model became increasingly accurate in mapping seagrass patches. In contrast, the year 2000 showed lower metrics, possibly due to insufficient training data from that period. However, it should be noted that the recall rate dropped in 2020, likely due to the scarcity of seagrass making it harder for the model to find positive examples. The F1-Score, which combines precision and recall, varied, with the lowest score in 2020, again likely due to seagrass scarcity. Focusing solely on binary classification (seagrass vs. non-seagrass) simplified the model and maximized accuracy. This approach was effective due to the minimal discrepancy between the classes.

There are certain limitations in this study, such as the complex, often opaque nature of DNNs, the need for more comprehensive training data, and the potential for misidentifying seagrass patches, especially when similar features like oyster reefs are present. Despite these challenges, GeoAI, particularly through the use of DNNs, is a practical and effective tool for identifying seagrass in Mosquito Lagoon. This research underscores the urgent need for conservation actions to curb the decline of seagrass and its wider implications on the ecosystem. This study not only advances an understanding of seagrass ecosystems but also opens up new possibilities for applying GeoAI in other environmental contexts, such as monitoring other critical habitats, like coral reefs or wetlands. These techniques could also be instrumental in tracking changes due to climate impacts or human activities, providing vital data for conservation strategies. Moreover, the methods employed here have broader applications in land use planning and environmental monitoring. For instance, they could aid in assessing the health of forests or in mapping urban expansion and its impact on natural habitats. This research serves as a foundation for future studies, where integrating field data with advanced AI models could lead to more efficient and sustainable management of natural resources. This development represents a meaningful advancement in efforts to reconcile human needs with ecological integrity, illustrating the significant role that technological innovation can play in supporting environmental conservation.

Case Study 2: Understanding urban air pollution and vegetation change using GeoAI in Dhaka, Bangladesh

Mashoukur Rahaman, Jane Southworth, Amobichukwu Chukwudi Amanambu, Bewuket B. Tefera, Ali R. Alruzuq, Mohammad Safaei, Md Muyeed Hasan, Audrey Culver Smith, 2025. Combining deep learning and machine learning techniques to track air pollution in relation to vegetation cover utilizing remotely sensed data. ***Journal of Environmental Management, 376***, https://doi.org/10.1016/j.jenvman.2025.124323.

Urban growth, air pollution, and vegetation loss

Rapid urbanization is reshaping cities across the globe, bringing economic opportunity while simultaneously intensifying environmental pressures. At present, more than half of the world's population lives in urban areas, a proportion projected to rise substantially over the coming decades (Han et al., 2022; Lu et al., 2024). One of the most severe consequences of this transformation is declining air quality. Fine particulate matter, particularly PM2.5 and PM10, is now considered one of the leading environmental health risks worldwide, contributing to millions of premature deaths each year (Diener & Mudu, 2021; Rai, 2016). Urban vegetation plays a critical role in moderating these impacts. Trees and green spaces can reduce ambient particulate concentrations through adsorption and interception, although their effectiveness varies by species, canopy structure, and spatial configuration (Li et al., 2023; Su et al., 2022). At the same time, chronic exposure to air pollution can impair vegetation health, altering plant physiology, morphology, and long-term survival (Rai, 2016). Understanding the spatial relationship between air pollution and vegetation is therefore central to sustainable urban planning and environmental management.

Dhaka, the capital of Bangladesh, exemplifies these challenges. Rapid population growth, industrial expansion, and increasing vehicular traffic have transformed much of the city from predominantly rural land into dense urban fabric over only a few decades (Hossain & Huggins, 2021; Kamal Pasha et al., 2022). Dhaka now ranks among the most polluted cities globally, with particulate concentrations frequently exceeding international air quality standards, particularly during the winter season (EPA, 2024; Rana et al., 2016). Simultaneously, urban green spaces have declined and become increasingly fragmented.

Why GeoAI is needed in complex urban environments

Traditional approaches to studying air pollution and vegetation rely heavily on the national level ground-based air quality monitoring stations and field surveys. While these methods provide accurate localized measurements, they are limited in spatial coverage and often lack the temporal depth needed to examine long-term urban change. In rapidly growing cities such as Dhaka, monitoring networks are sparse and unevenly distributed, making it difficult to assess environmental conditions across the entire urban landscape. Satellite remote sensing offers a powerful alternative by providing consistent, city wide observations over long time periods. However, the increasing volume and complexity of satellite data require advanced analytical tools capable of extracting meaningful patterns. GeoAI addresses this challenge by combining geospatial data with artificial intelligence methods that can learn from large, complex datasets and reveal spatial relationships that are not easily captured using traditional techniques.

This case study demonstrates how GeoAI can be applied to investigate the relationship between air pollution and vegetation change across Dhaka using an integrated approach that combines machine learning and deep learning with satellite remote sensing data (Rahaman et al., 2025).

Integrating satellite data with machine learning and deep learning

This study draws on multi decadal time-series satellite imagery from the Landsat program, to examine vegetation patterns across Dhaka from 1990 to 2022. Landsat 5, 7 and 8 data provide a consistent and widely used record of land surface conditions, making them well suited for long term environmental analysis. To complement these spatial data, ground-based measurements of PM2.5 and PM10 were obtained from air quality monitoring stations distributed across the city. A defining feature of this case study is the deliberate use of different GeoAI methods for different analytical tasks.

Machine learning models were used to predict areas of low, medium, and high air pollution across the city. Because air quality measurements are limited to a small number of monitoring stations, machine learning provides a means of extrapolating pollution levels across space by identifying relationships between measured pollution and other spatial variables derived from satellite data. These variables include indicators of vegetation, built surfaces, land surface temperature, and proximity to major transportation corridors. Several machine learning algorithms were applied and evaluated, allowing for comparison of model performance and spatial consistency. This multi model approach reduces reliance on any single algorithm and increases confidence in the resulting pollution susceptibility maps (Rahaman et al., 2025).

Deep learning models were used to map dense vegetation cover directly from satellite imagery. These models are particularly effective for image segmentation tasks, where the goal is to classify each pixel based on its land cover characteristics. Multiple deep learning architectures were trained and applied, producing vegetation maps for different time periods and enabling analysis of long-term vegetation change across the city.

Patterns of air pollution and vegetation change

The machine learning results revealed clear spatial patterns in air pollution across Dhaka. Areas predicted to experience high pollution were strongly associated with dense road networks, extensive built-up land cover, and limited vegetation. Although individual models varied slightly in their predictions, there was strong agreement on the locations of the most polluted zones, demonstrating the robustness of the GeoAI approach. The deep learning analysis revealed a substantial decline in urban vegetation between 1990 and 2022. While pockets of vegetation persisted in some areas, large portions of the city experienced significant losses, particularly in zones undergoing rapid urban expansion. Importantly, different deep learning models produced broadly consistent vegetation patterns, reinforcing confidence in the observed trends (Rahaman et al., 2025).

When the outputs of the machine learning and deep learning models were examined together, a clear spatial relationship emerged (Case Study Figure 2.1). Areas predicted to have high levels of air pollution frequently overlapped with areas experiencing vegetation loss. Statistical analysis confirmed a negative relationship between air pollution intensity and vegetation abundance. Although this relationship is influenced by multiple factors beyond vegetation alone, including urban form and meteorological conditions, the results support previous findings that highlight the role of vegetation in mitigating particulate pollution and the vulnerability of green spaces to degraded air quality (Su et al., 2022; Zhang et al., 2024).

Broader implications for GeoAI and urban sustainability

This case study highlights a key principle underlying GeoAI applications. Machine learning and deep learning are not interchangeable tools but complementary approaches that serve different analytical purposes. Machine learning excels at prediction and classification when explanatory variables are available, while deep learning is particularly powerful for extracting spatial patterns directly from imagery. Integrating these methods allows researchers to develop a more comprehensive understanding of complex urban environmental systems. The implications of this work extend beyond

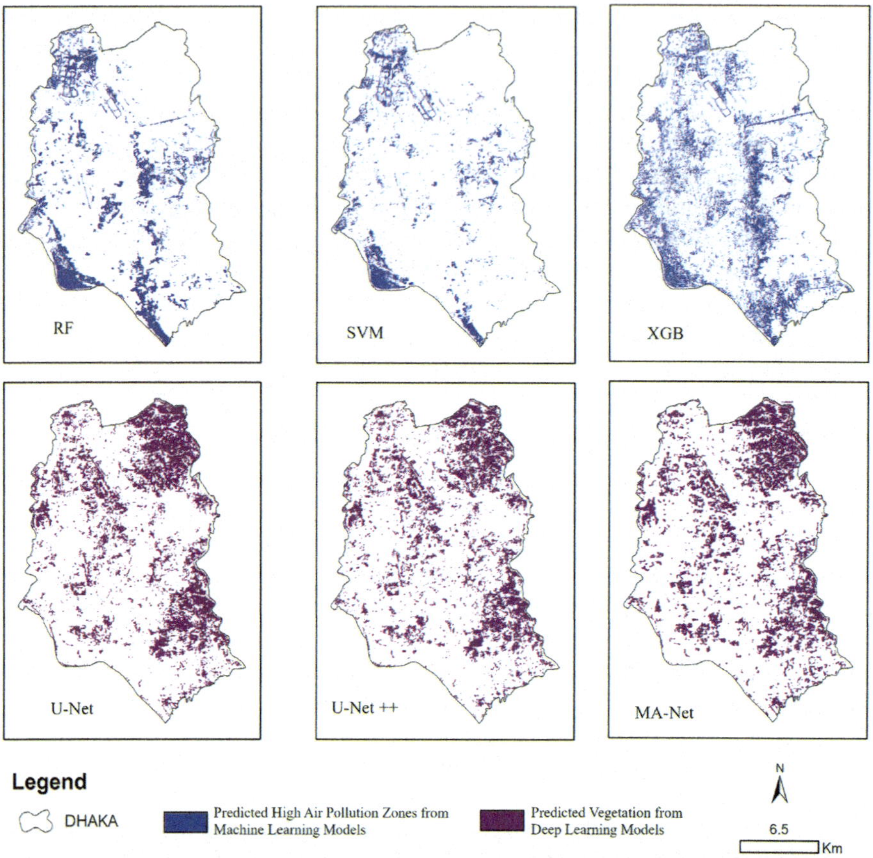

Case Study Figure 2.1 Spatial predictions made by three Machine Learning (ML) models and three Deep Learning (DL) models for Dhaka, Bangladesh. The displayed maps are delineated into two distinct categories based on their predictions: ML model output of areas with predicted high air pollution (depicted in red) and DL model output of areas with predicted high vegetation cover (shown in green).
Adapted from: Rahaman, M., Southworth, J., Amanambu, A. C., Tefera, B. B., Alruzuq, A. R., Safaei, M., Hasan, M. M., & Culver Smith, A. (2025). Combining deep learning and machine learning techniques to track air pollution in relation to vegetation cover utilizing remotely sensed data. Journal of Environmental Management, 376, 124323. https://doi.org/10.1016/j.jenvman.2024.124323.

Dhaka. Many cities worldwide face similar challenges related to rapid growth, declining green space, and deteriorating air quality. GeoAI approaches such as those demonstrated here offer a scalable and transferable framework for monitoring environmental change, identifying priority areas for intervention, and supporting data driven urban planning, particularly in regions where traditional monitoring networks are limited. As an introductory case study, this example illustrates how GeoAI can bridge theory and practice. It shows how big data and artificial intelligence can be applied to pressing real world challenges in ways that are scientifically rigorous, spatially explicit, and accessible to a broad audience. In doing so, it sets the stage for the more technical methods, applications, and challenges explored in subsequent chapters of this book.

References

Abbasi, E. (2025). Application of remote sensing and geospatial technologies in predicting vector-borne disease outbreaks. *Royal Society Open Science, 12*(10), Article 250536. https://doi.org/10.1098/rsos.250536

Alemohammad, H. (2023). Incorporating training data uncertainty in machine learning models for satellite imagery. In *EGU general assembly conference abstracts, EGU-10528*. https://ui.adsabs.harvard.edu/abs/2023EGUGA..2510528A/abstract.

American Association of Geographers. (2018). *Annual meeting of the American association of geographers: GeoAI sessions and workshops.*

Arashpour, M. (2023). AI explainability framework for environmental management research. *Journal of Environmental Management, 342*, Article 118149.

Arora, R., Basu, A., Mianjy, P., & Mukherjee, A. (2018). Understanding deep neural networks with rectified linear units. In *Presented at the international conference on learning representations, 30 Apr–3 May 2018.*

Aziz, G., Minallah, N., Saeed, A., Frnda, J., & Khan, W. (2024). Remote sensing based forest cover classification using machine learning. *Scientific Reports, 14*(1), 69.

Barnes, T. J. (2004). The place of locational analysis: A selective and interpretive history. *Progress in Human Geography, 28*(1), 69–95. https://doi.org/10.1191/0309132504ph469oa

Chen, G., Chen, S., Li, D., & Chen, C. (2025). A hybrid deep learning air pollution prediction approach based on neighborhood selection and spatio-temporal attention. *Scientific Reports, 15*(1), 3685. https://doi.org/10.1038/s41598-025-88086-1

Chinnaswamy, A., & Walton, N. (2022). AI and environmental health: A platform ecosystem perspective. In *Artificial intelligence and environmental sustainability: Challenges and solutions in the era of industry 4.0* (pp. 1–22). Singapore: Springer Nature Singapore.

Chui, M., Manyika, J., Miremadi, M., Henke, N., Chung, R., Nel, P., & Malhotra, S. (2018). *Notes from the AI frontier: Applications and value of deep learning*. McKinsey Global Institute. https://www.mckinsey.com.

Couclelis, H. (1986). Artificial intelligence in geography: Conjectures on the shape of things to come. *The Professional Geographer, 38*(1), 1–11. https://doi.org/10.1111/j.0033-0124.1986.00001.x

Diener, A., & Mudu, P. (2021). How can vegetation protect us from air pollution? A critical review on green infrastructure mitigation of air pollution in urban environments. *International Journal of Environmental Research and Public Health, 18*(4), 1–25. https://doi.org/10.3390/ijerph18042100

Dobson, J. E. (1983). Automated geography. *The Professional Geographer, 35*(2), 135–143. https://doi.org/10.1111/j.0033-0124.1983.00135.x

EPA. (2024). *National ambient air quality standards (NAAQS) for particulate matter*. United States Environmental Protection Agency. https://www.epa.gov/naaqs.

Fourqurean, J. W., Duarte, C. M., Kennedy, H., Marbà, N., Holmer, M., Mateo, M. A., Apostolaki, E. T., Kendrick, G. A., Krause-Jensen, D., McGlathery, K. J., & Serrano, O.

(2012). Seagrass ecosystems as a globally significant carbon stock. *Nature Geoscience, 5* (7), 505–509. https://doi.org/10.1038/ngeo1477

Han, L., Zhou, W., Li, W., & Li, L. (2022). Urbanization strategy and environmental changes: An insight with relationship between population change and fine particulate matter. *Science of the Total Environment, 806*, Article 150647. https://doi.org/10.1016/j.scitotenv.2021.150647

Harvey, D. (1972). *Explanation in geography*. Edward Arnold.

Hossain, M. S., & Huggins, R. (2021). Urbanization, environmental degradation and sustainable development in Bangladesh. *Environmental Development, 39*, Article 100651. https://doi.org/10.1016/j.envdev.2021.100651

Hughes, A. R., Williams, S. L., Duarte, C. M., Heck, K. L., & Waycott, M. (2009). Associations of concern: Declining seagrasses and threatened dependent species. *Frontiers in Ecology and the Environment, 7*(5), 242–246. https://doi.org/10.1890/080041

Janowicz, K., Gao, S., McKenzie, G., Hu, Y., & Bhaduri, B. (2020). GeoAI: Spatially explicit artificial intelligence techniques for geographic knowledge discovery and beyond. *International Journal of Geographical Information Science, 34*(4), 625–636. https://doi.org/10.1080/13658816.2019.1684500

Kamal Pasha, S. M., Islam, A. S., Rahman, M. M., & Shahid, S. (2022). Urban expansion and environmental change in Dhaka megacity using remote sensing and GIS. *Remote Sensing Applications: Society and Environment, 25*, Article 100668. https://doi.org/10.1016/j.rsase.2021.100668

Lavallin, E., & Downs, R. R. (2021). Geospatial artificial intelligence: Toward a research agenda. *Computers, Environment and Urban Systems, 86*, Article 101584. https://doi.org/10.1016/j.compenvurbsys.2020.101584

Li, X., Zhang, J., Li, H., & Chen, Y. (2023). Vegetation structure and particulate matter removal efficiency in urban environments. *Urban Forestry and Urban Greening, 82*, Article 127869. https://doi.org/10.1016/j.ufug.2023.127869

Lopez, C. B., Tilney, C. L., Muhlbach, E., Bouchard, J. N., Villac, M. C., Henschen, K. L., Markley, L. R., Abbe, S. K., Shankar, S., Shea, C. P., Flewelling, L., Garrett, M., Badylak, S., Phlips, E. J., Hall, L. M., Lasi, M. A., Parks, A. A., Paperno, R., Adams, D. H., … Hubbard, K. A. (2021). High resolution spatiotemporal dynamics of harmful algae in the Indian River Lagoon (Florida)—A case study of Aureoumbra lagunensis, Pyrodinium bahamense, and pseudo-nitzschia. *Frontiers in Marine Science, 8*. https://doi.org/10.3389/fmars.2021.769877

Lu, Y., Xu, J., Chen, Y., & Zhang, H. (2024). Urban air pollution and health risks under rapid urbanization: A global assessment. *Environmental Research, 235*, Article 116676. https://doi.org/10.1016/j.envres.2023.116676

Mao, Q., Tuia, D., Blaschke, T., Li, X., Soergel, U., & Gong, P. (2017). Deep learning in remote sensing: A review. *IEEE Geoscience and Remote Sensing Magazine, 5*(4), 8–36. https://doi.org/10.1109/MGRS.2017.2762307

Maxwell, A. E., Warner, T. A., & Fang, F. (2018). Implementation of machine-learning classification in remote sensing: An applied review. *International Journal of Remote Sensing, 39*(9), 2784–2817. https://doi.org/10.1080/01431161.2018.1433343

Openshaw, S. (1994). Computational human geography: A research agenda. *Environment and Planning A: Economy and Space, 26*(4), 499–508. https://doi.org/10.1068/a260499

Orth, R. J., Carruthers, T. J. B., Dennison, W. C., Duarte, C. M., Fourqurean, J. W., Heck, K. L., Hughes, A. R., Kendrick, G. A., Kenworthy, W. J., Olyyarnik, S., Short, F. T., Waycott, M., & Williams, S. L. (2006). A global crisis for seagrass ecosystems. *BioScience, 56*(12), 987. https://doi.org/10.1641/0006-3568(2006)56[987:agcfse]2.0.co;2

Phlips, E. J., Badylak, S., Lasi, M. A., Chamberlain, R., Green, W. C., Hall, L. M., Hart, J. A., Lockwood, J. C., Miller, J. D., Morris, L. J., & Steward, J. S. (2015). From red tides to green and brown tides: Bloom dynamics in a restricted subtropical lagoon under shifting climatic conditions. *Estuaries and Coasts, 38*(3), 886–904.

Radanliev, P. (2025). AI ethics: Integrating transparency, fairness, and privacy in AI development. *Applied Artificial Intelligence, 39*(1), Article 2463722. https://doi.org/10.1080/08839514.2025.2463722

Rahaman, M., Southworth, J., Amanambu, A. C., Tefera, B. B., Alruzuq, A. R., Safaei, M., Hasan, M. M., & Culver Smith, A. (2025). Combining deep learning and machine learning techniques to track air pollution in relation to vegetation cover utilizing remotely sensed data. *Journal of Environmental Management, 376*, Article 124323. https://doi.org/10.1016/j.jenvman.2024.124323

Rai, P. K. (2016). Impacts of particulate matter pollution on plants: Implications for environmental biomonitoring. *Ecotoxicology and Environmental Safety, 129*, 120–136. https://doi.org/10.1016/j.ecoenv.2016.03.012

Rana, M. M., Mahmud, M., & Bari, M. A. (2016). Seasonal variation of air pollution in Dhaka city. *Atmospheric Pollution Research, 7*(3), 422–431. https://doi.org/10.1016/j.apr.2015.12.003

Rathi, D., Jain, S., & Sreedevi, I. (2017). Underwater fish species classification using convolutional neural network and deep learning. In *2017 ninth international conference on advances in pattern recognition, 27–30 December 2017*. https://doi.org/10.1109/ICAPR.2017.8593044

Reynolds, P. (2018). Seagrass and seagrass beds. *Smithsonian National Museum of Natural History*. https://ocean.si.edu/ocean-life/plants-algae/seagrassand-seagrass-beds. (Accessed 3 May 2023) (last date accessed.

Smith, N. (1984). *Uneven development: Nature, capital, and the production of space*. Basil Blackwell.

St. Johns River Water Management District. (2020). Seagrasses of the lagoon. https://www.sjrwmd.com/waterways/indian-river-lagoon/seagrass/.

Su, J. G., Dadvand, P., Nieuwenhuijsen, M. J., & Bartoll, X. (2022). Green space and particulate matter concentration reduction: A systematic review. *Environmental Research, 204*, Article 112350. https://doi.org/10.1016/j.envres.2021.112350

Tan, J. W., Chang, S.-W., Abdul-Kareem, S., Yap, H. J., & Yong, K.-T. (2020). Deep learning for plant species classification using leaf vein morphometric. *IEEE/ACM Transactions on Computational Biology and Bioinformatics, 17*(1), 82–90. https://doi.org/10.1109/TCBB.2018.2848653

Valentine, J. F., & Duffy, J. E. (2006). The central role of grazing in seagrass ecology. In *Seagrasses: Biology, ecology and conservation* (pp. 463–501). London, UK: Springer Nature.

Wang, J., Li, B., Zhou, Y., Meng, Q., Rende, S. F., & Rocco, E. (2020). Real-time and embedded compact deep neural networks for seagrass monitoring. In *2020 IEEE international conference on systems, man, and cybernetics, 11–14 October 2020, Toronto, ON, Canada*. https://doi.org/10.1109/smc42975.2020.9283372

Wang, J., Yang, S., & Zhao, H. (2023). Crisis event summary generative model based on hierarchical multimodal fusion. *Pattern Recognition, 144*, Article 109890. https://doi.org/10.1016/j.patcog.2023.109890

Weidmann, F., Jager, J., Reus, G., Schultz, S. T., Kruschel, C., Wolff, V., & Fricke-Neuderth, K. (2019). A closer look at seagrass meadows: Semantic segmentation for visual coverage estimation. In *OCEANS 2019 - Marseille, 17–20 June 2019, Marseille, France* (pp. 1–6). https://doi.org/10.1109/oceanse.2019.8867064

Yadav, V., Yadav, A. K., Singh, V., & Singh, T. (2024). Artificial neural network an innovative approach in air pollutant prediction for environmental applications: A review. *Results in Engineering, 22*, Article 102305. https://doi.org/10.1016/j.rineng.2024.102305

Zhang, Q., Han, Y., Li, V. O., & Lam, J. C. (2022). Deep-AIR: A hybrid CNN-LSTM framework for fine-grained air pollution estimation and forecast in metropolitan cities. *IEEE Access, 10*, 55818–55841.

Zhang, J., Li, X., Chen, Y., & Lu, Y. (2024). Urban vegetation structure and air pollution mitigation: Implications for sustainable city planning. *Landscape and Urban Planning, 240*, Article 104856. https://doi.org/10.1016/j.landurbplan.2023.104856

Foundations of Remote Sensing and Earth Observation

2

Cerian Gibbes[1], *Jane Southworth*[2], *Erin Bunting*[3], *Mashoukur Rahaman*[2], *Mohammad Safaei*[2], *Torit Chakraborty*[2], *Md Muyeed Hasan*[2], *Anusha Chaudhary*[2] *and Audrey Culver Smith*[2]

[1]Department of Geography & Environmental Studies, University of Colorado Colorado Springs, Colorado Springs, CO, United States; [2]Department of Geography, University of Florida, Gainesville, FL, United States; [3]Department of Geography, Environment, and Spatial Sciences, Michigan State University, East Lansing, MI, United States

Key terms and definitions

Foundations, platforms, and infrastructure

Remote sensing (RS): The science of collecting information about Earth's surface without direct contact, typically using airborne- or satellite-based sensors that measure reflected or emitted **electromagnetic** energy.

Earth observation (EO): The collection, analysis, and interpretation of data from spaceborne, airborne, or ground-based sensors to monitor Earth's physical, chemical, and biological systems.

Aerial photography: An early precursor to modern remote sensing involving photographs taken from aircraft or balloons, providing foundational approaches for mapping and spatial interpretation.

GIScience (Geographic Information Science): The discipline that studies the concepts and methods behind geographic information systems (GIS), including spatial data models, analysis, and visualization.

Geospatial data science: An interdisciplinary domain that combines spatial analysis, statistics, computing, and AI to extract insight from large-scale geographic data.

Cloud computing: The use of distributed computing infrastructure to store, manage, and analyze large datasets. In remote sensing, platforms such as Google Earth Engine and commercial cloud environments (e.g., AWS) have dramatically increased the scalability of analysis.

High-performance computing (HPC): Advanced computing systems and techniques that enable fast processing of large, complex remote sensing datasets and computationally intensive models.

Environmental monitoring: The systematic use of remote sensing and related observation methods to track environmental conditions and change over time to support management and policy decisions.

Remote Sensing, Big Data, and GeoAI. https://doi.org/10.1016/B978-0-443-26741-3.00015-5

Big data in remote sensing: Remotely sensed datasets characterized by large **volume**, high **velocity**, and increasing **variety** (and often **veracity** challenges), requiring scalable storage, processing, and analytics—frequently supported by cloud-based systems.

Sensors, data types, and major missions

Multispectral imaging: The collection of image data in multiple discrete spectral bands, supporting land cover mapping and material/vegetation discrimination based on spectral reflectance.

Hyperspectral imaging: A technique that captures data in many narrow, contiguous spectral bands (often hundreds), allowing detailed identification of materials and subtle spectral differences.

LiDAR (Light Detection and Ranging): An active remote sensing technology that uses laser pulses to measure distance and generate high-precision 3D information about terrain, vegetation structure, and built environments.

SAR (Synthetic Aperture Radar): An active radar system that collects information regardless of daylight or many weather conditions, widely used for mapping, surface structure, and deformation monitoring.

UAV (Unmanned Aerial Vehicle): Drone-based platforms equipped with optical, thermal, LiDAR, or multispectral sensors that collect flexible, very high-resolution data at local scales.

Landsat program: A series of NASA–USGS Earth-observing satellites launched since 1972, providing the longest continuous multispectral record of Earth's land surface.

MODIS (Moderate Resolution Imaging Spectroradiometer): A sensor on NASA's Terra and Aqua satellites that collects global observations in 36 spectral bands, widely used for vegetation, land surface, atmosphere, and ocean studies.

ASTER (Advanced Spaceborne Thermal Emission and Reflection Radiometer): A sensor on NASA's Terra satellite designed for mapping surface temperature, emissivity, reflectance, and elevation.

Sentinel missions (ESA/Copernicus): A European Space Agency satellite series under the Copernicus program providing global observations, including Sentinel-1 (radar), Sentinel-2 (optical), and Sentinel-3 (ocean/atmosphere), with broad open access.

AVHRR (Advanced Very High Resolution Radiometer): A long-running NOAA sensor series providing coarse-resolution imagery used for global vegetation and climate applications since the late 1970s.

SPOT (Satellite Pour l'Observation de la Terre): A French satellite program launched in 1986 providing medium-resolution multispectral imagery for land use and environmental monitoring.

IKONOS: One of the first commercial high-resolution satellites (1999–2015), providing sub-5 m imagery used in detailed mapping and urban studies.

QuickBird: A commercial high-resolution satellite providing imagery at approximately 0.61–2.44 m resolution, widely used for mapping and urban/environmental applications.

WorldView: A commercial satellite series offering very high spatial resolution (as fine as ~30 cm in later sensors) and frequent revisit times, used for precision mapping, monitoring, and change analysis.

GOES (Geostationary Operational Environmental Satellites): NOAA's geostationary weather and environmental satellites providing near–real-time atmospheric observations.
Terra and Aqua satellites: NASA Earth Observing System platforms hosting multiple instruments (including MODIS; Terra also hosts ASTER), supporting integrated observation of land, ocean, and atmospheric process

Core methods and analytics

Land cover classification: The process of assigning pixels or image objects to categories (e. g., forest, water, urban) using spectral, spatial, and/or temporal information.
Change detection: Methods used to identify and map changes in land cover or surface conditions over time by comparing multi-temporal remote sensing datasets.
Time series analysis: The analysis of remotely sensed observations collected repeatedly over time to characterize seasonal patterns, long-term trends, and abrupt disturbances.
Object-based image analysis (OBIA): An approach that segments imagery into meaningful objects prior to classification, using not only spectral information but also shape, texture, and spatial context.
Spectral index: A mathematical combination of two or more spectral bands designed to highlight specific surface properties, such as vegetation greenness (e.g., NDVI) or water presence (e.g., NDWI).
Data fusion: The combination of multiple datasets (which may differ by time, resolution, viewpoint, or sensor) to produce information that is more consistent, accurate, or useful than any single dataset alone.
Multi-sensor fusion: A specific type of data fusion that integrates observations from different sensor types (e.g., optical, radar, LiDAR) to leverage complementary strengths and improve mapping or interpretation.
Multi-sensor image fusion: Techniques that merge imagery from multiple sensors to enhance spatial detail, spectral richness, temporal frequency, or robustness to missing data.
Machine learning (ML): A subset of AI that uses algorithms to learn patterns from data and perform prediction or classification without being explicitly programmed with rules.
Deep learning (DL): A branch of machine learning based on multi-layer neural networks that can learn complex features from large datasets, often highly effective for image classification, object detection, and segmentation.
Convolutional neural network (CNN): A deep learning model architecture designed for image analysis that learns spatial features through convolutional filters and hierarchical representations.
GeoAI (Geospatial Artificial Intelligence): An interdisciplinary field combining AI and spatial science to analyze and model geographic data for mapping, monitoring, prediction, and decision support.
Cognitive remote sensing: An emerging approach in which AI systems move beyond detection to incorporate learning and reasoning about geospatial processes, often linking data-driven models with domain knowledge and physical understanding.
Explainable AI (XAI): Methods that increase transparency and interpretability of AI models, helping users understand how predictions are made and supporting trust, validation, and accountability.

1. Foundations of remote sensing and early earth observation

1.1 Conceptual origins of remote sensing

The conceptual foundations of Earth observation (EO) and remote sensing (RS) predate the advent of satellite-based technologies by several centuries. Introductory accounts of remote sensing often begin with early aerial photography, including the frequently reproduced images captured using pigeons equipped with miniature cameras in the early twentieth century (Fig. 2.1). While these images illustrate creative experimentation with overhead observation, they do not mark the true beginning of EO. Instead, the intellectual origins of remote sensing are more appropriately traced to earlier scientific developments that established the physical basis for observing the Earth from a distance.

Two foundational advances are particularly important. The development of optical instruments, most notably Galileo's telescope in the seventeenth century, demonstrated that distant objects could be systematically observed and measured. Later, the discovery of radiant energy by William Herschel in the early nineteenth century expanded scientific understanding beyond the visible portion of the electromagnetic spectrum, establishing the principle that information about physical objects could be inferred from emitted or reflected energy (Cracknell, 2018). Together, these developments provided the conceptual and physical framework upon which modern RS systems would later be built.

1.2 Aerial photography and military reconnaissance (1850s–1950s)

Prior to the satellite era, EO relied almost exclusively on balloon- and aircraft-based platforms to collect imagery of the Earth's surface. One of the earliest documented examples occurred in 1858, when Gaspard-Félix Tournachon, known as Nadar, captured aerial photographs of Paris from a tethered balloon at an altitude of approximately 1600 feet. Nadar's work demonstrated the potential of overhead imagery for mapping and interpretation, and similar efforts soon followed, including James Wallace Black's balloon-based photographs of Boston in 1860 (Cosgrove & Fox, 2010).

In the United States, aerial observation was adopted for military reconnaissance during the mid-nineteenth century. Balloon-based surveillance was used by the Union Army during the American Civil War to gather intelligence on enemy troop movements, representing one of the first systematic applications of RS for warfare (Cosgrove & Fox, 2010; Green, 2025). By the late nineteenth century, substantial investment had been directed toward the development of the U.S. Army Balloon Corps, underscoring the strategic value of aerial observation (Shock, 1997).

Figure 2.1 Collage of pigeons with Julious Neubronner's cameras and three aerial photos taken by the methods. The photos were taken in the early 1900s in Germany (by Julius Neubronner (1852–1932 [public domain], via wikimedia commons).

The role of aerial photography expanded dramatically during the twentieth century. During World War I, imagery collected from aircraft became central to battlefield mapping and tactical planning, while World War II saw extensive use of aerial photographs for strategic bombing, infrastructure assessment, and post-conflict damage evaluation (Campbell, 2008; Pearson, 2002). These practices continued into the Cold War, when high-altitude reconnaissance aircraft such as the U-2 enabled imaging of sensitive military installations at unprecedented spatial detail (Hinsman, 2014; Hopkins, 1998). Collectively, these military-driven developments established many of the technical and analytical foundations that later informed civilian and scientific Earth observation.

1.3 The emergence of satellite earth observation (1960s–1970s)

The launch of meteorological satellites in the 1960s marked a decisive transition from episodic aerial campaigns to continuous, space-based observation of the Earth. Early missions such as the Television Infrared Observation Satellite (TIROS) series demonstrated the feasibility of routine global monitoring, while the later development of geostationary satellites established temporal cadence and spatial coverage patterns that remain fundamental to EO today (Cracknell, 2018).

These early satellite systems fundamentally altered how geographers and Earth scientists approached environmental observation. Satellite EO enabled repeated, standardized measurements across large spatial extents, providing a consistent observational framework that was not possible with aircraft-based methods alone. As scientific interest expanded beyond meteorology to include land-surface processes, the potential of satellite RS for environmental monitoring became increasingly apparent.

In response, NASA and the U.S. Geological Survey (USGS) collaborated during the late 1960s and early 1970s to develop satellite systems designed explicitly to monitor the Earth's land surface. This collaboration laid the groundwork for a new era of EO focused on terrestrial processes, culminating in the launch of the Landsat program.

1.4 The landsat program as a transformational pivot

Launched in 1972, Landsat was the first satellite series designed specifically for systematic, long-term observation of the Earth's land surface at continental to global scales (Goward et al., 2021; Loveland et al., 2022; Ojima et al., 2022). The initial Landsat mission carried the Multispectral Scanner (MSS), which provided four spectral bands at a spatial resolution of 80 m, enabling global studies of vegetation, water bodies, and land use that were previously infeasible (Jensen, 1986). Subsequent Landsat missions expanded these capabilities substantially. Landsat 4 and Landsat 5 introduced the Thematic Mapper (TM), adding shortwave and thermal infrared bands and improving spatial resolution to 30 m for most spectral bands. Later missions, including Landsat seven with the Enhanced Thematic Mapper Plus (ETM+) and Landsat 8 and nine with the Operational Land Imager (OLI and OLI-2), further enhanced spectral fidelity, radiometric resolution, and continuity (Table 2.1; Ojima et al., 2022; Loveland et al., 2022).

Jointly managed by NASA and the USGS, the Landsat program became one of the longest-running and most influential EO initiatives in history. Its consistent, multispectral record has supported long-term analyses of deforestation, urban expansion, agricultural change, and other land-surface processes across diverse regions of the world (Goward et al., 2021). Despite early challenges related to data cost and access, Landsat established the empirical backbone for time-series analysis in remote sensing.

1.5 Constraints of the Pre–Open data era (1970s–2000s)

Although satellite EO represented a major technological advance, early RS research was constrained by significant practical limitations. During the 1980s and 1990s, satellite imagery was expensive, and access was restricted. In the 1990s, a single Landsat scene typically cost approximately $4,400, limiting most studies to one or two observation dates (USGS). As a result, many early RS analyses focused on relatively small spatial extents and could not adequately capture temporal dynamics. Computational constraints further shaped the scope of early RS research. Data storage and processing resources were costly relative to available computing infrastructure, limiting the feasibility of large-scale or multi-temporal analyses (Fig. 2.2; McCallum, 2023). Software environments such as ERDAS IMAGINE, PCI Geomatica, and ENVI enabled increasingly sophisticated image processing, but analyses remained bounded by hardware and data availability (Cracknell, 2018). Together, these constraints influenced the methodological character of RS studies prior to the open-data era. Research designs typically relied on limited temporal sampling, moderate spatial resolution, and single-sensor datasets. While these studies established many foundational principles of RS analysis, they also underscore the importance of later policy and technological shifts that enabled broader access to EO data.

2. From flagship missions to a global earth observation ecosystem

2.1 The USGS open-archive decision as an inflection point

A major structural transformation in remote sensing occurred in 2008, when the USGS adopted an open-access policy and released the entire Landsat archive free of charge. As a result, many studies were constrained to a small number of images and limited temporal coverage. The decision to make the Landsat archive freely available was motivated by the recognition that maximizing data accessibility would substantially increase the scientific and societal value of Earth observation data. This

Table 2.1 Review of most common sensors and satellite platforms.

Satellite	Launch year	Sensor	Spectral bands wavelength (μm)	Spatial resolution (m)	Satellite series	Operating agency
Landsat 1	1972	Multispectral Scanner (MSS)	4 bands (Green: 0.5-0.6, Red: 0.6-0.7, NIR: 0.7-0.8, NIR: 0.8-1.1)	80	Landsat 1, 2, 3	USGS/ NASA
Landsat 2	1975	Multispectral Scanner (MSS)	Same as Landsat 1	80	Landsat 1, 2, 3	USGS/ NASA
Landsat 3	1978	Multispectral Scanner (MSS)	Same as Landsat 1	80	Landsat 1, 2, 3	USGS/ NASA
Landsat 4	1982	Thematic Mapper (TM)	7 bands (Blue: 0.45-0.52, Green: 0.52-0.60, Red: 0.63-0.69, NIR, SWIR, TIR)	30 (120 for Thermal)	Landsat 4, 5	USGS/ NASA
Landsat 5	1984	Thematic Mapper (TM)	Same as Landsat 4	30 (120 for Thermal)	Landsat 4, 5	USGS/ NASA
Landsat 7	1999	Enhanced Thematic Mapper Plus (ETM +)	8 bands (incl. Panchromatic 0.52-0.90)	30 (15 for Pan, 60 for Thermal)	Landsat 7	USGS/ NASA
Landsat 8	2013	Operational Land Imager (OLI)	11 bands (Coastal, Blue, Green, Red, NIR, SWIR1, SWIR2, Cirrus, Thermal IR)	30 (15 for Pan, 100 for Thermal)	Landsat 8	USGS/ NASA
Landsat 9	2021	Operational Land Imager (OLI-2)	Same as Landsat 8	30 (15 for Pan, 100 for Thermal)	Landsat 9	USGS/ NASA
AVHRR	1978-present	Advanced Very High-Resolution Radiometer (AVHRR)	6 bands (Visible to Thermal IR, 0.58–12.5 μm)	1100	NOAA Satellites	NOAA

Terra (EOS AM-1)	1999	MODIS (Moderate Resolution Imaging Spectroradiometer)	36 bands (0.41–14.38 μm)	250/500/1000	Terra, Aqua, Aura	NASA
Aqua (**EOS PM-1**)	2002	MODIS (Moderate Resolution Imaging Spectroradiometer)	Same as Terra	250/500/1000	Terra, Aqua, Aura	NASA
Sentinel-2A	2015	Multispectral Instrument (MSI)	13 bands (Coastal, Blue, Green, Red, NIR, SWIR, Red Edge, Water Vapor)	10/20/60	Sentinel-2A, 2B	ESA
Sentinel-2B	2017	Multispectral Instrument (MSI)	Same as Sentinel-2A	10/20/60	Sentinel-2A, 2B	ESA
IKONOS	1999	IKONOS Sensor	4 bands (Blue: 0.45-0.52, Green: 0.52-0.60, Red: 0.63-0.69, NIR: 0.76-0.90)	1 (Panchromatic), 4 (Multispectral)	IKONOS	GeoEye/ Maxar
QuickBird	2001	QuickBird Sensor	5 bands (Blue: 0.45-0.52, Green: 0.52-0.60, Red: 0.63-0.69, NIR: 0.76-0.90)	0.61 (Panchromatic), 2.44 (MS)	QuickBird	DigitalGlobe/ Maxar
WorldView-1	2007	WorldView Sensor	Panchromatic (0.45–0.80 μm)	0.50	WorldView-1	DigitalGlobe/ Maxar
WorldView-2	2009	WorldView Sensor	8 bands (Coastal, Blue, Green, Yellow, Red, Red Edge, NIR1, NIR2)	0.46 (Pan), 1.85 (Multispectral)	WorldView-1, 2, 3, 4	DigitalGlobe/ Maxar
WorldView-3	2014	WorldView Sensor	29 bands (Visible to SWIR)	0.31 (Panchromatic)	WorldView-1, 2, 3, 4	DigitalGlobe/ Maxar
WorldView-4	2016	WorldView Sensor	Same as WorldView-3	0.31 (Panchromatic)	WorldView-1, 2, 3, 4	DigitalGlobe/ Maxar

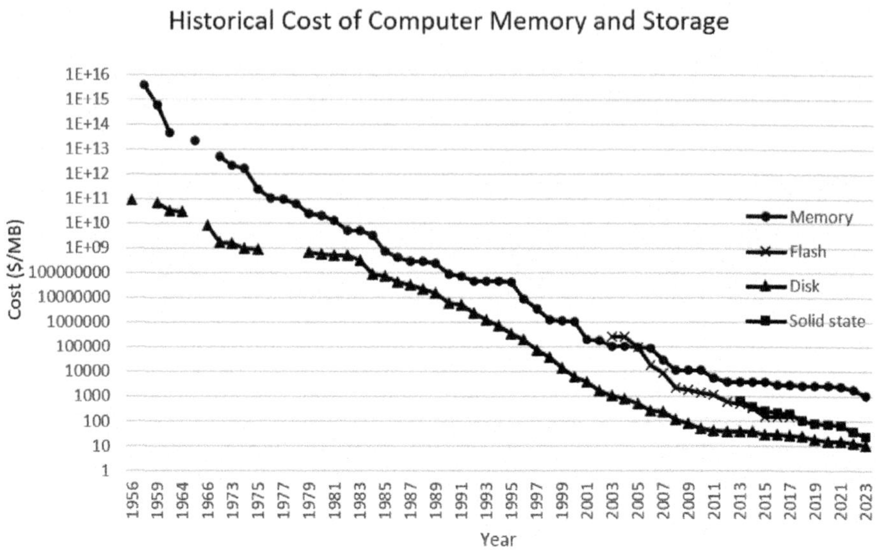

Figure 2.2 Historical cost of computer memory and storage. Data is expressed in USD per MB using a logarithmic scale.
Adapted from McCallum (2023).

policy shift enabled researchers to conduct multi-decadal analyses of land-surface processes, facilitated regional-to-global scale studies, and accelerated innovation across disciplines including climate science, land-use planning, disaster response, and agricultural monitoring (Cracknell, 2018). Importantly, open access to Landsat data also catalyzed the development of cloud-based analytical platforms such as Google Earth Engine, which lowered computational barriers and enabled scalable processing of large EO datasets. Together, these changes transformed RS from a field characterized by data scarcity into one defined by data abundance, setting the stage for the methodological shifts examined later in this chapter.

2.2 Government-funded global monitoring missions

Alongside Landsat, a series of government-funded satellite missions expanded the scope and capabilities of global Earth observation. One of the most influential has been the Advanced Very High Resolution Radiometer (AVHRR), which has operated continuously on NOAA satellites since 1978. Despite its relatively coarse spatial resolution of approximately 1.1 km, AVHRR provides a long-term, consistent global record that has been essential for climate monitoring, vegetation dynamics, and sea surface temperature analysis (Cracknell, 2018).

NASA's Earth Observing System further strengthened global EO capacity with the launch of the Terra satellite in 1999 and Aqua in 2002, both carrying the Moderate Resolution Imaging Spectroradiometer (MODIS). MODIS captures data in 36 spectral bands ranging from the visible to the thermal infrared, with spatial resolutions of 250 m, 500 m, and 1 km, enabling frequent global coverage suited to monitoring dynamic Earth system processes such as deforestation, primary productivity, and atmospheric conditions (Table 2.1). More recently, the European Space Agency's Sentinel missions under the Copernicus program have continued the trend toward freely available, high-quality EO data. Sentinel-2A and Sentinel-2B, launched in 2015 and 2017 respectively, provide multispectral imagery across 13 spectral bands at spatial resolutions of 10, 20, and 60 m, with a combined revisit time of approximately 5 days. These missions have become central to applications such as agricultural monitoring, land-use mapping, and disaster response (Table 2.1).

2.3 Commercial satellite proliferation and high-resolution EO

While government-funded missions have historically dominated Earth observation, commercial satellite programs have played an increasingly important role in expanding data availability and spatial resolution. The launch of IKONOS in 1999 marked a significant milestone, as it was among the first commercial satellites to provide sub-5 m multispectral imagery and 1 m panchromatic data suitable for detailed mapping and urban analysis (Ryan et al., 2003; Toutin & Cheng, 2002).

Subsequent commercial platforms, including QuickBird and the WorldView series operated by DigitalGlobe (now Maxar Technologies), further improved spatial resolution and spectral diversity. WorldView-3, for example, offers panchromatic imagery at approximately 30 cm resolution and expanded spectral coverage extending into the shortwave infrared (Table 2.1). These systems have enabled high-precision applications in urban planning, infrastructure monitoring, and security, complementing the broader coverage provided by public missions.

2.4 Increasing data density: Sensors, revisit, and scale

The combined expansion of public and commercial EO missions has led to a rapid increase in the number of active satellites, sensor diversity, and temporal revisit frequency. Belward and Skøien (2015) document a steady growth in the number of Earth observation satellites, driven by increased launch rates and extended mission lifespans. At the same time, technological advances have supported both continuity and innovation in sensor design. While certain spatial resolutions, such as the 30 m grid used by Landsat, have been maintained to ensure long-term comparability, an increasing number of sensors now provide much higher spatial and temporal resolution. Modern constellations, including small-satellite systems, can deliver daily or near-real-time coverage, enabling applications that require frequent monitoring such as precision agriculture and disaster management.

2.5 Methodological implications prior to the GeoAI era

The structure of the early RS data ecosystem had direct implications for research design and methodological choices. Studies conducted during the 1990s and early 2000s typically relied on a limited number of observations, moderate or coarse spatial resolution, and single-sensor data sources. Pixel-based classification, spectral indices, and simple change detection approaches dominated the literature, reflecting both data and computational constraints. As computational capacity increased and data availability slowly expanded, methodological diversity grew. Researchers began exploring non-parametric classifiers, including neural networks, particularly for studies integrating multisource data (Lu & Weng, 2007). Calls for systematic evaluation of classification approaches led to comparative assessments of methods and greater attention to analytical rigor (Khatami et al., 2016). These developments represent important precursors to later machine learning and GeoAI frameworks.

2.6 From expanding archives to data-driven science

By the late 2000s and early 2010s, the convergence of open data policies, expanding satellite constellations, and improving computational infrastructure fundamentally reshaped remote sensing. RS research increasingly shifted toward multi-temporal, multi-sensor analyses and larger spatial extents, moving beyond per-scene studies toward systematic monitoring of Earth systems. These structural changes in the EO data ecosystem directly underpin the empirical patterns documented in the next section. The citation analysis that follows demonstrates how methodological advances, particularly the rise of machine learning, time-series analysis, and GeoAI, emerged gradually in response to increasing data availability and scale, rather than appearing as abrupt paradigm shifts.

3. Tracing the evolution of remote sensing: Insights from three decades of top-cited research

The historical evolution of RS and EO outlined in Sections 1 and 2 establishes the technological, institutional, and data foundations that shape contemporary remote sensing research. From early conceptual advances and aerial reconnaissance, through the development of satellite-based EO and the central role of the Landsat program, the field has progressed from data scarcity toward an era defined by dense archives, multisensor coverage, and global accessibility. The 2008 USGS decision to make the Landsat archive freely available represents a critical inflection point, enabling systematic time-series analysis and large-scale synthesis that were previously infeasible (Wulder et al., 2012).

Building directly on this context, this section examines how structural shifts in data availability, sensor diversity, and computing capacity are reflected in the intellectual record of the discipline. Through a systematic review of top-cited papers published between 1990 and 2024 in three flagship remote sensing journals, we trace the transformation of research questions, data complexity, and analytical methods across 3 decades. Rather than viewing the emergence of machine learning and GeoAI as abrupt methodological ruptures, the citation analysis reveals them as the culmination of incremental change driven by expanding EO capabilities.

3.1 Selection criteria and analytical approach

The literature review spans calendar years 1990 through 2024. For each year within this period, the single most-cited research article from each of the three journals was identified. Citation counts were determined in October 2025, ensuring a consistent reference point for comparison across all years. Selection was conducted by reviewing all articles published within each calendar year and identifying the paper with the highest total citation count as of October 2025. Review articles were excluded from consideration, as their inherently retrospective and synthetic nature typically results in elevated citation counts that reflect cumulative influence rather than representing the primary research directions emerging in a given year. The objective of this analysis was to capture leading empirical and methodological contributions shaping the field at the time of publication.

In rare cases, papers that were highly cited but did not represent substantive advances in remote sensing research (e.g., narrowly scoped technical notes or content not indicative of broader methodological or thematic trajectories) were also excluded. These instances were minimal and applied consistently across journals to maintain comparability. Where applicable, journal-designated "best papers" were separately flagged but retained within the analytical pool. For each selected article, a standardized analytical framework was utilized to document (Tables 2.2–2.4):

- primary research questions or objectives
- data sources and sensor types
- number of observation dates and temporal structure
- spatial scale and study design
- analytical methods employed

This structured extraction enabled systematic comparison across decades. Particular attention was given to trends in temporal depth (e.g., single-date vs. multitemporal analyses), data diversity (e.g., single-sensor vs. multi-sensor integration), and methodological evolution (e.g., statistical classification, machine learning, deep learning, hybrid approaches). By documenting these characteristics consistently across years, we were able to assess how data cost, archive accessibility, sensor proliferation, and computational capacity influenced the trajectory of remote sensing research.

The citation lists for the three journals collectively provide a compact history of EO from 1990 to 2024 (Tables 2.2–2.4). *IEEE Transactions on Geoscience and Remote Sensing* (IEEE TGRS) tilts toward methodological innovation in sensors, signal processing, and algorithms (Table 2.2). *Remote Sensing of Environment* (RSE) emphasizes environmental and land surface applications, with iconic work on vegetation indices, biophysical retrievals, and later global products and time series (Table 2.3).

Table 2.2 Top-cited papers in IEEE transactions on geoscience and remote sensing identified for each calendar year (1990–2024) based on citation counts as of october 2025; review articles were excluded.

Year	In-text citation	Topic	Method family	Data complexity
1990	Benediktsson et al. (1990)	Deep Learning, Neural Networks	Deep Learning	Single-sensor
1990	Lopes et al. (1990)	SAR analysis	Classical RS	Single-sensor
1991	Bamler (1991)	Remote sensing study, Doppler	Classical RS	Single-sensor
1992	King et al. (1992)	Water/aerosol mapping	Classical RS	Multi-sensor
1993	Kealy and Hook (1993)	Climate monitoring	Classical RS	Single-sensor
1994	Harsanyi and Chang (1994)	Remote sensing study, Hyperspectral	Classical RS	Single-sensor
1995	Dubois et al. (1995)	SAR analysis	Classical RS	Multi-sensor Time-series
1996	Wan and Dozier (1996)	Climate monitoring, Simulation	Classical RS	Multi-sensor
1997	Cloude and Pottier (1997)	SAR analysis	Classical RS	Single-sensor
1998	Freeman and Durden (1998)	SAR analysis	Classical RS	Single-sensor
1999	Soh and Tsatsoulis (1999)	SAR analysis	Statistical ML	Time-series
2000	Ferretti et al. (2000)	SAR analysis	Classical RS	Time-series
2001	Ferretti et al. (2001)	SAR analysis	Classical RS	Time-series
2002	Berardino et al. (2002)	SAR analysis	Classical RS	Time-series
2003	Platnick et al. (2003)	Cloud detection	Classical RS	Multi-sensor Time-series
2004	Melgani and Bruzzone (2004)	Machine learning, Hyperspectral	Statistical ML	Single-sensor
2005	Nascimento and Dias (2005)	Remote sensing study, Hyperspectral	Classical RS	Single-sensor
2006	Gao et al. (2006)	Data fusion	Classical RS	Multi-sensor Time-series
2007	Krieger et al. (2007)	SAR analysis	Classical RS	Multi-sensor Time-series
2008	Fauvel et al., (2008)	Data fusion	Statistical ML	Single-sensor

2009	Berni et al. (2009)	Vegetation analysis	Classical RS	Single-sensor Time-series
2010	Pal & Foody (2010)	Data fusion	Statistical ML	Multi-sensor
2011	Ferretti et al. (2011)	SAR analysis	Classical RS	Time-series
2012	Yokoya et al. (2012)	Data fusion	Classical RS	Multi-sensor Time-series
2013	King et al. (2013)	Climate monitoring	Classical RS	Multi-sensor Time-series
2014	Chen et al. (2014)	Remote sensing study	Classical RS	Single-sensor
2015	Vivone et al. (2014)	Remote sensing study	Classical RS	Multi-sensor
2016	Chen, Chen et al. (2016) and Chen, Jiang et al. (2016)	Deep learning	Deep Learning	Time-series
2017	Xia et al. (2017)	Remote sensing study, Google Earth	Classical RS	Time-series
2018	Zhong et al. (2018)	Deep learning, Hyperspectral	Deep Learning	Single-sensor
2019	Li et al. (2019)	Deep learning, Hyperspectral	Deep Learning	Multi-sensor Time-series
2020	Wan et al. (2021)	Deep learning, Hyperspectral	Deep Learning	Multi-sensor
2021	Yu et al. (2021)	SAR analysis, Data fusion	Statistical ML	Multi-sensor
2022	Han et al. (2022)	Remote sensing study	Classical RS	Multi-sensor Time-series
2023	Sun et al. (2023)	Deep learning	Deep Learning	Multi-sensor
2024	Chen et al. (2024)	SAR analysis	Classical RS	Single-sensor

Table 2.3 Top-cited papers in remote sensing of environment identified for each calendar year (1990–2024) based on citation counts as of october 2025; review articles were excluded.

Year	In-text citation	Topic	Method family	Data complexity
1990	Smith et al. (1990)	Vegetation analysis	Classical RS	Time-series
1991	Hall et al. (1991)	Data fusion	Classical RS	Multi-sensor Time-series
1992	Chappelle et al. (1992)	Remote sensing study	Classical RS	Time-series
1993	Kruse et al. (1993)	Software development	Classical RS	Single-sensor
1994	Qi et al. (1994)	Vegetation analysis	Classical RS	Multi-sensor
1995	Roujean and Breon (1995)	Vegetation analysis	Classical RS	Single-simulation
1996	Gao (1996)	Vegetation analysis	Classical RS	Single-sensor
1997	Huete et al. (1997)	Vegetation analysis	Classical RS	Multi-sensor
1998	Holben et al. (1998)	Aerosol monitoring	Classical RS	Time-series
1999	Wagner et al. (1999)	Soil moisture analysis	Classical RS	Time-series
2000	Smirnov et al. (2000)	Remote sensing study	Classical RS	Time-series
2001	Broge and Leblanc (2001)	Vegetation analysis	Classical RS	Single-simulation
2002	Sims and Gamon (2002)	Remote sensing study	Classical RS	Single-sensor
2003	Hansen and Schjoerring (2003)	Vegetation analysis	Classical RS	Single-sensor
2004	Haboudane et al. (2004)	Vegetation analysis	Classical RS	Time-series
2005	Liu et al. (2005)	Agricultural mapping	Classical RS	Multi-sensor Time-series
2006	Chen et al. (2006)	Urban analysis	Classical RS	Multi-sensor Time-series
2007	Yuan and Bauer (2007)	Urban analysis	Classical RS	Multi-sensor Time-series
2008	Jiang et al. (2008)	Vegetation analysis	Classical RS	Time-series
2009	Hu (2009)	Remote sensing study	Classical RS	Time-series
2010	Kennedy et al. (2010)	Forest monitoring	Classical RS	Time-series

2011	Myint et al. (2011)	Land cover mapping	Classical RS	Single-sensor
2012	Zhu and Woodcock (2012)	Remote sensing study	Classical RS	Time-series
2013	Jin et al. (2013)	Land cover mapping	Classical RS	Time-series
2014	Feyisa et al. (2014)	Water/flood mapping	Classical RS	Time-series
2015	Estel et al. (2015)	Vegetation analysis	Statistical ML	Time-series
2016	Dong et al. (2016)	Remote sensing study	Classical RS	Time-series
2017	Veloso et al. (2017)	Vegetation analysis	Classical RS	Multi-sensor Time-series
2018	Belgiu and Csillik (2018)	Vegetation analysis	Classical RS	Time-series
2019	Zhong et al. (2019)	Vegetation analysis	Deep Learning	Time-series
2020	Tong et al. (2020)	Land cover mapping	Deep Learning	Multi-sensor Time-series
2021	Potapov et al. (2021)	Forest monitoring	Classical RS	Multi-sensor Time-series
2022	Wei et al. (2022)	Ozone monitoring	Deep Learning	Multi-sensor Time-series
2023	Wu et al. (2023)	Urban analysis	Statistical ML	Multi-sensor Time-series
2024	Yu et al. (2024)	Urban analysis	Classical RS	Multi-sensor Time-series

Table 2.4 Top-cited papers in ISPRS journal of photogrammetry and remote sensing identified for each calendar year (1990–2024) based on citation counts as of october 2025; review articles were excluded.

Year	In-text citation	Topic	Method family	Data complexity
1990	Sobrino et al. (1990)	Climate monitoring	Classical RS	Single-sensor
1991	Ehlers (1991)	SAR analysis	Classical RS	Multi-sensor
1992	Delécolle et al. (1992)	Agricultural mapping	Classical RS	Multi-sensor Time-series
1993	Meyer et al. (1993)	Forest monitoring	Classical RS	Single-sensor
1994	Maselli et al. (1994)	Change detection	Classical RS	Multi-sensor Time-series
1995	Weidner and Förstner (1995)	Remote sensing study	Classical RS	Single-sensor
1996	Yuan and Elvidge (1996)	Land cover mapping	Classical RS	Time-series
1997	Naesset (1997)	Forest monitoring	Classical RS	Single-sensor
1998	Kraus and Pfeifer (1998)	Vegetation analysis	Classical RS	Single-sensor
1999	Axelsson (1999)	Vegetation analysis	Classical RS	Single-sensor
2000	Pollefeys et al. (2000)	Remote sensing study	Classical RS	Multi-sensor Time-series
2001	Elvidge et al. (2001)	Remote sensing study	Classical RS	Time-series
2002	Kääb (2002)	Climate monitoring	Classical RS	Multi-sensor Time-series
2003	Rabus et al. (2003)	SAR analysis	Classical RS	Multi-sensor
2004	Benz et al. (2004)	Data fusion	Classical RS	Multi-sensor
2005	Gruen and Akca (2005)	Remote sensing study	Classical RS	Single-sensor
2006	Wagner et al. (2006)	Remote sensing study	Classical RS	Single-sensor
2007	Höfle and Pfeifer (2007)	Climate monitoring	Classical RS	Single-sensor
2008	Darvishzadeh et al. (2008)	Vegetation analysis	Classical RS	Single-sensor
2009	Reitberger et al. (2009)	Forest monitoring	Classical RS	Single-sensor
2010	Awrangjeb et al. (2010)	Urban analysis	Classical RS	Multi-sensor

2011	Yu et al. (2011)	Forest monitoring	Statistical ML	Single-sensor
2012	Shao and Lunetta (2012)	Forest monitoring	Statistical ML	Time-series
2013	Frampton et al. (2013)	Vegetation analysis	Classical RS	Single-sensor
2014	Cheng et al. (2014)	Data fusion	Statistical ML	Single-sensor
2015	Vo et al. (2015)	Urban analysis	Classical RS	Single-sensor
2016	Shean et al. (2016)	Remote sensing study	Classical RS	Time-series
2017	Zhou et al. (2017)	Vegetation analysis	Classical RS	Time-series
2018	Marmanis et al. (2018)	Deep learning	Deep Learning	Single-sensor
2019	Huang and Wang (2019)	Urban analysis	Classical RS	Multi-sensor
2020	Diakogiannis et al. (2020)	Data fusion	Deep Learning	Single-sensor
2021	Li et al. (2021)	Change detection	Deep Learning	Multi-sensor Time-series
2022	Wang et al. (2022)	Urban analysis	Deep Learning	Multi-sensor Time-series
2023	Wang et al., (2023)	Vegetation analysis	Classical RS	Multi-sensor Time-series
2024	Dong et al. (2024)	Change detection	Deep Learning	Multi-sensor Time-series

ISPRS Journal of Photogrammetry and Remote Sensing (ISPRS) sits at the interface of photogrammetry, three-dimensional reconstruction, and mapping, with a more explicit focus on objects, buildings, and topography that gradually incorporates machine learning and deep learning (Table 2.4).

Across all three journals, the early 1990s are dominated by physical modeling, radiometric and atmospheric correction, and the development of new indices and metrics. The 2000s saw more complex multi-sensor and multi-temporal data, and more systematic fusion of RS with models and ancillary data. From about 2010 onward, there is a clear rise in machine learning, time series analysis, and global scale mapping. In the 2020s deep learning and GeoAI become more central, particularly in IEEE TGRS and ISPRS, with RSE showing strong adoption of machine learning and deep learning for large area monitoring (Tables 2.5–2.7). The decadal distribution in these tables captures this shift quite neatly, with only one deep learning or neural network paper among the IEEE TGRS top cited list in the 1990s, increasing to five from 2016-2024, and similar jumps in ISPRS (Tables 2.8 and 2.9, Fig. 2.3). Taken together, the three sets of top cited papers show a progression that mirrors this chapter's historical story. The field moves from algorithm and sensor centric work with carefully curated scenes, to large multi-year and multi-sensor datasets, through to methods that assume big data as the starting point and use GeoAI to extract information at scale (Tables 2.5–2.9).

3.2 IEEE TGRS: From signal processing to deep feature learning

The IEEE TGRS articles are heavily method-oriented (Tables 2.2 and 2.5). In the 1990s, the top cited papers include early comparisons of neural networks and statistical classifiers, adaptive speckle filtering for synthetic aperture radar, entropy-based classification, and scattering models for polarimetric SAR. These papers mostly rely on single date or simulated datasets, often with synthetic signals or carefully chosen test scenes, and the focus is on theoretical performance, estimation bounds, and algorithm design. SAR and microwave systems are very prominent in this decade, and hyperspectral data appear as a specialized but growing area. In the 2000s, the emphasis shifts toward multi-temporal interferometry, subsidence and deformation monitoring, global cloud products, and more advanced signal processing for SAR stacks. Many of the flagship papers from this decade introduce or refine concepts like permanent scatterers, time series interferometry, and global algorithms for MODIS cloud properties. The data are still often limited in space compared to later decades, but there is a clear move beyond single images toward temporal stacks and systematic monitoring.

The 2010s and early 2020s show an expansion of machine learning and deep learning within IEEE TGRS. Top cited papers now focus on hyperspectral image classification with deep networks, benchmark datasets for aerial scene classification, spectral spatial residual networks, and dynamic graph convolutional networks for spatiotemporal data. Hyperspectral imagery becomes a dominant data source in the deep learning group, and the number of papers using neural networks or deep architectures rises sharply compared to earlier decades. Multi-sensor fusion, interferometric data integration, and cross calibration of large satellite constellations also feature prominently. Methodological complexity increases, and many papers assume

Table 2.5 Decadal summary of top cited papers in IEEE transactions on geoscience and remote sensing.

Decade	Representative themes	Dominant data sources	Dominant methods	GeoAI and big data signals
1990 to 1999	SAR speckle filtering, scattering models, early neural network classification, thermal emissivity and cloud properties	Airborne and spaceborne SAR, simulated signals, early hyperspectral scenes	Physical and statistical models, entropy-based classifiers, early neural networks, maximum likelihood	First experiments with neural networks, mostly single date scenes, focus on algorithm design rather than scale
2000 to 2009	Interferometric SAR, subsidence and deformation, global cloud and aerosol products	Multi date SAR stacks, MODIS, ENVISAT, scatterometers	Time series interferometry, permanent scatterers, physical retrieval algorithms, some machine learning	Move from single scenes to temporal stacks, early large-scale products, growing emphasis on automation
2010 to 2019	Hyperspectral image classification, small target detection, benchmark datasets, global cloud climatologies	Hyperspectral imagery, global satellite archives, mixed sensor datasets	Spectral spatial feature extraction, nonnegative matrix factorization, graph and patch-based methods, early deep networks	Clear increase in machine learning and deep learning, hyperspectral deep feature learning, larger training datasets
2020 to 2024	Deep networks for hyperspectral and spatiotemporal data, graph convolutionalnetworks, multi system fusion, sensor cross calibration	Hyperspectral and multispectral constellations, interferometric stacks, cross sensor time series	Convolutional neural networks, deep residual networks, graph neural networks, automated cross calibration methods	GeoAI is central. Methods assume large data, complex constellations, and high-performance computing as the default context

Table 2.6 Decadal summary of top cited papers in remote sensing of environment.

Decade	Representative themes	Dominant data sources	Dominant methods	GeoAI and big data signals
1990 to 1999	Vegetation and water indices, PAR, bidirectional reflectance, radiometric rectification, biophysical retrievals	Landsat TM, AVIRIS, field spectra and tower measurements	Index design and evaluation, physical models, radiometric correction, ratio analysis	Foundations for global monitoring. Single sites and small networks dominate, but several studies already explore global coverage concepts
2000 to 2009	Advanced vegetation indices, and pigment retrievals, global and multi-site monitoring, cloud screening and quality control	Multiyear multi-site satellite records, simulated data	Inversion and retrieval algorithms, radiative transfer, regression and comparative model evaluation	Strong move to multiyear and multi-site data, start of truly global products and operational quality control pipelines
2010 to 2019	Multi-temporal crop monitoring, paddy rice mapping, farmland abandonment and recultivation, automated water extraction and change detection	Landsat and MODIS time series, Sentinel 2, high resolution commercial imagery	Object-based image analysis, change detection methods, initial machine learning and deep learning for classification	Time series and automation are central. Many studies use long archives and test machine learning at regional and continental scales
2020 to 2024	Global canopy height, high resolution land cover mapping, deep learning-based crop classification and dense time series analysis	Global multi-sensor archives, high resolution optical, microwave and ancillary data, optimized networks	Deep learning for multi-temporal classification, data fusion, statistical and physical hybrid models	RSE becomes a key outlet for big data applications. GeoAI techniques are used to build global or near global products that link directly to climate and land system questions

Table 2.7 Decadal summary of top cited papers in ISPRS journal of photogrammetry and remote sensing.

Decade	Representative themes	Dominant data sources	Dominant methods	GeoAI and big data signals
1990 to 1999	Thermal inertia and soil moisture, multisensor image fusion, crop models, topographic correction, early automatic building extraction, laser scanning for terrain	High resolution aerial imagery, early satellite data, first airborne lidar, multi-sensor optical data	Photogrammetric methods, image fusion, radiometric correction, basic classification and feature extraction	Foundations for object-based mapping and three-dimensional analysis. GeoAI is not yet visible, but scene complexity and three-dimensional structure are already central
2000 to 2009	Three-dimensional reconstruction from image sequences, nighttime lights, high mountain deformation, SRTM based topography, object-oriented image analysis	Lidar, SAR, nighttime lights, global DEMs, high resolution optical imagery	Multi resolution segmentation, object-oriented analysis, time series deformation models, automated reconstruction	Rapid diversification of sensors. Early signs of big data through global DEMs and large nighttime light datasets
2010 to 2019	Automatic building detection, comparison of machine learning classifiers, Sentinel 2 capability, multi class object detection, point cloud segmentation, crop yield prediction	Airborne and terrestrial lidar, high resolution optical imagery, Sentinel 2, multi-temporal datasets	Support vector machines, neural networks, random forests, object detection pipelines, open-source processing chains	Machine learning increases. Papers compare algorithms and propose pipelines that are designed to scale and to be reused by others
2020 to 2024	Deep learning frameworks for semantic segmentation and change detection, transformer-based architectures, CLIP style models, vegetation trait estimation	Multi-sensor imagery, three-dimensional point clouds, dense time series, pre-trained vision language models	ResUNet, generative adversarial networks, transformer enhanced U Nets, CLIP based feature extraction	GeoAI is fully visible. Deep learning and foundation style models are at the heart of highly cited work, and many methods are explicitly designed for large scale and cross domain transfer

Table 2.8 Data complexity by decade showing the growth of multi-sensor and time-series studies among the most cited papers in three flagship remote sensing journals: Remote Sensing of Environment, ISPRS Journal of photogrammetry and remote sensing, and IEEE transactions on geoscience and remote sensing (1990–2024).

Decade (s)	Single-sensor	Multi-sensor	Time-series	Multi-sensor time-series
1990	16	5	6	4
2000	11	2	8	9
2010	8	4	14	4
2020	2	3	0	10

Table 2.9 Counts by method family and data paradigm: method–data matrix for top cited papers across the three journals: remote Sensing of Environment, ISPRS journal of photogrammetry and remote sensing, and IEEE transactions on geoscience and remote sensing (1990–2024).

Method family	Single-sensor	Multi-sensor	Time-series	Multi-sensor time-series
Classical RS	29	10	23	20
Statistical ML	4	2	3	1
Deep learning	4	2	2	6

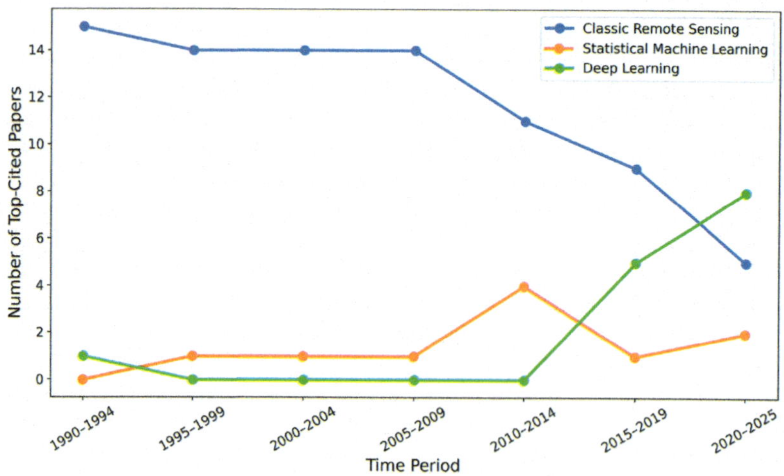

Figure 2.3 Evolution of methodological families in the most-cited papers (1990–2024) across three flagship remote sensing journals: Remote sensing of environment, ISPRS journal of photogrammetry and remote sensing, and IEEE transactions on geoscience and remote sensing.

access to sizable datasets and high-performance computing. Within this journal, GeoAI appears first as experiments with neural networks in the early 1990s, then resurfaces with much greater maturity once big data and GPU computing become widely available.

3.3 Remote sensing of environment: From indices and retrievals to global time series

The RSE list provides a complementary view that is more application centered (Tables 2.3 and 2.6). In the 1990 to 1999 decade, the highly cited papers introduce and refine many of the vegetation and water indices that became standard in the field. Examples include the modified soil adjusted vegetation index, the normalized difference water index, and global comparisons of vegetation indices. There are also influential studies on bidirectional reflectance, absorption of photosynthetically active radiation, and radiometric rectification. Many of these papers rely on data from Landsat TM, AVHRR, AVIRIS, field spectra, and relatively modest time series. Several still use simulated or experimental datasets, but the goal is already to support global or regional monitoring. During the 2000s, the top cited RSE papers continue to refine indices and retrievals but also begin to emphasize extended time series and global or multi-site analyses. There are studies that draw on 6 years or more of satellite data, multiple AERONET stations, and multi-year collections for vegetation monitoring. The methods are still largely physical or statistical rather than explicitly machine learning based, but the datasets grow larger and more continuous. This is the decade where RSE becomes the venue for high quality global products, with careful validation and uncertainty analysis.

From 2010 onward, there is a marked shift toward multi-temporal crop monitoring, mapping paddy rice, tracking farmland abandonment and recultivation, and automated water extraction. Object based classification, comprehensive change detection, cloud and cloud shadow detection, and per pixel versus object-based comparisons all appear in the top cited set. Many of these studies use multiyear Landsat, MODIS, or Sentinel time series and explicitly focus on automation over large extents. Later RSE papers in the 2010s and early 2020s use deep learning and machine learning for multi-temporal crop classification, high resolution land cover mapping, and global canopy height or biomass estimation. All five of the 2020–24 highly cited RSE papers use multiple sensors and are explicitly focused on time series. In this journal, we can see the transition from index design and single site experiments to time series analysis, global coverage, and data driven mapping that is very close in spirit to GeoAI, even when the methods are not always branded in that way.

3.4 ISPRS journal: From photogrammetry and 3D mapping to semantic deep learning

The ISPRS Journal series traces the evolution of photogrammetry and mapping in a similar way (Tables 2.4 and 2.7). In the 1990s the top cited papers focus on thermal inertia, multi-sensor image fusion, RS and crop production models, radiometric topographic correction, and one of the earliest highly cited pieces on automatic building extraction from high resolution imagery. There are also important contributions to determination of tree height, terrain models in wooded areas, and processing of laser scanner data for digital terrain models. These papers show a strong emphasis on three-dimensional structure, radiometric normalization, and early laser scanning. In the 2000s, ISPRS' top cited papers move decisively into three-dimensional reconstruction and global nighttime lights, as well as radar-based terrain deformation. The Shuttle Radar Topography Mission is a key milestone in this period, providing a global digital elevation model that underpins much of later geospatial analysis. Object oriented and multi resolution image analysis for high resolution data appears, along with work on monitoring high mountain terrain deformation, and automated reconstruction from image sequences. Data diversity increases markedly in this decade, with lidar, SAR, optical imagery, and extensive ground data often combined within the same study.

From 2010 onward, the ISPRS Journal became one of the central venues for machine learning and deep learning in photogrammetry and RS. Highly cited papers include automatic detection of residential buildings in complex environments, comparison of support vector machines, neural networks, and random forests, Sentinel two capability assessments, multi class geospatial object detection, point cloud segmentation, open source pipelines for mass processing of point clouds, crop yield prediction with multi-temporal data, and several deep network architectures such as ResUNet, generative adversarial network based change detection, UNet like transformers, and CLIP based change detection. The number of top cited papers that use neural networks or deep architectures increases sharply from 2010 to 2024, with four out of five in the 2020 to 2024 group using deep learning explicitly. This journal therefore provides a clear view of the move from geometric and radiometric problems to semantic segmentation, object detection, and representation learning, all of which are central to GeoAI.

3.5 Synthesis relative to GeoAI and big data

Across the three journals, the decadal patterns in these top cited lists support a simple story (Tables 2.5–2.7). In the 1990s, the field is learning how to correct and calibrate images, design robust indices, and derive basic structural and biophysical variables from relatively sparse data. In the 2000s large global missions like SRTM and MODIS, and the growth of multiyear archives, push researchers toward more systematic products and multi-sensor integration, but methods are still mostly rooted in classical statistics and physical models. In the 2010s both data and computing capacity grow, cloud platforms and open archives make global analysis practical, and machine learning and time series methods become normative. By the early 2020s deep learning, graph networks, and transformer architectures are routinely used on multi-sensor, multi-temporal datasets, and the journals highlight benchmark datasets, open pipelines, and models designed for reuse and transfer learning (Tables 2.8 and 2.9, Fig. 2.3).

These patterns align with the narrative in the earlier part of the chapter. The historical shift from analog to digital, from image processing to big data RS, and from computational to cognitive RS is mirrored in the citation records of these three flagship journals. The top cited papers show that GeoAI did not appear suddenly. Instead, it emerges gradually from decades of method development, index design, radiometric correction, and multi-sensor fusion, and it becomes dominant only once large image archives, time series, and cloud computing make data driven learning both possible and necessary.

Fig. 2.3 quantifies how methodological emphasis within top-cited remote-sensing research has shifted from 1990 to 2024. The early period from 1990 through the mid-2000s is dominated by Classical Remote Sensing approaches, including physically based radiometric corrections, spectral indices, and empirical or regression-based models, which account for the overwhelming majority of influential papers. During this time, statistical learning methods are largely absent, and early neural network applications appear only sporadically and remain isolated from mainstream practice. In the late 2000s and early 2010s, Statistical Machine Learning methods such as support vector machines and random forests begin to emerge more consistently, though they remain a minority relative to classical approaches. These methods function primarily as extensions of established empirical workflows rather than a wholesale methodological shift, and the field continues to rely heavily on sensor physics, calibrated reflectance, and targeted study areas.

A clear inflection occurs after 2015, when Deep Learning methods rapidly gain prominence across image classification, object detection, and large-area mapping applications. This shift coincides with the expansion of global open data archives (e.g., Landsat, MODIS, Sentinel), advances in cloud computing, and the availability of scalable training frameworks. By the 2020–25 period, deep learning surpasses both classical remote sensing and statistical machine learning among the most-cited papers. The relatively modest growth of statistical ML highlights its role as a transitional phase, while the sharp rise of deep learning reflects a structural transformation in the field. Fig. 2.3 captures the discipline's evolution from handcrafted features and per-scene analysis toward automated, data-intensive, learning-based EO characteristic of GeoAI.

Table 2.8 summarizes how data complexity in highly cited remote sensing studies has evolved from 1990 to 2024. In the 1990s, single-sensor and single-date analyses remain common, reflecting data availability and computational constraints of the period. However, multi-sensor studies are already present, and time-series approaches appear in a substantial subset of the most cited papers, indicating early interest in temporal dynamics despite limited archives. During the 2000s, time-series approaches expand noticeably, driven by the increasing availability of continuous AVHRR and MODIS observations and growing access to multi-date Landsat data. At the same time, multi-sensor time-series studies begin to emerge as researchers combine optical, thermal, and radar observations to address land-cover change, vegetation dynamics, and climate-related questions. Single-sensor studies remain important but no longer dominate the most influential work.

The 2010s mark a clear inflection point in data complexity. Time-series analyses become the most prevalent data paradigm among top-cited papers, with nearly half of the studies in this decade relying on multi-temporal observations. Multi-sensor approaches continue to appear, though often as targeted integrations rather than fully fused time-series frameworks. Single-date, single-sensor studies are now a clear minority among highly cited work. In the 2020s, the most complex category, multi-sensor time-series, becomes the largest single group among top-cited papers. Most influential studies in this period integrate both multi-sensor data streams and dense temporal archives, reflecting the maturation of cloud-based platforms and scalable analytical workflows. The absence of studies in the "time-series only" category during the 2020s reflects this increasing integration rather than a decline in temporal analysis; by this stage, nearly all time-series studies among the most cited papers also incorporate multi-sensor data and are therefore classified within the multi-sensor time-series category. This progression illustrates a decisive shift from image-based analysis toward dynamic, multi-source monitoring, providing the data foundation upon which contemporary GeoAI methods operate.

The method–data matrix (Table 2.9) highlights how analytical methods and data paradigms co-evolve in highly cited remote sensing research. Classical Remote Sensing methods still account for the largest share of top-cited papers overall; however, they are no longer confined to single-sensor, single-date analyses. Instead, classical approaches are frequently applied to time-series and multi-sensor time-series datasets, reflecting the long-standing role of physically based models and empirical indices in interpreting increasingly complex data archives. Statistical machine learning appears as a transitional methodological family within the matrix. Its presence is modest across all data paradigms, with applications spanning single-sensor, multi-sensor, and time-series studies. This distribution reflects the adoption of support vector machines, random forests, and related algorithms primarily as enhancements to established workflows during the 2000s and early 2010s, rather than as a dominant analytical paradigm. Deep learning and GeoAI methods cluster most strongly in the higher-complexity regions of the matrix. The majority of deep learning papers rely on either time-series or multi-sensor time-series data, with the largest concentration occurring in the multi-sensor, multi-temporal category. This pattern underscores that the rise of GeoAI in remote sensing is not driven solely by advances in algorithms. Instead, it reflects the concurrent maturation of the data ecosystem, including long observational archives, diverse sensor constellations, and cloud-based infrastructures capable of supporting large-scale, computationally intensive analysis.

3.6 Three decades of transformation in remote sensing research

Taken together, the citation analysis across *Remote Sensing of Environment*, *ISPRS Journal of Photogrammetry and Remote Sensing*, and *IEEE Transactions on Geoscience and Remote Sensing* traces a clear and accelerating transformation in the discipline. The 1990s and early 2000s were defined by steady methodological refinement within a largely classical paradigm, dominated by per-pixel classification, physical modeling, and single-sensor observations. From the mid-2000s onward, the expansion of global archives and the proliferation of new sensors introduced a decisive shift toward multi-sensor and multi-temporal approaches, reflected in the growing proportion of top-cited studies exploiting long time-series and data fusion. The 2010s represent a critical transition period, in which machine-learning frameworks begin to integrate with these richer datasets, complementing established statistical and physically based approaches rather than fully replacing them. By the 2020s, deep learning and cloud-based processing have become central to the most influential work, enabling analysis at continental to global scales.

Across all three journals, this trajectory underscores a dual evolution: first in data complexity (Table 2.8), from isolated scenes to dynamic, heterogeneous, and continuous observations, and second in analytical sophistication (Table 2.9), moving from empirical and statistical methods toward integrated GeoAI systems. The convergence of these trends defines the current era of big-data remote sensing, in which methodological innovation and data abundance reinforce one another. Evidence from the citation record demonstrates that the field's intellectual center of gravity has shifted decisively toward GeoAI and data-intensive EO, positioning RS as a leading contributor to data science and AI in the environmental sciences.

3.7 Why this history matters

This historical framing is essential for understanding contemporary developments in RS and GeoAI. The analytical approaches, data structures, and research questions that dominate the field today are not simply products of recent advances in machine learning or computing power; they are the outcome of decades of technological, institutional, and policy decisions that shaped how EO data are collected, archived, and accessed. From the continuity of the Landsat program to the expansion of global satellite constellations and the adoption of open data policies, these historical choices established the conditions under which data-intensive and AI-enabled methods could emerge. Recognizing this lineage helps situate GeoAI not as a methodological rupture, but as a logical extension of long-standing efforts to observe, understand, and model Earth system dynamics at scale.

4. Conclusion

The preceding sections traced the historical evolution of RS from its conceptual and technological foundations (Sections 1 and 2), through empirical evidence of shifting research practices documented via citation analysis (Section 3) and illustrated these transitions through two case studies highlighting early analytical approaches and constraints. Together, these perspectives demonstrate that contemporary GeoAI methods did not emerge in isolation, but instead reflect a cumulative progression shaped by sensor development, data access policies, computational capacity, and evolving analytical frameworks.

RS has advanced through successive technological revolutions that have transformed how scientists observe and understand the Earth, a trajectory documented historically in this chapter and empirically confirmed through patterns in the published literature. From its origins in aerial photography and film-based mapping to today's cloud-enabled, data-intensive systems, each era has been marked by innovations in sensors, data storage, analytical methods, domains of application, and computation. The contemporary landscape is defined by the fusion of geospatial data science, artificial intelligence, and high-performance computing, an integrative framework increasingly described as GeoAI, which enables continuous, intelligent EO (Chen, Chen et al., 2016, Chen, Jiang et al, 2016; Cracknell, 2018; Li, 2020; Li et al., 2021; Song et al., 2023).

Early foundations of the field emerged with nineteenth-century aerial photography, which introduced new vantage points on the landscape and laid the groundwork for systematic overhead observation. By the mid-20th century, military demands accelerated advances in airborne cameras and stereoscopic interpretation, establishing technical and analytical practices that would later be adapted for scientific Earth observation. The launch of Landsat one in 1972 marked a defining shift, as multispectral satellite imagery enabled the first long-term, global record of land-cover dynamics.

Integration with GIScience during the 1980s and 1990s marked a critical transition, as remotely sensed data were increasingly analyzed within geographic information systems. This integration allowed RS data to be combined with socioeconomic, ecological, and infrastructural layers, expanding research beyond classification accuracy toward understanding the processes driving land-use and land-cover change.

The big-data transformation unfolded in the early 2000s, when remote sensing transitioned from an era of data scarcity to one defined by data abundance. Global sensors such as Landsat, MODIS, ASTER, and Sentinel generated rapidly growing archives with frequent temporal coverage, while cloud-based platforms, notably Google Earth Engine, enabled distributed computation across planetary-scale datasets. Statistical and machine-learning approaches, including regression trees, support-vector machines, and random forests, improved generalization across diverse landscapes. Big-data remote sensing is characterized not only by data volume, but also by velocity, variety, and veracity, requiring scalable algorithms capable of near–real-time processing and analysis. Critically, these transformations were enabled not only by sensor proliferation, but also by institutional decisions, most notably the USGS open-access Landsat policy, that reshaped who could participate in data-intensive remote sensing research.

The emergence of GeoAI reflects the convergence of artificial intelligence, remote sensing, and spatial analysis, enabling a new generation of intelligent EO. Deep-learning architectures, including convolutional neural networks and U-Net–based models, now underpin applications ranging from deforestation detection to disaster response. These models learn spatial and temporal dependencies directly from imagery, offering unprecedented scalability and reproducibility. As noted by Song et al. (2023), this shift represents a new era of geocomputation defined by automation, openness, and the integration of physical and data-driven insights. Each technological advancement has expanded both the spatial and temporal scales of analysis and the scope of scientific questions that can be addressed. Overall, the integration of GeoAI represents the most transformative phase to date, uniting data abundance with analytical sophistication to support real-time, predictive environmental intelligence.

5. Future directions and unifying definitions

As remote sensing enters the next decade, its future lies in deeper convergence across sensors, disciplines, and scales of analysis. Satellite constellations, unoccupied aerial vehicle (UAV) networks, and in situ Internet-of-Things (IoT) systems are increasingly creating a persistent, multi-modal view of the planet. Managing these continuous data streams will require adaptive, interpretable models that integrate physical process knowledge with AI-driven inference. The next frontier of GeoAI will emphasize reasoning as much as recognition. Hybrid physical–data models, graph-based learning, and causal AI approaches have the potential to move the field beyond pattern detection toward explanation and prediction.

Ethical and practical considerations have become increasingly prominent with the rise of GeoAI, as advances in automation and data integration introduce challenges related to transparency, equity, and governance. Dependence on commercial cloud infrastructures risks widening resource disparities among institutions, while the opacity of deep neural networks complicates trust, validation, and reproducibility. Efforts in explainable artificial intelligence (XAI) seek to enhance interpretability and accountability, while increasing spatial and temporal resolution raises concerns related to privacy and surveillance. Advances in explainable and ethical AI will be central to ensuring that these systems remain transparent, trustworthy, and accessible to scientists across diverse institutional and geographic contexts.

In this way, the evolution traced in this chapter, from early aerial imagery and constrained satellite records to open, intelligent EO systems, sets the foundation for future work in which GeoAI enables not only sharper observation, but deeper understanding of a rapidly changing Earth. As such, it seems only fitting that the chapter concludes with an attempted unifying definition for GeoAI nested in the efforts of organizations around the world and across various industries.

Although the term GeoAI is now widely used, there is no single universal definition adopted across governmental organizations or in the academic literature. Instead, agencies tend to define it through their own missions and operational needs (Table 2.10). Despite these differences in emphasis, the definitions across organizations converge around several core ideas. All recognize GeoAI as fundamentally involving the integration or *artificial intelligence with spatial data*, particularly large, complex, and often satellite-derived datasets and other such big datasets. They consistently highlight *automation*, *pattern detection*, *classification*, *prediction*, *workflow integration*, and *decision support at scale*. The primary distinctions lie not in what GeoAI is, but in why it is used, whether for advancing scientific knowledge, improving environmental forecasting, supporting sustainable development, enhancing operational efficiency, or enabling strategic intelligence (Table 2.10). Synthesizing the governmental, academic, and operational perspectives reviewed above, we propose the following unifying definition of GeoAI:

GeoAI is the integration of artificial intelligence methods with geospatial data, models, and technologies to discover and model spatial–temporal patterns, generate predictions, and enable scalable, data-driven decision-making across scientific, operational, and policy domains.

Table 2.10 Commonly employed definitions of GeoAI across governmental, research, and industry sources.

Organization/source	Definition
United States Geological Survey (USGS, n.d.)	Application of AI methods and techniques to geospatial data, processes, models, and applications.
National Geospatial-Intelligence Agency (GEOINT, n.d..)	AI integrated into geospatial intelligence workflows to process large spatial data volumes and support decision-making.
National Aeronautics and Space Administration (NASA, n,d.)	Uses AI to extract patterns and insights from massive Earth observation datasets rather than a formal GeoAI definition.
National Oceanic and Atmospheric Association (NOAA, nd.)	AI tools for analyzing geospatial Earth system data for forecasting and monitoring.
European Space Agency (ESA, n. d.)	Integrates machine learning with satellite data to automate extraction and classification of geospatial patterns.
United Nations (UN, nd.)	AI applied to geospatial data and Earth observations to support sustainable development and monitoring.
Food and Agriculture Organization (FAO, nd.)	AI and machine learning applied to geospatial and EO data for agriculture and food-security monitoring.
ESRI (ESRI, nd.)	Geospatial artificial intelligence (GeoAI) is the application of artificial intelligence (AI) fused with geospatial data, science, and technology to accelerate real-world understanding of business opportunities, environmental impacts, and operational risks.

Case studies

The following case studies highlight RS questions and methods representative of the foundational approaches taken in earlier RS studies designed to understand changes in both social and physical earth systems. RS facilitated measurements of environmental changes in one) wetland ecosystems in Trinidad and two) tree canopy in a savanna ecosystem. The case studies are intended to provide an accessible synthesis of work presented in greater depth in Gibbes et al. (2009, 2010).

Case Study 1: Traditional pixel-based classification

Gibbes, C., Southworth, J., & Keys, E. (2009). Wetland conservation: Change and fragmentation in Trinidad's protected areas. Geoforum, 40(1), 91–104.

This case study uses RS to examine the effectiveness of land conservation policies. Human activities like agriculture, logging, and urbanization significantly alter land cover, disrupting ecosystems, and threatening biodiversity. Many countries attempt to counteract these impacts by establishing protected areas that limit or restrict human activity. The tension between economic development and environmental conservation complicates the efficacy of protected areas. RS facilitates the spatio-temporal study of the extent and patterns of land cover change which can serve as an indicator of the effectiveness of protected areas. This study employs pixel-based classifications and vegetation indices to address the impact of protected area designation on two wetlands in Trinidad and Tobago.

This study of land cover change focuses on the two largest coastal protected areas in Trinidad, Caroni and Nariva. Nariva Swamp, located on the eastern side of Trinidad, is the nation's biggest freshwater wetland, covering 6234 ha. Its ecosystems include palm swamps, herbaceous swamps, and mangrove forests. Its rich biodiversity and significance as a habitat for multiple species make it important both locally and globally. Key species include the palmiste palm (*Roystonea oleracea*), red-bellied macaw (*Ara manilata*), cascadura (*Hoplosternum littorale*), and conch (*Pomacea urceus*). On the western side of the island, Caroni Swamp spans 5611 ha and is characterized by mangroves and herbaceous swamps. It is Trinidad's largest brackish water ecosystem, home to the scarlet ibis (*Eudocimus ruber*) and a variety of commercial marine species. Caroni also provides a habitat for the silky anteater (*Cyclopes didactylus*), over two dozen fish species, the crab-eating raccoon (*Procyon cancrivorous*), and oysters (*Crassostrea rhizophorae*). Like Nariva, Caroni's wetlands are vital to surrounding communities, who rely on their resources to support their livelihoods.

Traditional pixel-based land cover classifications of Caroni and Nariva wetlands and a surrounding buffer area were derived using the Gaussian Maximum Likelihood technique. The land cover in the study area was categorized into five distinct types: forest, wetland, built/bare, agricultural land, and water. To enhance the accuracy of the classification, in-situ training samples for each of the five categories were collected from the study sites. An accuracy assessment of the classification was conducted, yielding an overall accuracy of 84.5% and a kappa coefficient of 0.80. Transformed Divergence, a metric for class separability, was used to measure the distinction between classes, with the lowest separability found between agriculture and built/bare areas, as well as between forest and wetland. The study aimed to understand shifts between natural and human-modified land covers. Natural in this context refers to forest, wetland, and water classes. Human-modified land covers included agriculture and built/bare. Discriminant analysis was used to determine the capability of various vegetation indices to differentiate amongst the land cover classes defined in the classification. Change analyses were conducted on both continuous and discrete datasets to assess class-to-class transitions as well as alterations within classes

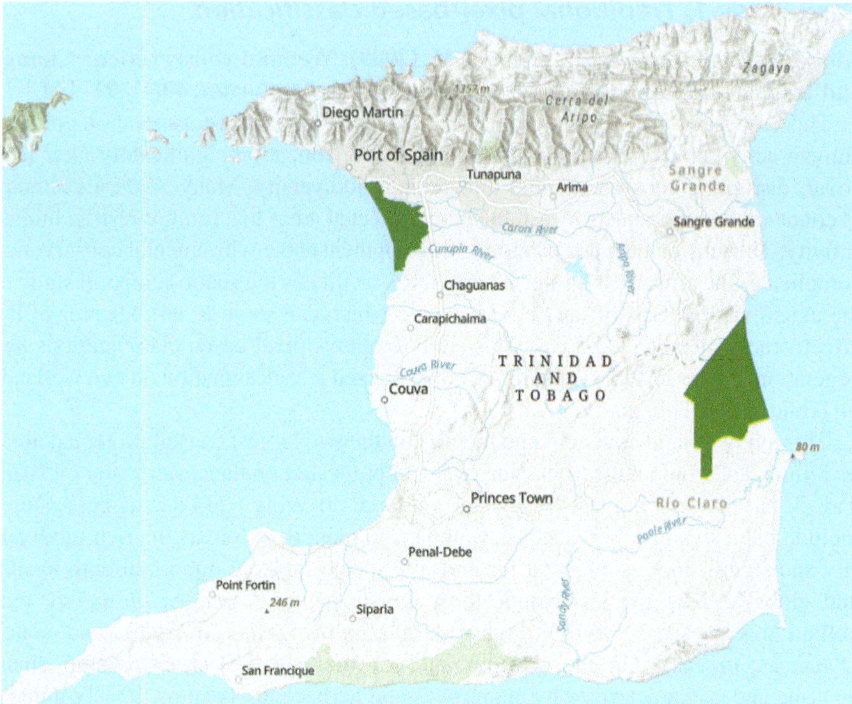

Case Study Figure 1.1 Study area across trinidad and tobago including the studied wetland locations.

represented by vegetation indices. Landscape metrics, quantitative measures of structure, composition, and spatial arrangement of the different land cover classes, were calculated for the protected areas and a surrounding buffer region so that within and outside park comparisons could be made (Case Study Figure 1.1).

The classification of Caroni shows an initial increase in forest/wetland areas, followed by a decrease (Case Study Figure 1.2). Meanwhile, the percentage of land classified as agriculture/bare-built decreases between 1980 and 1990, then increases from 1990 to 2000. In contrast, Caroni's buffer shows a steady increase in forest/wetland areas and a continued decline in agriculture/bare-built land. Nariva follows a similar pattern, with an initial increase in forest/wetland areas, followed by a decline by 2000, while agriculture/bare-built areas decrease before increasing between 1990 and 2000 (Case Study Figure 1.2). Nariv's buffer mirrors these trends, with forest/wetland areas first rising then falling, and agriculture/bare-built areas decreasing then increasing (Case Study Figure 1.2). If the conservation of these wetlands is successful, we would expect to observe limited change between land covers within the protected wetlands, particularly from forest/wetland to agriculture/bare-built. The nature of changes in these regions suggests otherwise and serves as an indicator of the effectiveness of the conservation areas.

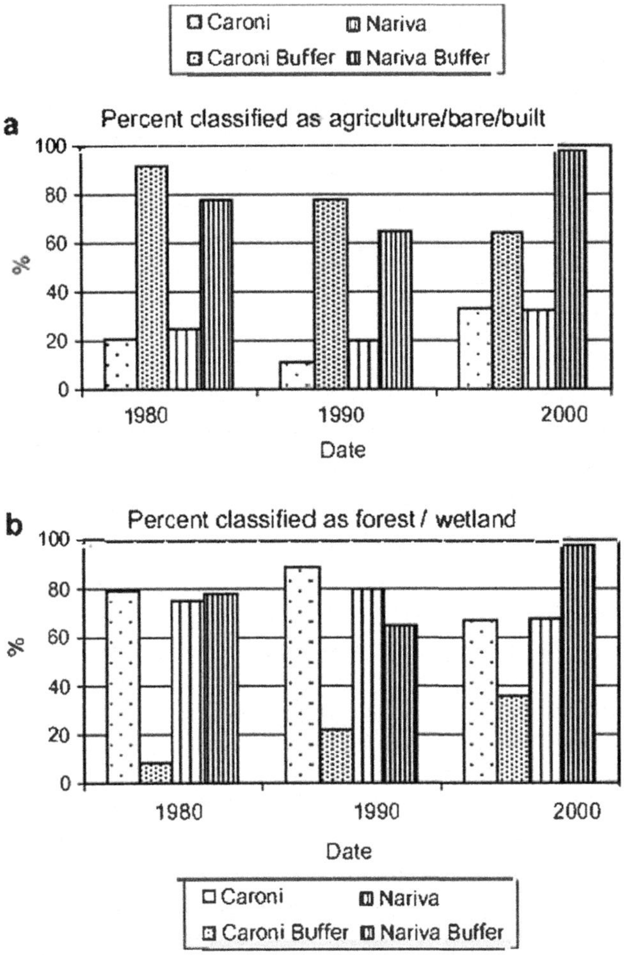

Case Study Figure 1.2 Results from remote sensing analysis presented in Gibbes et al., 2009. Results highlight the limited temporal frequency typical of early remote sensing studies.

This case study is typical of the trend in RS studies from the 1990s-2010, where studies relied on a limited number of observations of the landscape, often covering small spatial extents, and the methods included pixel-based classifications and indices derived from a single data source.

Case Study 2: Object oriented classification

Gibbes, C., Adhikari, S., Rostant, L., Southworth, J., & Qiu, Y. (2010). Application of object-based classification and high-resolution satellite imagery for savanna ecosystem analysis. Remote Sensing, 2(12), 2748–2772.

Savanna ecosystems are crucial to Earth system processes, such as global carbon cycling. Beyond their biophysical values, savannas support human livelihoods for communities across multiple continents. These landscapes are situated between forests and grasslands and characterized by a mix of herbaceous grasses and woody cover. The naturally heterogeneous nature of savannas presents challenges for the study of these ecosystems as it is difficult to measure and delineate attributes of the system due to high levels of spatial and temporal variation. RS enables systematic observation of important attributes, such as tree canopy. Tree cover is essential for savanna ecosystem functioning as it influences biodiversity, structural complexity, and resource availability by creating microhabitats. Reductions in tree cover can lead to habitat degradation and loss of resources for both wildlife and local human populations. Research shows that trees and grasses can coexist due to nitrogen fixation, while large herbivores can affect tree population dynamics by suppressing tree reproduction (Cramer et al., 2010; Goheen et al., 2007).

This research uses vegetation indices combined with an object-based classification in delineate tree canopy in a limited area within the Caprivi, Namibia. The study area encompasses approximately 278.6 km^2. The limited study area was in part determined by the cost and availability of high-resolution satellite imagery. The study area includes portions of Bwabwata National Park on the west side of the Kwando River (132.1 km^2) and communal conservancy areas on the east side (146.48 km^2). Communal conservancy is a legally recognized area of land where local communities manage and use natural resources for their benefit. This form of land tenure and management is common throughout much of southern Africa. Specifically, the study asks (1) How does tree distribution differ between protected areas and community managed land? (2) What is the current spatial distribution of large trees within the landscape? (3) Can the measures of tree presence be scaled up to the medium resolution of the Landsat TM imagery?

Using a combination of one high-resolution IKONOS imagery (4 $\times$ 4 m) and one medium-resolution Landsat TM imagery (30 $\times$ 30 m) the analysis focuses on identifying tree cover. The Landsat image was subset to match the spatial extent of the IKONOS image. An object-based analysis was applied to the IKONOS imagery, and a vegetation index was applied to characterize the landscape using Landsat TM imagery (Case Study Figure 2.1). Automated feature extraction was applied to the IKONOS image using ERDAS Imagine 9.3 software. The image was segmented into distinct, non-overlapping units based on predefined criteria and a supervised method trained with known tree crown locations as identified from field data was used to assign a probability to each pixel, indicating whether it represented a tree. The determination of tree/not tree was performed based on geometric features including area, perimeter-area ratio, and shadow. These features were trained using field observations. A probability filter was then applied to remove any objects with a probability of less than 0.9 of being tree crowns.

Focusing on trees with canopy diameters of 12 m or larger, tree distribution was assessed using point pattern analysis. Tree cover data from the high-resolution IKONOS imagery was linked to the medium-resolution Landsat imagery through an

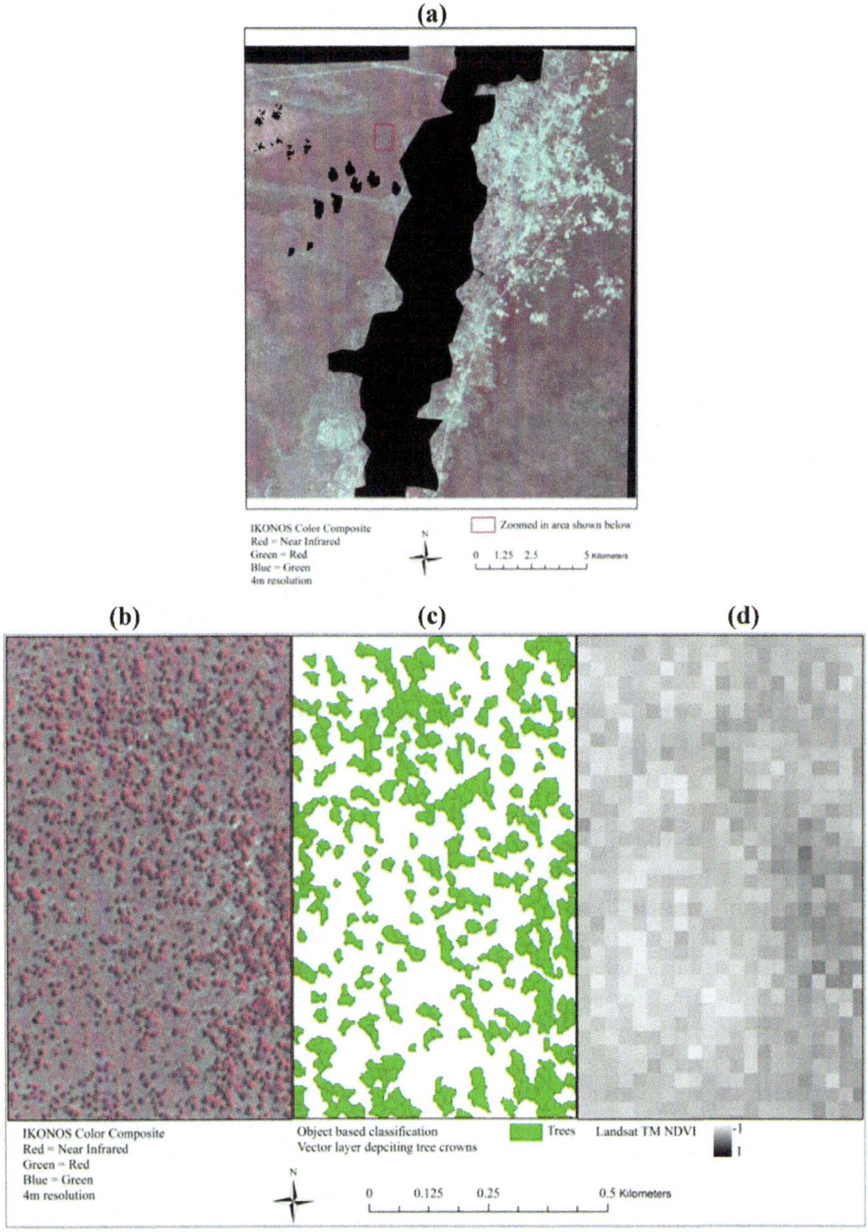

Case Study Figure 2.1 Results from the object-based classification. (a) IKONOS imagery covering the broader study area, (b) a detailed view of the focal area using the IKONOS color composite (RGB = near-infrared, red, and green, where vegetation appears red and exposed soil appears cyan), (c) object-based classification results showing individual tree crown polygons and (d) a landsat-derived NDVI image of the focal area, highlighting the difference in spatial resolution between datasets.

assessment of tree count within 30 × 30 m grid cells corresponding to the Landsat TM data. The number of trees per grid cell and the proportion of tree cover within each Landsat pixel were calculated and compared to the normalized difference vegetation index (NDVI) values from the Landsat image. A non-parametric statistical method was used to compare NDVI values across the tree cover classes, providing insights into the relationship between tree cover and vegetation health in the savanna ecosystem.

Object-based classification is more effective for this RS application than traditional pixel-based classification because it considers the shape and spatial relationships of tree crowns, improving accuracy. The classification achieved an 84% accuracy rate by comparing the detected tree locations with actual field observations. Results indicated a greater density of trees in the protected area compared to community conservation areas. This pattern likely results from differing land management practices, the land in the community conservation areas is multi-purpose, being used for both conservation purposes and daily livelihood activities (e.g., harvesting timber) while the land in the protected area is exclusively allocated for biodiversity conservation. Results also revealed significant clustering of larger trees in the southern portions of the study area, while smaller trees in the same size cohort were more evenly distributed. The number of trees per 30 × 30 m grid cell ranged from 0 to 5. Regarding the relationship between tree count and NDVI values, no statistically significant relationship was indicated between NDVI and tree coverage as measured in the high-resolution imagery.

The RS approach used here provided valuable data on tree cover and distribution. The integration of high- and medium-resolution imagery, along with statistical analysis, offered an early attempt to use multi-source, multi-scale image datasets and diverse RS methods to better quantify earth's surface. While many studies use vegetation indices like NDVI to assess savanna vegetation, this research highlighted the importance of incorporating high resolution spatial data to accurately characterize vegetation type, especially in heterogeneous savanna landscapes.

The study linked high-resolution imagery with medium-resolution Landsat data, offering an early example of multi-scale approaches to monitoring earth's surface. Though at the time not as commonly used, object-based classifications improved pixel-based methods by incorporating both spectral and spatial information. The study's generalizability was limited by the small spatial extent covered in the study area, which in turn limits the ability to replicate the land management types within the study. Thus, the study is limited to descriptive analysis like many other studies from this time period (e.g., Hartter & Southworth, 2009; Munroe et al., 2007; Nagendra et al., 2004). The successful application of an object-based analysis is an example of an early deviation from pixel-based classification to improve the representation of key ecosystem features that were not adequately captured when the pixel served as the minimum, and only unit of observation. The study demonstrates an application of object-based analysis using high-resolution satellite imagery limited by the early constraints of RS for earth observation.

References

Awrangjeb, M., Ravanbakhsh, M., & Fraser, C. S. (2010). Automatic detection of residential buildings using LiDAR data and multispectral imagery. *ISPRS Journal of Photogrammetry and Remote Sensing, 65*(5), 457–467. https://doi.org/10.1016/j.isprsjprs.2010.06.001

Axelsson, P. (1999). Processing of laser scanner data—Algorithms and applications. *ISPRS Journal of Photogrammetry and Remote Sensing, 54*(2–3), 138–147.

Bamler, R. (1991). Doppler frequency estimation and the Cramér–Rao bound. *IEEE Transactions on Geoscience and Remote Sensing, 29*(3), 385–390.

Belgiu, M., & Csillik, O. (2018). Sentinel-2 cropland mapping using pixel-based and object-based time-weighted dynamic time warping analysis. *Remote Sensing of Environment, 204*, 509–523.

Belward, A. S., & Skøien, J. O. (2015). Who launched what, when and why; trends in global land-cover observation capacity from civilian Earth observation satellites. *ISPRS Journal of Photogrammetry and Remote Sensing, 103*, 115–128.

Benediktsson, J. A., Swain, P. H., & Ersoy, O. K. (1990). Neural network approaches versus statistical methods in classification of multisource remote sensing data. *IEEE Transactions on Geoscience and Remote Sensing, 28*(4), 540–552.

Benz, U. C., Hofmann, P., Willhauck, G., Lingenfelder, I., & Heynen, M. (2004). Multi-resolution, object-oriented fuzzy analysis of remote sensing data for GIS-ready information. *ISPRS Journal of Photogrammetry and Remote Sensing, 58*(3–4), 239–258.

Berardino, P., Fornaro, G., Lanari, R., & Sansosti, E. (2002). A new algorithm for surface deformation monitoring based on small baseline differential SAR interferograms. *IEEE Transactions on Geoscience and Remote Sensing, 40*(11), 2375–2383. https://doi.org/10.1109/TGRS.2002.803792

Berni, J. A. J., Zarco-Tejada, P. J., Suárez, L., & Fereres, E. (2009). Thermal and narrowband multispectral remote sensing for vegetation monitoring from an unmanned aerial vehicle. *IEEE Transactions on Geoscience and Remote Sensing, 47*(3), 722–738.

Broge, N. H., & Leblanc, E. (2001). Comparing prediction power and stability of broadband and hyperspectral vegetation indices for estimation of green leaf area index and canopy chlorophyll density. *Remote sensing of environment, 76*(2), 156–172.

Campbell, J. B. (2008). Origins of aerial photographic interpretation, US Army, 1916 to 1918. *Photogrammetric Engineering & Remote Sensing, 74*(1), 77–93.

Chappelle, E. W., Kim, M. S., & McMurtrey III, J. E. (1992). Ratio analysis of reflectance spectra (RARS): An algorithm for the remote estimation of the concentrations of chlorophyll a, chlorophyll b, and carotenoids in soybean leaves. *Remote Sensing of Environment, 39*(3), 239–247.

Chen, Y., Chen, H., Gorkhali, A., Lu, Y., Ma, Y., & Li, L. (2016). Big data analytics and big data science: A survey. *Journal of Management Analytics, 3*(1), 1–42. https://doi.org/10.1080/23270012.2016.1141332

Chen, Y., Jiang, H., Li, C., Jia, X., & Ghamisi, P. (2016). Deep feature extraction and classification of hyperspectral images based on convolutional neural networks. *IEEE Transactions on Geoscience and Remote Sensing, 54*(10), 6232–6251.

Chen, C.-L., Li, H., Wei, Y., Xia, T., & Tang, Y. Y. (2014). A local contrast method for small infrared target detection. *IEEE Transactions on Geoscience and Remote Sensing, 52*(1), 574–581.

Chen, J., Xiong, R., Yu, H., Xu, G., & Xing, M. (2024). Nonparametric full-aperture autofocus imaging for microwave photonic SAR. *IEEE Transactions on Geoscience and Remote Sensing, 62*, 1–15.

Chen, X. L., Zhao, H. M., Li, P. X., & Yin, Z. Y. (2006). Remote sensing image-based analysis of the relationship between urban heat island and land use/cover changes. *Remote Sensing of Environment, 104*(2), 133–146.

Cheng, G., Han, J., Zhou, P., & Guo, L. (2014). Multi-class geospatial object detection and geographic image classification based on collection of part detectors. *ISPRS Journal of Photogrammetry and Remote Sensing, 98*, 119–132. https://doi.org/10.1016/j.isprsjprs.2014.10.002

Cloude, S., & Pottier, E. (1997). An entropy based classification scheme for land applications of polarimetric SAR. *IEEE Transactions on Geoscience and Remote Sensing, 35*(1), 68–78.

Cosgrove, D., & Fox, W. L. (2010). *Photography and flight*. Reaktion Books. https://books.google.com/books?hl=en&lr=&id=ZzKxrhj8Gr4C.

Cracknell, A. P. (2018). The development of remote sensing in the last 40 years. *International Journal of Remote Sensing, 39*(23), 8387–8427. https://doi.org/10.1080/01431161.2018.1550919

Cramer, M. D., Van Cauter, A., & Bond, W. J. (2010). Growth of N2-fixing African savanna Acacia species is constrained by below-ground competition with grass. *Journal of Ecology, 98*(1), 156–167. https://doi.org/10.1111/j.1365-2745.2009.01594.x

Darvishzadeh, R., Skidmore, A., Schlerf, M., Atzberger, C., Corsi, F., & Cho, M. (2008). LAI and chlorophyll estimation for a heterogeneous grassland using hyperspectral measurements. *ISPRS Journal of Photogrammetry and Remote Sensing, 63*(4), 409–426.

Delécolle, R., Maas, S. J., Guérif, M., & Baret, F. (1992). Remote sensing and crop production models: Present trends. *ISPRS Journal of Photogrammetry and Remote Sensing, 47*(2–3), 145–161.

Diakogiannis, F. I., Waldner, F., Caccetta, P., & Wu, C. (2020). ResUNet-a: A deep learning framework for semantic segmentation of remotely sensed data. *ISPRS Journal of Photogrammetry and Remote Sensing, 162*, 94–114. https://doi.org/10.1016/j.isprsjprs.2020.01.013

Dong, S., Wang, L., Du, B., & Meng, X. (2024). ChangeCLIP: Remote sensing change detection with multimodal vision-language representation learning. *ISPRS Journal of Photogrammetry and Remote Sensing, 208*, 53–69. https://doi.org/10.1016/j.isprsjprs.2024.01.004

Dong, J., Xiao, X., Menarguez, M. A., Zhang, G., Qin, Y., Thau, D., & Moore, B. (2016). Mapping paddy rice planting area in northeastern Asia with Landsat 8 images, phenology-based algorithm and google Earth engine. *Remote Sensing of Environment, 185*, 142–154.

Dubois, P. C., Van Zyl, J., & Engman, T. (1995). Measuring soil moisture with imaging radars. *IEEE Transactions on Geoscience and Remote Sensing, 33*(4), 915–926.

Ehlers, M. (1991). Multisensor image fusion techniques in remote sensing. *ISPRS Journal of Photogrammetry and Remote Sensing, 46*(1), 19–30.

Elvidge, C. D., Imhoff, M. L., Baugh, K. E., Hobson, V. R., Nelson, I., Safran, J., & Tuttle, B. T. (2001). Night-time lights of the world: 1994–1995. *ISPRS Journal of Photogrammetry and Remote Sensing, 56*(2), 81–99.

Environmental Systems Research Institute (ESRI). (n.d.). https://www.esri.com/en-us/capabilities/geoai/overview. (Accessed 28 April 2026).

Estel, S., Kuemmerle, T., Alcántara, C., Levers, C., Prishchepov, A., & Hostert, P. (2015). Mapping farmland abandonment and recultivation across Europe using MODIS NDVI time series. *Remote Sensing of Environment, 163*, 312–325.

European Space Agency (ESA). (n.d.). https://www.esa.int/Enabling_Support/Preparing_for_the_Future/Discovery_and_Preparation/Artificial_intelligence_in_space. (Accessed 28 April 2026).

Fauvel, M., Benediktsson, J. A., Chanussot, J., & Sveinsson, J. R. (2008). Spectral and spatial classification of hyperspectral data using SVMs and morphological profiles. *IEEE Transactions on Geoscience and Remote Sensing, 46*(11), 3804–3814.

Ferretti, A., Fumagalli, A., Novali, F., Prati, C., Rocca, F., & Rucci, A. (2011). A new algorithm for processing interferometric data-stacks: SqueeSAR. *IEEE Transactions on Geoscience and Remote Sensing, 49*(9), 3460–3470.

Ferretti, A., Prati, C., & Rocca, F. (2000). Nonlinear subsidence rate estimation using permanent scatterers in SAR interferometry. *IEEE Transactions on Geoscience and Remote Sensing, 38*(5), 2202–2212.

Ferretti, A., Prati, C., & Rocca, F. (2001). Permanent scatterers in SAR interferometry. *IEEE Transactions on Geoscience and Remote Sensing, 39*(1), 8–20.

Feyisa, G. L., Meilby, H., Fensholt, R., & Proud, S. R. (2014). Automated water extraction index: a new technique for surface water mapping using landsat imagery. *Remote Sensing of Environment, 140*, 23–35.

Food and Agriculture Organization (FAO). (n.d.). https://www.fao.org/geospatial/en. (Accessed 28 April 2026).

Frampton, W. J., Dash, J., Watmough, G., & Milton, E. J. (2013). Evaluating the capabilities of Sentinel-2 for quantitative estimation of biophysical variables in vegetation. *ISPRS Journal of Photogrammetry and Remote Sensing, 82*, 83–92. https://doi.org/10.1016/j.isprsjprs.2013.04.007

Freeman, A., & Durden, S. (1998). A three-component scattering model for polarimetric SAR data. *IEEE Transactions on Geoscience and Remote Sensing, 36*(3), 963–973.

Gao, B. C. (1996). NDWI—A normalized difference water index for remote sensing of vegetation liquid water from space. *Remote Sensing of Environment, 58*(3), 257–266.

Gao, F., Masek, J., Schwaller, M., & Hall, F. (2006). On the blending of the landsat and MODIS surface reflectance: Predicting daily landsat surface reflectance. *IEEE Transactions on Geoscience and Remote Sensing, 44*(8), 2207–2218.

GEOINT. (n.d.). https://www.nga.mil/news/GEOINT_Artificial_Intelligence_.html. (Accessed 28 April 2026).

Gibbes, C., Adhikari, S., Rostant, L., Southworth, J., & Qiu, Y. (2010). Application of object based classification and high-resolution satellite imagery for savanna ecosystem analysis. *Remote Sensing, 2*(12), 2748–2772.

Gibbes, C., Southworth, J., & Keys, E. (2009). Wetland conservation: Change and fragmentation in Trinidad's protected areas. *Geoforum, 40*(1), 91–104.

Goheen, J. R., Young, T. P., Keesing, F., & Palmer, T. M. (2007). Consequences of herbivory by native ungulates for the reproduction of a savanna tree. *Journal of Ecology, 95*(1), 129–138. https://doi.org/10.1111/j.1365-2745.2006.01196.x

Goward, S. N., Masek, J. G., Loveland, T. R., et al. (2021). Semi-centennial of landsat observations & pending landsat 9 launch. *Photogrammetric Engineering & Remote Sensing, 87*(8), 533–539.

Green, J. L. (2025). The role of balloons in union military intelligence in the civil war. In *Secrets on display: Stories and spycraft from the International Spy Museum* (p. 108).

Gruen, A., & Akca, D. (2005). Least squares 3D surface and curve matching. *ISPRS Journal of Photogrammetry and Remote Sensing, 59*(3), 151–174.

Höfle, B., & Pfeifer, N. (2007). Correction of laser scanning intensity data: data and model-driven approaches. *ISPRS Journal of Photogrammetry and Remote Sensing, 62*(6), 415–433.

Haboudane, D., Miller, J. R., Pattey, E., Zarco-Tejada, P. J., & Strachan, I. B. (2004). Hyperspectral vegetation indices and novel algorithms for predicting green LAI of crop canopies: Modeling and validation in the context of precision agriculture. *Remote Sensing of Environment, 90*(3), 337–352.

Hall, F. G., Strebel, D. E., Nickeson, J. E., & Goetz, S. J. (1991). Radiometric rectification: Toward a common radiometric response among multidate, multisensor images. *Remote Sensing of Environment, 35*(1), 11–27.

Han, J., Tao, Z., Xie, Y., Li, H., Guan, X., Yi, H., et al. (2022). Radiometric cross-calibration of GF-6/WFV sensor using MODIS images with different BRDF models. *IEEE Transactions on Geoscience and Remote Sensing, 60*, 1–11.

Hansen, P. M., & Schjoerring, J. K. (2003). Reflectance measurement of canopy biomass and nitrogen status in wheat crops using normalized difference vegetation indices and partial least squares regression. *Remote Sensing of Environment, 86*(4), 542–553.

Harsanyi, J. C., & Chang, C.-I. (1994). Hyperspectral image classification and dimensionality reduction: An orthogonal subspace projection approach. *IEEE Transactions on Geoscience and Remote Sensing, 32*(4), 779–785.

Hartter, J., & Southworth, J. (2009). Dwindling resources and fragmentation of landscapes around parks: Wetlands and forest patches around kibale national park, Uganda. *Landscape Ecology, 24*(5), 643–656.

Hinsman, A. (2014). Undetected media: intelligence and the U-2 spy plane. *The Velvet Light Trap, 73*(1), 19–38.

Holben, B. N., Eck, T. F., Slutsker, I. A., Tanre, D., Buis, J. P., Setzer, A., … Smirnov, A. (1998). AERONET—A federated instrument network and data archive for aerosol characterization. *Remote Sensing of Environment, 66*(1), 1–16.

Hopkins, R. S., III. (1998). *United States strategic aerial reconnaissance and the Cold War, 1945–1961*. Doctoral dissertation, University of Virginia. https://search.proquest.com/openview/fb53f8ed698c8f6a88d830870f5bc2d7/1.

Hu, C. (2009). A novel ocean color index to detect floating algae in the global oceans. *Remote Sensing of Environment, 113*(10), 2118–2129.

Huang, X., & Wang, Y. (2019). Investigating the effects of 3D urban morphology on the surface urban heat island effect in urban functional zones by using high-resolution remote sensing data: A case study of Wuhan, Central China. *ISPRS Journal of Photogrammetry and Remote Sensing, 152*, 119–131. https://doi.org/10.1016/j.isprsjprs.2019.04.010

Huete, A. R., Liu, H. Q., Batchily, K. V., & Van Leeuwen, W. J. D. A. (1997). A comparison of vegetation indices over a global set of TM images for EOS-MODIS. *Remote Sensing of Environment, 59*(3), 440–451.

Jensen, J. R. (1986). *Introductory digital image processing: A remote sensing perspective*. University of South Carolina. http://www.osti.gov/scitech/biblio/5166368.

Jiang, Z., Huete, A. R., Didan, K., & Miura, T. (2008). Development of a two-band enhanced vegetation index without a blue band. *Remote Sensing of Environment, 112*(10), 3833–3845.

Jin, S., Yang, L., Danielson, P., Homer, C., Fry, J., & Xian, G. (2013). A comprehensive change detection method for updating the national land cover database to circa 2011. *Remote Sensing of Environment, 132*, 159–175.

Kääb, A. (2002). Monitoring high-mountain terrain deformation from repeated air- and spaceborne optical data: Examples using digital aerial imagery and ASTER data. *ISPRS Journal of Photogrammetry and Remote Sensing, 57*(1–2), 39–52.

Kealy, P. S., & Hook, S. J. (1993). Separating temperature and emissivity in thermal infrared multispectral scanner data: Implications for recovering land surface temperatures. *IEEE Transactions on Geoscience and Remote Sensing, 31*(6), 1155–1164.

Kennedy, R. E., Yang, Z., & Cohen, W. B. (2010). Detecting trends in forest disturbance and recovery using yearly landsat time series: 1. LandTrendr—Temporal segmentation algorithms. *Remote Sensing of Environment, 114*(12), 2897–2910.

Khatami, R., Mountrakis, G., & Stehman, S. V. (2016). A meta-analysis of remote sensing research on supervised pixel-based land-cover image classification processes: General guidelines for practitioners and future research. *Remote Sensing of Environment, 177*, 89–100. https://doi.org/10.1016/j.rse.2016.02.028

King, M. D., Kaufman, Y. J., Menzel, W. P., & Tanré, D. (1992). Remote sensing of cloud, aerosol, and water vapor properties from the moderate resolution imaging spectrometer (MODIS). *IEEE Transactions on Geoscience and Remote Sensing, 30*(1), 2–27. https://doi.org/10.1109/36.124212

King, M. D., Platnick, S., Menzel, W. P., Ackerman, S. A., & Hubanks, P. A. (2013). Spatial and temporal distribution of clouds observed by MODIS onboard the terra and aqua satellites. *IEEE Transactions on Geoscience and Remote Sensing, 51*(7), 3826–3852.

Kraus, K., & Pfeifer, N. (1998). Determination of terrain models in wooded areas with airborne laser scanner data. *ISPRS Journal of Photogrammetry and Remote Sensing, 53*(4), 193–203.

Krieger, G., Moreira, A., Fiedler, H., Hajnsek, I., Werner, M., Younis, M., & Zink, M. (2007). TanDEM-X: A satellite formation for high-resolution SAR interferometry. *IEEE Transactions on Geoscience and Remote Sensing, 45*(11), 3317–3341.

Kruse, F. A., Lefkoff, A. B., Boardman, J. W., Heidebrecht, K. B., Shapiro, A. T., Barloon, P. J., & Goetz, A. F. (1993). The spectral image processing system (SIPS)—Interactive visualization and analysis of imaging spectrometer data. *Remote Sensing of Environment, 44*(2–3), 145–163.

Li, W. (2020). GeoAI: Where machine learning and big data converge in GIScience. *Journal of Spatial Information Science*, (20), 71–77.

Li, X., Du, Z., Huang, Y., & Tan, Z. (2021). A deep translation (GAN) based change detection network for optical and SAR remote sensing images. *ISPRS Journal of Photogrammetry and Remote Sensing, 179*, 14–34. https://doi.org/10.1016/j.isprsjprs.2021.07.007

Li, S., Song, W., Fang, L., Chen, Y., Ghamisi, P., & Benediktsson, J. A. (2019). Deep learning for hyperspectral image classification: An overview. *IEEE Transactions on Geoscience and Remote Sensing, 57*(9), 6690–6709.

Liu, J., Liu, M., Tian, H., Zhuang, D., Zhang, Z., Zhang, W., … Deng, X. (2005). Spatial and temporal patterns of China's cropland during 1990–2000: An analysis based on landsat TM data. *Remote Sensing of Environment, 98*(4), 442–456.

Lopes, A., Touzi, R., & Nezry, E. (1990). Adaptive speckle filters and scene heterogeneity. *IEEE Transactions on Geoscience and Remote Sensing, 28*(6), 992–1000. https://doi.org/10.1109/36.62623

Loveland, T. R., Anderson, M. C., Huntington, J. L., et al. (2022). Seeing our planet anew: Fifty years of Landsat. *Photogrammetric Engineering & Remote Sensing, 88*(7), 429–436.

Lu, D., & Weng, Q. (2007). A survey of image classification methods and techniques for improving classification performance. *International Journal of Remote Sensing, 28*(5), 823–870. https://doi.org/10.1080/01431160600746456

Marmanis, D., Schindler, K., Wegner, J. D., Galliani, S., Datcu, M., & Stilla, U. (2018). Classification with an edge: Improving semantic image segmentation with boundary detection. *ISPRS Journal of Photogrammetry and Remote Sensing, 135*, 158–172. https://doi.org/10.1016/j.isprsjprs.2017.11.009

Maselli, F., Conese, C., & Petkov, L. (1994). Use of probability entropy for the estimation and graphical representation of the accuracy of maximum likelihood classifications. *ISPRS Journal of Photogrammetry and Remote Sensing, 49*(2), 13–20.

McCallum, J. (2023). *Price and performance changes of computer technology with time*. U.S. Bureau of Labor Statistics.

Melgani, F., & Bruzzone, L. (2004). Classification of hyperspectral remote sensing images with support vector machines. *IEEE Transactions on Geoscience and Remote Sensing, 42*(8), 1778–1790.

Meyer, P., Itten, K. I., Kellenberger, T., Sandmeier, S., & Sandmeier, R. (1993). Radiometric corrections of topographically induced effects on landsat TM data in an alpine environment. *ISPRS Journal of Photogrammetry and Remote Sensing, 48*(4), 17–28.

Munroe, D. K., Nagendra, H., & Southworth, J. (2007). Monitoring landscape fragmentation in an inaccessible mountain area: Celaque national park, western Honduras. *Landscape and Urban Planning, 83*(2–3), 154–167.

Myint, S. W., Gober, P., Brazel, A., Grossman-Clarke, S., & Weng, Q. (2011). Per-pixel vs. object-based classification of urban land cover extraction using high spatial resolution imagery. *Remote Sensing of Environment, 115*(5), 1145–1161.

Naesset, E. (1997). Determination of mean tree height of forest stands using airborne laser scanner data. *ISPRS Journal of Photogrammetry and Remote Sensing, 52*(2), 49–56.

Nagendra, H., Munroe, D. K., & Southworth, J. (2004). From pattern to process: Landscape fragmentation and the analysis of land use/land cover change. *Agriculture, Ecosystems & Environment, 101*(2–3), 111–115. https://doi.org/10.1016/j.agee.2003.09.003

Nascimento, J. M. P., & Dias, J. M. B. (2005). Vertex component analysis: A fast algorithm to unmix hyperspectral data. *IEEE Transactions on Geoscience and Remote Sensing, 43*(4), 898–910.

National Aeronautics and Space Administration (NASA). (n.d.). https://www.earthdata.nasa.gov/learn/earth-observation-data-basics/artificial-intelligence. (Accessed 28 April 2026).

National Oceanic and Atmospheric Administration (NOAA). (n.d.). https://www.noaa.gov/ai/about. (Accessed 28 April 2026).

Ojima, D. S., DeFries, R. S., Goward, S. N., et al. (2022). Landsat@ 50. *Frontiers in Ecology and the Environment, 20*(6), 340–342.

Pal, M., & Foody, G. M. (2010). Feature selection for classification of hyperspectral data by SVM. *IEEE Transactions on Geoscience and Remote Sensing, 48*(5), 2297–2307.

Pearson, A. W. (2002). Allied military model making during world war II. *Cartography and Geographic Information Science, 29*(3), 227–242. https://doi.org/10.1559/152304002782008468

Platnick, S., King, M. D., Ackerman, S. A., Menzel, W. P., Baum, B. A., Riédi, J. C., & Frey, R. A. (2003). The MODIS cloud products: Algorithms and examples from terra. *IEEE Transactions on Geoscience and Remote Sensing, 41*(2), 459–473.

Pollefeys, M., Koch, R., Vergauwen, M., & Van Gool, L. (2000). Automated reconstruction of 3D scenes from sequences of images. *ISPRS Journal of Photogrammetry and Remote Sensing, 55*(4), 251–267.

Potapov, P., Li, X., Hernandez-Serna, A., Tyukavina, A., Hansen, M. C., Kommareddy, A., et al. (2021). Mapping global forest canopy height through integration of GEDI and landsat data. *Remote Sensing of Environment, 253.*

Qi, J., Chehbouni, A., Huete, A. R., Kerr, Y. H., & Sorooshian, S. (1994). A modified soil adjusted vegetation index. *Remote Sensing of Environment, 48*(2), 119–126.

Rabus, B., Eineder, M., Roth, A., & Bamler, R. (2003). The shuttle radar topography mission—A new class of digital elevation models acquired by spaceborne radar. *ISPRS Journal of Photogrammetry and Remote Sensing, 57*(4), 241–262.

Reitberger, J., Schnörr, C., Krzystek, P., & Stilla, U. (2009). 3D segmentation of single trees exploiting full waveform LiDAR data. *ISPRS Journal of Photogrammetry and Remote Sensing, 64*(6), 561–574. https://doi.org/10.1016/j.isprsjprs.2009.04.002

Roujean, J. L., & Breon, F. M. (1995). Estimating PAR absorbed by vegetation from bidirectional reflectance measurements. *Remote Sensing of Environment, 51*(3), 375–384.

Ryan, R., Baldridge, B., Schowengerdt, R. A., Choi, T., Helder, D. L., & Blonski, S. (2003). IKONOS spatial resolution and image interpretability characterization. *Remote Sensing of Environment, 88*(1–2), 37–52.

Shao, Y., & Lunetta, R. S. (2012). Comparison of support vector machine, neural network, and CART algorithms for the land-cover classification using limited training data points. *ISPRS Journal of Photogrammetry and Remote Sensing, 70*, 78–87. https://doi.org/10.1016/j.isprsjprs.2012.04.001

Shean, D. E., Alexandrov, O., Moratto, Z. M., Smith, B. E., Joughin, I. R., Porter, C., & Morin, P. (2016). An automated, open-source pipeline for mass production of digital elevation models (DEMs) from very-high-resolution commercial stereo satellite imagery. *ISPRS Journal of Photogrammetry and Remote Sensing, 116*, 101–117. https://doi.org/10.1016/j.isprsjprs.2016.03.012

Shock, J. R. (1997). *The US Army barrage balloon program, 315.* Merriam Press. https://books.google.com/books?hl=en&lr=&id=-JOLFUuamsC.

Sims, D. A., & Gamon, J. A. (2002). Relationships between leaf pigment content and spectral reflectance across a wide range of species, leaf structures and developmental stages. *Remote Sensing of Environment, 81*(2–3), 337–354.

Smirnov, A., Holben, B. N., Eck, T. F., Dubovik, O., & Slutsker, I. (2000). Cloud-screening and quality control algorithms for the AERONET database. *Remote Sensing of Environment, 73*(3), 337–349.

Smith, M. O., Ustin, S. L., Adams, J. B., & Gillespie, A. R. (1990). Vegetation in deserts: I. A regional measure of abundance from multispectral images. *Remote Sensing of Environment, 31*(1), 1–26.

Sobrino, J. A., Caselles, V., & Becker, F. (1990). Significance of the remotely sensed thermal infrared measurements obtained over a citrus orchard. *ISPRS Journal of Photogrammetry and Remote Sensing, 44*(6), 343–354.

Soh, L. K., & Tsatsoulis, C. (1999). Texture analysis of SAR sea ice imagery using gray level co-occurrence matrices. *IEEE Transactions on Geoscience and Remote Sensing, 37*(2), 780–795.

Song, Y., Kalacska, M., Gašparović, M., Yao, J., & Najibi, N. (2023). Advances in geocomputation and geospatial artificial intelligence (GeoAI) for mapping. *International Journal of Applied Earth Observation and Geoinformation, 120.* https://doi.org/10.1016/j.jag.2023.103300

Sun, X., Wang, P., Lu, W., Zhu, Z., Lu, X., He, Q., Li, J., Rong, X., Yang, Z., Chang, H., He, Q., Yang, G., Wang, R., Lu, J., & Fu, K. (2023). RingMo: A remote sensing foundation model with masked image modeling. *IEEE Transactions on Geoscience and Remote Sensing, 61*, Article 5612822. https://doi.org/10.1109/TGRS.2022.3194732

Tong, X.-Y., Xia, G.-S., Lu, Q., Shen, H., Li, S., You, S., & Zhang, L. (2020). Land-cover classification with high-resolution remote sensing images using transferable deep models. *Remote Sensing of Environment, 237.*

Toutin, T., & Cheng, P. (2002). QuickBird—A milestone for high resolution mapping. *Earth Observation Magazine, 11*(4), 14–18.

United Nations (UN). (n.d.). https://unopengis.org/DWG5.html. (Accessed 28 April 2026).

United States Geological Survey (USGS). (n.d.). https://www.usgs.gov/publications/geoai-spatial-image-processing. (Accessed 28 April 2026).

Veloso, A., Mermoz, S., Bouvet, A., Le Toan, T., Planells, M., Dejoux, J.-F., & Ceschia, E. (2017). Understanding the temporal behavior of crops using Sentinel-1 and Sentinel-2-like data for agricultural applications. *Remote Sensing of Environment, 199*, 415–426.

Vivone, G., Alparone, L., Chanussot, J., Dalla Mura, M., Garzelli, A., Licciardi, G. A., et al. (2014). A critical comparison among pansharpening algorithms. *IEEE Transactions on Geoscience and Remote Sensing, 53*(5), 2565–2586.

Vo, A.-V., Truong-Hong, L., Laefer, D. F., & Bertolotto, M. (2015). Octree-based region growing for point cloud segmentation. *ISPRS Journal of Photogrammetry and Remote Sensing, 104*, 88–100. https://doi.org/10.1016/j.isprsjprs.2015.01.011

Wagner, W., Lemoine, G., & Rott, H. (1999). A method for estimating soil moisture from ERS scatterometer and soil data. *Remote Sensing of Environment, 70*(2), 191–207.

Wagner, W., Ullrich, A., Ducic, V., Melzer, T., & Studnicka, N. (2006). Gaussian decomposition and calibration of a novel small-footprint full-waveform digitising airborne laser scanner. *ISPRS Journal of Photogrammetry and Remote Sensing, 60*(2), 100–112.

Wan, Z., & Dozier, J. (1996). A generalized split-window algorithm for retrieving land-surface temperature from space. *IEEE Transactions on Geoscience and Remote Sensing, 34*(4), 892–905.

Wan, S., Gong, C., Zhong, P., Pan, S., Li, G., & Yang, J. (2021). Hyperspectral image classification with context-aware dynamic graph convolutional network. *IEEE Transactions on Geoscience and Remote Sensing, 59*(1), 597–612. https://doi.org/10.1109/TGRS.2020.2994205

Wang, L., Li, R., Zhang, C., Fang, S., Duan, C., Meng, X., & Atkinson, P. M. (2022). UNetFormer: A UNet-like transformer for efficient semantic segmentation of remote sensing urban scene imagery. *ISPRS Journal of Photogrammetry and Remote Sensing, 190*, 196–214. https://doi.org/10.1016/j.isprsjprs.2022.06.008

Wang, Q., Moreno-Martínez, Á., Muñoz-Marí, J., Campos-Taberner, M., & Camps-Valls, G. (2023). Estimation of vegetation traits with kernel NDVI. *ISPRS Journal of Photogrammetry and Remote Sensing, 195*, 408–417. https://doi.org/10.1016/j.isprsjprs.2022.12.019

Wei, J., Li, Z., Li, K., Dickerson, R. R., Pinker, R. T., Wang, J., et al. (2022). Full-coverage mapping and spatiotemporal variations of ground-level ozone (O3) pollution from 2013 to 2020 across China. *Remote Sensing of Environment, 270.*

Weidner, U., & Förstner, W. (1995). Towards automatic building extraction from high-resolution digital elevation models. *ISPRS Journal of Photogrammetry and Remote Sensing, 50*(4), 38–49.

Wu, W.-B., Ma, J., Banzhaf, E., Meadows, M. E., Yu, Z.-W., Guo, F.-X., et al. (2023). A first Chinese building height estimate at 10 m resolution (CNBH-10 m) using multi-source earth observations and machine learning. *Remote Sensing of Environment, 291.*

Wulder, M. A., Masek, J. G., Cohen, W. B., Loveland, T. R., & Woodcock, C. E. (2012). Opening the archive: How free data has enabled the science and monitoring promise of landsat. *Remote Sensing of Environment, 122*, 2–10.

Xia, G.-S., Hu, J., Hu, F., Shi, B., Bai, X., Zhong, Y., et al. (2017). AID: A benchmark data set for performance evaluation of aerial scene classification. *IEEE Transactions on Geoscience and Remote Sensing, 55*(7), 3965–3981.

Yokoya, N., Yairi, T., & Iwasaki, A. (2012). Coupled nonnegative matrix factorization unmixing for hyperspectral and multispectral data fusion. *IEEE Transactions on Geoscience and Remote Sensing, 50*(2), 528–537.

Yu, H., Cao, N., Lan, Y., & Xing, M. (2021). Multisystem interferometric data fusion framework: A three-step sensing approach. *IEEE Transactions on Geoscience and Remote Sensing, 59*(10), 8501–8509.

Yu, Z., Chen, J., Chen, J., Zhan, W., Wang, C., Ma, W., et al. (2024). Enhanced observations from an optimized soil-canopy-photosynthesis and energy flux model revealed evapotranspiration-shading cooling dynamics of urban vegetation during extreme heat. *Remote Sensing of Environment, 305.*

Yu, X., Hyyppä, J., Vastaranta, M., Holopainen, M., & Viitala, R. (2011). Predicting individual tree attributes from airborne laser point clouds based on the random forests technique. *ISPRS Journal of Photogrammetry and Remote Sensing, 66*(1), 28–37. https://doi.org/10.1016/j.isprsjprs.2010.08.003

Yuan, F., & Bauer, M. E. (2007). Comparison of impervious surface area and normalized difference vegetation index as indicators of surface urban heat island effects in landsat imagery. *Remote Sensing of Environment, 106*(3), 375–386.

Yuan, D., & Elvidge, C. D. (1996). Comparison of relative radiometric normalization techniques. *ISPRS Journal of Photogrammetry and Remote Sensing, 51*(3), 117–126.

Zhong, L., Hu, L., & Zhou, H. (2019). Deep learning based multi-temporal crop classification. *Remote Sensing of Environment, 221*, 430–443.

Zhong, Z., Li, J., Luo, Z., & Chapman, M. (2018). Spectral–spatial residual network for hyperspectral image classification: A 3-D deep learning framework. *IEEE Transactions on Geoscience and Remote Sensing, 56*(2), 847–858.

Zhou, X., Zheng, H. B., Xu, X. Q., He, J. Y., Ge, X. K., Yao, X., Cheng, T., Zhu, Y., Cao, W. X., & Tian, Y. C. (2017). Predicting grain yield in rice using multi-temporal vegetation indices from UAV-based multispectral and digital imagery. *ISPRS Journal of Photogrammetry and Remote Sensing, 130*, 246–255. https://doi.org/10.1016/j.isprsjprs.2017.05.003

Zhu, Z., & Woodcock, C. E. (2012). Object-based cloud and cloud shadow detection in landsat imagery. *Remote Sensing of Environment, 118*, 83–94.

Further reading

Ash, M. G. (2019). Interdisciplinarity in historical perspective. *Perspectives on Science, 27*(4), 619–642.

Chan, J. C.-W., & Paelinckx, D. (2008). Evaluation of random forest and adaboost tree-based ensemble classification and spectral band selection for ecotope mapping using airborne hyperspectral imagery. *Remote Sensing of Environment, 112*(6), 2999–3011.

Council, National Research. (1998). *People and pixels: Linking remote sensing and social science.* National Academies Press. https://books.google.com/books?hl=en&lr=&id=I5mcAgAAQBAJ.

European Space Agency. (2020). *European space Agency—Esa. In Organization.* https://www.taylorfrancis.com/chapters/edit/10.4324/9781003138587-37/european-space-agency%E2%80%94esa.

Li, W. (2021). GeoAI and deep learning. In *International encyclopedia of geography: People, the Earth, environment and technology.* https://doi.org/10.1002/9781118786352.wbieg2083

Li, J., Zheng, K., Gao, L., Han, Z., Li, Z., & Chanussot, J. (2025). Enhanced deep image prior for unsupervised hyperspectral image super-resolution. *IEEE Transactions on Geoscience and Remote Sensing, 63*, 1–18.

Quattrochi, D. A., Walsh, S. J., Jensen, J. R., & Ridd, M. K. (2003). Remote sensing and its relationship to geography. In *Geography in America at the dawn of the 21st century* (p. 376).

Roughgarden, J., Running, S. W., & Matson, P. A. (1991). What does remote sensing do for ecology? *Ecology, 72*(6), 1918–1922.

Sun, X., Wang, P., Lu, W., Zhu, Z., Lu, X., He, Q., et al. (2022). RingMo: A remote sensing foundation model with masked image modeling. *IEEE Transactions on Geoscience and Remote Sensing, 61*, 1–22.

Wan, S., Gong, C., Zhong, P., Du, B., Zhang, L., & Yang, J. (2019). Multiscale dynamic graph convolutional network for hyperspectral image classification. *IEEE Transactions on Geoscience and Remote Sensing, 58*(5), 3162–3177.

Getting Started in Big Data Remote Sensing

3

Erin Bunting[1], *Dan Wanyama*[2] *and Vasily Tolmanov*[1]
[1]Department of Geography, Environment, and Spatial Sciences, Michigan State University, East Lansing, MI, United States; [2]Department of Geography, Sustainability, Community, and Urban Studies, University of Connecticut, Storrs, CT, United States

Key terms and definitions

Remote sensing: The acquisition of information about Earth's surface without direct contact, typically using satellite, airborne, or Unmanned Aerial Vehicle (UAV) sensors to measure reflected or emitted energy.
Satellite platforms: Spaceborne systems equipped with sensors that collect Earth observation data (e.g., Landsat, SPOT, MODIS).
Preprocessing: A set of correction and normalization steps applied to raw remote sensing data to improve quality and make images suitable for analysis.
Geometric correction: A preprocessing technique that corrects distortions caused by Earth's curvature, sensor orientation, and platform movement to ensure accurate spatial alignment.
Atmospheric correction: The removal or reduction of atmospheric effects (e.g., aerosols, clouds, water vapor) to obtain accurate surface reflectance values.
Radiometric correction: Adjustments applied to correct calibration errors, sensor degradation, or radiance inconsistencies to ensure consistent brightness values across time and space.
BRDF correction (bidirectional reflectance distribution function): A correction that accounts for changes in surface reflectance caused by varying sensor viewing angles and illumination conditions.
Destriping: A noise-reduction technique used to remove striping artifacts caused by sensor inconsistencies, often in older satellite sensors.
Line drop correction: A method used to fill in missing scan lines in imagery—commonly associated with Landsat 7 SLC-off data.
Cloud masking: The process of identifying and removing cloud-covered pixels from optical satellite imagery to ensure accurate surface analysis.
Mosaicking: Combining multiple adjacent or overlapping images into a single, seamless composite.
Subsetting: Cropping or trimming a larger satellite image to a specific geographic area of interest.
Google earth engine (GEE): A cloud-based geospatial platform that performs large-scale data storage, preprocessing, and analysis of global satellite datasets.
Interoperability: The ability of software platforms, tools, or datasets to work together seamlessly through standardized data formats and protocols.
Scalability: The capacity of tools, workflows, or infrastructure to handle increasing data volumes or expand to large geographic areas without performance loss.

Remote Sensing, Big Data, and GeoAI. https://doi.org/10.1016/B978-0-443-26741-3.00006-4

Noise reduction: Techniques applied to remove random or systematic noise from imagery while preserving important features.
Terrain correction: A preprocessing step that accounts for topographic effects on illumination or viewing geometry, improving reflectance accuracy in complex landscapes.
Petabyte (PB): a unit of digital storage equal to 1000 terabytes.
Exabyte: a unit of digital storage equal to 1000 petabytes representing the largest scale of data storage.

1. Introduction

The growth of remote sensing (RS) data over recent decades has transformed the way we observe and analyze the Earth's landscapes. This shift is driven by technological advances from satellite platforms to aerial surveys and sensor development, all of which have significantly expanded the scope and scale of data available for environmental monitoring, resource management, and scientific research. The advent of new analytical platforms, such as Google Earth Engine (GEE), has simplified access to this data, democratizing the ability to work with large, diverse datasets. However, these developments also introduce a range of complexities in terms of data management and preprocessing, especially as the volume, variety, and velocity of these data continue to increase (Chi et al., 2016; Chen 1996).

One of the fundamental challenges in handling big data in RS is the management of the massive volumes of data generated from multiple sensors and platforms. This is increasingly challenging, especially as data fusion processes become standard in RS applications. Traditional data storage and retrieval systems are often ill-suited to managing data at such scales, prompting the need for scalable solutions. The transition from traditional RS practices to big data-driven workflows necessitates a shift in how researchers approach data acquisition and processing. In the past, practitioners dealt with relatively small datasets that could be easily stored and processed on local servers or personal computers. Today, RS data volumes are generated at a scale that is orders of magnitude larger, requiring scalable storage solutions, cloud-based processing, and sophisticated metadata management systems to ensure the efficient retrieval and integration of information (Wang et al., 2019; Yan et al., 2018). Efficient metadata management also becomes crucial as researchers require sophisticated methods to organize, catalog, and retrieve relevant datasets for analysis (Lü et al., 2011). Without proper metadata and organization, it becomes difficult to track the edits, errors, accuracy, and quality of datasets, all of which are essential for ensuring the validity of subsequent analyses.

The sheer volume and variety of RS data available today also raises questions about the infrastructure required to support its analysis. With data being collected by multiple satellites, drones, and ground-based sensors, the challenge is no longer just accessing the data but also ensuring that it can be processed effectively and efficiently. As the volume of data grows, the risk of data bottlenecks and inefficient processing pipelines increases, potentially hindering the timely use of critical information. In this context, scalable infrastructure capable of supporting large-scale data processing has become a priority for both public and private sectors (Lisjak et al., 2024).

Preprocessing, an essential step in preparing RS data for accurate analysis, is another area that has become increasingly complex in the context of big data. As datasets grow, maintaining the accuracy and consistency of preprocessing techniques (e.g., geometric correction, atmospheric correction, and radiometric normalization) becomes more challenging. The use of automated tools, like those available in GEE, simplifies many preprocessing steps but can also obscure the nuance and complexity of these methods, leading to a potential loss of expertise in manual data preparation (Eschenbach, 2021; Janga et al., 2023). This presents a paradox: while automation increases efficiency, it may reduce practitioners' understanding of the fundamental methodologies, which could affect the interpretation of results. Moreover, the integration of data from diverse sources, ranging from satellite sensors to aerial platforms and ground-based networks, introduces additional challenges in data harmonization. For instance, discrepancies in spatial resolution, temporal coverage, and spectral characteristics need to be resolved before datasets can be effectively combined for analysis (Dalla Mura et al., 2015; Schmitt & Xiao, 2016; Zhang 2010). Harmonizing such diverse datasets requires sophisticated preprocessing workflows that can accommodate the unique characteristics of each data source while ensuring compatibility and synchronization (Zhang 2010). The scalability of infrastructure for handling these tasks is a critical factor that affects the ability of researchers and organizations to leverage big data in RS. Large-scale processing requires not only vast computational resources but also specialized expertise in managing cloud-based systems, high-performance computing environments, and distributed data processing workflows (Lee et al., 2011). As demand for big data analysis grows, these infrastructure challenges will only intensify, making it imperative for organizations to invest in scalable and robust processing systems.

A key aspect of modern RS is its integration into diverse fields, ranging from agriculture and forestry to urban planning and public health. With the rise of global environmental challenges related to climate change, deforestation, urbanization, and disaster management, there is an increasing demand for accurate, high-resolution RS data that can inform policy, scientific research, and on-the-ground decision-making. In agriculture, RS enables precision farming by providing near real-time information on crop health, soil moisture levels, and weather conditions (Sabir et al., 2024; Sishodia et al., 2020; Surendran et al., 2024). In urban environments, RS is used to monitor infrastructure development and manage natural resources (Miller & Small, 2003). These applications highlight the growing relevance of RS data, as it allows stakeholders to make data-driven decisions (Avtar et al., 2020; Bachmann et al., 2022; Wang et al., 2019). However, this reliance on RS data also brings to the forefront the need for robust preprocessing frameworks that ensure the quality and reliability of the data. RS provides this unique perspective, offering continuous, large-scale observations of the Earth's surface, enabling the detection of changes over time. As such, the role of big data in RS has expanded dramatically, and with it, the need for effective data management and preprocessing techniques to ensure that the data can be meaningfully applied.

This chapter seeks to unpack the complexity of big data remote sensing, focusing on the critical phases of data acquisition and preprocessing. As the field evolves with an unprecedented influx of data volumes, propelled by platforms like GEE, researchers and practitioners face new challenges and opportunities (Chi et al. 2016). This chapter provides a comprehensive overview of how to manage and preprocess vast amounts of RS data effectively through case studies and tested approaches. We will delve into the nuances of data management issues, from scalable storage solutions to efficient metadata handling, and explore how innovations in new analytical platforms have revolutionized data access and analysis. However, we also address concerns that arise from the increasing complexity of these tools, such as the potential loss of hands-on understanding of image preprocessing techniques. Through two case studies that showcase different approaches to data management and preprocessing, we aim to expand knowledge on how to navigate and harness the power of big data in RS, ensuring that they can leverage these advancements while maintaining a robust grasp of fundamental methodologies. Throughout the chapter we will focus on the following topics:

- **Data management:** Handling vast volumes and diverse types of remote sensing data requires scalable storage solutions and sophisticated metadata management to ensure efficient retrieval and integration.
- **Scalability and infrastructure**: Developing and maintaining infrastructure capable of supporting large-scale data processing and analysis demands significant resources and expertise, posing a challenge for many organizations and researchers.
- **Preprocessing difficulties**: As the amount of data grows, maintaining the accuracy and consistency of preprocessing techniques becomes increasingly complex, with the risk of losing a nuanced understanding of how preprocessing affects analysis.
- **Integration of diverse data sources**: Combining data from various platforms and sensors, such as satellites, aerial surveys, and sensor networks, presents challenges in terms of data compatibility, synchronization, and harmonization.
- **Tool proficiency and usability**: With advanced platforms like GEE simplifying data access and processing, there is a risk of practitioners becoming overly reliant on these tools, potentially diminishing their ability to perform detailed manual data preprocessing and analysis.

2. Theoretical foundations

2.1 Data management

The management of big data, including RS data and other geospatial products, presents a set of theoretical challenges that can be explained using three key defining properties: Volume, Velocity, and Variety, commonly known as the 3 V's of big data (Gupta 2014). As Earth Observation (EO) technology advances, increasing both the abundance and collection frequency of these data, managing these massive datasets requires a fundamental understanding of the computational, methodological, and epistemological challenges inherent in most types of geospatial big data. Within the context of geospatial big data, RS, and GeoAI, the 3 V's can be broken down as follows:

- **Volume** refers to the sheer magnitude of (remotely sensed) data generated from the numerous EO systems, including multiple sources such as satellites, UAVs, planes, and flux towers.

- **Variety** refers to the heterogeneity of geospatial big data and encompasses multiple sensor types (e.g., optical, radar, LiDAR, hyperspectral, etc.) and data formats (e.g., raster, vector, point clouds). Additionally, datasets are available at varying spatial and temporal extents. Given the now relatively long history of the field of RS, there is consistent imagery that dates back as early as the 1980s (e.g., AVHRR and Landsat) but early imagery from the CORONA reconnaissance satellite dates back to the 1960s and were declassified for public use in the 1990s with additional images released in the early 2000s. Additionally, with the evolution of technology, mostly in terms of sensors and platforms, we are seeing rapidly evolving changes in data availability at higher spatial resolutions.
- **Velocity** represents the rapid pace at which such data is acquired, transmitted, and processed.

2.1.1 Data volume and scalability

One of the central challenges in geospatial big data management, resulting from the vast data volumes, is scalability. Within the context of this book chapter, scalability refers to the ability of geospatial big data management approaches and infrastructure to efficiently manage the increasing volumes, varieties, and velocities of RS data.

The increasing spatial and temporal resolution of satellite imagery generates petabytes of data, requiring robust storage infrastructure. For instance, NASA's Earth Science Data and Information Systems (ESDIS) as of early 2025 held more than 100 petabytes (PB) of data, with 300 terabytes (TB) of data distributed daily (ESDIS Project Website https://www.earthdata.nasa.gov/about/esdis) (Janga et al., 2023). Similarly, GEE, Google's cloud-based RS platform, stored roughly 50 PB of imagery as of 2022, a volume that has skyrocketed to upwards of 90 PB as of late 2025 (https://earthengine.google.com/).

2.1.2 Data variety and integration

Geospatial big data encompasses diverse sources, including optical, radar, LiDAR, and hyperspectral imagery, each with unique characteristics. Theoretical frameworks such as data fusion models and ontologies facilitate integration across heterogeneous data types. Semantic interoperability (i.e., the ability of different systems or technologies to not only exchange data but also understand it), driven by standardized metadata schemas (e.g., OGC standards), is a critical piece as it ensures consistency across datasets and also provides a structure for systems to work together meaningfully (Harvey et al., 1999). However, the dynamic nature of RS data presents ongoing challenges in maintaining temporal and spatial coherence in multi-source integration. For example, integrating Landsat and Sentinel imagery requires careful alignment of spatial resolution and spectral response to ensure compatibility (Claverie et al., 2018). Additionally, combining MODIS time series with higher-resolution datasets, such as Planet imagery, can introduce distortions in trend analyses due to temporal gaps and differences in revisit frequency.

2.1.3 Data velocity and real-time processing

The rapid acquisition of RS data necessitates efficient data streaming and real-time analytics. Traditional batch processing models are insufficient for handling high-velocity data, leading to the adoption of event-driven architectures and edge computing paradigms. Theoretical advancements in stream processing and decentralized computing are essential for optimizing real-time geospatial analytics, particularly for applications such as disaster response and environmental monitoring (e.g., Ahmad 2024; Liang et al., 2023).

2.2 Computational complexity and algorithmic efficiency

RS data processing involves computationally intensive tasks such as atmospheric correction, radiometric calibration, and feature extraction (Schowengerdt 2006). Theoretical developments in high-performance computing (HPC) and artificial intelligence (AI) have improved efficiency, yet challenges remain in algorithmic scalability and explainability. Machine learning models, for instance, excel in pattern recognition but lack theoretical transparency in decision-making, raising concerns regarding model interpretability and reproducibility (Tursunalieva et al., 2024).

2.3 The epistemology of geospatial big data

Beyond technical challenges, geospatial big data raises epistemological questions regarding the nature of knowledge production in RS. The shift from hypothesis-driven research to data-driven discovery necessitates a critical examination of biases in data representation, model assumptions, and the implications of automated decision-making. Theoretical debates on the role of AI and machine learning in scientific inference underscore the need for transparent methodologies that maintain scientific rigor.

The theoretical foundations of geospatial big data management in RS span multiple domains, from computational theory to epistemology. Addressing these challenges requires interdisciplinary approaches that integrate advances in cloud computing, AI, HPC, and data governance. By establishing robust theoretical frameworks, researchers can enhance the scalability, reliability, and interpretability of geospatial big data analytics, ultimately advancing the field of RS in an era of unprecedented data availability.

3. Acquisition barriers and challenges

The acquisition of RS data has evolved significantly over the past few decades as work was done to expand this technology and remove barriers to use such as cost, accessibility, and processing needs (Gorelick et al., 2017; Liu et al., 2018; Wang et al., 2019). The advent of new sensing platforms, processing algorithms, and cloud functionality was not just an attempt to remove barriers to data utilization but also a push to expand into the world of big data applications, time series analysis, and even the use of modeling approaches common to GeoAI. The field of RS has far surpassed the dates of single-date image analysis, as was common early on. Now, we are in an analytical world where such cloud systems enable unique data fusion approaches, the integration of ancillary data, and work from the local to global scale with dense time series. Such transformations have resulted in unprecedented opportunities and applications.

3.1 Traditional acquisition approaches

RS data are acquired using satellites, aircrafts, or ground-based systems. Each platform may carry multiple sensors, resulting in differences in spatial, temporal, spectral, and radiometric resolutions. Satellite missions, such as the Landsat program and Moderate Resolution Imaging Spectroradiometer (MODIS), have played a central role in vegetation RS studies at local to regional scales over the past several decades. In contrast, the Advanced Very High Resolution Radiometer (AVHRR), onboard NOAA satellites, was primarily designed for weather monitoring but has been widely used to assess environmental change at regional to continental scales (Cracknell 2024; Ehrlich et al., 1994). By today's standards, these datasets may be considered relatively low resolution; however, they represent extensive and highly utilized long-term archives. For example, Landsat 5, during its 29-year mission, collected roughly 2.5 million images in total (https://www.usgs.gov/landsat-missions/landsat-5). Since 2008, when the USGS adopted an open data policy making Landsat imagery freely available, cumulative downloads from the Earth Resources Observation and Science (EROS) Center have reached multiple petabytes, highlighting the extraordinary scale and impact of the archive (https://www.usgs.gov/landsat-missions/news/fifteen-years-open-data-allows-advancements-landsat-use-and-research). Similarly, MODIS products represent a massive and continuously expanding data record, although both MODIS Terra and Aqua platforms are currently approaching the end of their operational lifetimes, both missions are scheduled to end in late 2026-early 2027.

Aircraft-based RS has also played a vital role in capturing high-resolution data, often used for more localized studies, such as agriculture, urban planning, and land-use mapping. For instance, the State of Michigan over the last 15 years has heavily invested in such aerial imagery to help with land management and conservation efforts across multiple land use/cover types through a program called MiSAIL (Rachol 2024). The use of aerial photography has many pros and cons. For instance, across much of the US, early aerial images are available long before satellite data, enabling stakeholders to have time series dating back to the 1930s or earlier (dependent on the area). However, there are cons to using such data, including: (1) the older sets of aerial images are typically not digital and therefore require significant processing for analysis, (2) the imagery is not collected on a consistent basis as with satellite data, (3) due to variable flights heights there is not a consistent spatial scale across all flight lines, adding a barrier to image comparison and analysis.

As RS applications expanded to address dynamic processes such as deforestation, urban expansion, agricultural monitoring, and disaster response (e.g., Ahmad 2024; Miller & Small, 2003; Surendran et al., 2024), the need for more frequent and higher-resolution data became increasingly evident. The limitations of these long-used data volumes became particularly apparent as the scale and complexity of environmental and land management issues grew. For example, large-scale deforestation monitoring requires frequent, high-resolution imagery to detect small-scale changes

(Albamonte et al., 2025), yet many early satellites had long revisit periods and coarse spatial resolution, making it difficult to track deforestation in near real-time. Similarly, urban planners seeking to monitor rapid urban growth faced challenges due to the lack of high-resolution imagery capable of capturing fine-scale details such as individual buildings or roads (Metzler et al., 2023). In agricultural monitoring, low-resolution imagery made it difficult to assess crop health and detect localized stress factors such as drought or pest infestations (Adedeji et al., 2025).

As data volumes increased, traditional manual processing methods struggled to keep pace. The sheer amount of imagery being generated by newer satellites, aerial platforms, and UAVs required more efficient processing techniques. Traditional RS approaches, which relied heavily on manual classification and interpretation, became impractical as datasets grew larger and more complex. This challenge underscored the need for automated data processing techniques, leading to the rise of AI and machine learning in RS. These advancements have helped address the shortcomings of traditional methods, enabling faster, more accurate, and more scalable analysis of RS data.

RS technologies have revolutionized our ability to observe and understand Earth's dynamic systems. From monitoring climate variability to tracking urban development, satellite imagery and geospatial data play a crucial role in scientific research, policy-making, and resource management (Pei et al., 2021; Scott & Rajabifard, 2017). However, the rapid expansion of RS data presents significant acquisition barriers and challenges. The remainder of Section 3 of this chapter explores the complexities associated with acquiring, processing, and managing vast volumes of RS data, examining barriers and how they have been overcome.

3.2 Significance of acquisition barriers

Acquisition barriers in RS refer to the obstacles encountered in gathering, processing, and utilizing Earth observation data. These barriers stem from technological, financial, logistical, and analytical constraints. Addressing these challenges is critical, as RS data fuels scientific research, supports decision-making processes, and drives advancements in geospatial technologies. As IBM reported, 2.5 quintillion bytes of data are generated daily, and 90% of the world's data has been created within the past 2 years (Marr, 2021). The sheer volume, variety, and velocity of data introduce substantial acquisition barriers that must be addressed to ensure the effective use of RS technologies.

3.3 Key barriers to remote sensing data acquisition

3.3.1 Data volume and storage constraints

The ever-growing volume of RS data, reaching petabyte (PB) and exabyte (EB) levels, poses major storage and computational challenges. The European Space Agency (ESA) alone has accumulated over 1.5 PB of Earth observation data ("Earth Science in Action for Tomorrow's World—ESA Earth Observation Science Strategy," n.d.), and NASA's Earthdata Cloud was projected to exceed 250 PB by the end of 2025 (https://www.earthdata.nasa.gov/about/competitive-programs/access/eos-data-cloud). Efficient storage solutions, HPC, and cloud-based infrastructures such as Amazon Web Services (AWS), Microsoft Azure, and GEE have emerged to address these constraints (Amazon Web Services, Inc. 2024; Gorelick et al., 2017; Microsoft Corporation 2024). Each of these technologies requires sufficient funds, internet speeds, and infrastructure to tap into such data resources. While each technology is available in most developed countries, this is not always the case for developing countries, resulting in a significant barrier to big data RS data utilization. Specifically, there are barriers because of a lack of supporting infrastructure, not a lack of desire to understand or use such technologies.

3.3.2 Data variety and heterogeneity

RS data are diverse, comprising multiple formats (GeoTIFF, HDF, ASCII) and varying spatial, spectral, temporal, and radiometric resolutions. The sheer volume and heterogeneity of data sources pose a significant challenge in data acquisition, integration, and management. Integrating disparate datasets requires advanced data fusion techniques and standardized storage frameworks, which can be difficult to implement without specialized knowledge and computational resources. Efforts to harmonize multi-source satellite data, such as the Landsat-Sentinel-2 composite (Claverie et al., 2018), exemplify strategies for overcoming these barriers by creating temporally dense, spatially continuous datasets that support diverse analytical applications.

However, beyond the technical challenge of integrating data, there is a fundamental knowledge barrier associated with big data RS. To effectively utilize such diverse datasets, an understanding of core RS principles is essential; otherwise, there is a high risk of misusing the data or misinterpreting results. A notable concern in the RS field is the increasing reliance on cloud-based platforms like GEE. While GEE provides immense capabilities for big data analysis and algorithm development, its repository consists of preprocessed, analysis-ready datasets, which may create the illusion that traditional RS understanding and preprocessing knowledge are unnecessary (Amani et al., 2020; Sharma et al., 2022; Suman et al., 2025; Xiaona et al., 2022). This could lead to a decline in critical thinking regarding data preprocessing steps, error sources, and sensor-specific nuances. For instance, a common pitfall is misunderstanding the distinction between top-of-atmosphere (TOA) and surface reflectance (SR) data, both of which are available in GEE. Without a clear grasp of the atmospheric correction processes that differentiate these datasets, users may misinterpret reflectance patterns, leading to inaccurate conclusions in applications such as vegetation health assessment or land cover classification. Similarly, a lack of awareness regarding data fusion principles can compromise the validity of multi-sensor integration efforts. RS datasets are collected under different spectral, spatial, temporal, and radiometric conditions, and without a firm understanding of these characteristics, data fusion may produce misleading results (Dalla Mura et al., 2015; Ghamisi et al., 2019).

The acquisition barrier extends beyond individual dataset challenges to include accessibility issues, computational limitations, and institutional constraints. While open-access repositories such as NASA's Earthdata and ESA's Copernicus Hub have improved data availability, the financial burden has shifted from purchasing imagery to covering the cost of storage, computation, and network infrastructure required to manage and process massive volumes of freely available data (Chi et al., 2016). As such, even when data is freely available, the computational resources required to process and analyze large-scale imagery remain a limiting factor for many researchers and institutions, particularly those in resource-constrained settings (Chi et al., 2016).

Addressing the acquisition barrier in big data RS requires a multi-faceted approach. Educational initiatives should emphasize foundational RS principles alongside training in cloud-based platforms to ensure that users can critically assess and preprocess data appropriately. Additionally, advancements in data standardization, metadata documentation, and interoperability between platforms can facilitate smoother data integration and reduce the challenges associated with data acquisition. Without such efforts, the full potential of big data RS cannot be realized, as barriers to access and understanding will continue to hinder its effective application.

3.3.3 Data acquisition costs

In the era of big data RS, the cost of acquiring high-resolution satellite imagery and geospatial data remains a significant barrier. Despite advances in open-access satellite programs like NASA's Landsat and ESA's Sentinel missions, high-resolution (<1 meter pixel size) commercial datasets continue to be prohibitively expensive. Proprietary datasets from platforms such as WorldView, Ikonos, and Planet Scope offer superior spatial resolution and, in some cases, higher or more flexible revisit capabilities, which are valuable for applications such as precision agriculture, urban planning, and disaster response (Breunig et al., 2020; Frazier & Hemingway, 2021; Huang et al., 2022). However, the financial burden associated with purchasing these datasets limits accessibility, particularly for researchers, non-profit organizations, and policymakers in resource-constrained settings.

Cloud-based platforms and collaborative initiatives have played a pivotal role in mitigating some of these financial barriers (Amani et al., 2020; Suman et al., 2025; Wang et al., 2024). GEE, NASA's Earthdata, and ESA's Copernicus Hub provide access to vast repositories of publicly available satellite imagery, significantly expanding opportunities for large-scale geospatial analysis (Aschbacher 2017; Earth Science Data Systems 2026; Gorelick 2013; Jutz & Pilar Milagro-Perez, 2020). These platforms integrate HPC resources with extensive RS archives, enabling users to perform complex spatial analyses without the need for local storage or processing power (Aschbacher 2017; Earth Science Data Systems 2026; Gorelick 2013; Jutz & Pilar Milagro-Perez, 2020). However, while such initiatives enhance accessibility, they do not always include the highest-resolution commercial imagery, which remains restricted behind paywalls. This exclusion can be particularly challenging for GeoAI applications that require fine-scale training data to improve model accuracy and predictive capabilities.

Beyond the direct costs of acquiring imagery, significant financial and technical investments are required for data storage, processing, and analysis (Chi et al., 2016).

The growing volume of RS data, particularly from high-frequency and hyperspectral sensors, demands robust computational infrastructure. Organizations without access to cloud computing or dedicated HPC clusters often struggle to manage and process these large datasets, which can hinder the development and deployment of GeoAI models. The cost of training deep learning models on high-resolution RS data is further compounded by licensing fees for commercial geospatial software, AI platforms, and the need for specialized expertise in data preprocessing and model optimization.

Addressing the challenge of data acquisition costs requires a multi-pronged approach. Potential solutions include expanding governmental and private-sector partnerships to subsidize access to high-resolution imagery, implementing tiered pricing models that make commercial data more affordable for academic and non-profit research, and promoting open-access initiatives that support data sharing agreements. Additionally, advancements in data compression techniques, on-the-fly cloud processing, and federated learning could help reduce the financial burden associated with managing large geospatial datasets.

Ultimately, ensuring equitable access to RS data is critical for the advancement of GeoAI and its applications in addressing global challenges. By lowering financial barriers and investing in scalable data infrastructure, the scientific community can harness the full potential of big data RS to drive innovation and inform evidence-based decision-making across a wide range of disciplines.

3.3.4 The power of cloud computing

For years users of remotely sensed images, especially those looking to work with extensive time series or big datasets, struggled with acquiring the data and processing it to complete any sort of analysis. The struggles associated with this process came from multiple factors, including the need to store and manage vast data quantities, the bandwidth and internet connection needed to download such data, and the computing power needed to analyze such data volumes (Ma et al., 2015). Traditional RS workflows often required researchers to download imagery from sources such as the USGS and NASA, and then locally store large datasets, necessitating powerful computational resources for processing. This approach posed significant limitations, particularly when dealing with high-resolution, multi-temporal, and multi-sensor datasets. While limiting, this workflow was necessary because the data had already undergone corrections or low-level preprocessing by the original providers, highlighting the importance of government data repositories such as USGS Earth Explorer, USGS GLOVIS, and NASA AppEEARS.

Cloud computing has fundamentally reshaped the landscape of RS data acquisition, storage, and analysis, providing a scalable and efficient solution to handle the growing volume of data. The emergence of cloud-based platforms, such as AWS, Microsoft Azure, and GEE, has mitigated these challenges, offering unprecedented access to RS archives while harnessing high-performance computing capabilities for large-scale analysis (Lee et al., 2011). Cloud-based RS infrastructures are designed to address key challenges associated with big data, including data volume, velocity, and variety. These platforms provide users with on-demand access to extensive datasets, often in petabyte-scale repositories, without requiring local storage. GEE, for example, hosts over 40 years of Landsat imagery in addition to MODIS, Sentinel, and numerous other RS products, allowing researchers to perform analyses directly in the cloud using Google's powerful computational resources (Gorelick et al., 2017). AWS and Azure, meanwhile, provide storage solutions like Amazon S3 and Azure Blob Storage, which enable users to access and manage RS data seamlessly, with options for integrating custom machine learning workflows and scalable parallel processing pipelines.

A major advantage of cloud computing in RS is its ability to streamline data processing and analysis. Cloud-based infrastructure enables researchers to run complex algorithms on massive datasets without computational bottlenecks associated with traditional desktop-based processing (Amani et al., 2020; Xu et al., 2022). GEE, for instance, allows users to conduct time-series analysis, change detection, and classification without requiring extensive local computation. Its built-in functions, including image preprocessing, feature extraction, and spatial statistics, reduce the technical burden on researchers and facilitate more efficient workflows. AWS and Azure offer similar capabilities, supporting the deployment of machine learning frameworks such as TensorFlow, PyTorch, and Scikit-learn for geospatial data analysis (Géron, 2022).

Another transformative aspect of cloud computing is the democratization of RS analysis. Traditionally, large-scale geospatial processing was restricted to institutions with access to HPC clusters or dedicated data centers. Cloud platforms have removed these barriers by offering flexible models, allowing researchers, governments, and even small organizations to leverage cutting-edge computational power without investing in costly hardware. This accessibility is particularly valuable for researchers in developing regions, where local infrastructure limitations previously restricted engagement with large-scale RS studies.

Despite its advantages, the adoption of cloud computing in RS is not without challenges. Issues such as data privacy, storage costs, and dependency on internet connectivity can serve as barriers or limitations, particularly for organizations operating in bandwidth-constrained environments. Additionally, the proprietary nature of some cloud platforms raises concerns about long-term accessibility and data sovereignty, as users may become reliant on specific service providers for their analytical workflows. Some studies note that cloud computing platforms such as GEE are used predominantly by institutions in developed countries while comparatively fewer studies originate from less developed nations (Kumar & Mutanga, 2018). However, ongoing advancements in open-source cloud solutions, such as the Open Data Cube initiative and cloud-optimized GeoTIFF formats, are helping to address some of these concerns by promoting greater interoperability and long-term data accessibility (Iosifescu Enescu et al., 2021).

Overall, the integration of cloud computing into RS workflows has revolutionized how geospatial data is processed, analyzed, and shared. By leveraging scalable computational resources, vast data archives, and advanced machine learning capabilities, cloud-based platforms have unlocked new possibilities for RS applications across disciplines. As technology continues to evolve, further advancements in cloud-based geospatial analysis will likely enhance our ability to monitor and understand Earth's dynamic systems at unprecedented scales.

3.3.5 Computational challenges

Processing high-resolution, multi-temporal satellite imagery requires substantial computational power, posing a challenge for researchers and organizations. Traditional desktop-based processing is often inadequate due to memory constraints, limited processing speeds, and the sheer volume of data generated by modern EO missions (Mohr et al., 2025). The exponential growth in spatial, spectral, and temporal resolutions of satellite data necessitates more advanced computational approaches to handle storage, preprocessing, and analytical workflows effectively.

Scalable cloud computing solutions and parallel processing architectures have emerged as essential tools in overcoming these challenges. Cloud platforms, such as GEE, leverage distributed computing to significantly reduce processing time, enabling users to perform large-scale geospatial analyses without the need for extensive local computational infrastructure. These platforms allow researchers to access and process PB of RS data using cloud-based servers, facilitating near-real-time analytics for applications such as land cover change detection, disaster monitoring, and environmental modeling (Gorelick et al., 2017; Xiao & Xiao, 2012).

Despite these advancements, computational barriers remain, particularly for users with limited access to cloud resources or HPC environments. While cloud platforms offer powerful processing capabilities, they often come with associated costs for large-scale data storage, API usage, and premium analytics services. Additionally, the integration of GeoAI techniques, such as deep learning and machine learning algorithms, further amplifies computational demands. Training complex neural networks on high-resolution satellite imagery requires specialized Graphics Processing Units (GPUs), which can be cost-prohibitive for many institutions.

Another key challenge is data transfer and accessibility. High-resolution RS datasets can be massive in size, making traditional download-and-process workflows not just inefficient but archaic in regard to current standards. The need for real-time or near-real-time processing has driven interest in edge computing and federated learning, where processing occurs closer to the data source, reducing latency and bandwidth requirements. However, implementing these technologies requires infrastructure investments and expertise in distributed computing frameworks.

To address these computational challenges, several strategies can be explored. Open-source tools (GEE, R, Python) and optimized algorithms, such as GPU-accelerated computing and compression techniques, can help reduce processing times and storage requirements. Additionally, collaborative initiatives promoting shared computing resources, such as NASA's Open Science initiative or ESA's Data Cube approach, can provide broader access to computational power for RS and GeoAI applications (Giuliani et al., 2019; Sheffer et al., 2018). Advances in cloud-based AI frameworks, including pre-trained models for geospatial applications, can also help reduce the computational burden associated with training and deploying machine learning models.

Ultimately, overcoming computational challenges in big data RS and GeoAI requires a combination of technological innovation, cost-effective cloud solutions, and increased accessibility to high-performance computing resources. As data volumes continue to grow, investments in scalable infrastructure and efficient processing methodologies will be crucial in unlocking the full potential of satellite-derived insights for scientific research and decision-making.

3.3.6 Data security and privacy concerns

With the increasing reliance on cloud-based RS platforms, concerns regarding data security and privacy have emerged ("Terms of Service—Google Earth Engine," n.d.). While platforms like GEE provide extensive global datasets, they do not support the processing of private datasets with users' own computing resources. The development of hybrid cloud solutions and encryption protocols aims to address these concerns while maintaining accessibility and security (Solymosi et al., 2023).

3.4 Overcoming acquisition barriers

3.4.1 Cloud computing and big data platforms

The rise of cloud-based platforms, such as AWS, Microsoft Azure, and GEE, has revolutionized RS data processing. These platforms provide scalable storage, rapid processing capabilities, and easy accessibility, reducing computational bottlenecks that have historically limited the ability to analyze large-scale geospatial data (Suman et al., 2025). By enabling server-side processing and parallel computing, these solutions allow users to process PB of RS data efficiently (Suman et al., 2025; Wang et al., 2024). Additionally, cloud-based platforms reduce dependency on high-end local hardware, democratizing access to advanced geospatial analytics.

3.4.2 AI and machine learning enhancements

AI-driven algorithms, such as deep learning models (e.g., YOLO, Faster R-CNN, and U-Net), have significantly enhanced image classification, object detection, and pattern recognition in large datasets (Li & Hsu, 2022; Wang et al., 2024). Machine learning techniques automate feature extraction, allowing for rapid and accurate interpretation of complex satellite imagery. These advancements not only streamline data processing but also improve the precision of land cover classification, anomaly detection, and environmental monitoring applications (Janowicz et al., 2020; Wang et al., 2024). Furthermore, AI-powered automation reduces manual intervention, making it easier for researchers and analysts to process vast amounts of data with minimal effort (Janowicz et al., 2020; Wang et al., 2024).

3.4.3 Open data initiatives

Government agencies and international organizations have increasingly adopted open-data policies to promote accessibility to high-quality RS datasets (Zhu et al.,

2019). Programs such as the ESA Copernicus initiative, NASA's Earth Observing System (EOS), and the USGS Landsat archive exemplify efforts to make RS data freely available to the global community. Open data repositories provide valuable resources for researchers, policymakers, and businesses, enabling them to leverage high-quality satellite imagery without financial constraints (Radočaj et al., 2020). These initiatives also support collaboration between academia, government, and industry to drive innovation in RS and GeoAI applications.

3.4.4 Interoperability standards and data fusion techniques

One additional challenge in RS is integrating data from multiple sources, including optical, radar, and LiDAR sensors (Kahraman & Bacher, 2021). Standardization efforts, such as those led by the Open Geospatial Consortium (OGC), promote interoperability by establishing common data formats, metadata standards, and web service protocols (Chen et al., 2025). These standards facilitate seamless data exchange and integration, ensuring compatibility across different platforms and software environments. Advances in data fusion techniques further enhance the usability of RS data by combining complementary datasets, improving the accuracy of geospatial analyses, and enabling comprehensive multi-sensor applications.

In addition to satellite imagery, ancillary data such as weather, climate, and topographic information play a crucial role in enhancing RS analyses. RS datasets, particularly those collected at a global scale, often need to be complemented with other types of data to improve accuracy and understanding. For example, combining RS imagery with weather data from NOAA or climate models can provide insights into environmental changes, such as droughts, floods, and forest cover dynamics (Aryal & Lakshmi, 2025).

In recent years, advances in data fusion techniques have allowed for the integration of heterogeneous data sources. The fusion of high-resolution imagery with environmental and socio-economic data from sources such as census surveys and land use records has enabled a more comprehensive understanding of geographic phenomena. As RS data itself has grown in volume and complexity, the ability to seamlessly integrate and analyze such diverse data sets has become an essential capability. Cloud platforms, like those mentioned earlier, facilitate these integrations, allowing users to merge satellite imagery with real-time data streams and historical records to derive actionable insights. The combination of imagery and ancillary data is not only useful for mapping and monitoring purposes but is also central to the growing field of GeoAI (Boutayeb et al., 2025). By incorporating external datasets with RS images, machine learning models can be trained to detect patterns and make predictions about various phenomena, from deforestation to urban growth and biodiversity loss (Boutayeb et al., 2025). The integration of different data types ensures that analyses are not limited to one source but are enriched by contextual information, enhancing their accuracy and predictive power.

As RS technology advances, the process of data acquisition has evolved from traditional methods to cloud-based, scalable platforms that handle vast amounts of geospatial data. The rise of cloud computing has enabled the processing of large datasets in ways that were previously impossible, empowering researchers and policymakers to tackle complex environmental and social challenges. Furthermore, the integration of ancillary data with RS imagery has enhanced the depth of analysis, leading to more informed decision-making. The intersection of RS, big data, and GeoAI represents an exciting frontier for geographic sciences, where innovative tools and techniques continue to unlock new possibilities for understanding and managing our planet.

3.4.5 Improved sensor technologies

Next-generation satellites equipped with enhanced spatial, spectral, and temporal resolutions help overcome acquisition challenges. Advances in hyperspectral imaging, Synthetic Aperture Radar (SAR), and real-time data acquisition technologies enable more detailed and timely EO (Kahraman & Bacher, 2021; Lu et al., 2020). Higher-resolution sensors provide finer-scale imagery, while improved revisit rates ensure more frequent updates for dynamic monitoring applications such as disaster response, deforestation tracking, and urban expansion analysis (Veraverbeke et al., 2018). Additionally, miniaturized satellite technologies and constellations of small satellites (e.g., CubeSats) are expanding data availability and accessibility, offering cost-effective solutions for global environmental monitoring (Saeed et al., 2020; Veraverbeke et al., 2018).

RS has unlocked unprecedented opportunities for EO, yet acquisition barriers persist. The challenges of data volume, variety, computational constraints, and accessibility are being addressed through cloud computing, AI advancements, and open-data initiatives (Chi et al., 2016). Overcoming these barriers is essential to harnessing the full potential of RS big data for scientific research, environmental monitoring, and decision-making. As technology evolves, continuous efforts in data standardization, machine learning integration, and scalable computing infrastructures will be vital in shaping the future of RS data acquisition.

By leveraging cloud computing, AI advancements, open data initiatives, interoperability standards, and sensor technology improvements, the RS community is steadily overcoming acquisition barriers. These innovations are transforming the way geospatial data is acquired, processed, and utilized, paving the way for more effective and scalable applications of big data RS and GeoAI.

4. Preprocessing

4.1 Traditional steps

Preprocessing is a critical step in RS that ensures data accuracy, consistency, and usability for downstream analyses. Traditional preprocessing workflows include several key steps ranging from geometric correction to cloud masking and/or mosaicking (Table 3.1). These techniques correct for errors and distortions that can arise during data acquisition, making the data ready for interpretation.

Of the preprocessing methods introduction in Table 3.1, it's important to highlight a few key steps. First, geometric correction looks to reduce image distortions caused by the Earth's curvature, sensor orientation, and/or platform motion (Shlien, 1979b). Without geometric correction, the spatial accuracy of RS data would be compromised, especially when integrating data from multiple sources or platforms. Geometric correction is crucial for ensuring that different datasets align spatially. Next, sensors capture reflectance values that can be influenced by atmospheric conditions, such as clouds, aerosols, and water vapor (Li et al., 2010). Atmospheric correction removes these effects, providing more accurate surface reflectance values. This is particularly important in multi-temporal studies, where changes in the Earth's surface need to be distinguished from variations caused by atmospheric conditions. Third, sensors can suffer calibration errors or degradation over time, leading to inconsistencies in the radiometric values they capture. Radiometric correction adjusts for these sensor-specific errors, ensuring that the data is consistent across time and space (TEILLET, 1986). This consistency is critical for quantitative analyses that rely on accurate brightness measurements. Other preprocessing methods, include destriping, which removes striping artifacts caused by sensor inconsistencies; line drop correction, which fills in missing data gaps particularly for Landsat 7 imagery; and cloud masking, which identifies and removes cloud cover from optical imagery (Table 3.1) (Sun et al., 2019; Tsai & Chen, 2008; Zhu & Woodcock, 2012). Each of these methods helps ensure that RS data is of high quality and suitable for accurate analysis.

The rapid evolution of RS technology has fundamentally transformed the preprocessing of image datasets, shifting from labor-intensive, manual corrections to largely automated, cloud-based workflows. Traditionally, preprocessing methods such as geometric correction, atmospheric correction, radiometric normalization, and noise reduction required significant computational resources and technical expertise. Researchers often had to download raw satellite imagery, apply various correction algorithms, and manage extensive metadata to ensure data consistency. However, with the advent of cloud computing and AI, preprocessing has become more efficient, scalable, and accessible to a wider range of users.

4.2 Preprocessing difficulties

Preprocessing challenges have historically posed significant hurdles to effective RS analysis. Further, incorrect or improper preprocessing has led to incorrect outcomes or misuse of remotely sensed images. Some of the key difficulties include:

- **Inconsistent Data Quality**: Variations in sensor calibrations, atmospheric conditions, and acquisition times can introduce inconsistencies. The solution, development of automated preprocessing pipelines, such as GEE's atmospheric correction functions, has significantly reduced manual processing time and errors.
- **Computational Constraints**: Processing large RS datasets requires significant computing power. The solution, adoption of cloud computing platforms like AWS and GEE, has enabled scalable parallel processing.
- **Limited Access to Standardized Algorithms**: Traditional preprocessing workflows often require specialized knowledge and access to proprietary software. The solution, utilizing open-source software and tools like QGIS, SNAP, and R-based RS packages, has democratized access to high-quality preprocessing tools.

Table 3.1 A comprehensive list and description of common remote sensing data preprocessing methodologies.

Preprocessing method	Description	Platforms	Challenges	Relevance
Geometric correction	Corrects image distortions caused by the Earth's curvature, sensor orientation, and platform motion.	Satellite (e.g., Landsat, Sentinel), aerial	Ensuring accurate georeferencing for multi-source data integration.	Essential for aligning data across time and sensors for consistent spatial analysis (Bannari et al., 1995; Shlien 1979).
Atmospheric correction	Removes the effects of the atmosphere on the reflectance values captured by the sensor.	Satellite, drone, aerial	Compensating for varying atmospheric conditions across different regions and times.	Crucial for accurate surface reflectance measurements, especially in multi-temporal studies (Hadjimitsis et al., 2004; Kondratyev et al., 2013; Lyapustin et al., 2012).
Radiometric correction	Adjusts for sensor-specific errors like calibration differences and sensor degradation.	Satellite, UAV, aerial	Requires access to calibration parameters; challenging across diverse sensors.	Ensures consistency in brightness values across datasets for quantitative analysis (Franklin & Giles, 1995; Hill & Sturm, 1991; TEILLET 1986; Young et al., 2017)
BRDF (Bidirectional reflectance distribution function) correction	Corrects for the angular dependence of surface reflectance.	Satellite, aerial, drone	Complex to implement, as it requires accurate models of surface and atmospheric interactions.	Important for analyzing surface properties, especially in forest or urban landscapes (Ngan et al., 2005; F. Li et al., 2010).

Destriping	Removes striping artifacts caused by sensor inconsistencies during data acquisition.	Satellite (e.g., Landsat, MODIS)	Requires careful filtering without losing spatial detail.	Vital for producing visually consistent images, especially in older satellite datasets (Hill & Sturm, 1991; Horn & Woodham, 1979; Sun et al., 2019).
Line drop correction	Interpolates missing scan lines or data gaps that occur during image acquisition.	Satellite (e.g., Landsat, SPOT), aerial	Filling gaps while maintaining data integrity can be challenging.	Important for continuity in time-series data and ensuring no critical information is lost (Tsai & Chen, 2008).
Cloud masking	Identifies and removes cloud cover from optical imagery.	Satellite (e.g., Landsat, Sentinel-2), UAV	Distinguishing between clouds, cloud shadows, snow, and bright surfaces can be difficult.	Improves the usability of optical imagery for land surface analysis (Z. Li, Shen, et al., 2022; Lyapustin et al., 2012; Zhu & Woodcock, 2012, 2014).
Mosaicking	Combines multiple images from different orbits and passes into a single, seamless image.	Satellite, UAV, aerial	Matching radiometric and geometric properties between images.	Useful for generating large-area composite images from multiple scenes (Li et al., 2019).
Subsetting	Extracts a specific region of interest from a larger dataset.	Satellite, aerial, drone	Risk of losing context when subsetting from larger datasets.	Essential for focusing analysis on relevant areas, reducing processing time (Goslee 2011).
Spectral normalization	Adjusts for differences in spectral characteristics between sensors or over time.	Satellite, UAV	Harmonizing datasets from different sensors or time periods.	Enables accurate multi-source and multi-temporal data analysis (Leach et al., 2019).

Continued

Table 3.1 A comprehensive list and description of common remote sensing data preprocessing methodologies.—cont'd

Preprocessing method	Description	Platforms	Challenges	Relevance
Terrain correction	Adjusts for topographic effects that cause distortions in imagery, especially in hilly areas.	Satellite (e.g., Sentinel-1, radar), UAV, aerial	Requires accurate digital elevation models (DEMs).	Essential for radar and optical images in regions with complex terrain (Proy et al., 1989; Richter et al., 2009).
Noise reduction (Denoising)	Reduces sensor noise or interference in the imagery.	Satellite, drone, aerial	Risk of blurring important image features during the noise reduction process.	Important for enhancing image quality, especially in high-noise environments (Chang et al., 2016; Rasti et al., 2021).
Resampling	Changes the spatial resolution of the data to match that of another dataset.	Satellite (e.g., Landsat, MODIS), UAV, aerial	Risk of losing spatial detail or introducing artifacts.	Important for integrating data from multiple sources with different resolutions (Lyons et al., 2018).

- **Data Storage and Management**: Storing and managing vast amounts of RS data presents logistical challenges. The solution, advances in hierarchical data storage systems, such as the use of Cloud Optimized GeoTIFFs (COG), allow for efficient data retrieval and storage.

4.3 Difficulties in dealing with big data/long time series

With the increasing availability of high-resolution imagery and the proliferation of long-term satellite missions, new challenges arise in handling big data and long time series RS datasets:

- **Scalability Issues**: Traditional preprocessing pipelines struggle to keep up with petabyte-scale datasets. The solution, distributed computing frameworks such as Apache Spark, allow parallel processing across large-scale RS archives.
- **Automated Anomaly Detection**: Long time series datasets are prone to sensor drift, missing data, and other inconsistencies. The solution, AI-driven anomaly detection models, such as deep learning-based temporal outlier detection, are being developed to automatically identify and correct errors.
- **Harmonization Across Sensors**: Long-term analysis requires data consistency across different satellite generations. The solution, cross-sensor harmonization efforts, such as the Harmonized Landsat-Sentinel (HLS) project, provide consistent surface reflectance products from multiple sensors.
- **Efficient Data Retrieval and Processing**: Querying and processing long-term data can be time-consuming. The solution, use of cloud-based analysis-ready data (ARD) formats, such as NASA's Earthdata, allows for on-demand access to preprocessed datasets.

By leveraging advancements in cloud computing, AI, and automated preprocessing workflows, researchers can now process and analyze vast amounts of RS data more efficiently than ever before. These innovations continue to reshape the field, enabling large-scale environmental monitoring, land cover change analysis, and GeoAI applications at unprecedented spatial and temporal scales. The evolution of RS has been accompanied by the rise of GeoAI, which combines geographic data with AI, deep learning (DL), and machine learning (ML) techniques being used to derive deeper insights, understand both pattern and process, and for predictive modeling. Together, RS and GeoAI are shaping the future of Earth observation, offering new opportunities for understanding and managing complex environmental challenges.

4.4 The integration of GeoAI into remote sensing

The introduction of GeoAI marks a significant leap in the evolution of RS methods. GeoAI blends geographic data with AI algorithms to analyze and interpret large datasets. This integration has revolutionized how RS data is processed, interpreted, and used for decision-making. For example, machine learning algorithms can be applied to classify land cover or detect changes in the environment over time. These algorithms can learn from training data and improve their performance as more data becomes available. This is particularly valuable in applications like land use and land cover classification, where traditional methods might struggle to handle the vast amounts of data generated by modern RS platforms. One of the key advantages of GeoAI is its ability to process and analyze big data rapidly (Dritsas & Trigka, 2025; Li et al., 2022). With the massive amounts of data generated by satellites, drones, and UAVs, traditional manual analysis would be too slow and inefficient. By using AI, large datasets can be processed quickly, enabling faster and more accurate decision-making in areas like high-resolution mapping (Li et al., 2024), disaster response (Hanny et al., 2025), agricultural monitoring (Sagan et al., 2024), and environmental management (Dritsas & Trigka, 2025).

4.5 Challenges and future directions

Despite the remarkable progress made in RS and GeoAI, several challenges remain. One of the main challenges is the integration of data from multiple sources with varying resolutions, sensor types, and times of acquisition (Zhu et al., 2018). Preprocessing methods like resampling and spectral normalization help address these challenges by harmonizing datasets to ensure compatibility (Saunier et al., 2022; Tsai & Chen, 2008). However, achieving seamless integration across diverse datasets still requires continuous research and innovation. Another challenge is dealing with noisy or incomplete data. Noise reduction and destriping are common techniques used to address sensor noise and artifacts, but balancing noise removal with the preservation of important image details can be tricky (Chang et al., 2016; Rasti et al., 2021). Similarly, ensuring that terrain correction is accurately applied to data in complex topographies remains an area for improvement. As RS and GeoAI continue to evolve, there will be an increasing emphasis on developing more advanced algorithms and preprocessing techniques that can handle the complexity of big data. The future of RS lies in further enhancing the synergy between these technologies, enabling more precise, efficient, and timely insights for managing Earth's resources and addressing global challenges.

5. Discussion

The evolution of RS, driven by big data and GeoAI, has revolutionized how Earth observation data is acquired and processed. Advances in sensor technologies, cloud computing, and AI, such as the use of platforms like GEE, enables the handling of vast geospatial datasets with improved efficiency and accuracy. However, these advancements also bring challenges related to tool proficiency, scalability, and infrastructure.

While AI tools and big data platforms have simplified data acquisition and preprocessing, they have also introduced additional complexity into RS workflows. As automation expands and usability improves through cloud-based platforms and AI-driven systems, users may increasingly rely on preprocessed datasets without fully understanding the underlying correction, calibration, and preprocessing steps. This reduced awareness raises the risk of misinterpretation and compromised data reliability, making it essential for practitioners to understand these fundamental processes to maintain scientific rigor. Standardized and well-documented workflows improve usability while maintaining transparency and accuracy. In addition, with the rise of automated RS tools, ensuring proficiency in using these tools has become essential. Training programs, workshops, and educational courses must emphasize both theoretical and practical knowledge to bridge the gap between automation and fundamental understanding. Researchers and professionals should engage with open-source tools, participate in collaborative projects, and develop best practices to enhance their expertise in GeoAI and RS. Furthermore, as RS applications scale globally, ensuring equitable access to computational resources and cloud-based platforms is crucial. Investment in scalable infrastructure and cost-effective cloud solutions will help democratize access to HPC for RS applications. Lastly, incorporating multiple data sources, integrating cloud and edge computing, and developing resilient infrastructure are key to managing and processing large-scale RS data efficiently. Standardized data formats, robust storage solutions, and interoperability frameworks are needed to ensure seamless data management and analysis.

6. Conclusion

6.1 Summary

RS has undergone a remarkable transformation, from early platforms and coarse-resolution sensors to today's sophisticated constellation of satellites, drones, and cloud-based analytical platforms. This chapter traces this evolution, emphasizing how advances in data acquisition, preprocessing, and computational infrastructure have fundamentally reshaped Earth observation workflows.

Key preprocessing techniques, including geometric, radiometric, atmospheric, and BRDF corrections, remain foundational to high-quality RS. Their increasing automation through platforms like GEE has, in some ways, reduced barriers to entry while simultaneously elevating expectations for scale, efficiency, and standardization. That said, these systems have also created new barriers due to the high computational needs they demand, power that many regions of the world don't currently have the capacity to support. This preprocessing and analysis shift underscores the need for continued transparency and user understanding of these essential steps. This technological progress also brings challenges. Tool proficiency, workflow scalability, data interoperability, and infrastructure requirements remain central concerns. Cloud-based systems increase accessibility but require robust training, computational literacy, and awareness of methodological limitations. As RS expands into a truly big-data science, the importance of rigorous preprocessing, reproducible workflows, and scientifically grounded interpretation becomes even more critical.

The rise of big data and GeoAI represents the next major milestone. Integrating machine learning, deep learning, and AI-driven automation with RS has enabled unprecedented capabilities, from rapid change detection and predictive modeling to the handling of massive, multi-sensor time series. These innovations extend the reach and impact of RS across domains such as environmental monitoring, land cover classification, hydrology, agriculture, and hazard assessment.

6.2 Closing thoughts and future directions

Looking ahead, the convergence of RS, AI, and cloud computing will continue to accelerate innovation across all fields from geography to environmental sciences and geospatial technologies. Several trends are poised to shape the next era:

- AI-native preprocessing pipelines will increasingly automate correction, harmonization, and quality control, enabling near-real-time readiness of multi-sensor datasets.
- Interoperable, analysis-ready data will become standard, reducing the burden on individual researchers and supporting cross-platform, global-scale applications.
- Multi-modal data fusion: combining optical, SAR, LiDAR, hyperspectral, UAV, and socio-environmental datasets will enable richer insights into pattern, process, and change.
- Edge computing and on-board AI will allow satellites and drones to process data prior to downlink, improving efficiency for rapid response situations such as wildfires, floods, and extreme heat events. We are already seeing such drone technology on the market.
- Increased use of vibe-coding (i.e., using AI to help develop code decks) will rise in popularity with both positive and negative repercussions. This is a positive trend in that it reduces the barrier to get users into RS analysis and big data analysis in general. This is a negative trend in that, if users cannot correct or validate their code, errors might be propagated in results. This is already being seen across academia in multiple outlets.
- Training and education must keep pace, ensuring researchers and practitioners understand both the power and the limitations of increasingly automated tools.

Ultimately, the future of RS will be shaped by how effectively the community balances innovation with scientific rigor, accessibility, and responsible use.

6.3 Closing remarks

RS and GeoAI now operate in a landscape defined by data abundance, computational power, and global connectivity. These tools offer extraordinary potential for understanding and addressing environmental challenges, from climate variability and land-use change to biodiversity loss and resource management. Yet their impact depends on thoughtful implementation, transparent workflows, and sustained investment in training, infrastructure, and interdisciplinary collaboration. By embracing both the opportunities and responsibilities that accompany big-data RS, researchers, educators, and practitioners are well positioned to advance a new era of EO, an era that is more scalable, integrated, equitable, and aligned with the complex environmental realities of our changing world.

Case studies

The case studies selected for this chapter highlight the various stages, decision-making, and approaches utilized in big data RS from preprocessing through the analysis stage. The first case study, with a study site across the US, highlights the use of dense time series remotely sensed data in conjunction with ancillary variables related to socioecological and climate trends. This work, funded by the National Science

Foundation (Award Number: 2203235) ultimately looks to understand the relationship between heat wave shape, persistence, and duration as it impacts or is impacted by environmental and human factors. The second case study highlights the use of multiple remotely sensed data products in conjunction with one another. A common practice in RS is the use of not just multiple image time series but other data products in conjunction to characterize, for example, the rate of vegetation change because of human utilization of the landscape. This case study highlights how such data fusion can be completed from start to finish. Further, it highlights the importance of data fusion for developing meaningful inferences across the landscape.

Case study 1: Spatiotemporal trends in heat waves across the United States and their impact on environmental health

Background: Heat waves are extreme weather events that pose significant risk to human health, ecosystems, and infrastructure. Understanding the environmental drivers of heat wave occurrence and persistence is critical for anticipating their impacts and supporting effective mitigation strategies. This case study focuses on analyzing heat wave patterns across the contiguous US from 1981 to 2018, with a particular emphasis on the role of soil moisture and vegetation condition in extreme heat events. To achieve this, we integrate multiple sources of environmental data including remotely sensed, modeled, and field observations into a flexible geospatial workflow. This case study illustrates how environmental research leveraging geospatial technologies is inherently a dynamic and iterative process. Here we show just one part of the research conducted at Michigan State University (MSU) Department of Geography, Environment, and Spatial Sciences under the supervision of Dr. Bunting (Bunting et al., 2024; Wanyama et al., 2023).

Implementation: Working with geospatial data includes using the most convenient software, where convenience is defined by several interconnected factors. (A) Data upload and access speed is often the initial limiting step in geospatial workflows. For example, we relied on GEE at the outset because it provides immediate, cloud-based access to a vast array of satellite products without the need to manually download, store, and reupload datasets into a separate analysis environment (Gorelick et al., 2017b). (B) Availability of computational power, processing efficiency, and memory resources to perform large-scale analysis locally versus using more convenient, but less transparent cloud computing (Xu et al., 2022; Zhao et al., 2016). This topic is particularly important once analyses scale in complexity or spatial/temporal extent. Cloud environments, such as GEE, can dramatically speed up large-scale image processing or time series analysis, yet they do so through server-side operations that obscure how resources are allocated and how intermediate steps are handled. Conversely, using a local computer (e.g., using R or Python) offers greater transparency, reproducibility, and the ability to profile and optimize memory use, but may require breaking large spatial datasets into tiles, aggregating smaller pixels into larger ones, or using parallelization strategies to avoid system limitations (Gorelick et al., 2017b; Kumar & Mutanga, 2018; Xu et al., 2022b). (C) Transparency of the algorithm and the ability to fine-tune analyses is another important consideration. Black-box tools, often found in cloud environments or proprietary platforms, can accelerate workflows, but they may hide assumptions related to preprocessing (e.g., radiometric corrections and resampling) or algorithm function (e.g., classification parameterization). Therefore, in this work, we prioritized tools that allow users to modify parameters, inspect intermediate outputs, and fully document analytical decisions. (D) Lastly, the capacity to integrate GIS operations, statistical modeling, and effective visualization strongly influences software choice. In our workflows, we used R and RStudio to perform detailed statistical and spatial analyses locally, ensuring clear documentation and reproducibility, while ArcGIS Pro served as a powerful environment for geovisualization and interpretation of results.

Preprocessing data is a crucial step. Complex geospatial analyses often include multiple steps of environmental data processing. It includes transforming raw data into formats suitable for research purposes, and may include data cleaning, filtering, or even transforming the data into trends. Using multiple variables for the analysis complicates the process. If the analysis requires data to be in the same format or scale, then an extra step, such as resampling the data, is required.

This case study represents the next step in analyzing changes in heat wave drivers and understanding their consequences across the contiguous US over the period from 1981 to 2018. After assessing the persistence of heat waves across various climate regions and identifying areas of extreme occurrences (Wanyama et al., 2023), we now focus on exploring the key factors associated with heat wave persistence. Specifically, we aim to determine whether areas experiencing extreme heat waves exhibit distinct characteristics related to soil moisture and vegetation conditions.

The methodology for studying heat wave events across the US climate regions (Keellings et al., 2018; Keellings & Hamid, 2020; Wanyama et al., 2023) included several key components, the first being the defining of heat wave extents. Daily PRISM (Parameter elevation Regression on Independent Slopes Model) temperature data, with a spatial resolution of 4 km, for the continental US from 1981 to 2018 was used (Daly et al., 2002; Keellings et al., 2018). Analysis was conducted for each NOAA climate region across the continental US. Heat waves were defined using a threshold equal to the 95th percentile of the local daily maximum temperature for each extended summer month (May through September) for at least three consecutive days (Keellings et al., 2018). A hierarchical clustering algorithm with indicator semivariograms identified heat clusters and delineated heat wave events (Keellings et al., 2018). Bounding polygons were drawn around heat clusters belonging to a single heat wave (Keellings et al., 2018; Keellings & Hamid, 2020). A GIS model was developed using ArcPy and implemented in ArcGIS Pro to compute heat wave persistence by calculating overlapping polygons and counting the number of overlaps at each location. The generalized overlapping polygons with the most persistent heat waves inside of the NOAA climate region were named "Persistent heat wave regions" (Case Study Figure 1.1).

The above analysis enabled the definition of persisting heat waves in several fashions. In addition to quantifying the total number of overlapping heat waves, we identified areas that were consistently affected by long-duration, most extreme (99th percentile), and highly fragmented events within each climate region. The research

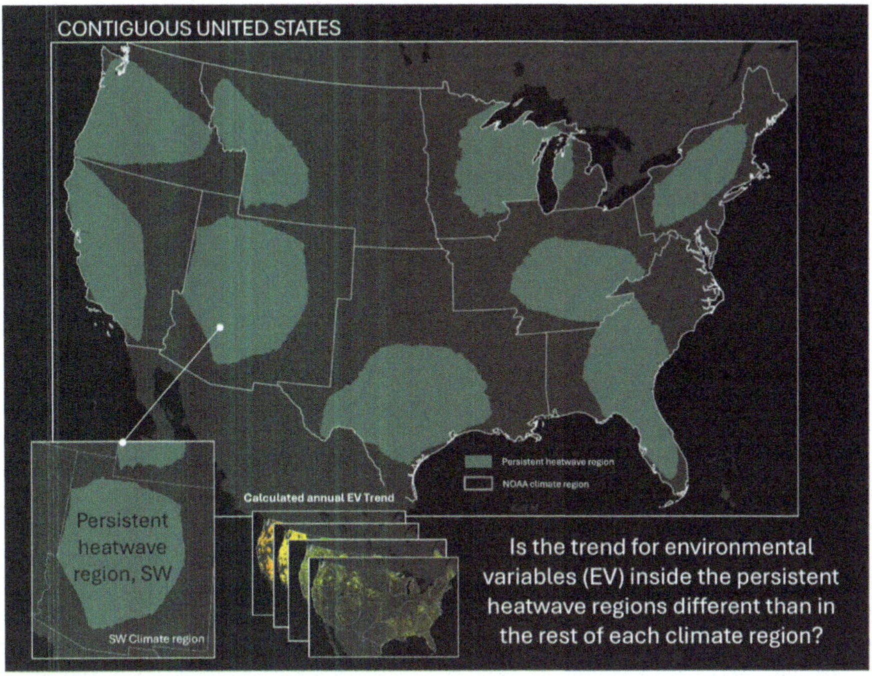

Case Study Figure 1.1 Persistent heat wave locations across the United States as denoted in Wanyama et al. (2023).

question asks: "How do the patterns of environmental characteristics and underlying drivers compare inside versus outside the most extreme heat wave events? We hypothesize that some variables such as soil moisture and vegetation trends will differ more than others. Specifically, spectral indices that act as proxies for vegetation characteristics (e.g., the normalized difference vegetation index, NDVI) are expected to be significantly lower within extreme heat wave boundaries compared to areas outside them, with more pronounced declines in regions where other environmental covariates (e.g., soil moisture and high impervious surface cover) are both affected by extreme heat and contribute to variation in spectral index values. In other words, NDVI declines during extreme heat waves are expected to be greatest in areas where extreme heat intensifies low soil moisture and where high impervious surface cover exacerbates conditions that suppress vegetation.

As a first step, after using the literature to compile the list of environmental variables required, a script was developed in GEE to generate, stack, and extract all dataset for random points across the US. The attributes included, among others, elevation, slope, water and urban area proximity, and soil moisture indices. GEE was chosen as it allows us to upload data we generated previously and uses already existing GEE datasets to create a joint dataset. The goal is to prepare a dataset for Random Forest modeling and subsequent statistical analysis inside and outside the areas of heat waves boundaries.

The script is divided into multiple categories:

1. Data Import and Country Selection: First, the script loads country boundaries, filters to the US, and generates 100,000 random points within the country for sampling purposes. As such we cover all the study area but in a random approach to minimize bias.
2. Digital Elevation Model (DEM): Using the 30-m Shuttle Radar Topography Mission (SRTM) DEM we derived elevation data and calculated slope/aspect.
3. Population and Protected Areas: Layers for population density (GPWv4) and protected areas (GAP PAD-US v2.0) were utilized to contextualize human presence and conservation priorities. Both datasets were provided in raster format, which allowed for streamlined spatial overlay, zonal statistics, and region to region comparison.
4. Calculation of Distance to Water Bodies and Urban Areas: Distances to permanent water bodies (JRC/GSW1_2) and urban areas were calculated to capture the spatial gradients that may influence environmental condition and human exposure. These distance rasters were generated using Euclidean distance functions, ensuring continuous surfaces that quantify proximity.
5. Drought Indices: Standardized Precipitation Evapotranspiration Index (SPEI) for various temporal windows (3, 6, 12, and 24 months) were included to capture drought sensitivity. Temporally, these indices cover the same years as the heat wave data. Sen's Slope analysis was used to reveal the annual trend in these time series data. Next, binary masks were applied to denote areas with significant trends, where 1 represents significant change (positive or negative) and 0 represents non-significant trends.
6. Vegetation Indices: Landsat derived seasonal median and mean composites of NDVI were used to assess the relationship between heat waves and vegetation condition. Temporally these composites cover the same years as the heat wave data. Like SPEI above, Sen's Slope analysis was used to reveal the annual trend in these time series data and a binary mask was applied to identify areas with significant trends.

GEE was used to merge the raster outputs into a single image collection. We then extracted values at each random point location for the aforementioned variables, creating a comprehensive dataset ready for both machine learning and statistical analysis. For each point we also denoted its climate region, state, and developed a binary variable of inside/outside persistent heat wave regions. ArcGIS Pro was used for such data extraction. For the final stage of the analysis, we used the R package ggstatsplot (Patil, 2021) to evaluate statistical differences between groups and to visualize the results.

Results: We present only results disaggregated by climate regions, specifically the southwest (SW) NOAA climate region. Results indicate that between 2001 and 2018, SPEI values and trends for all windows, except one (SPEI-3 months), within extreme heat wave regions were consistently lower than those outside these regions (Case Study Figure 1.2). The magnitude of this difference increased across the windows from 6 to 24 months. NDVI trends showed the similar behavior. Changes in value and trend of NDVI inside heat waves were of slightly lower magnitude than outside the polygons. Other variables revealed statistically significant differences between areas inside and outside heat wave zones. Elevation was notably higher within persistent heat wave regions, with mean values around 2000 m compared to approximately 1400 m outside. This elevation difference may reflect topographic gradients and how they shape local climate dynamics. Distance-to-water also showed a marked contrast.

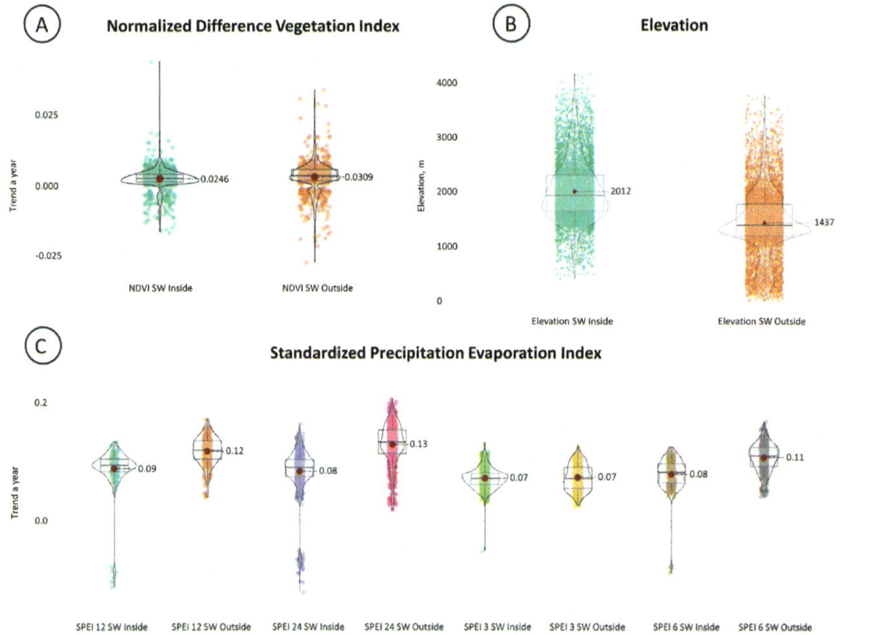

Case Study Figure 1.2 Variables across the SW NOAA climate region highlighting the differences inside and outside heat wave persistent regions for key environmental aspects. (A) NDVI, (B) elevation, and (C) SPEI at across different windows 3–24 months.

Points inside heat wave polygons were, on average, closer to permanent water bodies (~6000 m) than points outside (~10,000 m). Although both distances indicate generally dry landscapes, the shorter proximity inside heat wave regions may indicate the presence of localized hydrologic features within broadly arid zones, or alternatively, that more distant-from-water environments outside the polygons exhibit different land-surface characteristics that mitigate heat wave persistence. Together, these results illustrate that the most persistent heat wave areas are not only climatologically distinct but also differ systematically in their physical, hydrologic, and demographic characteristics, providing important context for interpreting environmental responses to extreme heat.

Using the SW climate region as an example, the results illustrate how persistent heat wave conditions align with distinct environmental characteristics (Case Study Figure 1.2). Both NDVI and SPEI trends indicate that areas experiencing the most extreme and long-duration heat waves also exhibit notably drier conditions. SPEI values were consistently lower inside persistent heat wave polygons across multiple timescales, signaling sustained moisture deficits. Differences in trend values for SPEI increase from shorter to longer time scales, which highlights the shorter-term fluctuations. Correspondingly, NDVI trends were slightly lower inside these boundaries, suggesting reduced vegetation productivity or rebound rates in areas repeatedly exposed to extreme heat. Together, these patterns reinforce the hypothesis that persistent heat stress is closely linked to ecological drying and diminished vegetation health.

Impact: Assessment of changes in environmental variables in the context of a changing climate and the development of heat waves is an important research topic. Several decades of observations, including RS data, climate records, and other sources, allow us to analyze the relationships between these parameters, quantify their mutual contributions, and explore potential causal links between heat wave dynamics and changes in environmental variables. This analysis can be conducted at a regional scale and subsequently generalized to the national level. Such an approach enables us to understand how the system responds and adapts to shifts in thermal regimes and increasing parameters of the heat waves across the US, as well as to identify changes in the functioning of key parameters such as soil moisture, overall vegetation health and related vegetation trends.

Data and Study Limitations: This short case study integrates datasets from multiple sources, each characterized by distinct spatial resolutions, temporal frequencies, and preprocessing histories. PRISM temperature data (4 km), SRTM elevation data (30 m), Landsat-derived NDVI (30 m), gridded population data, drought indices, and proximity raster were all harmonized into a single workflow; however, these differences in scale introduce potential uncertainties. Spatial mismatches may lead to smoothing or exaggeration of local environmental gradients, particularly when coarser-resolution datasets are used to interpret fine-scale vegetation or soil moisture conditions. In addition, several datasets, especially those accessed through cloud-based platforms like GEE, carry preprocessing assumptions (e.g., resampling, cloud masking, atmospheric corrections) that may influence derived trends.

As persistent heat wave boundaries were delineated at one resolution and environmental covariates were extracted at others, scale differences may influence statistical relationships. Further, the reliance on randomly sampled points introduces an element of spatial averaging that may obscure localized effects. While these limitations do not undermine the identified patterns, they highlight the need for caution when interpreting fine-scale ecological or biophysical patterns/mechanisms.

Case study 2: Agricultural expansion and natural vegetation decline in Eastern Africa

Background: This case study leverages readily available time series Earth observation data to comprehensively assess the changing dynamics in agricultural lands and forests (and/or natural vegetation) in Eastern Africa for the period 2000–21. By integrating geospatial datasets with environmental variables, this study aims to provide insights into the mechanisms shaping land-use transitions and inform strategies for sustainable land management in vulnerable regions. The study also demonstrates key image preprocessing practices employed in these kinds of analyses.

Implementation, Data: Data utilized in this case study include fractional tree cover from GLOBMAP (Liu et al., 2024) and MODIS vegetation continuous fields (VCF) (DiMiceli et al., 2022), slope computed from resampled elevation data, and gridded precipitation (Funk et al., 2015). Data preprocessing involved eliminating poor quality pixels from the VCF data, resampling all data to the GLOBMAP resolution, and generating annual composites over the 2000 to 2021 period. We defined forest areas in two steps. First, utilizing a common forest fractional cover dataset (Hansen et al., 2013), we identified forest cover in the year 2000 as 30 m pixels with a canopy cover of 30% or more. Second, we aggregated the output, defining forest cover areas as the larger GLOBMAP pixels that contained at least 50% of the smaller tree cover pixels.

Implementation, Methods: We analyzed annual tree cover data from GLOBMAP along with percent tree cover, percent non-tree cover, and percent non-vegetated cover from MODIS VCF for the period from 2000 to 2021 to identify changes and spatial patterns. We assessed trends in these variables along with annual precipitation using Mann-Kendall and Sen's Slope. We summarized gains and losses in the four vegetation measurements for four ecoregions and examined their associations with elevation, slope, and precipitation. We also calculated the annual area of land under each of the four vegetation measurements, normalized by forest area in 2000, and created temporal plots for each country. Finally, we conducted correlation analyses between the vegetation measurements and elevation, slope, and precipitation. Through our robust fusion of these datasets, we were able to more thoroughly examine the changing dynamics of agricultural lands and natural vegetation in Eastern Africa, a region where agricultural and forest systems are particularly susceptible to the increasing effects of climate change and variability.

Results: Spatial trends in forest cover

In 2000, Uganda had 76,967 km^2 of forest cover (approximately 31.5% of its land area), Tanzania had 261,132 km^2 (27.5%), Burundi had 4923 km^2 (17.5%), Rwanda had 570 km^2 (17.4%), and Kenya had 30,575 km^2 (5.2%). Across the five countries, there was a net increase of 6015 km^2 (1.6% of the forest area) in fractional tree cover, a net loss of 50,114 km^2 (13.3%) in MODIS percent tree cover, and a net increase of 32,997 km^2 (8.7%) in percent non-tree cover. These changes were spatially concentrated. Significant losses in tree cover and gains in non-tree cover were observed in western Uganda, indicating a shift from forest to non-tree vegetation. Similar patterns were observed in western and southern Tanzania, especially near protected areas. Gains in tree cover were noted in Tanzania's Miombo woodlands and major protected areas in Uganda, as well as in Kenya's Mau Forest complex. Some increases in non-vegetated cover were also observed where tree cover was lost (Case Study Figure 2.1).

Results: Temporal trends in vegetation cover

Case Study Figure 2.2 shows temporal plots of the four vegetation measurements for each country. In Uganda, there was a significant increase in non-tree cover and a decrease in fractional tree cover which peaked around 2015 before reversing. Similar, though not statistically significant, trends were observed in Tanzania, Burundi, and Rwanda, with non-tree cover increasing until about 2013 before reversing. Fractional tree cover generally followed the opposite trend. Trends in Tanzania and Uganda, which had the most forest cover in 2000, greatly influenced the overall regional trend, which showed a significant increase in non-tree cover and an overall, non-significant decrease in tree cover.

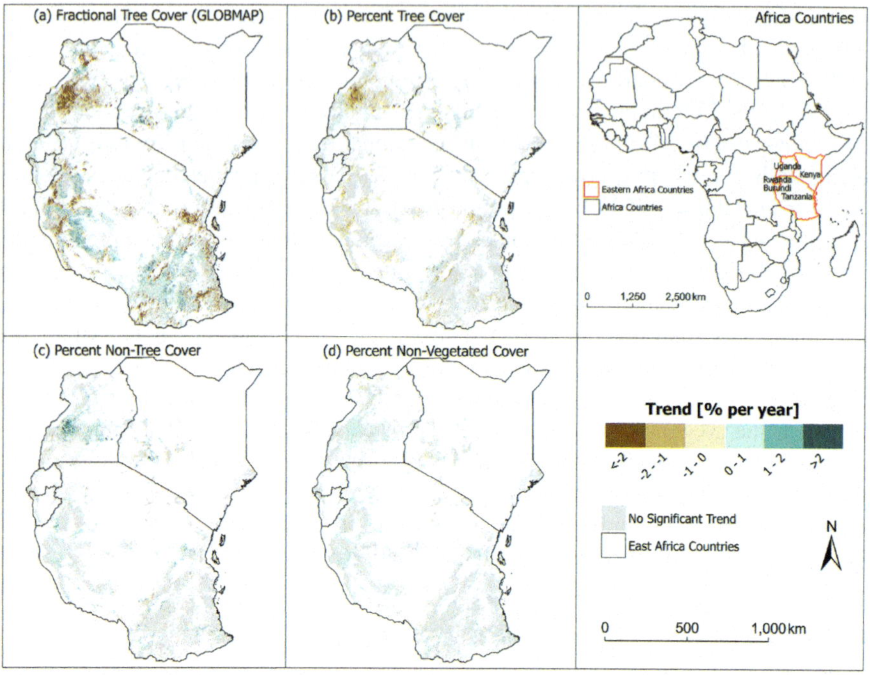

Case Study Figure 2.1 Trends from 2000 to 2021 in the four vegetation measurements: Fractional tree cover from GLOBMAP (a), percent tree cover (b), percent non-tree cover (c) and percent non-vegetated cover (d), all from MODIS. Only statistically significant trends at $P = .05$ are shown.

The MODIS percent tree cover significantly underestimated tree cover in all five countries, particularly in Uganda, Tanzania, Burundi, and Rwanda. In these countries, areas with $\geq$50% tree cover represented less than 20% of the forest area in 2000 throughout the 2000–21 period, while this proportion was relatively higher in Kenya (>25%). In addition, the amount of non-vegetated cover was negligible across the five countries.

Results: Linking forest cover trends to environmental drivers

Trend analysis of annual precipitation revealed a significant increase in some locations. Additionally, fractional tree cover and percent non-tree cover had significant associations with precipitation in some of these locations. This suggests that the increase in precipitation may have influenced the observed trends in tree cover and non-tree cover in these locations. Although there were no strong associations between vegetation measurements and slope, elevation, or long-term mean precipitation, we found that tree cover loss increased with elevation up to about 1000 m, then decreased. This trend was especially noticeable in the montane forest ecoregion, highlighting the tendency of farming communities to convert intact high-elevation forests into new agricultural lands (Wanyama et al., 2020).

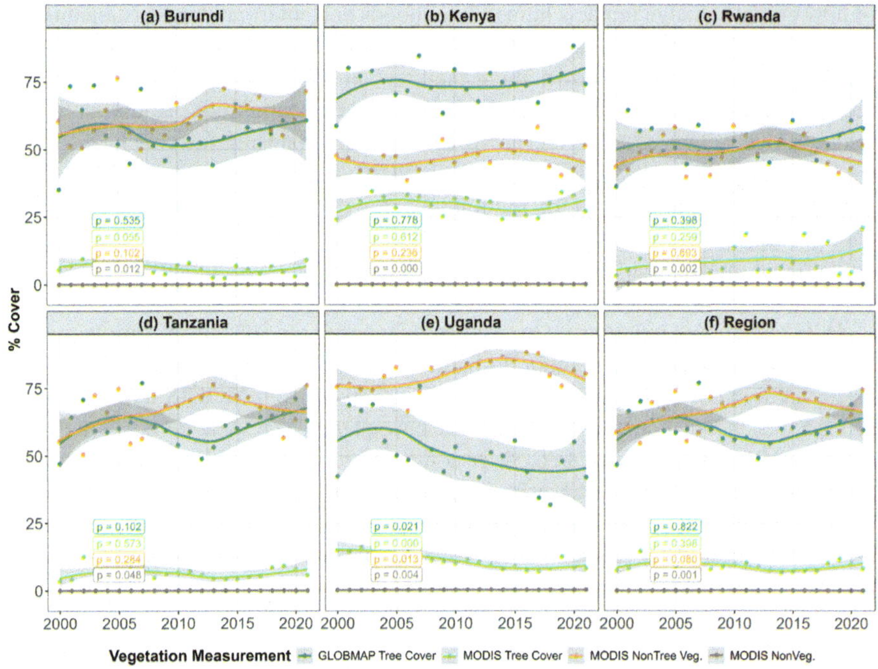

Case Study Figure 2.2 Graphs of annual percent proportion of the four vegetation measurements per country (a–e) and for the whole region (f) using data from GLOBMAP and MODIS vegetation continuous fields (VCF). Areas with tree cover, non-tree cover, and non-vegetated cover were defined to be pixels with $\geq$50% percent proportion tree cover, non-tree cover, and non-vegetated cover, respectively. For reference, Uganda had 76,967 km^2 of forest cover in 2000, Tanzania had 261,132 km^2, Burundi had 4923 km^2, Rwanda had 570 km^2, while Kenya had 30,575 km^2. *P*-values were obtained from the Mann-Kendall test.

Conclusions: This case study identified overall trends of increasing tree cover loss from 2000 to 2021, consistent with findings from other research (e.g., Vancutsem et al., 2021; Ciobotaru et al., 2021; Liu et al., 2025). We also observed that significant tree cover losses were accompanied by major increases in non-tree vegetated cover, while non-vegetated areas generally remained low. This indicates a widespread shift from forest to non-tree vegetation (agriculture) in the region. These changes were highly spatially heterogeneous, with most of the transitions occurring in western Uganda, parts of eastern and southern Tanzania, and the greater Mau Forest complex in Kenya. The changes were associated with precipitation and elevation, and to a lesser extent with slope, although the significance, strength, and nature of these relationships varied considerably by location and ecoregion type.

Implications: This study holds significant importance for several reasons. Firstly, the results highlight how data selection and source can significantly influence research outcomes. The high variability in outputs observed in this case study underscores the importance of carefully considering data inputs when developing RS research and interpreting results. Second, this work addresses existing gaps in our understanding of how climate change impacts human livelihoods and environmental sustainability at both local and regional scales. Eastern Africa, in particular, has not been extensively studied, leaving much unknown about the ways in which climate change influences the interactions between nature and nature-reliant communities. Additionally, the long-term consequences of widespread agricultural expansion (as concluded in this study, Akanga et al., 2022; Wanyama et al., 2020) and the loss of natural vegetation remain unclear for this region. By pinpointing areas frequently targeted for conversion and detailing their characteristics, this study equips local and national governments with crucial information to develop policies that balance environmental conservation with the livelihood needs of dependent populations.

Data and Study Limitations: Our case study incorporated data from various sources with different spatial and temporal characteristics. Although both the GLOBMAP fractional tree cover data and VCF data were derived from MODIS data, they were produced using distinct methodologies. Consequently, MODIS percent tree cover significantly underestimated tree cover compared to GLOBMAP and higher resolution data. Since MODIS VCF is computed so that the sum of percent tree cover, non-tree cover, and non-vegetated cover equals 100% for each pixel, the underestimation of tree cover likely led to an overestimation of non-tree cover. Nevertheless, integrating these datasets allowed us to more thoroughly investigate the evolving dynamics of agricultural lands and forests (and/or natural vegetation) in Eastern Africa.

References

Adedeji, O., Sun, Y., Li, S., & Guo, W. (2025). Effects of spatial resolution on assessing cotton water stress using unmanned aerial system imagery. *Remote Sensing, 17*(24), 4018. https://doi.org/10.3390/rs17244018

Ahmad, R. (2024). Smart remote sensing network for disaster management: An overview. *Telecommunication Systems, 87*(1), 213–237.

Akanga, D. O., Dahlin, K. M., & Moore, N. J. (2022). Accelerating agricultural expansion in the greater mau forest complex, Kenya. *Remote Sensing Applications: Society and Environment, 28*(November). https://doi.org/10.1016/j.rsase.2022.100860

Albamonte, G., Falcone, G., Monaco, M., & Senatore, S. (2025). Constructing a knowledge base from remote sensing indicators for deforestation assessment. *Applied Intelligence, 55*(15), 1014. https://doi.org/10.1007/s10489-025-06896-2

Amani, M., Ghorbanian, A., Ali Ahmadi, S., et al. (2020). Google Earth engine cloud computing platform for remote sensing big data applications: A comprehensive review. *IEEE Journal of Selected Topics in Applied Earth Observations and Remote Sensing, 13*, 5326–5350.

Amazon Web Services, Inc. (2024). *Amazon web services*. https://aws.amazon.com/.

Aryal, A., & Lakshmi, V. (2025). Assessing antecedent climatic and hydrological conditions and anthropogenic impacts to drive catastrophic flooding in the northeastern United States. *Frontiers in Remote Sensing, 6*(September). https://doi.org/10.3389/frsen.2025.1639845

Aschbacher, J. (2017). ESA's Earth observation strategy and copernicus. In *Satellite Earth observations and their impact on society and policy*. Singapore: Springer. https://library.oapen.org/bitstream/handle/20.500.12657/27724/1002282.pdf?sequence=1#page=96.

Avtar, R., Komolafe, A. A., Kouser, A., et al. (2020). Assessing sustainable development prospects through remote sensing: A review. *Remote Sensing Applications: Society and Environment, 20*(November). https://doi.org/10.1016/j.rsase.2020.100402

Bachmann, N., Tripathi, S., Brunner, M., & Herbert, J. (2022). The contribution of data-driven technologies in achieving the sustainable development goals. *Sustainability, 14*(5), 2497. https://doi.org/10.3390/su14052497

Bannari, A., Morin, D., Bénié, G. B., & Bonn, F. J. (1995). A theoretical review of different mathematical models of geometric corrections applied to remote sensing images. *Remote Sensing Reviews, 13*(1–2), 27–47. https://doi.org/10.1080/02757259509532295

Boutayeb, A., Lahsen-Cherif, I., & Ahmed, El K. (2025). When machine learning meets geospatial data: A comprehensive GeoAI review. *IEEE Journal of Selected Topics in Applied Earth Observations and Remote Sensing, 18*, 13135–13191. https://doi.org/10.1109/JSTARS.2025.3568715

Breunig, F. M., Soares Galvão, L., Dalagnol, R., et al. (2020). Delineation of management zones in agricultural fields using cover–crop biomass estimates from PlanetScope data. *International Journal of Applied Earth Observation and Geoinformation, 85*.

Bunting, E. L., Tolmanov, V., & Keellings, D. (2024). What is a weat wave: A survey and literature synthesis of heat wave definitions across the United States. *PLOS Climate, 3*(9).

Chang, Y., Yan, L., Wu, T., & Zhong, S. (2016). Remote sensing image stripe noise removal: from image decomposition perspective. *IEEE Transactions on Geoscience and Remote Sensing, 54*(12), 7018–7031.

Chen, Jung, Y., Li, W., Losier, L. M., et al. (2025). *Thirty years of the open geospatial consortium-history, present, and future*. https://eartharxiv.org/repository/view/8623/.

Chen, J. M. (1996). Evaluation of vegetation indices and a modified simple ratio for boreal applications. *Canadian Journal of Remote Sensing, 22*(3), 229–242.

Chi, M., Plaza, A., Atli Benediktsson, J., Sun, Z., Shen, J., & Zhu, Y. (2016a). Big data for remote sensing: challenges and opportunities. *Proceedings of the IEEE, 104*(11), 2207–2219.

Ciobotaru, A.-M., Patel, N., & Pintilii, R.-D. (2021). Tree cover loss in the mediterranean region—an increasingly serious environmental issue. *Forests, 12*(10), 1341.

Claverie, M., Ju, J., Masek, J. G., et al. (2018). The harmonized landsat and Sentinel-2 surface reflectance data set. *Remote Sensing of Environment, 219*, 145–161.

Cracknell, A. P. (2024). *The advanced very high resolution radiometer AVHRR*. https://www.taylorfrancis.com/books/mono/10.1201/9781003572817/advanced-high-resolution-radiometer-avhrr-arthur-cracknell.

Daly, C., Gibson, W. P., Taylor, G. H., Johnson, G. L., & Pasteris, P. (2002). A knowledge-based approach to the statistical mapping of climate. *Climate Research, 22*, 99–113.

DiMiceli, C., Sohlberg, R., & Townshend, J. (2022). *"MODIS/Terra Vegetation Continuous Fields Yearly L3 Global 250m SIN Grid V061." NASA EOSDIS land processes distributed Active Archive Center (DAAC) data set, MOD44B-061.*

Dritsas, E., & Trigka, M. (2025). Remote sensing and geospatial analysis in the big data era: a survey. *Remote Sensing, 17*(3), 550.

Earth Science Data Systems, NASA. (2026). *Your gateway to NASA Earth observation data | NASA earthdata*. Earth Science Data Systems, NASA. January 23 https://www.earthdata.nasa.gov/.

Earth science in action for tomorrow's world – ESA Earth observation science strategy., n.d. Accessed November 13, 2025. https://zenodo.org/records/13973612.

Ehrlich, D., Estes, J. E., & Singh, A. (1994). Applications of NOAA-AVHRR 1 Km data for environmental monitoring. *International Journal of Remote Sensing, 15*(1), 145–161. https://doi.org/10.1080/01431169408954056

Enescu, I., Ionuț, L.de E., Haas-Artho, D., et al. (2021). Cloud optimized raster encoding (core): A web-native streamable format for large environmental time series. *Geomatics, 1*(3), 369–382.

Eschenbach, W. J. von (2021). Transparency and the black box problem: Why we do not trust AI. *Philosophy & Technology, 34*(4), 1607–1622. https://doi.org/10.1007/s13347-021-00477-0

Franklin, S. E., & Giles, P. T. (1995). Radiometric processing of aerial and satellite remote-sensing imagery. *Computers & Geosciences, 21*(3), 413–423.

Frazier, A. E., & Hemingway, B. L. (2021). A technical review of planet smallsat data: Practical considerations for processing and using planetscope imagery. *Remote Sensing, 13*(19), 3930.

Funk, C., Peterson, P., Landsfeld, M., et al. (2015). The climate hazards infrared precipitation with Stations—A new environmental record for monitoring extremes. *Scientific Data, 2* (1). https://doi.org/10.1038/sdata.2015.66

Géron, A. (2022). *Hands-on machine learning with Scikit-Learn, Keras, and TensorFlow: Concepts, tools, and techniques to build intelligent systems* (3rd ed.). O'Reilly Media, Inc.

Ghamisi, P., Rasti, B., Yokoya, N., et al. (2019). Multisource and multitemporal data fusion in remote sensing: A comprehensive review of the state of the art. *IEEE Geoscience and Remote Sensing Magazine, 7*(1), 6–39.

Giuliani, G., Camara, G., Killough, B., & Minchin, S. (2019). Earth observation open science: Enhancing reproducible science using data cubes. *Data, 4*(4), 147. https://doi.org/10.3390/data4040147

Gorelick, N., Hancher, M., Dixon, M., Ilyushchenko, S., Thau, D., & Moore, R. (2017a). Google Earth engine: Planetary-scale geospatial analysis for everyone. *Remote Sensing of Environment, 202*, 18–27. https://doi.org/10.1016/j.rse.2017.06.031

Gorelick, N. (2013). Google Earth engine. *EGU General Assembly Conference Abstracts, 15*.

Goslee, S. C. (2011). Analyzing remote sensing data in R: The landsat package. *Journal of Statistical Software, 43*, 1–25.

Gupta, M. (2014). Organizational culture and the three v's of big data. *MWAIS 2014 Proceedings, May 15*. https://aisel.aisnet.org/mwais2014/10.

Hadjimitsis, D. G., Clayton, C. R. I., & Hope, V. S. (2004). An assessment of the effectiveness of atmospheric correction algorithms through the remote sensing of some reservoirs. *International Journal of Remote Sensing, 25*(18), 3651–3674. https://doi.org/10.1080/01431160310001647993

Hanny, D., Schmidt, S., Gandhi, S., Granitzer, M., & Resch, B. (2025). A multimodal GeoAI approach to combining text with spatiotemporal features for enhanced relevance classification of social media posts in disaster response. *Big Earth Data, 23*, 1–45. https://doi.org/10.1080/20964471.2025.2572140. October.

Hansen, M. C., Potapov, P. V., Moore, R., Hancher, M., Turubanova, S. A., Tyukavina, A., Thau, D., Stehman, S. V., Goetz, S. J., Loveland, T. R., Kommareddy, A., Egorov, A., Chini, L., Justice, C. O., & Townshend, J. R. G. (2013). High-resolution global maps of 21st-century forest cover change. *Science, 342*(6160), 850–853. https://doi.org/10.1126/science.1244693

Harvey, F., Werner, K., Hardy, P., Bishr, Y., & Catharina, R. (1999). Semantic interoperability: A central issue for sharing geographic information. *The Annals of Regional Science, 33*(2), 213–232.

Hill, J., & Sturm, B. (1991). Radiometric correction of multitemporal thematic mapper data for use in agricultural land-cover classification and vegetation monitoring. *International Journal of Remote Sensing, 12*(7), 1471–1491. https://doi.org/10.1080/01431169108955184

Horn, B. K. P., & Woodham, R. J. (1979). Destriping landsat MSS images by histogram modification. *Computer Graphics and Image Processing, 10*(1), 69–83.

Huang, D., Yang, T., & Qin, R. (2022). An evaluation of PlanetScope images for 3D reconstruction and change detection – Experimental validations with case studies. *GIScience and Remote Sensing, 59*(1), 744–761. https://doi.org/10.1080/15481603.2022.2060595

Janga, B., Asamani, G. P., Sun, Z., & Cristea, N. (2023). A review of practical AI for remote sensing in Earth sciences. *Remote Sensing, 15*(16), 4112. https://doi.org/10.3390/rs15164112

Janowicz, K., Gao, S., McKenzie, G., Hu, Y., & Bhaduri, B. (2020). GeoAI: Spatially explicit artificial intelligence techniques for geographic knowledge discovery and beyond. *International Journal of Geographical Information Science, 34*(4), 625–636. https://doi.org/10.1080/13658816.2019.1684500

Jutz, S., & Pilar Milagro-Perez, M. (2020). Copernicus: The European Earth observation programme. *Revista de Teledetección, 56* (V–XI).

Kahraman, S., & Bacher, R. (2021). A comprehensive review of hyperspectral data fusion with lidar and sar data. *Annual Reviews in Control, 51*(January), 236–253. https://doi.org/10.1016/j.arcontrol.2021.03.003

Keellings, D., & Hamid, M. (2020). Spatiotemporal evolution of heat wave severity and coverage across the United States. *Geophysical Research Letters, 47*(9). https://doi.org/10.1029/2020GL087097

Keellings, D., Bunting, E., & Engström, J. (2018). Spatiotemporal changes in the size and shape of heat waves over North America. *Climatic Change, 147*(1–2), 165–178. https://doi.org/10.1007/s10584-018-2140-3

Kondratyev, K. Y., Kozoderov, V. V., & Smokty, O. I. (2013). *Remote sensing of the Earth from space: Atmospheric correction.* Springer Science & Business Media. https://books.google.com/books?hl=en&lr=&id=x5j6CAAAQBAJ&oi=fnd&pg=PA1&dq=Kondratyev,+K.Y.,+Kozoderov,+V.V.+and+Smokty,+O.I.,+2013.%C2%A0Remote+sensing+of+the+Earth+from+space:+atmospheric+correction.+Springer+Science+%26+Business+Media.&ots=10vKqCMR9I&sig=bXoAEAkLdEe33STIPciTpqSQfQ8.

Kumar, L., & Mutanga, O. (2018). Google Earth engine applications since inception: Usage, trends, and potential. *Remote Sensing, 10*(10), 1509.

Leach, N., Coops, N. C., & Obrknezev, N. (2019). Normalization method for multi-sensor high spatial and temporal resolution satellite imagery with radiometric inconsistencies. *Computers and Electronics in Agriculture, 164*.

Lee, C. A., Gasster, S. D., Plaza, A., Chang, C.-I., & Huang, B. (2011a). Recent developments in high performance computing for remote sensing: A review. *IEEE Journal of Selected Topics in Applied Earth Observations and Remote Sensing, 4*(3), 508–527.

Li, W., & Hsu, C.-Y. (2022). GeoAI for large-scale image analysis and machine vision: Recent progress of artificial intelligence in geography. *ISPRS International Journal of Geo-Information, 11*(7), 385. https://doi.org/10.3390/ijgi11070385

Li, F., Jupp, D. L. B., Reddy, S., et al. (2010). An evaluation of the use of atmospheric and BRDF correction to standardize landsat data. *IEEE Journal of Selected Topics in Applied Earth Observations and Remote Sensing, 3*(3), 257–270.

Li, X., Feng, R., Guan, X., Shen, H., & Zhang, L. (2019). Remote sensing image mosaicking: Achievements and challenges. *IEEE Geoscience and Remote Sensing Magazine, 7*(4), 8–22.

Li, W., Hsu, C. Y., Wang, S., Witharana, C., & Liljedahl, A. (2022). Real-time GeoAI for high-resolution mapping and segmentation of arctic permafrost features: The case of ice-wedge polygons. In *Proceedings of the 5th ACM SIGSPATIAL international workshop on AI for geographic knowledge discovery, November* (pp. 62–65). https://doi.org/10.1145/3557918.3565869

Li, Z., Shen, H., Weng, Q., Zhang, Y., Dou, P., & Zhang, L. (2022). Cloud and cloud shadow detection for optical satellite imagery: Features, algorithms, validation, and prospects. *ISPRS Journal of Photogrammetry and Remote Sensing, 188*, 89–108.

Liang, H., Zhang, Z., Hu, C., Gong, Y., & Cheng, D. (2023). A survey on spatio-temporal big data analytics ecosystem: Resource management, processing platform, and applications. *IEEE Transactions on Big Data, 10*(2), 174–193.

Lisjak, J., Petrinović, M., & Keleminec, S. (2024). Harnessing remote sensing technologies for successful large-scale projects. *Tehnički Glasnik, 18*(si1), 104–109.

Liu, P., Di, L., Du, Q., & Wang, L. (2018). Remote sensing big data: Theory, methods and applications. *Remote Sensing, 10*(5), 711. https://doi.org/10.3390/rs10050711

Liu, Y., Liu, R., Chen, J., Wei, X., Lin, Q., & Lei, Z. (2024). A global annual fractional tree cover dataset during 2000–2021 generated from realigned MODIS seasonal data. *Scientific Data, 11*(1), 832. https://doi.org/10.1038/s41597-024-03671-9

Liu, W., Zhang, X., Zhao, T., Wang, J., Li, Z., & Liu, L. (2025). Revealing the proximate drivers behind global tree cover loss using multisourced remote sensing products during 2000–2020. *Forest Ecology and Management, 579.*

Lü, X.F., C.Q. Cheng, J.Y. Gong, and L. Guan. 2011. "Review of data storage and management technologies for massive remote sensing data." Science China Technological Sciences 54 (12): 3220–3232. 10.1007/s11431-011-4549-z.

Lu, B., Dao, P. D., Liu, J., He, Y., & Shang, J. (2020). Recent advances of hyperspectral imaging technology and applications in agriculture. *Remote Sensing, 12*(16), 2659.

Lyapustin, A. I., Wang, Y., Laszlo, I., et al. (2012). Multi-angle implementation of atmospheric correction for MODIS (MAIAC): 3. Atmospheric correction. *Remote Sensing of Environment, 127*, 385–393.

Lyons, M. B., Keith, D. A., Phinn, S. R., Mason, T. J., & Elith, J. (2018). A comparison of resampling methods for remote sensing classification and accuracy assessment. *Remote Sensing of Environment, 208*, 145–153.

Ma, Y., Wu, H., Wang, L., et al. (2015). Remote sensing big data computing: Challenges and opportunities. *Future Generation Computer Systems, 51*, 47–60.

Marr, B. (2021). *How much data do we create every day? The mind-blowing stats everyone should read.* https://www.forbes.com/sites/bernardmarr/2018/05/21/how-much-data-do-we-create-every-day-the-mind-blowing-stats-everyone-should-read/.

Metzler, A. B., Nathvani, R., Sharmanska, V., et al. (2023). Phenotyping urban built and natural environments with high-resolution satellite images and unsupervised deep learning. *Science of the Total Environment, 893*(October). https://doi.org/10.1016/j.scitotenv.2023.164794

Microsoft Corporation. (2024). *Microsoft azure*. https://azure.microsoft.com/.

Miller, R. B., & Small, C. (2003). Cities from space: Potential applications of remote sensing in urban environmental research and policy. *Environmental Science & Policy, 6*(2), 129–137.

Mohr, M., Pebesma, E., Dries, J., Lippens, S., Janssen, B., Thiex, D., … Griffiths, P. (2025). Federated and reusable processing of Earth observation data. *Scientific Data, 12*(1), 194.

Mura, D., Mauro, Prasad, S., Pacifici, F., Gamba, P., Chanussot, J., & Benediktsson, J. A. (2015). Challenges and opportunities of multimodality and data fusion in remote sensing. *Proceedings of the IEEE, 103*(9), 1585–1601.

Ngan, A., Durand, F., & Matusik, W. (2005). Experimental analysis of BRDF models. *Rendering Techniques, 2005*(16th), 2.

Patil, I. (2021). Visualizing statistical analyses: The "ggstatsplot" ecosystem. *Journal of Open Source Software, 6*(61), 3167. https://doi.org/10.21105/joss.03167

Pei, T., Xu, J., Liu, Y., et al. (2021). GIScience and remote sensing in natural resource and environmental research: Status quo and future perspectives. *Geography and Sustainability, 2*(3), 207–215.

Proy, C., Tanre, D., & Deschamps, P. Y. (1989). Evaluation of topographic effects in remotely sensed data. *Remote Sensing of Environment, 30*(1), 21–32.

Rachol, C. M. (2024). The 3D elevation program—supporting michigan's economy. In *Fact sheet, nos* (pp. 2024–3031). U.S. Geological Survey. https://doi.org/10.3133/fs20243031

Radočaj, D., Obho\djaš, J., Jurišić, M., & Gašparović, M. (2020). Global open data remote sensing satellite missions for land monitoring and conservation: A review. *Land, 9*(11), 402.

Rasti, B., Chang, Y., Dalsasso, E., Denis, L., & Ghamisi, P. (2021). Image restoration for remote sensing: Overview and toolbox. *IEEE Geoscience and Remote Sensing Magazine, 10*(2), 201–230.

Richter, R., Kellenberger, T., & Kaufmann, H. (2009). Comparison of topographic correction methods. *Remote Sensing, 1*(3), 184–196.

Sabir, R. M., Mehmood, K., Sarwar, A., et al. (2024). Remote sensing and precision agriculture: A sustainable future. In S. Kanga, S. Kumar Singh, K. Shevkani, V. Pathak, & B. Sajan (Eds.), *Transforming agricultural management for a sustainable future*. Switzerland: World Sustainability Series. Springer Nature. https://doi.org/10.1007/978-3-031-63430-7_4

Saeed, N., Ahmed, E., Almorad, H., Dahrouj, H., Al-Naffouri, T. Y., & Mohamed-Slim, A. (2020). CubeSat communications: Recent advances and future challenges. *IEEE Communications Surveys & Tutorials, 22*(3), 1839–1862.

Sagan, V., Coral, R., Bhadra, S., et al. (2024). Hyperfidelis: A software toolkit to empower precision agriculture with GeoAI. *Remote Sensing, 16*(9), 1584.

Saunier, S., Pflug, B., Moletto Lobos, I., et al. (2022). Sen2Like: Paving the way towards harmonization and fusion of optical data. *Remote Sensing, 14*(16), 3855.

Schmitt, M., & Xiao, X. Z. (2016). Data fusion and remote sensing: an ever-growing relationship. *IEEE Geoscience and Remote Sensing Magazine, 4*(4), 6–23.

Schowengerdt, R. A. (2006). *Remote sensing: Models and methods for image processing*. Elsevier. https://books.google.com/books?hl=en&lr=&id=KQXNaDH0X-IC&oi=fnd&pg=PP1&dq=remote+sensing,+data+processing+involves+computationally+intensive+tasks+such+as+atmospheric+correction,+radiometric+calibration,+and+feature+extraction&ots=spm0ELAcPF&sig=NfqIgVbwpb2ugz5hGdIqiyj8RdI.

Scott, G., & Rajabifard, A. (2017). Sustainable development and geospatial information: A strategic framework for integrating a global policy agenda into national geospatial capabilities. *Geo-Spatial Information Science, 20*(2), 59–76. https://doi.org/10.1080/10095020.2017.1325594

Sharma, R., Mishra, D. R., Levi, M. R., & Sutter, L. A. (2022). Remote sensing of surface and subsurface soil organic carbon in tidal wetlands: A review and ideas for future research. *Remote Sensing, 14*(12), 12. https://doi.org/10.3390/rs14122940

Sheffer, A., Gentemann, C. L., & Parsons, M. A. (2018). *Best practices for a future open code policy for NASA space science*. AGU Fall Meeting.

Shlien, S. (1979a). Geometric correction, registration, and resampling of landsat imagery. *Canadian Journal of Remote Sensing, 5*(1), 74–89. https://doi.org/10.1080/07038992.1979.10854986

Sishodia, R. P., Ray, R. L., & Singh, S. K. (2020). Applications of remote sensing in precision agriculture: A review. *Remote Sensing, 12*(19), 3136.

Solymosi, R., Buil-Gil, D., Ceccato, V., Kim, E., & Jansson, U. (2023). Privacy challenges in geodata and open data. *Area, 55*(4), 456–464. https://doi.org/10.1111/area.12888

Suman, S., Maurya, S., Pandey, V., Srivastava, P. K., & Gupta, D. K. (2025). Cloud computing platforms–based remote sensing big data applications. *Google Earth Engine and Artificial Intelligence for Earth Observation*, 77–88.

Sun, Y.-J., Huang, T.-Z., Ma, T.-H., & Chen, Y. (2019). Remote sensing image stripe detecting and destriping using the joint sparsity constraint with iterative support detection. *Remote Sensing, 11*(6), 608. https://doi.org/10.3390/rs11060608

Surendran, U., Nagakumar, K. C. V., & Samuel, M. P. (2024). Remote sensing in precision agriculture. In P. M. Priyadarshan, S. M. Jain, S. Penna, M. Jameel, & Al-Khayri (Eds.), *Digital agriculture*. Springer International Publishing. https://doi.org/10.1007/978-3-031-43548-5_7

Teillet, P. M. (1986). Image correction for radiometric effects in remote sensing. *International Journal of Remote Sensing, 7*(12), 1637–1651. https://doi.org/10.1080/01431168608948958

Terms of Service – Google Earth engine., n.d. Accessed March 2, 2026. https://earthengine.google.com/terms/.

Tsai, F., & Chen, W. W. (2008). Striping noise detection and correction of remote sensing images. *IEEE Transactions on Geoscience and Remote Sensing, 46*(12), 4122–4131.

Tursunalieva, A., Alexander, D. L. J., Dunne, R., Li, J., Riera, L., & Zhao, Y. (2024). Making sense of machine learning: A review of interpretation techniques and their applications. *Applied Sciences, 14*(2), 496.

Vancutsem, C., Achard, F., Pekel, J. F., et al. (2021). Long-term (1990–2019) monitoring of forest cover changes in the humid tropics. *Science Advances, 7*(10). https://doi.org/10.1126/sciadv.abe1603

Veraverbeke, S., Dennison, P., Gitas, I., et al. (2018). Hyperspectral remote sensing of fire: state-Of-The-art and future perspectives. *Remote Sensing of Environment, 216*, 105–121.

Wang, L., Yan, J., & Ma, Y. (2019). *Cloud computing in remote sensing*. Chapman and Hall/CRC. https://www.taylorfrancis.com/books/mono/10.1201/9780429488764/cloud-computing-remote-sensing-lizhe-wang-jining-yan-yan-ma.

Wang, L., Yan, J., Ma, Y., et al. (2024). Cloud computing in remote sensing: A comprehensive assessment of state of the arts. In *Remote sensing handbook* (Vol. I)CRC Press. https://www.taylorfrancis.com/chapters/edit/10.1201/9781003541141-21/cloud-computing-remote-sensing-lizhe-wang-jining-yan-yan-ma-xiaohui-huang-jiabao-li-sheng-wang-haixu-ao-long-xiaohan-zhang.

Wang, S., Huang, X., Liu, P., et al. (2024). Mapping the landscape and roadmap of geospatial artificial intelligence (GeoAI) in quantitative human geography: An extensive systematic review. *International Journal of Applied Earth Observation and Geoinformation, 128* (April). https://doi.org/10.1016/j.jag.2024.103734

Wanyama, D., Moore, N. J., & Dahlin, K. M. (2020). Persistent vegetation greening and browning trends related to natural and human activities in the Mount Elgon ecosystem. *Remote Sensing, 12*(13), 2113. https://doi.org/10.3390/rs12132113

Wanyama, D., Bunting, E. L., Weil, N., & Keellings, D. (2023). Delineating and characterizing changes in heat wave events across the United States climate regions. *Climatic Change, 176*(2), 6. https://doi.org/10.1007/s10584-022-03476-y

Xiao, Z., & Xiao, Y. (2012). Security and privacy in cloud computing. *IEEE Communications Surveys & Tutorials, 15*(2), 843–859.

Xiaona, W., Jinyan, T., Xiaojuan, L., et al. (2022). Benefits of google Earth engine in remote sensing. *National Remote Sensing Bulletin, 26*(2), 299–309.

Xu, C., Du, X., Fan, X., et al. (2022a). Cloud-based storage and computing for remote sensing big data: A technical review. *International Journal of Digital Earth, 15*(1), 1417–1445. https://doi.org/10.1080/17538947.2022.2115567

Yan, J., Ma, Y., Wang, L., Raymond Choo, K.-K., & Jie, W. (2018). A cloud-based remote sensing data production system. *Future Generation Computer Systems, 86*, 1154–1166.

Young, N. E., Anderson, R. S., Chignell, S. M., Vorster, A. G., Lawrence, R., & Evangelista, P. H. (2017). A survival guide to landsat preprocessing. *Ecology, 98*(4), 920–932. https://doi.org/10.1002/ecy.1730

Zhang, J. (2010). Multi-source remote sensing data fusion: Status and trends. *International Journal of Image and Data Fusion, 1*(1), 5–24. https://doi.org/10.1080/19479830903561035

Zhao, L., Chen, L., Ranjan, R., Choo, K.-K. R., & He, J. (2016). Geographical information system parallelization for spatial big data processing: A review. *Cluster Computing, 19*(1), 139–152.

Zhu, Z., & Woodcock, C. E. (2012). Object-based cloud and cloud shadow detection in landsat imagery. *Remote Sensing of Environment, 118*, 83–94.

Zhu, Z., & Woodcock, C. E. (2014). Automated cloud, cloud shadow, and snow detection in multitemporal landsat data: An algorithm designed specifically for monitoring land cover change. *Remote Sensing of Environment, 152*, 217–234.

Zhu, X., Cai, F., Tian, J., & Williams, T. K.-A. (2018). Spatiotemporal fusion of multisource remote sensing data: Literature survey, taxonomy, principles, applications, and future directions. *Remote Sensing, 10*(4), 527. https://doi.org/10.3390/rs10040527

Zhu, Z., Wulder, M. A., Roy, D. P., et al. (2019). Benefits of the free and open landsat data policy. *Remote Sensing of Environment, 224*, 382–385.

Big Data Analytics for Geospatial Applications

4

Erin Bunting[1], *Jane Southworth*[2], *Hannah Herrero*[3] *and Cerian Gibbes*[4]
[1]Department of Geography, Environment, and Spatial Sciences, Michigan State University, East Lansing, MI, United States; [2]Department of Geography, University of Florida, Gainesville, FL, United States; [3]Department of Geography & Sustainability, University of Tennessee, Knoxville, TN, United States; [4]Department of Geography & Environmental Studies, University of Colorado Colorado Springs, Colorado Springs, CO, United States

Key terms and definitions

Ancillary data: Supporting datasets such as climate variables, elevation, or socioeconomic layers used to complement remote sensing imagery.
Anomaly mapping: The identification of deviations from long-term averages or baseline conditions to detect extreme values.
Black box model: A predictive model whose internal processes are difficult to interpret despite high performance.
Breakpoint detection: Analytical techniques used to identify abrupt structural changes in time-series data.
Change detection: Methods used to identify differences in land surface conditions by comparing imagery across multiple time periods.
Cloud computing: The use of remote servers to store and process large geospatial datasets at scale.
Cloud masking: Automated identification and removal of cloud-affected pixels from imagery.
Convolutional Neural Network (CNN): A deep learning architecture specialized for recognizing spatial patterns in images.
Data harmonization: The standardization of datasets from different sensors or time periods to ensure consistency.
Data fusion: The process of integrating multiple data sources, sensors, or indices to improve model accuracy and capture complementary information.
Deep Learning (DL) in Geospatial Analysis: Advanced machine learning approaches, such as neural networks, that automatically extract features and patterns from large geospatial datasets.
Earth Observation (EO): The collection of information about the Earth's surface using satellite, airborne, or ground-based sensing technologies.
Geometric correction: Spatial adjustment of imagery to ensure accurate geographic alignment.
Geographically Weighted Regression (GWR): A regression technique allowing relationships between variables to vary spatially.
Harmonic regression: A method for modeling seasonal cycles in time-series data using sinusoidal functions.

Remote Sensing, Big Data, and GeoAI. https://doi.org/10.1016/B978-0-443-26741-3.00002-7

High dimensionality: Data characterized by many variables, such as multiple spectral bands, indices, and time steps.
Hotspot analysis: A spatial statistical technique used to identify clusters of unusually high or low values within a geographic area.
Landscape metrics: Quantitative measures that describe spatial patterns such as fragmentation, patch size, and connectivity within landscapes.
Local Indicators of Spatial Association (LISA): Localized spatial statistics used to identify clustering and spatial outliers.
Mann–Kendall test: A non-parametric statistical test for detecting monotonic trends in time-series data.
Model interpretability (Explainability): The ability to understand and explain how a model produces its predictions, which is critical for trust and decision-making.
Modifiable Areal Unit Problem (MAUP): The sensitivity of statistical results to spatial aggregation units.
Modifiable Temporal Unit Problem (MTUP): The sensitivity of analytical results to temporal aggregation or segmentation.
Moran's I: A global measure of spatial autocorrelation indicating clustering or dispersion.
Multi-sensor integration: The combined use of data from different sensing technologies (e.g., optical, radar, LiDAR) to improve environmental monitoring and modeling.
Neighborhood structure: The rules defining spatial relationships among observations.
Normalized Burn Ratio (NBR): A spectral index used to assess burn severity and vegetation recovery.
Normalized Difference Vegetation Index (NDVI): A widely used vegetation index calculated from red and near-infrared reflectance to measure vegetation health and density.
Phenology: The study of seasonal biological cycles, such as vegetation growth and senescence, often analyzed using time-series remote sensing data.
Queen adjacency: Spatial contiguity defined by shared boundaries or corners.
Recurrent Neural Network (RNN): A neural network used for sequential and time-series data.
Rook adjacency: Spatial contiguity defined by shared boundaries.
Sensor noise: Measurement error introduced by sensing instruments.
Single-date classification: A traditional remote sensing approach that analyzes one image at a single point in time to categorize land cover types.
Spatial autocorrelation: The principle that nearby geographic observations tend to be more similar than distant ones.
Spatial dependence: The influence of nearby observations on one another.
Spatial nonstationarity: Variation in statistical relationships across space.
Spatial Statistics: Analytical methods used to quantify spatial patterns, relationships, and dependencies in geographic data.
Temporal alignment: Synchronizing datasets to comparable time periods.
Temporal Convolutional Network (TCN): A neural network architecture modeling long-range temporal patterns.
Theil–Sen slope: An estimator of time-series trend magnitude.
Time-Series Remote Sensing: The analysis of sequences of remotely sensed data collected over time to monitor trends, seasonal cycles, disturbances, and long-term environmental change.
Trajectory segmentation: Dividing time-series data into distinct change phases.
Vegetation Index (VI): A mathematical combination of spectral bands designed to enhance vegetation signals and provide proxies for properties such as greenness, biomass, or moisture.

1. Introduction

Over the past several decades, Earth observation (EO) has undergone a profound transformation driven by the rapid expansion of sensing technologies, satellite platforms, and data availability. Early remote sensing (RS) systems were limited by coarse spatial resolution, infrequent revisit times, and relatively narrow spectral capabilities. For example, early Landsat missions, Landsat MSS 1 to 3, in the 1970 and 1980s provided expensive global imagery but at spatial resolutions of 80m and revisit intervals over a couple weeks (18 days), making it difficult to capture rapid environmental changes such as crop growth or short-term disturbance (Chen et al., 2020; Hemati et al., 2021; Wulder et al., 2008). Many early RS applications, researchers often relied on only one or two images per year because satellite revisit times were long, cloud-free observations were limited in many parts of the world, and the cost of each image scene was high (Ju & Roy, 2008; Wulder et al., 2012). Today, a wide range of satellite and airborne (e.g., unoccupied aerial vehicles or airplanes) sensors continuously collect data at increasingly fine spatial, temporal, and spectral scales, producing an unprecedented volume of geospatial information. Modern EO systems also include diverse sensor types such as optical imagers, synthetic aperture radar (SAR), hyperspectral instruments, and LiDAR systems, each providing unique perspectives on environmental conditions and processes. In addition, the increasing frequency of observations, from daily revisit cycles to near-real-time monitoring, has enabled scientists to move beyond static snapshots of the Earth toward continuous tracking of landscape dynamics (Fu et al., 2024; Kuenzer et al., 2015). The result is a data environment characterized by high dimensionality, multi-sensor integration, and massive data volumes, often described as "big geospatial data." This shift has altered not only the types of questions researchers can ask, but also the analytical approaches employed/required to answer them. However, as GeoAI workflows evolve, it remains essential to understand and further develop foundational methods in time series RS, geospatial analysis, and spatial statistics, since these approaches are increasingly integrated into new workflows and are critical for model development and explainability.

As sensing technologies have evolved, geospatial analysis methods have advanced in parallel. Historically, RS analysis relied heavily on manual interpretation and expert-defined classification rules, where analysts used spectral thresholds or decision trees based on known physical relationships. These approaches gradually gave way to rule-based automation and statistical classification techniques, including supervised and unsupervised methods (Bárdossy & Samaniego, 2002; Wang et al., 2020). More recently, however, the field has experienced a major shift toward data-driven analytics, often referred to as GeoAI, in which machine learning and artificial intelligence techniques are used to automatically extract patterns from large and complex geospatial datasets.

This transition reflects several key changes in both data characteristics and analytical goals. First, modern machine learning methods enable the identification of nonlinear relationships and complex spatiotemporal interactions/trends that are difficult to capture using traditional statistical approaches due to the multidimensionality (Camps-Valls, 2009, pp. 1–6; Holloway & Mengersen, 2018). Second, these methods can leverage large volumes of data, helping address challenges related to limited or noisy reference information that frequently arise in geospatial applications (Camps-Valls, 2009, pp. 1–6; Zhu et al., 2017). Third, advances in deep learning have further expanded analytical capabilities by allowing automated feature extraction directly from raw imagery rather than relying solely on predefined spectral or spatial metrics (Li et al., 2024).

Deep learning techniques such as convolutional neural networks (CNNs) have proven particularly effective for spatial pattern recognition tasks, including land cover mapping, object detection, and urban feature extraction (Song et al., 2019; Zhu et al., 2017). For example, a CNN can be trained on high-resolution satellite imagery to automatically identify and map unique landscape features such as building damage after a hurricane or flood event, enabling for rapid post disaster assessment across large spatial extents (e.g., Li et al., 2023; Munawar et al., 2021). In other examples CNNs are used for vegetation segmentation, like Parker et al. under review who used NAIP imagery within a GeoAI workflow to develop a high-resolution classification of Michigan that particularly identifies continuous woody allotments of all sizes. At the same time, time-series and hybrid modeling approaches, including neural networks and temporal convolutional models, have improved the ability to analyze dynamic processes such as vegetation phenology, land use change, and environmental disturbances. For example, a temporal convolutional network can be used to analyze a dense time series of imagery to track seasonal vegetation cycles and pinpoint early signs of stress from drought or other drivers (e.g., Das Choudhury et al., 2023). These developments have been supported by advances in cloud computing and geospatial data platforms, which now enable scalable processing and analysis across massive multi-sensor datasets (Gorelick et al., 2017; Kumar & Mutanga, 2018).

Another defining characteristic of modern geospatial analytics is the growing emphasis on data integration and/or fusion. Contemporary workflows increasingly combine information from multiple sensor types, including optical imagery, radar data, LiDAR measurements, and ancillary environmental datasets such as climate or socioeconomic variables. Data fusion or joint analysis of these diverse data sources allows researchers to capture complementary or covarying information about surface conditions, improving both accuracy and robustness of geospatial models. For example, as highlighted in Case study 2, with an early data integration workflow, combining both RS imagery and shapefiles (discrete and continuous) into a unified modeling methodology.

As these capabilities have expanded, the focus of geospatial analysis has shifted significantly. Traditional RS workflows often centered on single-date image classification and static map production. In contrast, modern approaches emphasize dynamic landscape modeling, continuous monitoring, and predictive analytics. Time-series RS, which itself was a revolution to the field 15–20 years ago, enables researchers to track vegetation dynamics, detect disturbances, and assess long-term environmental trends rather than simply mapping conditions at one point in time (Bunting et al., 2018; Campo-Bescós et al., 2013; Fu et al., 2024). The combination of time series RS with GeoAI has further transformed these capabilities by enabling automated pattern discovery, scalable change detection across dense time series, and the integration of multi-source data streams to model complex environmental processes in near real time. This shift reflects a broader transition from descriptive mapping toward understanding processes, patterns, and change.

This chapter explores these RS and GeoAI developments with a focus on three interconnected themes: time-series RS, vegetation indices, and spatial statistical analysis. Together, these areas illustrate how modern geospatial analytics is moving beyond single-date classification toward integrated, multi-scale approaches for understanding change across landscapes through the use of big data and GeoAI. This chapter is organized to guide readers through key components of modern geospatial big data analytics. It begins by examining the shift from single-date analysis toward time-series RS, highlighting how dense time series enable monitoring of environmental dynamics and long-term change. The chapter then explores vegetation indices as a foundational tool for summarizing spectral information into interpretable measures of vegetation condition, productivity, and stress. Next, it introduces spatial statistical approaches that support pattern detection, clustering, and analysis of spatial relationships across landscapes. Throughout the chapter emphasis is placed on how advances in machine learning, cloud computing, and multi-source data integration are reshaping geospatial workflows. The chapter concludes by discussing challenges related to model evaluation, interpretability, and the growing importance of explainable GeoAI in supporting transparent and reliable decision-making.

2. The methodological evolution from single-date classification to time series remote sensing

As introduced above, early RS analysis was largely organized around a single-date image (e.g., Langley et al., 2001; Saadat et al., 2011). A researcher would acquire one or two cloud-free scenes, apply radiometric and geometric corrections, derive a set of spectral features, and then classify the image using supervised or unsupervised models to create thematic categories such as forest, cropland, urban, or water. This workflow reflected both the technical constraints and analytical priorities. Satellite revisit intervals were long, storage and processing capacity were limited, and image acquisition costs were high. As a result, classification accuracy and map production for a specific point in time were the central objectives (Langley et al., 2001).

Single-date classification proved effective for many foundational applications. National land cover inventories, urban extent mapping, wetland delineation, and post-event damage assessments were often conducted using one or two well-timed images. For example, a Landsat scene acquired during peak growing season could be used to distinguish cropland from forest or grassland based on spectral differences in near-infrared (NIR) reflectance or in the ratios of reflectance in the red versus NIR (Ding et al., 2014; Dusseux et al., 2014; Xu & Guo, 2014). Similarly, a single post-fire image could map burn scars using indices such as the Normalized Burn Ratio (NBR) (Chen et al., 2011; Escuin et al., 2008; Miller & Thode, 2007). In urban contexts, a high-resolution optical image could support building footprint extraction or impervious surface mapping.

However, the limitations of this approach became increasingly apparent as researchers sought to analyze dynamic processes rather than static conditions. Landscapes are driven by seasonal, annual, and interannual processes that dynamically shape large-scale and small-scale changes or fluctuations. For instance, vegetation greenness fluctuates seasonally. Agricultural fields transition from bare soil to canopy closure to harvest residue within months. Forest disturbances may unfold gradually through pest outbreaks or abruptly through wildfire. A single-date snapshot may misclassify these transient states. A cropland field imaged early in the growing season may appear similar to bare soil and could be misclassified as fallow land. A drought-stressed forest may exhibit spectral characteristics resembling recently harvested timber, leading to potential confusion in land cover classification. Temporary flooding can also create challenges, as inundated agricultural fields may be misidentified as permanent water bodies in a single-date image.

These issues reflect a broader conceptual shift: from asking *"What is here?"* to asking *"What is happening over time?"*

2.1 Emergence of multi-date and change detection approaches

As data availability improved, analysts began incorporating multi-date imagery into their workflows. Rather than relying on a single image, researchers compared two or more scenes to detect change. This approach brought on the use of change trajectories within the field (Boori & Vozenilek, 2014; Jianya et al., 2008). Image differencing, post-classification comparison, and spectral index trajectory analysis became common techniques (Dianat & Kasaei, 2009; Gómez et al., 2014; Storey et al., 2016; Xu et al., 2009). For example, pre- and post-fire images can be compared to quantify burn severity, providing insight into the extent and intensity of wildfire impacts (Chu & Guo, 2013). Similarly, two images captured several years apart can be used to map urban expansion, revealing patterns of land cover change and development over time. Seasonal changes in vegetation can be assessed by subtracting NDVI values across different dates, allowing researchers to identify growth cycles, periods of stress, or long-term trends in vegetation cover. These methods marked a transition toward temporal thinking but still treat time discretely. Change was detected between two fixed points rather than modeled as a continuous process.

2.2 Dense time series and continuous monitoring

The launch of higher-frequency satellite systems, such as MODIS, Sentinel-2, and harmonized Landsat collections, enabled the routine construction of dense time series, with observations every few days in many parts of the world. This fundamentally altered the analytical landscape. Instead of comparing isolated images, researchers could now model continuous trajectories, trends, cycles, and anomalies across months or years.

Time-series RS allows for:

- Phenological analysis, such as identifying the start of season (SOS), end of season (EOS), and peak greenness in agricultural systems.

- Disturbance detection, including abrupt forest loss, gradual degradation, or insect outbreaks.
- Crop type discrimination, where seasonal growth curves differentiate corn, soybeans, and wheat even when single-date imagery cannot.
- Drought monitoring, using deviations from long-term vegetation index averages.
- Urban growth tracking, by modeling incremental increases in impervious surface signals over time.

For example, instead of classifying a forest in a single year, a time-series approach might reveal a gradual decline in canopy vigor over 5 years followed by abrupt clearing in year six. Similarly, a dense NDVI time series can distinguish double-cropping systems from single-harvest systems by identifying multiple seasonal peaks.

2.3 From classification to modeling temporal dynamics

This evolution reflects not only a change in data volume but also in methodological orientation. Traditional single-date classification emphasized categorical outputs: forest versus non-forest, urban versus rural. Time-series approaches increasingly emphasize continuous modeling and process inference. Examples of advanced time-series analysis include fitting harmonic regression models to vegetation indices in order to extract key seasonal parameters such as amplitude, phase, and timing of peak greenness (Atkinson et al., 2012; Hermance et al., 2007; Jakubauskas et al., 2001). Breakpoint detection algorithms can be applied to identify abrupt structural changes in land cover trajectories, such as sudden deforestation or disturbance events (Bunting et al., 2017, 2018; Munson et al., 2019). Autoregressive models allow researchers to forecast vegetation productivity under future climate scenarios by leveraging past trends and seasonal patterns. More complex approaches, including recurrent neural networks (RNNs) and temporal convolutional networks (TCNs), are increasingly used to model non-linear temporal dependencies across multi-sensor datasets, capturing intricate interactions between vegetation dynamics, climate, and land management practices. RNNs are a class of deep learning models specifically designed to process sequential data by maintaining a hidden state that captures information from previous time steps, enabling them to learn temporal dependencies and dynamic patterns over time (Jeyavathana, 2022; Kazanskiy et al., 2025). TCNs, in contrast, use convolutions to model long-range temporal relationships in sequence data while allowing for parallel computation and stable gradient propagation (Fan et al., 2023; Osibo et al., 2024). Together, these architectures enhance the ability to represent complex, non-linear interactions in multi-sensor time series, making them particularly well suited for analyzing evolving relationships among vegetation dynamics, climate variability, and land management practices. In this context, classification, change detection and trajectories becomes one component of a broader modeling framework. The focus shifts from static map production to understanding rates of change, resilience, recovery, and system stability.

2.4 Integration with multi-sensor data

Time-series analysis has also benefited from multi-sensor integration. With such integration researchers can not only begin to more easily breakdown complex multidimensional datasets but also to better understand and articulate complex drivers of landscape change. Radar sensors such as Sentinel-1 provide consistent observations regardless of cloud conditions. LiDAR datasets contribute structural information about canopy height and vertical complexity. Climate reanalysis data add temperature and precipitation context. For example, a forest disturbance monitoring system might integrate NDVI trends with radar backscatter changes to detect both canopy loss and structural damage, providing a more complete picture of ecosystem impacts (e.g., Fernández-Guisuraga et al., 2023; Schellenberg et al., 2023). In agricultural applications, a crop monitoring workflow could combine satellite-derived phenology metrics with rainfall anomalies to improve predictions of yield variability across fields and seasons. Similarly, coastal wetland monitoring may fuse optical vegetation indices with tidal data and SAR-derived inundation patterns, enabling accurate assessment of vegetation health and hydrological dynamics in highly dynamic environments.

2.5 Implications for analytical practice

The methodological transition from single-date classification to time-series remote sensing carries several important implications:

1. Data preprocessing becomes more complex, requiring temporal smoothing, cloud masking, gap filling, and harmonization across sensors.
2. Feature extraction expands, incorporating trend coefficients, seasonal amplitudes, anomaly metrics, and trajectory shapes.
3. Model validation must account for time, including temporal cross-validation and forecasting accuracy assessment.
4. Interpretability becomes more critical, as complex temporal models may obscure causal relationships without careful explanation.

Importantly, time-series methods do not replace single-date analysis entirely. Rapid disaster response may still rely on a single post-event image. Urban infrastructure mapping at very high resolution may prioritize spatial detail over temporal depth. Rather than a linear replacement, the field has evolved toward a hierarchical and context-dependent toolkit, where single-date, multi-date, and dense time-series approaches coexist. Ultimately, the evolution from single-date classification to time-series remote sensing represents a broader epistemological shift in EO. The goal is no longer merely to map the Earth's surface, but to understand how it changes, why it changes, and what those changes imply for ecosystems, climate, and society. By enabling continuous monitoring, trajectory analysis, and predictive modeling, time-series remote sensing transforms EO from static cartography into dynamic environmental intelligence.

3. Vegetation indices as high-dimensional time series features

3.1 Overview of vegetation indices

Vegetation indices (VIs) are mathematical combinations of spectral reflectance measurements designed to enhance the signal of vegetation properties while reducing background noise from soil, atmosphere, or illumination effects (Jensen, 1996, 2009). Conceptually, they transform raw spectral bands, most often red or near-infrared (NIR), into interpretable proxies of biophysical characteristics such as greenness, biomass, canopy structure, chlorophyll content, and water status. The foundational logic behind most vegetation indices stems from two well-established spectral properties of healthy vegetation. First, strong chlorophyll absorption in the red portion of the spectrum (Jensen, 2009). Second, high reflectance in the near-infrared due to internal leaf structure (Jensen, 2009). The contrast between low red reflectance and high NIR reflectance creates a distinctive spectral signature that can be exploited to quantify vegetation vigor and density (Fig. 4.1). By mathematically combining these two regions of the spectrum, often through ratios, vegetation indices amplify, in this example, differences between vegetated and non-vegetated surfaces while minimizing the influence of external factors such as variable illumination, sensor geometry, and background soil reflectance. This contrast forms the basis of the most widely used index, the Normalized Difference Vegetation Index (NDVI), which expresses vegetation condition as a normalized ratio between NIR and red reflectance (Rouse, 1973; Tucker, 1980; Tucker et al., 2005). By exploiting the contrast between red absorption and NIR reflectance, early researchers developed simple yet powerful indices capable of mapping vegetation vigor across large regions.

Beyond NDVI, numerous vegetation indices have been developed to address specific limitations or target particular biophysical properties (Table 4.1). For example, soil-adjusted indices reduce sensitivity to exposed soil in sparsely vegetated areas, while enhanced vegetation indices incorporate additional spectral bands to improve performance in high-biomass conditions and reduce atmospheric effects (Huete, 1988). Other indices integrate shortwave infrared (SWIR) bands to assess plant water content or stress, reflecting the strong absorption of water in those wavelengths.

Collectively, vegetation indices provide a scalable and computationally efficient means of monitoring vegetation dynamics across spatial and temporal scales. From field-level crop assessment to regional and global ecosystem monitoring, VI's enable consistent, repeatable quantification of vegetation condition, supporting applications in agriculture, forestry, drought monitoring, carbon modeling, and land cover analysis. Table 4.1 highlights some of the commonly utilized VI's and provides a key citation for follow up.

3.2 Purpose and biophysical interpretation

VI's serve a wide range of analytical purposes in RS, Earth system science, ecology, landscape ecology, conservation, management, and many other fields of study. They are commonly seen as proxy measures and are used to monitor vegetation greenness and vigor, with indices such as NDVI and EVI providing proxies for photosynthetic activity and canopy density. Many indices also support the estimation of biophysical parameters, including leaf area index (LAI) and the fraction of absorbed photosynthetically active radiation (fAPAR), which are critical for modeling ecosystem productivity. LAI quantified the total one-sided leaf surface area per unit of ground area and serves as a key indicator of canopy density, light interception capacity, and potential

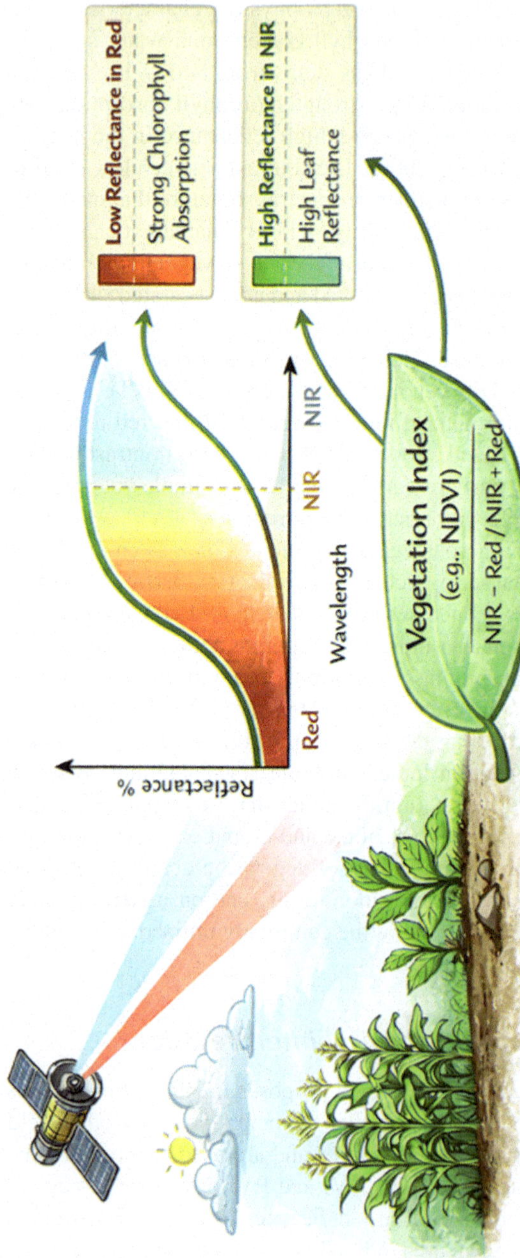

Figure 4.1 Spectral contrast between red absorption and near-infrared reflectance underlying vegetation indices.

Table 4.1 Common vegetation indices and their derivations.

Vegetation index name	Acronym	Formula	Key citation	Role in big data/GeoAI workflows
Normalized difference vegetation index	NDVI	(NIR − red)/(NIR + red)	Rouse (1973)	Treated as a dense temporal feature vector rather than a single metric; used to model trajectory shape, trend, anomalies, and persistence across long time series
Enhanced vegetation index	EVI	2.5 × (NIR − Red)/(NIR + 6Red − 7.5Blue + 1)	Huete et al. (2002)	Preferred in high-biomass and global-scale analyses where NDVI saturation biases machine learning predictions
Soil-adjusted vegetation index	SAVI	((NIR − Red)/(NIR + red + L)) × (1 + L)	Huete (1988)	Improves model stability in sparse vegetation and drylands, especially in time-series regression and classification workflows
Modified SAVI	MSAVI	(2NIR +1 − $\sqrt{}$((2NIR +1)^2–8(NIR–Red)))/2	Qi et al. (1994)	Used as a noise-reduction feature in arid and semi-arid time series
Normalized difference water index	NDWI	(NIR − SWIR)/(NIR + SWIR)	Gao (1996)	Acts as a stress-sensitive temporal signal, frequently fused with climate variables for drought and water-stress modeling
Normalized burn ratio	NBR	(NIR − SWIR2)/(NIR + SWIR2)	García and Caselles (1991)	Used in time-series breakpoint detection and recovery modeling, not just static burn severity mapping

Continued

Table 4.1 Common vegetation indices and their derivations.—cont'd

Vegetation index name	Acronym	Formula	Key citation	Role in big data/GeoAI workflows
Green NDVI	GNDVI	(NIR – Green)/(NIR + green)	Gitelson et al. (1996)	Enhances chlorophyll sensitivity in crop-focused machine learning workflows
Red-edge NDVI	NDRE	(NIR – RedEdge)/(NIR + RedEdge)	Danson and Plummer (1995); Xie et al. (2018)	High-value predictor in precision agriculture and trait inference models
Atmospherically resistant vegetation index	ARVI	(NIR – (2Red – Blue))/(NIR + (2Red – blue))	Kaufman and Tanre (1992)	Reduces atmospheric noise in large-area, multi-sensor time series
Difference vegetation index	DVI	NIR – red	Lillesand et al. (2015)	Retained primarily for legacy comparison and interpretability
Global environmental monitoring index	GEMI	Complex nonlinear NDVI-based formula	Pinty and Verstraete (1992)	Early example of index complexity anticipating feature engineering
Optimized soil-adjusted vegetation index	OSAVI	(NIR – Red)/(NIR + red + 0.16)	Rondeaux et al. (1996)	Improves robustness in dense agricultural monitoring
Visible atmospherically resistant index	VARI	(Green – Red)/(Green + red – blue)	Gitelson et al. (2002)	Useful where NIR unavailable, including UAV and RGB-only workflows
Leaf area index proxy (NDVI-based)	LAI-NDVI	Empirical function of NDVI	Running et al. (2004)	Links VI time series to process-based ecosystem productivity models

photosynthetic productivity (Jensen, 2009). fAPAR represents the production of incoming solar radiation in the 400–700 nm range that is absorbed by vegetation canopies, directly linking canopy structure and chlorophyll content to photosynthetic activity and ecosystem uptake (Jensen, 2009). Other indices are designed to assess canopy moisture and vegetation water content, such as NDWI, making them valuable for drought monitoring and hydrological studies. Some indices, including the Photochemical Reflectance Index (PRI) and red-edge–based metrics, are particularly sensitive to physiological stress and changes in chlorophyll concentration. Such indices are becoming very common in agricultural work and developed through the use of unoccupied aerial vehicles (e.g., Sharifi & Felegari, 2023; Zhao et al., 2025). Additionally, indices such as the NBR are widely used for mapping burn severity and monitoring post-disturbance recovery (Alcaras et al., 2022; Saberi & Harvey, 2023). Together, these applications demonstrate the versatility of vegetation indices as interpretable and biophysically meaningful features in both static and time-series analyses (Fig. 4.2).

NDVI time series, for example, can be used to track seasonal crop development, capturing the progression from early emergence through peak growth and eventual senescence. NDWI provides insight into vegetation water content and can help

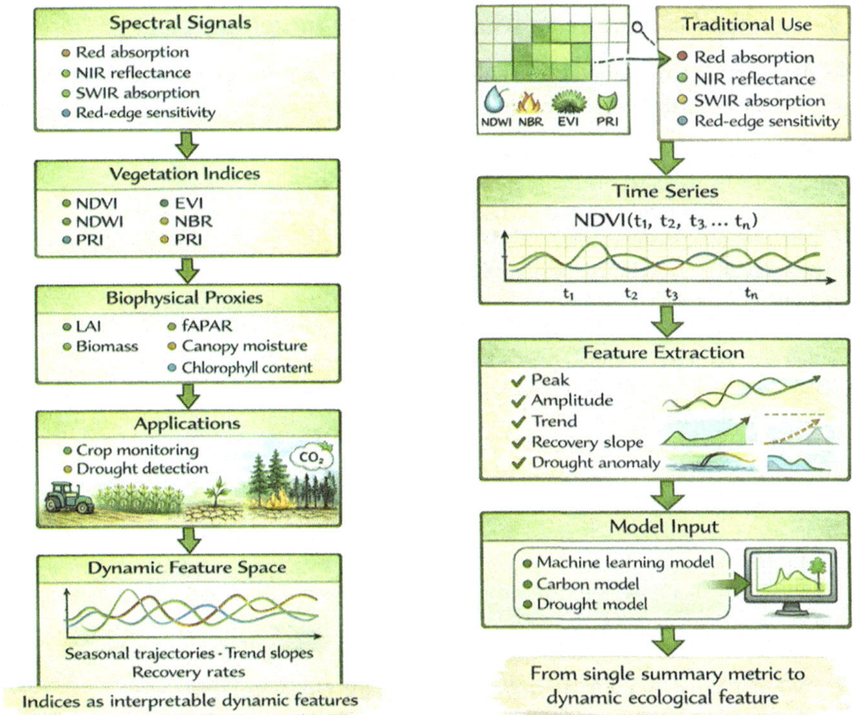

Figure 4.2 From spectral reflectance to dynamic ecological feature space: Expanding roles of vegetation indices.

identify periods of water stress during drought conditions by detecting declines in canopy moisture. NBR trajectories are widely applied to quantify wildfire burn severity and to monitor vegetation recovery in the years following a fire event. In densely vegetated tropical forests, where NDVI often saturates and loses sensitivity at high biomass levels, the Enhanced Vegetation Index (EVI) offers improved performance by reducing atmospheric and background influences and maintaining responsiveness in areas of dense canopy cover.

In modern workflows, vegetation indices are rarely used as single summary values. Instead, they are increasingly treated as high-dimensional dynamic temporal features, where each pixel is represented not by one index value, but by an index trajectory or trend across time. As illustrated in Fig. 4.2, vegetation indices provide a conceptual bridge between spectral reflectance and ecological interpretation. The left panel highlights how indices such as NDVI, EVI, NDWI, and NBR translate fundamental spectral contrasts into biophysical proxies and applied ecosystem metrics, while the right panel demonstrates how these indices are increasingly analyzed as temporal trajectories from which dynamic features such as seasonal amplitude, trend, recovery rate, and stress anomalies can be extracted. Together, the figure underscores the dual role of vegetation indices as both physically grounded proxies and high-dimensional temporal features in modern remote sensing and Earth system analysis.

3.3 Seminal works in vegetation index development

The development of vegetation indices has been shaped by several foundational studies. Key contributions include:

- Rouse et al. (1974)—Introduced the Normalized Difference Vegetation Index (NDVI) using ERTS (early Landsat) imagery (Rouse et al., 1974).
- Tucker (1979)—Demonstrated the utility of red and photographic infrared combinations for vegetation monitoring (Tucker, 1979).
- Huete (1988)—Developed the Soil-Adjusted Vegetation Index (SAVI) to reduce soil background effects (Huete, 1988).
- Asrar et al. (1984)—Linked spectral reflectance to absorbed photosynthetically active radiation (APAR) and leaf area index (Asrar et al., 1984).
- Gao (1996)—Proposed NDWI for vegetation liquid water content estimation (Gao, 1996).
- Huete et al. (1997)—Compared vegetation indices globally using TM data for EOS-MODIS applications (Huete et al., 1997).
- Huete et al. (2002)—Evaluated the radiometric and biophysical performance of MODIS vegetation indices (Huete et al., 2002).
- Bannari et al. (1995)—Provided a comprehensive review of vegetation indices (Bannari et al., 1995).
- Running et al. (2004)—Integrated vegetation index products into global productivity modeling (Running et al., 2004).

3.4 From single-value indices to index time series

An NDVI curve observed across a growing season captures multiple key stages of vegetation development. The initial rise in NDVI values reflects the onset of greening, marking the emergence of leaves and the beginning of active photosynthesis. The slope of the curve during early and mid-season indicates the rate of canopy development, as vegetation rapidly accumulates biomass and leaf area. The peak of the curve corresponds to maximum canopy density and biomass, often aligning with the period of highest productivity. Finally, the decline in NDVI values signals senescence timing, when vegetation begins to age, chlorophyll breaks down, and photosynthetic activity decreases (Fig. 4.3). Together, these phases illustrate how a seasonal NDVI trajectory provides a compact yet information-rich representation of plant growth dynamics (also known as Phenology).

The phenological curve contains far more ecological information than a single-date measurement. Dense image time series-based analysis leverages the temporal variability of vegetation index to reveal patterns that are not visible in single-date imagery. For example, double-cropping systems can be detected by identifying multiple NDVI peaks within a single growing year, reflecting two distinct planting and harvest cycles. In forested regions, recovery after wildfire can be monitored by modeling the rebound of indices such as the NBR over multiple years, allowing researchers to quantify regrowth rates and assess ecosystem resilience. Drought impacts can be identified by comparing current-season NDVI values to long-term averages, where sustained negative deviations indicate vegetation stress or reduced productivity (e.g., Bunting et al., 2017). Additionally, harmonic regression and related time-series modeling approaches can be applied to multi-year vegetation index records to extract seasonal amplitude, timing (phase), and trend components, providing quantitative metrics of ecosystem dynamics (Carrao et al., 2009; Wilson et al., 2018; Zhou et al., 2022). As shown in Fig. 4.3, vegetation index time series reveal distinct ecological signatures, including seasonal growth and senescence cycles, multiple cropping peaks within a year, and multi-year disturbance and recovery trajectories, underscoring the analytical power of temporal remote sensing.

In GeoAI workflows, vegetation index time series are commonly treated as structures, information-dense input features for machine learning models. With this approach the input data becomes a spatially and temporally varying surface into the

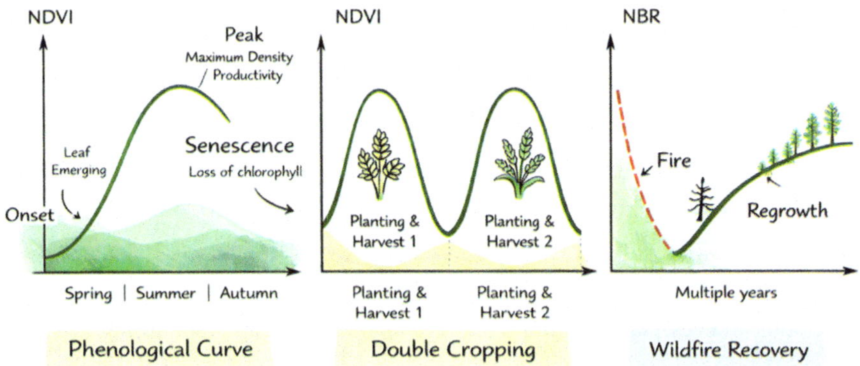

Figure 4.3 Phenological dynamics, cropping cycles, and post-disturbance recovery in vegetation time series.

model rather than simple summary statistics. Instead of relying on a single-date image, models ingest entire temporal stacks, sometimes called dense image collections in software like Google Earth Engine, that encode phenological dynamics from multiple seasons or years into model forecasts. For example, a temporal convolutional network may take multi-year sequences of NDVI, or any other index, as multivariate time series input data to aide in the classification of specific crop types, detect disturbance, or map regime change. With this approach each pixel's temporal depth is characterized by shape, time, amplitude, and frequency. These characteristics carry predictive value for landscape change, conditions, cover, and use.

3.5 Multi-index fusion/utilization

While no single date captures all aspects of vegetation condition, similarly no single vegetation index captures all aspects of vegetation condition. In many RS studies using multiple indices together and/or employing data fusion approaches with multiple indices represents related but conceptually distinct strategies for integrating spatial information. Using multiple indices together typically means stacking indices such as NDVI, EVI, SAVI, and NBR as separate predictor variables in a machine learning model. In this approach, each index preserves its physical interpretability, such as greenness, moisture status, canopy structure, or burn severity, and the model learns how to weight and combine them internally. In contrast, a fusion approach refers to mathematically combining multiple indices (or spectral data) into a new composite feature prior to modeling, often through techniques such as Principal Component Analysis. Data fusion can reduce the dimensionality, minimize noise, and capture index interactions in a compact data format. However, data fusion may also sacrifice interpretability. In GeoAI workflows, stacking multiple indices together is common when interpretability and/or explainability are priorities; whereas fusion is often favored in big data and sometimes deep learning where computation efficiency, redundancy reduction, and feature extraction are the key concerns.

Multi-index fusion combines complementary signals to improve discrimination and robustness. Examples of multi-index fusion include combining NDVI and NDWI to differentiate drought stress from normal seasonal senescence, allowing more precise detection of vegetation water deficits (Gu et al., 2007, 2008). Similarly, integrating NBR and NDVI can help distinguish fire disturbances from logging activities by capturing both canopy loss and changes in vegetation condition (Escuin et al., 2008). Fusing EVI with red-edge indices has been shown to improve crop yield predictions by leveraging complementary information about canopy structure and chlorophyll content.

3.6 Applications in time-series analysis

Vegetation indices play a central role in dynamic landscape monitoring across all global systems:

Crop Monitoring: VI time series has become a foundational tool for agricultural monitoring from field to regional scales. There are numerous examples in the literature in which VI's have been used for crop discrimination, health assessment, biomass abundance, and more. For instance, dense NDVI or EVI trajectories allow analysts to estimate planting and harvest dates by identifying key inflection points in seasonal greenness curves (Gao & Zhang, 2021; Rodigheri et al., 2023; Shammi & Meng, 2021). Other key phenological markers include:

- The onset of sustained greening often corresponds to crop emergence, while rapid declines signal senescence or harvest.
- The first detectable increase above baseline greening typically signals budburst or leaf out in perennial systems.
- Prolonged low greenness after planting may indicate poor germination or system level stress
- Subtle mid-season declines in greenness may coincide with flowering in crops as well.
- A premature decline in canopy vigor relative to growing degree day often indices heat stress.

These phenological markers can be extracted algorithmically across thousands of fields, supporting large-scale crop calendars and management assessments.

Beyond timing, cumulative greenness metrics, such as integrated NDVI over the growing season, are frequently correlated with biomass accumulation and final yield. By using summary statistics, VI values across critical growth stages can be combined enabling researchers to generate early-season yield forecasts that inform food security assessments, commodity markets, and policy planning.

Ecosystem Dynamics: At broader spatiotemporal scales, VI's support the analysis of ecosystem productivity and resilience. Multi-year time series allow researchers to model interannual variability in greenness and biomass, revealing how ecosystems respond to climate variability such as El Niño events, prolonged droughts, or unusually wet years. By examining trends and anomalies across decades, scientists can assess whether ecosystems are becoming more productive, more variable, or increasingly stressed. For instance, Southworth et al. (2023) used long term NDVI time series analysis to show patterns of global vegetation persistence. Specifically, the increased trend overall across all global land covers from the early 1980s, highlighting the variability in magnitude and consistency across different biomes and seasons. Further, this work showed seasonal and latitudinal patterns that had emerged, suggesting that climate drivers and land use/cover changes jointly influence how ecosystem greenness responds over multidecadal scales.

VI based trajectories or data products also provide a powerful means of tracking phenological shifts associated with climate variability or change. Earlier spring green-up, extended growing seasons, or delayed autumn senescence can be quantified using long-term NDVI or EVI records. These shifts have implications for carbon cycling, species interactions, and ecosystem services. Moreover, vegetation indices are integral to estimating gross primary productivity (GPP) through light-use efficiency models, which combine satellite-derived greenness metrics with climate variables to estimate carbon uptake at regional to global scales. In this context, vegetation indices serve as critical links between spectral observations and biogeochemical modeling.

Disturbance Detection: Time-series vegetation indices are equally central to disturbance detection and recovery monitoring. For example, in wildfire analysis, indices such as the NBR are used to quantify burn severity by comparing pre- and post-fire conditions (Alcaras et al., 2022; Escuin et al., 2008). When extended into a time series, NBR trajectories can track vegetation recovery over multiple years, distinguishing rapid regrowth from long-term ecosystem transformation. However, this is just one example of disturbance detection; deforestation can be detected through abrupt and sustained declines in NDVI or related indices, particularly in tropical forest regions. Breakpoint detection algorithms applied to vegetation index time series can identify the timing and magnitude of canopy removal, enabling near-real-time forest monitoring. In contrast, insect outbreaks or disease events often manifest as gradual canopy degradation rather than sudden loss. Subtle but persistent downward trends in greenness, reduced seasonal amplitudes, or altered phenological timing may signal stress before visible mortality occurs. These patterns demonstrate how vegetation index time series provide not only change detection, but also insight into the processes and severity of ecological disturbance (Fang et al., 2018; Hui & Cheng, 2026; Watts & Laffan, 2014; Zheng et al., 2025; Zhu & Woodcock, 2014).

For such disturbance detection, time-series VI analysis relies on a suite of statistical and geospatial approaches to quantify disturbance timing, magnitude, and recovery dynamics, which entails both parametric and nonparametric understanding. Simple change detection techniques like image differencing, percent change, or standardized anomaly mapping provide intuitive measures of departure from baseline conditions and are particularly effective for identifying abrupt events. More advanced trajectory-based approaches model the full temporal signal. Algorithms such as LandTrendr, BFAST (Breaks For Additive Season and Trend), and Continuous Change Detection and Classification (CCDC) segment time series into trend components and breakpoints, allowing researchers to detect both sudden canopy loss and gradual degradation (Fang et al., 2018; Hui & Cheng, 2026; Watts & Laffan, 2014; Zheng et al., 2025; Zhu & Woodcock, 2014). Harmonic regression and seasonal decomposition further separate long-term trends from phenological cycles, improving sensitivity to subtle disturbances such as insect outbreaks or drought-induced decline (Bunting et al., 2017; Carrao et al., 2009; Wilson et al., 2018; Zhou et al., 2022).

Beyond detection, there are key methods to support quantitative assessment of severity and recovery. For instance, non-parametric trend tests such as Mann–Kendall and Theil–Sen quantify persistent greening or browning trends over time (Bunting et al., 2018; Southworth et al., 2018). Recovery can be characterized using trajectory slopes, time-to-recovery metrics, and resilience indicators such as return time or post-disturbance greenness ratios (Bunting et al., 2017). Table 4.2 highlights key methods used in this area of research, but these methods also are common in other time series remote sensing work.

The continued expansion of satellite archives and cloud-based processing platforms has positioned these disturbance detection methods within a broader GeoAI and big data framework. With petabyte-scale datasets from sensors such as Landsat and Sentinel, traditional pixel-based analyses are increasingly complemented by deep learning, spatiotemporal data fusion, and automated feature extraction pipelines capable of processing decades of imagery at continental to global scales. As a result,

Table 4.2 Key methods in time series remote sensing vegetation index analysis.

Method category	Example techniques	Primary purpose	Best suited for	Scalability/Big data considerations
Simple change detection	Image differencing, percent change, anomaly mapping	Detect abrupt changes relative to baseline	Wildfire, clear-cutting	Computationally efficient; often used for rapid screening before advanced modeling
Trajectory segmentation	LandTrendr, BFAST, CCDC, piecewise regression	Identify breakpoints and long-term trends	Gradual degradation, mixed disturbances	Designed for dense satellite archives and scalable cloud execution
Seasonal decomposition	Harmonic regression, phenological metric extraction	Separate seasonal cycles from disturbance trends	Insect outbreaks, drought stress	Converts raw time series into interpretable features
Severity metrics	dNBR, RdNBR, burned area index	Quantify disturbance magnitude	Burn severity assessment	Enables standardized severity comparison across space and time
Trend analysis	Mann–Kendall test, Theil–Sen slope	Detect monotonic greening or browning trends	Long-term ecosystem change	Frequently applied pixel-wise at continental to global scales
Recovery & resilience metrics	Time to recovery, recovery slope, greenness ratio	Quantify ecosystem resilience	Post-fire or post-harvest recovery	Supports process-oriented modeling, not just detection
Classification approaches	Random forest, bayesian change-point detection, HMMs	Distinguish disturbance type and process	Operational monitoring systems	Often integrate time-series features into scalable ML pipelines

time-series VI analyses are evolving from retrospective change assessment toward near-real-time monitoring systems that combine scalable computation, cloud infrastructure, and AI to support decision-making in a wide variety of fields such as forestry, agriculture, and urban planning.

3.7 Limitations and uncertainties

Despite their widespread use, vegetation indices have several well-known limitations, even with our most prominently used indices such as NDVI (Gu et al., 2013; Tian et al., 2025). Saturation effects can occur in dense canopies, where NDVI and similar indices lose sensitivity to additional increases in biomass, making it difficult to distinguish high-biomass areas (Gu et al., 2013; Tian et al., 2025). In sparsely vegetated regions, the influence of the soil background can bias index values, potentially overstating or understating vegetation condition (Huete, 1988). The impact of soil background on VI values led to the development of the Soil Adjusted Vegetation Index (SAVI) (Huete, 1988). Atmospheric effects, including aerosols, haze, and clouds, can also alter reflectance measurements, introducing noise and uncertainty into the indices (Ren-hua et al., 1996; Xue & Su, 2017). Additionally, bidirectional reflectance effects, caused by variations in sun-sensor geometry, can influence the observed index values, introducing variability unrelated to actual vegetation changes (Petri & Galvão, 2019). These uncertainties highlight the importance of the user, and the requirement for them to have robust knowledge of the image data being used, including pre-processing steps undertaken and analytical approach used, in order to ensure the most appropriate use of the data, correct time-series workflows, and ultimately, appropriate image interpretation.

4. Spatial statistics in large-scale geospatial analytics

Spatial statistics provide a critical approaches for understanding patterns, relationships, and dependencies in geospatial data, enabling researchers to move beyond descriptive mapping toward quantitative analysis of spatial phenomena. At the core of spatial statistics are key concepts such as spatial autocorrelation and spatial dependence, which capture the idea that observations located near each other are often more similar than those farther apart (Geary, 1954; Moran, 1950). This principle underlies many geospatial analyses used in fields from ecosystem modeling to urban planning and highlights the importance of accounting for the inherent structure of spatial data. Scale and resolution are also fundamental considerations, as analyses can yield different results depending on the spatial grain or extent of the data. These issues are closely linked to the modifiable areal unit problem (MAUP), where the choice of spatial aggregation or spatial units can influence statistical outcomes and interpretations (Openshaw, 1984), emphasizing the need for careful study design and sensitivity analyses. Related to RS and GeoAI, these spatial statistical principles are essential for analyzing these massive, high-resolution, satellite datasets, where accounting for spatial autocorrelation, scale effects, and MAUP improves the reliability of machine learning models and enables robust big-data-driven insights.

Spatial statistical methods provide a diverse toolkit for quantifying, visualizing, and modeling spatial relationships. Global metrics, such as Moran's I or Geary's C, measure the degree of overall spatial autocorrelation in datasets, specifically revealing high or low value clusters across a region (Geary, 1954; Moran, 1950). Local metrics, including Local Indicators of Spatial Association (LISA) (Anselin, 1995), extend this approach to identify hotspots, cold spots, or outlier regions, highlighting areas of concentrated activity or anomaly. Hotspot and cluster analysis have been widely applied in environmental monitoring, epidemiology, and urban studies, providing actionable insights into spatial patterns of vegetation stress, disease incidence, or urban development (Getis, 1996, pp. 261–277; Getis & Ord, 1992). Spatial statistics increasingly integrate with GIS and RS, where large raster or vector datasets can be analyzed to uncover both broad trends and fine-scale variability (Goodchild et al., 2007; Longley et al., 2015). The synergy between GIS and RS has been particularly transformative, allowing the spatial structure of remotely sensed measurements to inform spatially explicit analyses and models.

A central mechanism enabling this integration is the explicit modeling of spatial neighborhoods. In spatial statistics, neighborhoods define how observations relate to one another, and they can be constructed in multiple ways depending on the spatial logic of the problem. Raster-based moving windows and kernel functions are common in continuous RS analysis. In contrast, contiguity relationships (rook or queen adjacency), distance-based buffers, k-nearest neighbors, and network- or graph-based connections are often used with discrete GIS features. Each neighborhood definition involves assumptions about spatial interaction and scale, allowing models to capture different forms of spatial dependence and heterogeneity. When neighborhood structures are translated into quantitative descriptors like spatial lag variables, local variance, texture metrics, hotspot statistics, or local indicators of spatial association (e. g., local Moran's I) they become powerful predictive features. These neighborhood-derived metrics embed contextual information directly into machine learning workflows. Continuous pixel-level measurements can be summarized within discrete zones, while discrete units can inherit contextual characteristics from surrounding raster surfaces. In this way, spatial statistics provide the mathematical bridge that links continuous and discrete data structures, allowing them to coexist within the same modeling framework.

Spatially informed models extend this integration further by incorporating spatial dependence directly into model structure. Approaches such as geographically weighted regression account for spatial non-stationarity by allowing relationships to vary locally (Brunsdon et al., 1998; Fotheringham et al., 2009). More recent developments such as spatial ensemble models, neural networks, and spatial deep learning include adjacency relationships or distance-decay functions during training. These methods reduce biases arising from clustered observations, mitigate spatial autocorrelation in residuals, and produce predictions that better reflect underlying geographic processes.

This synthesis is particularly important as geospatial datasets grow in size, resolution, and complexity. High-resolution, multi-sensor RS imagery generates massive continuous data streams, while GIS layers provide rich contextual information about land use, infrastructure, and socio-environmental boundaries. Embedding neighborhood-based spatial analysis within GeoAI workflows enables scalable modeling that preserves spatial context across scales. Ultimately, integrating spatial statistics with machine learning unifies the continuous and discrete traditions of geospatial analysis, advancing toward spatially aware predictive systems that more faithfully represent the structure and dynamics of geographic phenomena.

Computational considerations are critical when applying spatial statistics to large geospatial datasets. Traditional approaches, such as Fragstats and landscape metrics, quantify patch structure, fragmentation, and connectivity, but can become computationally intensive with fine-resolution multi-temporal data or when nested without other GeoAI workflows/models (Cushman et al., 2013; Cushman & McGarigal, 2019). Advances in parallel computing, cloud-based geospatial platforms, and optimized algorithms now allow spatial metrics to be calculated efficiently across entire regions, supporting analyses at continental or global scales. Modern tools also facilitate the combination of landscape metrics with GeoAI, such as using fragmentation indices as predictive features in land use classification or ecosystem modeling. Table 4.3, below, summarizes some of the most common spatial analytics methods, their definitions, and examples of typical applications:

Building on these concepts, spatial statistics methods such as Moran's I, hotspot analysis, and landscape metrics can be directly applied to remotely sensed time-series data to enhance GeoAI workflows. For instance, Moran's I can be used to quantify the spatial autocorrelation of seasonal NDVI or EVI values across a landscape, revealing regions where vegetation change is clustered or dispersed (e.g., Nejadrekabi et al., 2022). Hotspot analysis can identify areas of extreme vegetation stress, deforestation, or urban expansion by detecting statistically significant clusters in multi-temporal vegetation index data. Landscape metrics, calculated from classified imagery or continuous vegetation layers, allow researchers to quantify patch size, connectivity, and fragmentation, providing critical context for ecological modeling, habitat assessment, and disturbance monitoring (Cushman et al., 2013; Cushman & McGarigal, 2019; Nejadrekabi et al., 2022). By integrating these spatial statistics with machine learning models, GeoAI workflows gain the ability to account for spatial dependence, leverage neighborhood information, and improve predictive accuracy, creating a more robust and interpretable framework for analyzing large-scale geospatial datasets derived from modern remote sensing systems.

Closely paralleling MAUP in the spatial domain is the Modifiable Temporal Unit Problem, or MTUP, which refers to the sensitivity of analytical results to the way time is discretized, aggregated, or segmented (Cheng & Adepeju, 2014). In time series remote sensing, temporal resolution choices such as daily, 8-day, monthly, or annual composites, the selection of calendar versus phenological years, and the definition of breakpoint intervals in change detection algorithms can substantially alter detected trends, disturbance timing, and model coefficients (de Jong et al., 2012). A drought signal evident in high-frequency observations may be dampened or lost when aggregated to monthly means; similarly, long-term greening or browning trajectories can strengthen, weaken, or even reverse depending on the selected time window. In large-scale GeoAI workflows that leverage dense satellite archives, MTUP becomes structurally embedded in compositing strategies, smoothing procedures, and temporal

Table 4.3 Common spatial analysis methods and their definitions as it pertains to their use or combined use in GIS and remote sensing.

Method	Definition	Example of common use	Role in large-scale/ GeoAI analytics
Moran's I	A global measure of spatial autocorrelation, indicating whether similar values cluster or disperse across a study area	Assessing overall clustering of vegetation stress in a forest (Moran, 1950)	Diagnostic for spatial dependence and model bias in ML
Geary's C	Similar to Moran's I but more sensitive to local differences, measures spatial autocorrelation	Evaluating fine-scale patterns of urban temperature anomalies (Geary, 1954)	Complements global metrics in heterogeneous landscapes
Local indicators of spatial association (LISA)	Measures local spatial autocorrelation to identify hotspots, coldspots, and spatial outliers	Detecting localized disease clusters or vegetation anomaly zones (Anselin, 1995)	Used for feature generation and hotspot screening
Hotspot analysis (Getis-Ord Gi*)	Identifies statistically significant clusters of high or low values	Mapping urban crime hotspots or agricultural yield anomalies (Getis & Ord, 1992)	Inputs to risk mapping and prioritization models
Ripley's K function	Quantifies spatial point patterns across multiple scales to detect clustering or dispersion	Analyzing distribution of tree species or disease incidence points (Ripley, 1988)	Supports scale-explicit spatial inference
Spatial scan statistic (Kulldorff)	Detects clusters in space or space-time and tests statistical significance	Identifying disease outbreaks or deforestation hotspots (Kulldorff, 1997)	Integrates space-time logic in automated monitoring
Geographically weighted regression (GWR)	Spatially varying regression that accounts for local relationships between variables	Modeling how environmental factors affect crop yield differently across regions (Brunsdon et al., 1998)	Precursor to spatially adaptive ML models
Landscape metrics (Fragstats)	Quantifies spatial configuration, patch size, connectivity, and fragmentation	Assessing habitat fragmentation, forest structure, or wetland connectivity (Cushman et al., 2013)	Increasingly used as predictive features, not endpoints

modeling decisions. Just as spatial aggregation can bias inference if MAUP is ignored, temporal aggregation and segmentation can shape ecological interpretation if MTUP is not explicitly addressed (Cheng & Adepeju, 2014; de Jong et al., 2012). Recognizing and testing for MTUP through sensitivity analyses, transparent reporting of temporal design choices, and alignment of temporal units with ecological processes are therefore essential for producing reproducible and scientifically robust insights from big geospatial time series data. As shown in Fig. 4.4, both MAUP and MTUP originate in project design choices related to aggregation, scale, and segmentation, and their effects cascade into downstream modeling and policy-relevant conclusions (Cheng & Adepeju, 2014; de Jong et al., 2012).

4.1 Emerging big data analytics paradigms

As geospatial datasets have expanded in volume, dimensionality, and temporal depth, analytical emphasis has shifted from individual methods toward integrated paradigms for extracting information from large, heterogeneous data streams. Rather than replacing foundational remote sensing or spatial statistical techniques, these paradigms reframe how such methods are operationalized under big data conditions, emphasizing scalability, temporal structure, spatial context, and interpretability. Table 4.4 summarizes several emerging analytical paradigms that are increasingly shaping geospatial research in the era of big data. These approaches build directly on the time-series vegetation indices and spatial statistical methods discussed in earlier sections,

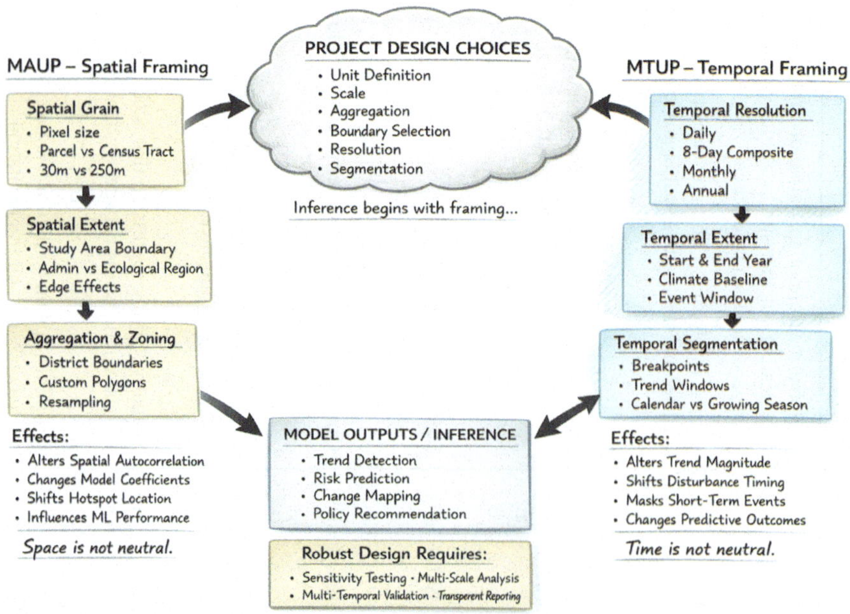

Figure 4.4 Conceptual framework of spatial and temporal unit sensitivities in geospatial project design.

Table 4.4 Emerging big data analytics paradigms in geospatial science.

Paradigm/ Technique	Core idea	What makes it "big data"	Relevance to this chapter	Representative reference
Dense time-series feature engineering	Treats time series as structured feature spaces rather than discrete observations	Leverages long, dense satellite archives across time	Builds directly on vegetation index and time-series methods	Zhu and Woodcock (2014)
Cloud-native EO analytics	Shifts computation to data rather than moving data to local systems	Enables planetary-scale processing and analysis	Supports scalability of Tables 4.1–4.3 and global applications	Gorelick et al. (2017)
Multi-sensor data fusion	Integrates complementary information from multiple sensor types	Increases dimensionality across sensors and data sources	Extends vegetation indices and spatial context	Camps-Valls et al. (2013)
Spatially aware machine learning	Explicitly incorporates spatial structure into modeling and validation	Addresses spatial dependence in large geospatial datasets	Bridges spatial statistics and GeoAI workflows	Meyer et al. (2019)
Interpretable machine learning for EO	Emphasizes transparency, uncertainty, and diagnostic insight	Addresses trust and bias in high-dimensional models	Aligns with model interpretability and validation discussion	Holloway and Mengersen (2018)

illustrating how traditional techniques are embedded within cloud-native processing environments, dense temporal feature spaces, multi-sensor fusion workflows, and spatially aware statistical learning frameworks. Together, these paradigms highlight the methodological evolution from static mapping toward scalable, process-oriented, and interpretable geospatial analytics.

5. Cross scale use of time series data

Time series data has become an essential input for a wide range of models that operate across local, regional, and global scales, providing dynamic information about environmental processes, human activities, and ecosystem responses. At the local scale, models often focus on site-specific applications such as crop yield prediction, soil moisture estimation, or phenology monitoring. For example, high-frequency NDVI or EVI time series can feed process-based crop models to estimate planting and harvest dates, cumulative biomass, and water stress impacts, enabling precision agriculture decisions (Dela Torre et al., 2021; Kooistra et al., 2023). Hydrological models can also leverage time series of rainfall, evapotranspiration, and vegetation indices to simulate water balance and runoff at field or watershed scales (Thoma et al., 2016; Thoma et al., 2019).

At regional and continental scales, time series data supports modeling of land use and land cover change, forest disturbance and regrowth, and ecosystem productivity (e.g., Hüttich et al., 2009; Jeyavathana, 2022; Zhao et al., 2025). RS-derived VI's over time can be integrated into hybrid statistical and machine learning models to estimate GPP, track drought impacts, or detect deforestation events (e.g., Munson et al., 2019; Sjöström et al., 2009). These models often produce outputs such as maps of biomass change, seasonal growth anomalies, or indicators of ecosystem health, which are critical for land management, conservation planning, and policy decision-making.

Global-scale models increasingly rely on aggregated time series from multiple satellite platforms, incorporating data from optical, radar, and LiDAR sensors. For example, climate–vegetation interaction models use global NDVI, fAPAR, or leaf area index time series to simulate carbon fluxes, biome productivity, and responses to climate variability. Earth system models and large-scale hydrological or biogeochemical models also incorporate time series inputs to provide projections of ecosystem function, land cover transitions, and climate feedbacks.

Despite these advances, the quality of model outputs is highly dependent on how time series data are used. Good practice involves accounting for temporal resolution, sensor differences, cloud contamination, and spatial heterogeneity. Poor use occurs when single-date or sparsely sampled time series are applied to processes that require continuous temporal information, when data gaps are not properly handled, or when models fail to account for seasonal or interannual variability. Misalignment of temporal scales between predictors and modeled processes can also introduce bias, reducing the reliability of predictions. Table 4.5, below, summarizes the types of models, typical outputs, and considerations for effective and poor use of time series data.

By linking time series data to models at multiple scales, researchers and practitioners can capture both the temporal dynamics and spatial variability of ecological

Table 4.5 Summary of model types/categories that can be used with remote sensing time series data.

Model type	Typical outputs	Good use	Poor use
Crop and phenology models	Planting/harvest dates, biomass, water stress indicators	Use high-frequency NDVI/EVI time series, account for seasonal trends	Single-date snapshots, ignore gaps or anomalies
Hydrological models	Runoff, soil moisture, evapotranspiration	Integrate continuous rainfall and vegetation time series, validate locally	Sparse temporal data, mismatched scales
Land cover change models	Maps of deforestation, urban expansion, vegetation degradation	Multi-year time series, sensor harmonization, account for clouds	Single-date images, ignore seasonal variability
Ecosystem productivity models	GPP, NPP, carbon flux estimates	Use consistent multi-sensor time series, align temporal resolution with process	Over-aggregation, ignoring sensor differences
Climate–Vegetation interaction models	Biome productivity, carbon cycle feedbacks	Global multi-sensor time series, gap-filling, interannual analysis	Infrequent sampling, failure to consider temporal autocorrelation

and human systems, improving predictions, understanding, and management of complex landscapes.

6. Discussion and future directions

As geospatial analytics enters an era defined by massive multi-sensor datasets, advanced modeling techniques, and real-time monitoring, it becomes essential to situate these innovations within a coherent analytical framework. Fig. 4.5 synthesizes the major components discussed throughout this chapter, illustrating how data harmonization, spatial and temporal modeling, uncertainty management, and interpretability interact within modern GeoAI workflows. By explicitly integrating considerations of MAUP and MTUP alongside model validation and application domains, the figure highlights both the power and the responsibility inherent in contemporary Earth system analytics.

Despite tremendous advances in RS, GeoAI, and spatial analytics, several challenges remain in the era of big geospatial data. One of the primary obstacles is managing and processing massive multi-sensor datasets that vary in spatial, temporal, and spectral resolution, while also considering issues of MAUP and MTUP in project design. Harmonizing data from satellites, airborne sensors, and ground-based measurements requires careful attention to calibration, atmospheric correction, and temporal alignment. Another persistent challenge is dealing with uncertainty, including sensor noise, cloud contamination, missing observations, and the effects of aggregation or scale mismatches on model outputs. Model interpretability and/or explainability remains of critical concern, as increasingly complex machine and deep learning

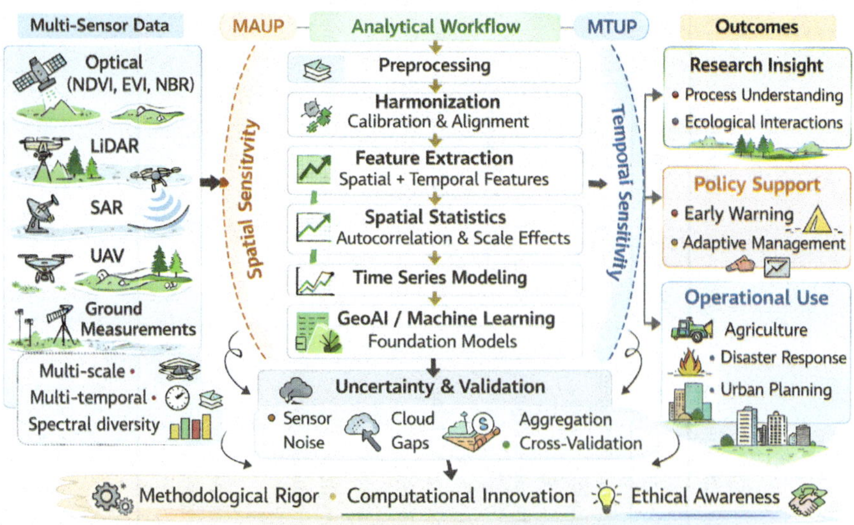

Figure 4.5 Integrated GeoAI workflow for trustworthy earth system analytics.

workflows can function as "black boxes," limiting users' ability to understand why certain predictions or classifications are made.

Emerging trends in geospatial analytics are beginning to address some of these challenges while also expanding the frontier of what is possible. Real-time analytics enable monitoring of dynamic processes such as vegetation stress, flood progression, or wildfire spread. Multi-sensor fusion, integrating optical imagery, LiDAR measurements, and ancillary environmental data (e.g., continuous, discrete, vector, or raster) allow for richer, more robust models that capture complementary aspects of the Earth and its dynamic or changing landscapes. Foundation models and self-supervised learning approaches are starting to enter geospatial workflows, providing flexible, pre-trained representations that can be adapted to multiple tasks, reducing the dependence on large, labeled datasets and accelerating the development of predictive models. These advancements carry significant implications for research, policy, and to integrate multi-scale, multi-sensor data opens opportunities to explore complex ecological and social–environmental interactions that were previously difficult to quantify or too multidimensional to assess. For policymakers and resource managers, improved monitoring and predictive capabilities can support evidence-based decision-making, early warning systems, and adaptive management strategies. Operational applications, from agriculture to disaster response and urban planning, stand to benefit from timely, accurate, and spatially explicit information, provided that the underlying models are transparent, interpretable, and appropriately validated.

7. Conclusions

The ongoing evolution of geospatial big data analytics reflects a fundamental shift in how we observe, model, and understand the Earth. Key takeaways include the importance of integrating time-series RS, vegetation indices, spatial statistics, and GeoAI approaches into coherent workflows that account for spatial and temporal dependencies, sensor characteristics, and environmental variability. Modern geospatial analytics move beyond static mapping toward dynamic, predictive, and process-oriented modeling, enabling continuous monitoring of ecosystems, human activities, and environmental change.

The role of big data analytics in geospatial science continues to expand, with emerging tools allowing researchers to handle unprecedented volumes of multi-sensor data, fuse information across scales, and leverage advanced machine learning methods. However, the power of these tools must be balanced with attention to interpretability, transparency, and trust in GeoAI systems. Ultimately, the future of geospatial analytics lies in its ability to combine methodological rigor, computational innovation, and ethical awareness to deliver actionable insights about the Earth system, from local fields to global landscapes. By grounding advanced GeoAI workflows in solid principles of spatial analysis, time-series interpretation, and model transparency, the field can continue to expand the frontiers of what is possible while maintaining confidence in the reliability and interpretability of its outputs.

Case studies

Case Study 1: Moving beyond land cover classes: Time series analysis of vegetation dynamics in Southern African Savannas

Gibbes, C., Southworth, J., Waylen, P. and B. Child (2014). Climate variability as a dominant driver of post-disturbance savanna degradation or recovery. *Applied Geography*, 53: 389–401.

Introduction and conceptual motivation

For much of the history of remote sensing (RS) based land analysis, changes in landscapes were primarily assessed through land cover classification. Discrete categories such as forest, grassland, agriculture, or shrubland were mapped at individual points in time and compared across years to identify change. This approach proved highly effective for documenting large scale transformations such as deforestation, urban expansion, and agricultural conversion. However, it struggled to represent ecosystems whose structure and function vary continuously through time rather than shifting abruptly between stable states. This limitation is especially evident in savanna and dryland systems. These environments are inherently heterogeneous and dynamic, shaped by interactions among climate variability, fire, herbivory, and soil conditions (Sankaran et al., 2005; Staver et al., 2011). Vegetation composition and productivity fluctuate seasonally and interannually, often without crossing clear categorical thresholds (Bégué et al., 2011; Carlson & Ripley, 1997; Jiang et al., 2006; Lambin & Ehrlich, 1997; Tucker, 1979; Wang et al., 2005). As a result, landscapes may appear stable in land cover maps while undergoing substantial ecological change.

This case study illustrates an early and influential shift away from static land cover classification toward time series analysis of continuous vegetation metrics using remotely sensed data. Drawing on work published by Gibbes et al. (2014), the study demonstrates how long-term vegetation index records can be used to understand ecosystem dynamics, resilience, and threshold behavior across large landscapes. Although the paper predates many recent developments in dense time series analysis and continuous vegetation products, it provides a conceptual foundation that aligns closely with the themes of this chapter. Rather than asking whether land cover has changed from one class to another, the study asks how vegetation productivity evolves through time, whether those changes exhibit vegetation persistence or recovery, and how they relate to shifts in precipitation patterns. This reframing represents a critical step toward modern geospatial analysis approaches that emphasize continuous variables and temporal trajectories over categorical snapshots (Waylen et al., 2014).

From land cover classification to continuous vegetation time series

Savanna ecosystems pose persistent challenges for land cover classification. These systems are composed of varying mixtures of grasses, shrubs, and trees whose relative abundance changes across seasons and years. Fire regimes, grazing pressure, and rainfall variability interact to produce landscapes that resist stable categorization (Andela et al., 2013; Campo-Bescós et al., 2013; Lehmann et al., 2014; Sankaran et al., 2005). A pixel classified as woodland or grassland in 1 year may not meaningfully differ in ecological function from its state a few years earlier, even if its categorical label changes. Recognizing these challenges, the study adopts a fundamentally different analytical strategy. Rather than mapping discrete land cover classes, vegetation dynamics are analyzed using the Normalized Difference Vegetation Index (NDVI) as a continuous proxy for vegetation productivity and condition. NDVI captures changes in photosynthetic activity and green biomass and can be derived consistently from satellite data over long time periods (Townshend & Justice, 1986).

Using a combined AVHRR and MODIS NDVI record spanning 1982 to 2009, a 28-year time series of peak growing season vegetation was constructed across southern Africa. April imagery was used to represent peak vegetation conditions following the rainy season, reducing phenological noise and emphasizing cumulative ecosystem responses to climate forcing. This approach allows vegetation to be analyzed as a continuous surface that varies over time rather than as a static set of land cover categories. This analytical shift exemplifies the movement away from classification-based land change analysis toward continuous vegetation products and time series thinking. Instead of identifying whether a pixel has changed class, the study evaluates how NDVI values evolve through time, whether trends are persistent, and whether observed changes exceed what would be expected from random variation (Southworth et al., 2016; Waylen et al., 2014) (Case Study Figure 1.1). Fig. 4.1 conceptualizes how shifts in both the mean and variability of precipitation generate distinct NDVI persistence outcomes, and how these responses depend upon a system's position along a savanna rainfall gradient. By separating climate forcing, ecological regime context, and vegetation trajectory, the model illustrates why continuous time series analysis is essential for detecting resilience or degradation in dynamic savanna systems.

Study region and environmental context

The study area spans approximately 683,000 square kilometers across portions of Angola, Namibia, Botswana, and Zambia, encompassing the Kavango, Kwando, and Zambezi river basins. This region covers a broad precipitation gradient, ranging from approximately 400 mm annually in the south to over 1400 mm in the north. Importantly, it crosses a widely cited ecological threshold near 650 mm of annual rainfall that separates precipitation limited savannas from systems increasingly regulated by fire and herbivory (Campo-Bescós et al., 2013; Sankaran et al., 2005; Staver et al., 2011). Vegetation types across the region include dry Kalahari woodlands, wetter miombo forests, and mixed grassland woodland mosaics. Human population density is relatively low across much of the study area, which reduces the confounding influence of intensive land use change and allows climate vegetation relationships to be examined more directly. The region experienced a well-documented climate shift in the mid to late 1970s associated with large scale changes in ocean atmosphere dynamics. This shift altered both the mean and variability of precipitation across southern Africa, creating a natural experiment for examining long-term ecosystem responses to climatic disturbance (Gibbes et al., 2014).

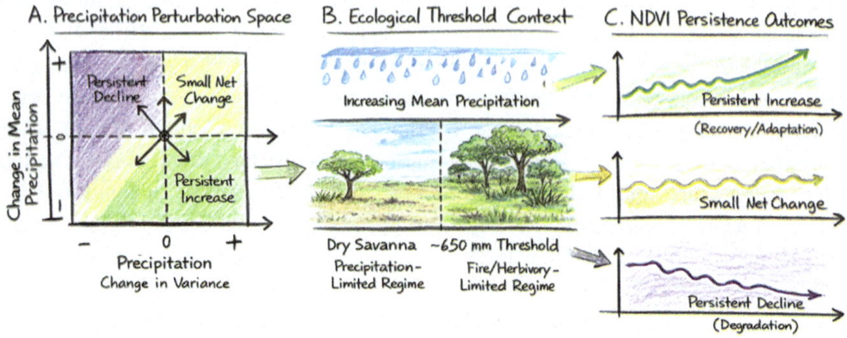

Case Study Figure 1.1 Regime-dependent vegetation persistence outcomes under shifts in precipitation mean and variability across a Savanna rainfall gradient.

Time series analysis and ecosystem resilience

A central contribution of the study lies in its explicit framing of vegetation dynamics through the lens of resilience and non-equilibrium ecosystem theory. Rather than assuming that ecosystems fluctuate around a stable equilibrium state, resilience theory is used to examine how systems respond to shocks and whether they return to previous conditions or transition toward new states (Carpenter & Brock, 2006; Holling, 1973). Remote sensing time series analysis is central to operationalizing this framework. By examining vegetation dynamics over nearly 3 decades, the study moves beyond snapshot comparisons and instead evaluates persistence, directionality, and variability in vegetation change (Case Study Figure 1.1). Three complementary persistence metrics are developed to characterize vegetation behavior. Directional persistence measures whether vegetation productivity has increased or decreased relative to an initial benchmark year. Relative directional persistence assesses whether vegetation tends to increase or decrease from 1 year to the next. Massive persistence captures the magnitude of year-to-year changes, accounting for both direction and size of NDVI shifts. These metrics transform the NDVI time series into interpretable measures of ecosystem behavior that can be mapped spatially and evaluated statistically. Importantly, observed trends are tested against a random walk process, allowing statistically significant changes to be identified rather than inferred subjectively (Gibbes et al., 2014; Southworth et al., 2016; Waylen et al., 2014). This emphasis on temporal structure, statistical rigor, and spatial patterning anticipates many contemporary approaches in GeoAI that focus on trend detection, anomaly identification, and regime shift analysis using dense Earth observation time series.

Vegetation responses to changes in climate variability

One of the most important findings of the study is that vegetation responses cannot be explained by changes in mean annual precipitation alone. While many vegetation studies focus primarily on average rainfall, this analysis demonstrates that changes in precipitation variability exert a strong and sometimes dominant influence on ecosystem behavior. Across much of the study region, mean annual precipitation declined following the climate shift of the late 1970s. However, NDVI time series reveals increasing vegetation productivity across large portions of the landscape in subsequent decades. This apparent contradiction highlights the limitations of static land cover approaches and simple climate correlations.

By examining both changes in precipitation mean and changes in interannual variability, distinct ecosystem responses emerge across the precipitation gradient. Drier savanna and Kalahari woodland systems exhibited increasing NDVI despite reduced rainfall, suggesting adaptation to high variability environments. In contrast, wetter miombo systems showed persistent declines in NDVI associated with increased precipitation variability, even where mean rainfall remained stable or increased (Gibbes et al., 2014). These findings underscore the value of continuous vegetation time series for understanding ecosystem processes. A land cover classification approach would struggle to capture such nuanced responses, particularly when productivity changes occur without clear shifts in vegetation type.

Spatial patterns, thresholds, and early signals of change

Mapping vegetation persistence metrics reveals spatially coherent patterns of vegetation change across the region. Areas of persistent NDVI increase dominate much of the southern and central portions of the study area, while clusters of declining NDVI are concentrated in wetter northern miombo regions. These patterns are interpreted through the lens of ecological thresholds. In some systems, changes in precipitation variability may push ecosystems beyond thresholds from which recovery is slow or incomplete. Importantly, these thresholds do not necessarily manifest as abrupt land cover transitions but instead appear as sustained changes in vegetation productivity and variability. This insight has significant implications for land change science and GeoAI applications. It suggests that meaningful ecosystem transitions may occur without obvious categorical change and that time series analysis of continuous variables is essential for detecting early warning signals of vulnerability. The study also highlights the importance of spatial heterogeneity as an indicator of ecosystem status. By examining both the mean and variance of NDVI across space and time, the analysis captures patterns of recovery, adaptation, and stress that would be invisible in single date land cover maps (Gibbes et al., 2014; Southworth et al., 2016; Waylen et al., 2014).

Implications for GeoAI

Although the study predates recent advances in GeoAI, its conceptual contributions align strongly with modern practice. The emphasis on continuous variables, long term time series, statistical significance, and spatially explicit analysis reflects the analytical mindset that underpins contemporary Earth observation and GeoAI approaches. Today, continuous vegetation products derived from sensors such as MODIS, Sentinel, and Landsat, combined with machine learning and deep learning methods, allow for even richer characterization of vegetation dynamics. However, the fundamental questions remain the same. How does vegetation respond through time to environmental drivers? What patterns indicate vegetation resilience versus vulnerability?

How can early signals of system change be detected before irreversible transitions occur? This case study illustrates that the move away from land cover classification is not simply a technical evolution but a conceptual one. Treating vegetation as a dynamic continuous process rather than a static categorical entity provides deeper insight into ecosystem functioning and change. This shift lies at the heart of this chapter and sets the stage for the GeoAI based time series methods discussed in this chapter.

Case Study 2: Understanding what drives vegetation change in Southern Africa using combined time-series analysis and machine learning

Southworth, J., Waylen, P., Muñoz-Carpena, R., Bunting, E., Zhu, L., Ryan, S. J., Herrero, H. V., Kaplan, D., & Campo-Bescós, M. A. (2018). Using a coupled dynamic factor—random forest analysis (DFRFA) to reveal drivers of spatio-temporal heterogeneity in the semi-arid regions of southern Africa. PLOS ONE, 13(12), e0208400.

Introduction

Vegetation change is one of the clearest signals that landscapes are responding to environmental pressures and human activity (Campo-Bescós et al., 2013; Lehmann et al., 2014). In semi-arid regions, these changes are especially important because vegetation supports food production, livestock grazing, wildlife habitat, and ecosystem stability. However, understanding why vegetation changes over time is challenging because many factors operate at once. Climate patterns such as rainfall and temperature influence plant growth, but so do human decisions, including land use, settlement expansion, and conservation policies.

Remote sensing (RS) has long been used to monitor vegetation through indices such as the Normalized Difference Vegetation Index (NDVI), which provides a consistent way to track plant greenness over time (Tucker, 1979). These datasets have made it possible to examine environmental change across large regions and longer time periods. However, many traditional studies focus primarily on mapping trends rather than explaining why they occur. This limitation becomes especially important in semi-arid landscapes, where vegetation patterns often reflect both environmental variability and human activities (Anyamba & Tucker, 2005; Nicholson, 2012). This study looks to move beyond simple trend mapping toward identifying single and interacting drivers of vegetation change. The research introduced a combined framework that integrates time-series modeling with machine learning to better explain spatial and temporal variability. This approach allows researchers to capture both large-scale environmental influences and local human and landscape factors shaping vegetation outcomes (Southworth et al., 2018).

Study area

The study focused on the Kwando, Okavango, and Zambezi catchments in southern Africa. This region spans multiple countries and includes diverse landscapes ranging from grasslands and savannas to wetlands and floodplains (Case Study Figure 2.1). Seasonal rainfall patterns dominate ecological processes, with vegetation growth closely tied to wet-season precipitation. The region is also characterized by significant human impacts across the landscape, some areas are protected for conservation while others support growing populations, agriculture, and infrastructure development. These differences create a complex mosaic of environmental and human influences. The combination of strong climate variability and rapid land-use change makes the region ideal for studying vegetation dynamics and their drivers (Southworth et al., 2018).

Results: Large-scale environmental drivers of vegetation change

The companion pieces to this study, Campo-Bescós et al. (2013), found that across this same study area climate variables strongly influence regional vegetation patterns. Analysis of satellite time-series data showed that precipitation was the most important driver of vegetation variability across the study area, but that the significance of its role varied depending on the mean annual precipitation (Campo-Bescós et al., 2013). Areas receiving higher and more consistent rainfall generally had denser and more stable vegetation and were likely not water limited. Across the study area but particularly in the non-water limited areas, temperature also played a significant role by influencing seasonal growth cycles and evapotranspiration processes. Soil moisture and atmospheric conditions further helped explain differences in vegetation

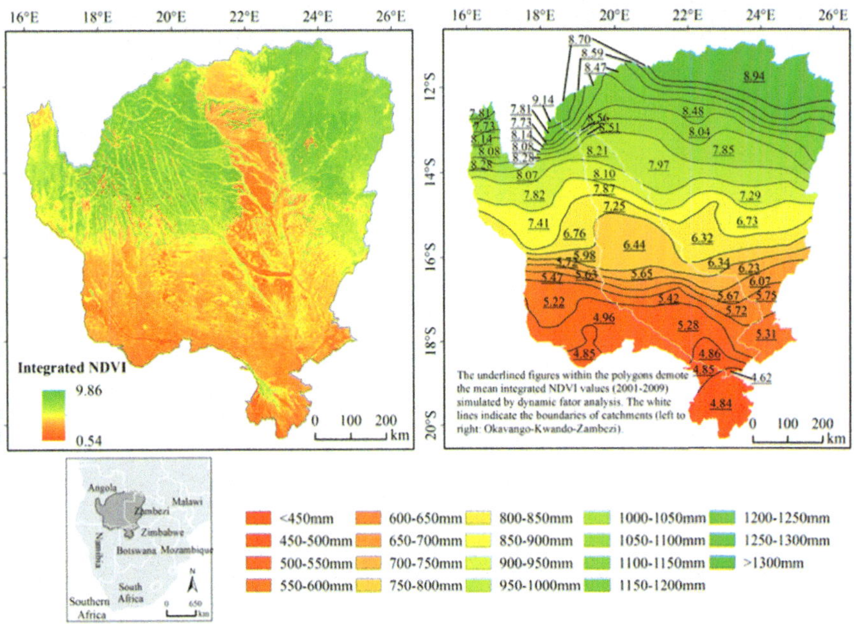

Case Study Figure 2.1 Study area across Botswana, Namibia, Zambia, Angola, and Zimbabwe including the patterns of greenness (high values = denser or greener vegetation, low values = sparse or brown downed vegetation) and mean annual precipitation.

response across landscapes. These findings align with decades of research demonstrating the dominant role of rainfall variability in shaping vegetation dynamics in semi-arid environments (Anyamba & Tucker, 2005; Nicholson, 2012). However, this article moved the field of savanna ecology forward in its understanding of the specific role of different climatic drivers. Specifically, these systems are often thought to be driven by precipitation alone, with much less emphasis placed on temperature, which was not found to be the case across the full study region. Despite the strong influence of climate, environmental variables alone could not fully explain observed vegetation patterns. Many locations show differences between predicted and actual vegetation conditions, indicating the presence of additional local drivers (Case Study Figure 2.2; Southworth et al., 2018).

Local variations and unexplained patterns

After accounting for environmental influences, the study identified areas where vegetation behaved differently than expected (Case Study Figure 2.2). Some locations

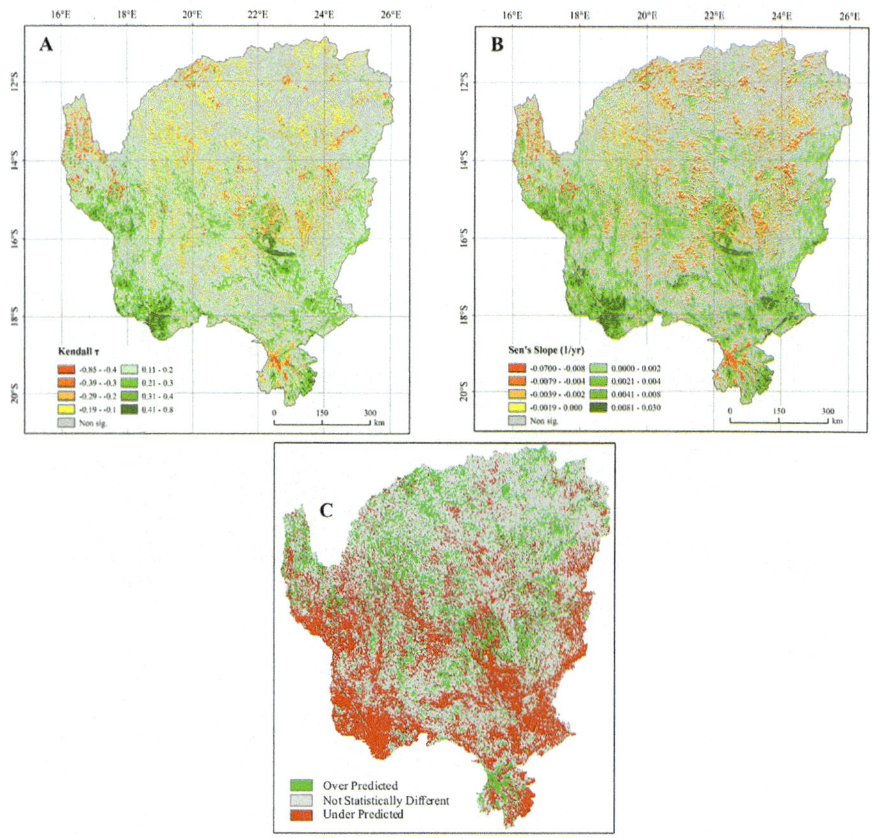

Case Study Figure 2.2 Trend analysis of NDVI time series across the study areas including (A) Kendall T, (B) sen slope, and (C) areas were Campo-Bescós et al., 2013(the companion piece to Southworth et al., 2018) under or over predicted NDVI based on continuous, mostly climate data.

showed stronger vegetation growth than predicted, while others exhibited weaker growth. These mismatches revealed the presence of localized factors influencing vegetation dynamics. The spatial distribution of these unexplained areas showed clear clustering rather than random patterns. A random forest analysis was used to determine the drivers of these landscape patterns, specifically focusing on the variables of key importance (Case Study Figure 2.3). Factors included in this random forest analysis included discrete and continuous variables such as distance metrics (from roads, towns and water), land cover, topography, population, and land management status (protected areas). These results highlight a key limitation of traditional environmental modeling approaches. While climate data can explain broad trends, they often fail to capture fine-scale variability caused by human decisions and land-use practices (Southworth et al., 2018). This study bridges this gap Case Study Figure 2.2.

Human and landscape drivers of heterogeneity

Machine learning analysis revealed that several local factors significantly influenced vegetation variability (Case Study Figure 2.3). Protected areas emerged as one of the strongest predictors of positive vegetation outcomes. These regions tended to maintain higher vegetation density due to reduced disturbance and controlled land use. Population density also played a major role. Areas with higher population concentrations often showed reduced vegetation cover, reflecting pressures from agriculture, fuelwood collection, and settlement expansion. Proximity to roads further influenced vegetation patterns by increasing access and facilitating land transformation (Case Study Figure 2.3).

Physical landscape features such as elevation, land cover, and soil type also contributed to vegetation differences. These factors affect water retention, nutrient availability, and microclimate conditions. The interaction between environmental,

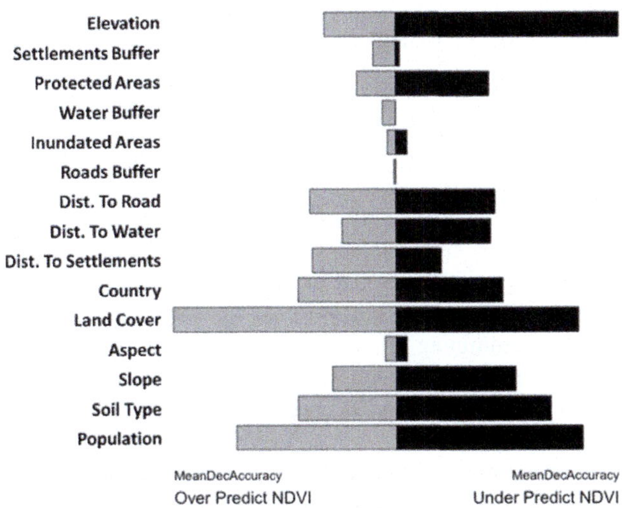

Case Study Figure 2.3 Random forest derived variable importance chart for continuous and discrete drivers of landscape variability within the study area.

human, and landscape factors created complex spatial patterns of vegetation change across the region (Southworth et al., 2018).

Implications for GeoAI

This study demonstrates the growing importance of integrating AI methods with RS, GIS, and geospatial analysis. The combined framework allowed researchers to move beyond simply describing patterns to identify their underlying causes. One key implication is the importance of multi-method integration. Time-series analysis effectively captures broad environmental trends, while machine learning can identify complex relationships among diverse variables. Together, they provide a more complete understanding of landscape dynamics. Another important insight is the need for multi-scale analysis. Environmental drivers often operate at regional scales, while human influences tend to be localized. GeoAI approaches must therefore be capable of analyzing patterns across multiple spatial and temporal scales simultaneously. The study also highlights the importance of interpretability in GeoAI. Rather than acting as a "black box," the machine learning approach identified which variables contributed most to vegetation variability. This transparency is critical for policy and management applications because it helps decision-makers understand the causes of environmental change. Finally, the findings demonstrate the potential of GeoAI to support sustainable land management. By identifying the drivers of vegetation change, planners can design targeted conservation strategies, anticipate environmental risks, and improve resource allocation.

Case Study 3: Assessing vegetation directional persistence in Southern African National Parks (2000–16)

Herrero, H., Southworth, J., Muir, C., Khatami, R., Bunting, E., & Child, B. (2020). An evaluation of vegetation health in and around Southern African National Parks during the 21st Century (2000–16). *Applied Sciences*, *10*(7), 2366.

Introduction

Understanding long-term vegetation resilience within savanna ecosystems is essential for conservation planning, particularly in regions where climate variability intersects with mounting human pressures. This case study explores directional persistence, a time-series metric derived from NDVI satellite data, to evaluate vegetation health and stability in and around 79 national parks across southern Africa between 2000 and 2016. By comparing trends inside protected areas with those in their surrounding buffer zones, and by examining differences across seasons and park sizes, this analysis provides a comprehensive understanding of vegetation responses to environmental conditions and management decisions in a landscape of global ecological importance.

Savannas dominate much of southern Africa and are characterized by high variability in precipitation, fire, herbivory, and human land use (Andela et al., 2013; Campo-Bescós et al., 2013; Lehmann et al., 2014; Sankaran et al., 2005). These factors, combined with other local scale drivers, contribute to the dynamic nature of savanna vegetation, making it critical for conservation scientists to monitor long-term ecological trends. Traditional standalone NDVI analyses provide snapshots of vegetation condition, but further analyses are needed to capture whether vegetation systems are significantly changing over extended periods (Bégué et al., 2011; Carlson & Ripley, 1997; Jiang et al., 2006; Lambin & Ehrlich, 1997; Tucker, 1979; Wang et al., 2005). The directional persistence approach was developed to address this limitation by quantifying both the direction and consistency of NDVI change over time (Southworth et al., 2016; Waylen et al., 2014). This method allows researchers to distinguish transient fluctuations from sustained ecological shifts, making it an especially powerful tool for evaluating the effectiveness of protected areas.

Directional persistence: Concept and analytical framework

Directional persistence utilizes big data and GeoAI techniques because this metric is calculated through a pixel-level random-walk statistic applied to seasonal NDVI data (derived monthly and then aggregated) from MODIS at 250m globally and subset to the study area. In this study, seasonal (early/late wet season and early/late dry season) NDVI values from a mean benchmark period (2000–04) were compared to values from each year of the subsequent comparison period (2005–16). When a pixel's NDVI in a given comparison year exceeded its benchmark value, it received a $+1$; when it fell below, it received a -1. The accumulated sum over 12 years produces the directional persistence value. Given this random walk is a two-tailed test, for a significance level of 0.05, the threshold of $+8$ or higher signifies significant long-term vegetation persistence in a positive direction, while -8 or lower signifies significant vegetation persistence in a negative direction. This approach incorporates statistical rigor uncommon in traditional vegetation trend analyses and produces results that are interpretable at fine spatial scales.

Spatial patterns of directional persistence across Southern Africa

The spatial patterns of directional persistence during the wet season (broken into early and late), because of its seasonal importance with vegetation across Southern Africa (with parks highlighted), are shown in Case Study Figure 3.1 below.

During the early wet season (December through February), strong positive persistence dominated the western portion of southern Africa, with Namibia and Angola exhibiting some of the most consistent greening across the region. A similar concentration of positive persistence appeared around the borders of South Africa, Zimbabwe, Botswana, and Zambia, corresponding with one of the region's largest and most established transfrontier conservation areas. These patterns suggest that certain regions have experienced sustained increases in vegetation greenness relative to the benchmark period. In contrast to these areas of improvement, the northernmost portion of the study area showed largely neutral persistence, indicating relatively stable vegetation conditions without major directional change. Small pockets of weak

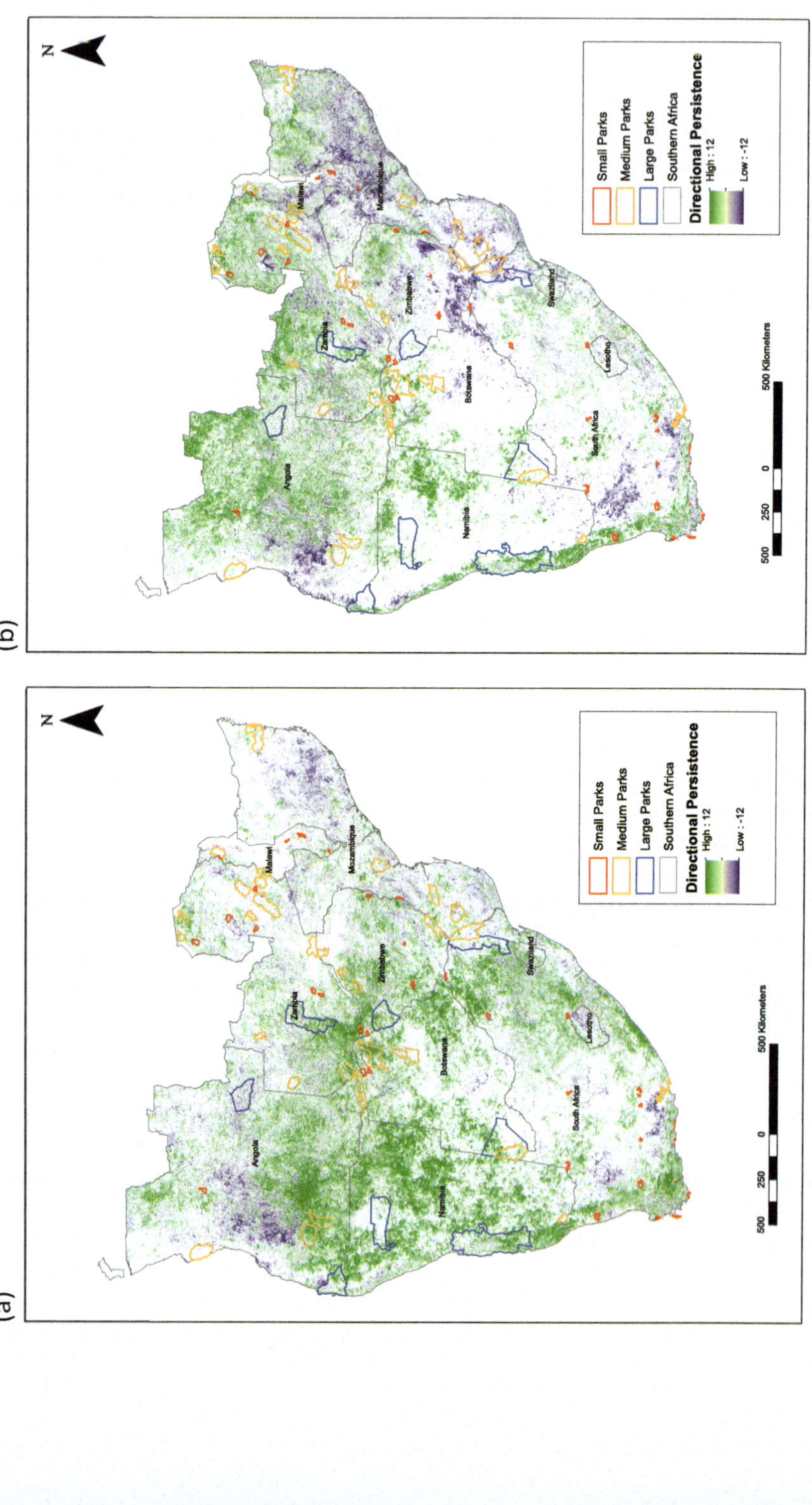

Case Study Figure 3.1 Directional persistence mapped across southern Africa with parks highlighted (a) significant values in December, January, and February and (b) significant values in March, April, and May.

negative persistence were observed in eastern Mozambique and western Angola, pointing to localized declines in vegetation health during this season.

The late wet season (March through May) displayed a more heterogeneous set of trends. Much of the region exhibited neutral persistence, likely reflecting the stabilization of vegetation after the peak rainfall period. Nevertheless, northeastern Namibia and areas of the Kalahari in both Namibia and South Africa showed notable positive persistence, suggesting continued greening even late into the wet season. Conversely, several regions demonstrated strong negative persistence, particularly the tri-border area among South Africa, Mozambique, and Zimbabwe, central-northern Mozambique, southwestern Angola, and parts of western central South Africa. These negative trends point to locations where vegetation has been under increasing stress or where ecological pressures have intensified relative to the early 2000s. For further analysis on the Greater Limpopo Transfrontier Conservation Area specifically, see Herrero et al. (2023).

Protected areas versus buffer zones

A central objective of this study was to assess whether national parks maintain healthier and more resilient vegetation than their surrounding buffer zones. Across all park sizes and during both wet seasons, national parks consistently exhibited higher levels of positive persistence and lower levels of negative persistence compared to their buffers (Case Study Figure 3.2). This pattern was particularly evident during the early wet season, when positive persistence in both parks and buffers ranged between 20%–31%, while negative persistence remained low at around 3%–4%. Small and medium parks outperformed their buffers with more positive and less negative persistence, indicating that even relatively smaller protected areas can contribute meaningfully to vegetation stability. Large parks, due to their extensive undisturbed interiors, often showed levels similar to or slightly better than their buffers.

When compared to the early wet season, the overall amount of positive persistence decreased in the late wet season, while negative persistence increased slightly, yet parks continued to outperform buffers across all size classes. Positive persistence ranged from 13%–26%, while negative persistence varied between 5%–10%. The seasonal reduction in positive persistence suggests the landscape experiences more mixed ecological signals following the main rainy period. However, the consistent difference between parks and buffers underscores the effectiveness of protected areas in maintaining long-term vegetation resilience.

Role of transfrontier conservation areas

Transfrontier conservation areas emerged as particularly influential features in shaping patterns of directional persistence. Regions located within or near these cross-border conservation complexes frequently showed strong positive persistence, indicating that ecological connectivity across political boundaries has a measurable positive effect on vegetation stability. These areas benefit from reduced fragmentation, standardized or cooperative management practices, and broader ecological gradients that allow species and ecological processes to function at landscape scales.

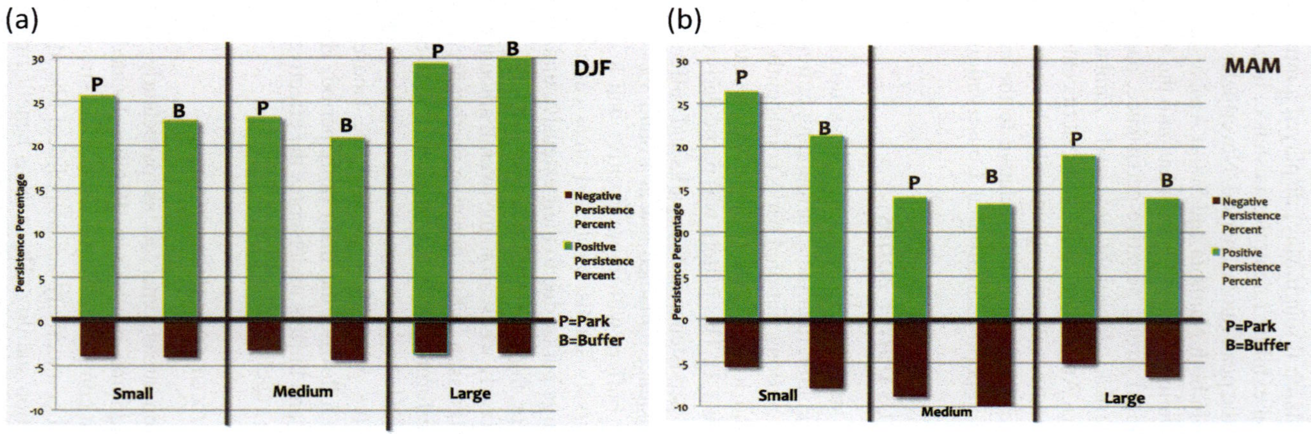

Case Study Figure 3.2 Percentage of significant positive and negative directional persistence values across small, medium, and large parks and buffer zones in (a) December, January, February, and (b) March, April, and May.

Ecological interpretation and implications

The widespread positive directional persistence observed in many parts of the study region aligns with global greening trends documented during the 21st century. Yet interpreting these changes requires caution. Increased NDVI may reflect shrub encroachment, shifts in herbivory, invasive species expansion, or atmospheric CO_2 fertilization, rather than improvements in ecosystem health (Bastin et al., 2017; Brandt et al., 2015; de Jong et al., 2011, 2012; Garonna et al., 2016; Guay et al., 2014; Mitchard et al., 2009; Olsson et al., 2005; Piao et al., 2011; Schimel et al., 2015; Zhang et al., 2015; Zhu et al., 2016). These findings also support literature on browning hotspots of Sub-Saharan Africa (Alcaraz-Segura et al., 2010; de Jong et al., 2012). Directional persistence provides a robust measure of long-term change, but it does not identify the exact ecological processes responsible. As climate change and human pressures continue to reshape these landscapes, incorporating directional persistence into regular monitoring frameworks will be invaluable for guiding adaptive management and supporting long-term ecosystem conservation.

Future directions and opportunities with GeoAI

Looking ahead, the integration of GeoAI, the fusion of large geospatial datasets with machine learning, offers powerful new avenues to advance directional persistence research. As remote sensing datasets grow in spatial and temporal resolution, GeoAI methods provide the analytical capacity to process and interpret these large, complex inputs more efficiently than traditional approaches. Machine learning models such as convolutional neural networks and transformer-based architectures could help detect subtle vegetation changes, identify emerging degradation hotspots, and refine the spatial detail at which persistence patterns are understood.

One of the most significant opportunities lies in using GeoAI to differentiate the ecological drivers behind positive or negative persistence. While directional persistence reveals long-term trends, it does not identify whether greening reflects shrub encroachment, invasive species, altered rainfall timing, or shifts in herbivore dynamics. By integrating field-based training data, high-resolution imagery, and biophysical variables, GeoAI models could help attribute observed changes to specific processes, providing clearer guidance for conservation interventions. GeoAI also offers potential for predictive modeling, allowing managers to anticipate how vegetation resilience may shift under future climate or land-use scenarios. Models trained on past persistence patterns, rainfall histories, and human impact layers could support scenario-based planning at park, buffer, and regional scales. Finally, the growing prominence of transfrontier conservation areas underscores the value of automated GeoAI pipelines for large scale, cross-border monitoring. These systems could streamline data ingestion, anomaly detection, and reporting, enabling coordinated and adaptive management across nations. Together, directional persistence and GeoAI create a powerful foundation for next generation conservation planning, offering both retrospective insight and forward-looking predictive capability.

References

Alcaras, E., Costantino, D., Guastaferro, F., Parente, C., & Pepe, M. (2022). Normalized burn ratio plus (NBR+): A new index for Sentinel-2 imagery. *Remote Sensing, 14*(7), 1727.

Alcaraz-Segura, D. O., Chuvieco, E., Epstein, H. E., Kasischke, E. S., & Trishchenko, A. (2010). Debating the greening vs. browning of the North American boreal forest: Di erences between satellite datasets. *Global Change Biology, 16*, 760–770.

Andela, N., Liu, Y. Y., van Dijk, A. I. J. M., de Jeu, R. A. M., & McVicar, T. R. (2013). Global changes in dryland vegetation dynamics (1988–2008) assessed by satellite remote sensing: Comparing a new passive microwave vegetation density record with reflective greenness data. *Biogeosciences, 10*, 6657–6676.

Anselin, L. (1995). Local indicators of spatial Association—LISA. *Geographical Analysis, 27* (2), 93–115. https://doi.org/10.1111/j.1538-4632.1995.tb00338.x

Anyamba, A., & Tucker, C. J. (2005). Analysis of Sahelian vegetation dynamics using NOAA-AVHRR NDVI data from 1981–2003. *Journal of Arid Environments, 63*(3), 596–614. https://doi.org/10.1016/j.jaridenv.2005.03.007

Asrar, G., Fuchs, M., Kanemasu, E. T., & Hatfield, J. L. (1984). Estimating absorbed photosynthetic radiation and leaf area index from spectral reflectance in Wheat[1]. *Agronomy Journal, 76*(2), 300–306. https://doi.org/10.2134/agronj1984.00021962007600020029x

Atkinson, P. M., Jeganathan, C., Dash, J., & Atzberger, C. (2012). Inter-comparison of four models for smoothing satellite sensor time-series data to estimate vegetation phenology. *Remote Sensing of Environment, 123*, 400–417.

Bannari, A., Morin, D., Bonn, F., & Huete, A. R. (1995). A review of vegetation indices. *Remote Sensing Reviews, 13*(1–2), 95–120. https://doi.org/10.1080/0275725950 9532298

Bárdossy, A., & Samaniego, L. (2002). Fuzzy rule-based classification of remotely sensed imagery. *IEEE Transactions on Geoscience and Remote Sensing, 40*(2), 362–374.

Bastin, J. F., Berrahmouni, N., Grainger, A., Maniatis, D., Mollicone, D., Moore, R., Patriarca, C., Picard, N., Sparrow, B., Abraham, E. M., et al. (2017). The extent of forest in dryland biomes. *Science, 356*, 635–638.

Bégué, A., Vintrou, E., Ruelland, D., Claden, M., & Dessay, N. (2011). Can a 25-year trend in Soudano-Sahelian vegetation dynamics be interpreted in terms of land use change? A remote sensing approach. *Global Environmental Change, 21*, 413–420.

Boori, M. S., & Vozenilek, V. (2014). Remote sensing and land use/land cover trajectories. *Journal of Geophysics & Remote Sensing, 3*(3), 0107.

Brandt, M., Mbow, C., Diouf, A. A., Verger, A., Samimi, C., & Fensholt, R. (2015). Ground- and satellite-based evidence of the biophysical mechanisms behind the greening Sahel. *Global Change Biology, 21*, 1610–1620.

Brunsdon, C., Fotheringham, S., & Charlton, M. (1998). Geographically weighted regression. *Journal of the Royal Statistical Society: Series D (The Statistician), 47*(3), 431–443. https://doi.org/10.1111/1467-9884.00145

Bunting, E. L., Munson, S. M., & Villarreal, M. L. (2017). Climate legacy and lag effects on dryland plant communities in the southwestern US. *Ecological Indicators, 74*, 216–229.

Bunting, E. L., Southworth, J., Herrero, H., Ryan, S. J., & Waylen, P. (2018). Understanding long-term savanna vegetation persistence across three drainage basins in Southern Africa. *Remote Sensing, 10*(7). https://doi.org/10.3390/rs10071013. Article 7.

Campo-Bescós, M. A., Muñoz-Carpena, R., Southworth, J., Zhu, L., Waylen, P. R., & Bunting, E. (2013). Combined spatial and temporal effects of environmental controls on long-term monthly NDVI in the Southern Africa savanna. *Remote Sensing, 5*(12), 6513–6538.

Camps-Valls, G., Tuia, D., Bruzzone, L., & Benediktsson, J. A. (2013). Advances in hyperspectral image classification: Earth monitoring with statistical learning methods. *IEEE Signal Processing Magazine, 31*(1), 45–54.

Camps-Valls, G. (2009). Machine learning in remote sensing data processing. In *2009 IEEE international workshop on machine learning for signal processing*. https://ieeexplore.ieee.org/abstract/document/5306233/?casa_token=PZd0etZQ6i8AAAAA:eQHSoHBgIS4JiY8tmQrJlngmIF0Cf1y3t4yOAICtkxHrewYN87ejxGB3T96veJuHYnB60Y4brw.

Carlson, T. N., & Ripley, D. A. (1997). On the relation between NDVI, fractional vegetation cover, and leaf area index. *Remote Sensing of Environment, 62*, 241–252.

Carpenter, S. R., & Brock, W. A. (2006). Rising variance: A leading indicator of ecological transition. *Ecology Letters, 9*(3), 311–318. https://doi.org/10.1111/j.1461-0248.2005.00877.x

Carrao, H., Gonçalves, P., & Caetano, M. (2009). A nonlinear harmonic model for fitting satellite image time series: Analysis and prediction of land cover dynamics. *IEEE Transactions on Geoscience and Remote Sensing, 48*(4), 1919–1930.

Chen, X., Vogelmann, J. E., Rollins, M., Ohlen, D., Key, C. H., Yang, L., Huang, C., & Shi, H. (2011). Detecting post-fire burn severity and vegetation recovery using multitemporal remote sensing spectral indices and field-collected composite burn index data in a ponderosa pine forest. *International Journal of Remote Sensing, 32*(23), 7905–7927. https://doi.org/10.1080/01431161.2010.524678

Chen, M., Fahrner, D., Arribas-Bel, D., & Rowe, F. (2020). A reproducible notebook to acquire, process and analyse satellite imagery: Exploring long-term urban changes. *The Region, 7* (2), R15–R46.

Cheng, T., & Adepeju, M. (2014). Modifiable temporal unit problem (MTUP) and its effect on space-time cluster detection. *PloS One, 9*(6), Article e100465.

Chu, T., & Guo, X. (2013). Remote sensing techniques in monitoring post-fire effects and patterns of forest recovery in boreal forest regions: A review. *Remote Sensing, 6*(1), 470–520.

Cushman, S. A., & McGarigal, K. (2019). Metrics and models for quantifying ecological resilience at landscape scales. *Frontiers in Ecology and Evolution, 7*, 440.

Cushman, S. A., McGarigal, K., McKelvey, K. S., Vojta, C. D., & Regan, C. M. (2013). Landscape analysis for habitat monitoring. Tech. In M. M. Rowland, & C. D. Vojta (Eds.), *A technical guide for monitoring wildlife habitat. Gen. Tech. Rep. WO-89*. Washington, DC: US Department of Agriculture, Forest Service. 18p., 6–1.

Danson, F. M., & Plummer, S. E. (1995). Red-edge response to forest leaf area index. *International Journal of Remote Sensing, 16*(1), 183–188. https://doi.org/10.1080/01431169508954387

Das Choudhury, S., Saha, S., Samal, A., Mazis, A., & Awada, T. (2023). Drought stress prediction and propagation using time series modeling on multimodal plant image sequences. *Frontiers in Plant Science, 14*, Article 1003150.

de Jong, R., de Bruin, S., de Wit, A., Schaepman, M. E., & Dent, D. L. (2011). Analysis of monotonic greening andbrowning trends from global NDVI time-series. *Remote Sensing of Environment, 115*, 692–702.

de Jong, R., Verbesselt, J., Schaepman, M. E., & De Bruin, S. (2012). Trend changes in global greening and browning: Contribution of short-term trends to longer-term change. *Global Change Biology, 18*, 642–655.

Dela Torre, D. M. G., Gao, J., & Macinnis-Ng, C. (2021). Remote sensing-based estimation of rice yields using various models: A critical review. *Geo-Spatial Information Science, 24* (4), 580–603. https://doi.org/10.1080/10095020.2021.1936656

Dianat, R., & Kasaei, S. (2009). Change detection in optical remote sensing images using difference-based methods and spatial information. *IEEE Geoscience and Remote Sensing Letters, 7*(1), 215–219.

Ding, Y., Zhao, K., Zheng, X., & Jiang, T. (2014). Temporal dynamics of spatial heterogeneity over cropland quantified by time-series NDVI, near infrared and red reflectance of landsat 8 OLI imagery. *International Journal of Applied Earth Observation and Geoinformation, 30*, 139–145.

Dusseux, P., Vertès, F., Corpetti, T., Corgne, S., & Hubert-Moy, L. (2014). Agricultural practices in grasslands detected by spatial remote sensing. *Environmental Monitoring and Assessment, 186*(12), 8249–8265. https://doi.org/10.1007/s10661-014-4001-5

Escuin, S., Navarro, R., & Fernández, P. (2008). Fire severity assessment by using NBR (normalized burn ratio) and NDVI (normalized difference vegetation index) derived from LANDSAT TM/ETM images. *International Journal of Remote Sensing, 29*(4), 1053–1073. https://doi.org/10.1080/01431160701281072

Fan, J., Zhang, K., Huang, Y., Zhu, Y., & Chen, B. (2023). Parallel spatio-temporal attention-based TCN for multivariate time series prediction. *Neural Computing & Applications, 35* (18), 13109–13118. https://doi.org/10.1007/s00521-021-05958-z

Fang, X., Zhu, Q., Ren, L., Chen, H., Wang, K., & Peng, C. (2018). Large-scale detection of vegetation dynamics and their potential drivers using MODIS images and BFAST: A case study in Quebec, Canada. *Remote Sensing of Environment, 206*, 391–402.

Fernández-Guisuraga, J. M., Suárez-Seoane, S., & Calvo, L. (2023). Radar and multispectral remote sensing data accurately estimate vegetation vertical structure diversity as a fire resilience indicator. *Remote Sensing in Ecology and Conservation, 9*(1), 117–132. https://doi.org/10.1002/rse2.299

Fotheringham, A. S., Brunsdon, C., & Charlton, M. (2009). Geographically weighted regression. *The Sage Handbook of Spatial Analysis, 1*, 243–254.

Fu, Y., Zhu, Z., Liu, L., Zhan, W., He, T., Shen, H., Zhao, J., Liu, Y., Zhang, H., Liu, Z., Xue, Y., & Ao, Z. (2024). Remote sensing time series analysis: A review of data and applications. *Journal of Remote Sensing, 4*, 0285. https://doi.org/10.34133/remotesensing.0285

Gao, F., & Zhang, X. (2021). Mapping crop phenology in near real-time using satellite remote sensing: Challenges and opportunities. *Journal of Remote Sensing*, 2021–8379391. https://doi.org/10.34133/2021/8379391

Gao, B.-C. (1996). NDWI—A normalized difference water index for remote sensing of vegetation liquid water from space. *Remote Sensing of Environment, 58*(3), 257–266.

García, M. J. L., & Caselles, V. (1991). Mapping burns and natural reforestation using thematic mapper data. *Geocarto International, 6*(1), 31–37. https://doi.org/10.1080/10106049109354290

Garonna, I., de Jong, R., & Schaepman, M. E. (2016). Variability and evolution of global land surface phenology over the past three decades (1982–2012). *Global Change Biology, 22*, 1456–1468.

Geary, R. C. (1954). The contiguity ratio and statistical mapping. *The Incorporated Statistician, 5*(3), 115–146.

Getis, A., & Ord, J. K. (1992). The analysis of spatial association by use of distance statistics. *Geographical Analysis, 24*(3), 189–206. https://doi.org/10.1111/j.1538-4632.1992.tb00261.x

Getis, A. (1996). *Local spatial statistics: An overview. Spatial analysis: Modelling in a GIS environment.*

Gibbes, C., Southworth, J., & Fogel, C. (2014). Using time series analysis to examine thresholds in vegetation response to precipitation variability in Southern Africa. *Applied Geography, 53*, 13–26. https://doi.org/10.1016/j.apgeog.2014.05.001

Gitelson, A. A., Kaufman, Y. J., & Merzlyak, M. N. (1996). Use of a green channel in remote sensing of global vegetation from EOS-MODIS. *Remote Sensing of Environment, 58*(3), 289–298. https://doi.org/10.1016/S0034-4257(96)00072-7

Gitelson, A. A., Stark, R., Grits, U., Rundquist, D., Kaufman, Y., & Derry, D. (2002). Vegetation and soil lines in visible spectral space: A concept and technique for remote estimation of vegetation fraction. *International Journal of Remote Sensing, 23*(13), 2537–2562. https://doi.org/10.1080/01431160110107806

Gómez, C., White, J. C., Wulder, M. A., & Alejandro, P. (2014). Historical forest biomass dynamics modelled with landsat spectral trajectories. *ISPRS Journal of Photogrammetry and Remote Sensing, 93*, 14–28.

Goodchild, M. F., Yuan, M., & Cova, T. J. (2007). Towards a general theory of geographic representation in GIS. *International Journal of Geographical Information Science, 21*(3), 239–260. https://doi.org/10.1080/13658810600965271

Gorelick, N., Hancher, M., Dixon, M., Ilyushchenko, S., Thau, D., & Moore, R. (2017). Google Earth engine: Planetary-scale geospatial analysis for everyone. *Remote Sensing of Environment, 202*, 18–27.

Gu, Y., Brown, J. F., Verdin, J. P., & Wardlow, B. (2007). A five-year analysis of MODIS NDVI and NDWI for grassland drought assessment over the central great plains of the United States. *Geophysical Research Letters, 34*(6), Article 2006GL029127. https://doi.org/10.1029/2006GL029127

Gu, Y., Hunt, E., Wardlow, B., Basara, J. B., Brown, J. F., & Verdin, J. P. (2008). Evaluation of MODIS NDVI and NDWI for vegetation drought monitoring using Oklahoma mesonet soil moisture data. *Geophysical Research Letters, 35*(22), Article 2008GL035772. https://doi.org/10.1029/2008GL035772

Gu, Y., Wylie, B. K., Howard, D. M., Phuyal, K. P., & Ji, L. (2013). NDVI saturation adjustment: A new approach for improving cropland performance estimates in the greater platte river basin, USA. *Ecological Indicators, 30*, 1–6.

Guay, K. C., Beck, P. S., Berner, L. T., Goetz, S. J., Baccini, A., & Buermann, W. (2014). Vegetation productivity patterns at high northern latitudes: A multi-sensor satellite data assessment. *Global Change Biology, 20*, 3147–3158.

Hemati, M., Hasanlou, M., Mahdianpari, M., & Mohammadimanesh, F. (2021). A systematic review of landsat data for change detection applications: 50 years of monitoring the Earth. *Remote Sensing, 13*(15), 2869.

Hermance, J. F., Jacob, R. W., Bradley, B. A., & Mustard, J. F. (2007). Extracting phenological signals from multiyear AVHRR NDVI time series: Framework for applying high-order annual splines with roughness damping. *IEEE Transactions on Geoscience and Remote Sensing, 45*(10), 3264–3276.

Herrero, H., Southworth, J., Muir, C., Khatami, R., Bunting, E., & Child, B. (2020). An evaluation of vegetation health in and around Southern African National parks during the 21st century (2000–2016). *Applied Sciences, 10*(7), 2366.

Herrero, H. V., Southworth, J., Khatami, R., Insalaco, S., & Muir, C. (2023). Examining the relationship between vegetation decline and precipitation in the national parks of the greater Limpopo transfrontier conservation area during the 21st century. *Frontiers in Environmental Science, 11*, Article 1106849.

Holling, C. S. (1973). Resilience and stability of ecological systems. *Annual Review of Ecology and Systematics, 4*, 1–23.

Holloway, J., & Mengersen, K. (2018). Statistical machine learning methods and remote sensing for sustainable development goals: A review. *Remote Sensing, 10*(9), 1365.

Huete, A. R., Liu, H. Q., Batchily, K., & van Leeuwen, W. (1997). A comparison of vegetation indices over a global set of TM images for EOS-MODIS. *Remote Sensing of Environment, 59*, 440–451.

Huete, A., Didan, K., Miura, T., Rodriguez, E. P., Gao, X., & Ferreira, L. G. (2002). Overview of the radiometric and biophysical performance of the MODIS vegetation indices. *Remote Sensing of Environment, 83*(1–2), 195–213. https://doi.org/10.1016/S0034-4257(02)00096-2

Huete, A. R. (1988). A soil-adjusted vegetation index (SAVI). *Remote Sensing of Environment, 25*(3), 295–309.

Hui, J., & Cheng, Y. (2026). Detecting disturbance and recovery in mining landscapes: A novel time-series framework based on improved LandTrendr and machine learning. *Environmental Research*, Article 123868.

Hüttich, C., Gessner, U., Herold, M., Strohbach, B. J., Schmidt, M., Keil, M., & Dech, S. (2009). On the suitability of MODIS time series metrics to map vegetation types in dry savanna ecosystems: A case study in the kalahari of NE Namibia. *Remote Sensing, 1*(4), 620–643. https://doi.org/10.3390/rs1040620

Jakubauskas, M., Mark, E., Legates, D. R., & Kastens, J. H. (2001). Harmonic analysis of time series AVHRR NDVI data. *Photogrammetric Engineering & Remote Sensing, 67*(4), 461–470.

Jensen, J. R. (1996). *Introductory digital image processing: A remote sensing perspective* (Ed.2). CABDirect2. xi + 318 pp.

Jensen, J. R. (2009). *Remote sensing of the environment: An Earth resource perspective.* Pearson Education.

Jeyavathana, R. B. (2022). Land use and land cover classification using landsat-8 multispectral remote sensing images and long short-term memory-recurrent neural network. *AIP Conference Proceedings, 2452*(1). https://pubs.aip.org/aip/acp/article-abstract/2452/1/070001/2824539.

Jiang, Z., Huete, A. R., Chen, J., Chen, Y., Li, J., Yan, G., & Zhang, X. (2006). Analysis of NDVI and scaled difference vegetation index retrievals of vegetation fraction. *Remote Sensing of Environment, 101*, 366–378.

Jianya, G., Haigang, S., Guorui, M., & Qiming, Z. (2008). A review of multi-temporal remote sensing data change detection algorithms. *The International Archives of the Photogrammetry, Remote Sensing and Spatial Information Sciences, 37*(B7), 757–762.

Ju, J., & Roy, D. P. (2008). The availability of cloud-free Landsat ETM+ data over the conterminous United States and globally. *Remote Sensing of Environment, 112*(3), 1196–1211.

Kaufman, Y. J., & Tanre, D. (1992). Atmospherically resistant vegetation index (ARVI) for EOS-MODIS. *IEEE Transactions on Geoscience and Remote Sensing, 30*(2), 261–270.

Kazanskiy, N., Khabibullin, R., Nikonorov, A., & Khonina, S. (2025). A comprehensive review of remote sensing and artificial intelligence integration: Advances, applications, and challenges. *Sensors, 25*(19), 5965.

Kooistra, L., Berger, K., Brede, B., Graf, L. V., Aasen, H., Roujean, J.-L., Machwitz, M., Schlerf, M., Atzberger, C., & Prikaziuk, E. (2023). Reviews and syntheses: Remotely sensed optical time series for monitoring vegetation productivity. *Biogeosciences Discussions, 2023*, 1–67.

Kuenzer, C., Dech, S., & Wagner, W. (2015). Remote sensing time series revealing land surface dynamics: Status quo and the pathway ahead. In C. Kuenzer, S. Dech, & W. Wagner (Eds.), *Remote sensing time series, 22* (pp. 1–24). Springer International Publishing. https://doi.org/10.1007/978-3-319-15967-6_1

Kulldorff, M. (1997). A spatial scan statistic. *Communications in Statistics - Theory and Methods, 26*(6), 1481–1496. https://doi.org/10.1080/03610929708831995

Kumar, L., & Mutanga, O. (2018). Google Earth engine applications since inception: Usage, trends, and potential. *Remote Sensing, 10*(10), 1509.

Lambin, E. F., & Ehrlich, D. (1997). Land-cover changes in Sub-Saharan Africa (1982–1991): Application of a change index based on remotely sensed surface temperature and vegetation indices at a continental scale. *Remote Sensing of Environment, 61*, 181–200.

Langley, S. K., Cheshire, H. M., & Humes, K. S. (2001). A comparison of single date and multitemporal satellite image classifications in a semi-arid grassland. *Journal of Arid Environments, 49*(2), 401–411.

Lehmann, C. E. R., Anderson, T. M., Sankaran, M., Higgins, S. I., Archibald, S., Ho mann, W. A., Hanan, N. P., Williams, R. J., Fensham, R. J., Felfili, J., et al. (2014). Savanna vegetation-fire-climate relationships differ among continents. *Science, 343*, 548–552.

Li, C., Zhang, Z., Liu, L., Kim, J. Y., & Sangaiah, A. K. (2023). A novel deep multi-instance convolutional neural network for disaster classification from high-resolution remote sensing images. *Ieee Journal of Selected Topics in Applied Earth Observations and Remote Sensing, 17*, 2098–2114.

Li, J., Cai, Y., Li, Q., Kou, M., & Zhang, T. (2024). A review of remote sensing image segmentation by deep learning methods. *International Journal of Digital Earth, 17*(1), Article 2328827. https://doi.org/10.1080/17538947.2024.2328827

Lillesand, T., Kiefer, R. W., & Chipman, J. (2015). *Remote sensing and image interpretation.* John Wiley & Sons. https://books.google.com/books?hl=en&lr=&id=AFHDCAAAQBAJ&oi=fnd&pg=PR12&dq=Remote+sensing,+lillis and&ots=0Eulf3dA9s&sig=2H3u00C7NCHQPbVKrRjciRYF9_o.

Longley, P. A., Goodchild, M. F., Maguire, D. J., & Rhind, D. W. (2015). *Geographic information science and systems.* John Wiley & Sons. https://books.google.com/books?hl=en&lr=&id=C_EwBgAAQBAJ&oi=fnd&pg=PR10&dq=Geographic+Information+Science+and+Systems,+4th+Edition+Paul+A.+Longley,+Michael+F.+Goodchild,+David+J.+Maguire,+David+W.+Rhind+&ots=r7NwWv21QG&sig=QWB8K1IOst5KZ7KmCU54DTeK_iQ.

Meyer, H., Reudenbach, C., Wöllauer, S., & Nauss, T. (2019). Importance of spatial predictor variable selection in machine learning applications–moving from data reproduction to spatial prediction. *Ecological Modelling, 411*, Article 108815.

Miller, J. D., & Thode, A. E. (2007). Quantifying burn severity in a heterogeneous landscape with a relative version of the Delta normalized Burn Ratio (dNBR). *Remote Sensing of Environment, 109*(1), 66–80.

Mitchard, E. T., Saatchi, S. S., Gerard, F. F., Lewis, S. L., & Meir, P. (2009). Measuring woody encroachment along a forest–sSavanna boundary in Central Africa. *Earth Interactions, 13*, 1–29.

Moran, P. A. (1950). Notes on continuous stochastic phenomena. *Biometrika, 37*(1/2), 17–23.

Munawar, H. S., Ullah, F., Qayyum, S., Khan, S. I., & Mojtahedi, M. (2021). UAVs in disaster management: Application of integrated aerial imagery and convolutional neural network for flood detection. *Sustainability, 13*(14), 7547.

Munson, S. M., Bunting, E. L., Bradford, J. B., Butterfield, B. J., & Gremer, J. R. (2019). Plant production responses to precipitation differ along an elevation gradient and are enhanced under extremes. *Ecosystems, 22*(4), 699–708. https://doi.org/10.1007/s10021-018-0296-3

Nejadrekabi, M., Eslamian, S., & Zareian, M. J. (2022). Spatial statistics techniques for SPEI and NDVI drought indices: A case study of Khuzestan Province. *International Journal of Environmental Science and Technology, 19*(7), 6573–6594. https://doi.org/10.1007/s13762-021-03852-8

Nicholson, S. E. (2012). LONG-TERM changes in african rainfall. *Weather, 44*(2), 46–56.

Olsson, L., Eklundh, L., & Ardö, J. (2005). A recent greening of the Sahel—Trends, patterns and potential causes. *Journal of Arid Environments, 63*, 556–566.

Openshaw, S. (1984). The modifiable areal unit problem. *Concepts and Techniques in Modern Geography*. https://cir.nii.ac.jp/crid/1570291225725496704.

Osibo, B. K., Ma, T., Bediako-Kyeremeh, B., Mamelona, L., & Darbinian, K. (2024). TCNT: A temporal convolutional network-transformer framework for advanced crop yield prediction. *Journal of Applied Remote Sensing, 18*(4), 044513–044513.

Petri, C. A., & Galvão, L. S. (2019). Sensitivity of seven MODIS vegetation indices to BRDF effects during the Amazonian dry season. *Remote Sensing, 11*(14), 1650.

Piao, S., Wang, X., Ciais, P., Zhu, B., Wang, T. A. O., & Liu, J. I. E. (2011). Changes in satellite-derived vegetation growth trend in temperate and boreal Eurasia from 1982 to 2006. *Global Change Biology, 17*, 3228–3239.

Pinty, B., & Verstraete, M. M. (1992). GEMI: A non-linear index to monitor global vegetation from satellites. *Vegetatio, 101*(1), 15–20. https://doi.org/10.1007/BF00031911

Qi, J., Chehbouni, A., Huete, A. R., Kerr, Y. H., & Sorooshian, S. (1994). A modified soil adjusted vegetation index. *Remote Sensing of Environment, 48*(2), 119–126.

Ren-hua, Z., Rao, N. X., & Liao, K. N. (1996). Approach for a vegetation index resistant to atmospheric effect. *Journal of Integrative Plant Biology, 38*(1). https://www.jipb.net/EN/abstract/abstract23925.shtml.

Ripley, B. D. (1988). *Statistical inference for spatial processes*. Cambridge [England]; New York: Cambridge University Press (with Internet Archive) http://archive.org/details/statisticalinfer0000ripl.

Rodigheri, G., Sanches, I. D., Richetti, J., Tsukahara, R. Y., Lawes, R., Bendini, H. do N., & Adami, M. (2023). Estimating crop sowing and harvesting dates using satellite vegetation index: A comparative analysis. *Remote Sensing, 15*(22), 5366.

Rondeaux, G., Steven, M., & Baret, F. (1996). Optimization of soil-adjusted vegetation indices. *Remote Sensing of Environment, 55*(2), 95–107. https://doi.org/10.1016/0034-4257(95)00186-7

Rouse, J. W., Haas, R. H., Schell, J. A., & Deering, D. W. (1974). Monitoring vegetation systems in the great plains with ERTS. *NASA Special Publication, 251*(309). https://scholar.google.com/scholar_lookup?title=Monitoring+vegetation+systems+in+the+Great+Plains+with+ERTS&author=Rouse,+J.W.&author=Haas,+R.H.&author=Schell,+J.A.&author=Deering,+D.W.&publication_year=1974&journal=NASA+Spec.+Publ.&volume=351&pages=309.

Rouse, J. W. (1973). *Monitoring the vernal advancement and retrogradation (green wave effect) of natural vegetation*. https://ntrs.nasa.gov/search.jsp?R=19730016613.

Running, S. W., Nemani, R. R., Heinsch, F. A., Zhao, M., Reeves, M., & Hashimoto, H. (2004). A continuous satellite-derived measure of global terrestrial primary production. *BioScience, 54*(6), 547–560.

Saadat, H., Adamowski, J., Bonnell, R., Sharifi, F., Namdar, M., & Ale-Ebrahim, S. (2011). Land use and land cover classification over a large area in Iran based on single date analysis of satellite imagery. *ISPRS Journal of Photogrammetry and Remote Sensing, 66* (5), 608–619.

Saberi, S. J., & Harvey, B. J. (2023). What is the color when Black is burned? Quantifying (re) burn severity using field and satellite remote sensing indices. *Fire Ecology, 19*(1), 24. https://doi.org/10.1186/s42408-023-00178-3

Sankaran, M., Hanan, N. P., Scholes, R. J., Ratnam, J., Augustine, D. J., Cade, B. S., Gignoux, J., Higgins, S. I., Roux, X. L., Ludwig, F., Ardo, J., Banyikwa, F., Bronn, A., Bucini, G., Caylor, K. K., Coughenour, M. B., Diouf, A., Ekaya, W., Feral, C. J., … Zambatis, N. (2005). Determinants of woody cover in African savannas. *Nature, 438* (7069), 846–849. https://doi.org/10.1038/nature04070

Schellenberg, K., Jagdhuber, T., Zehner, M., Hese, S., Urban, M., Urbazaev, M., Hartmann, H., Schmullius, C., & Dubois, C. (2023). Potential of Sentinel-1 SAR to assess damage in drought-affected temperate deciduous broadleaf forests. *Remote Sensing, 15*(4), 1004.

Schimel, D., Pavlick, R., Fisher, J. B., Asner, G. P., Saatchi, S., Townsend, P., Miller, C., Frankenberg, C., Hibbard, K., & Cox, P. (2015). Observing terrestrial ecosystems and the carbon cycle from space. *Global Change Biology, 21*, 1762–1776.

Shammi, S. A., & Meng, Q. (2021). Use time series NDVI and EVI to develop dynamic crop growth metrics for yield modeling. *Ecological Indicators, 121*, Article 107124.

Sharifi, A., & Felegari, S. (2023). Remotely sensed normalized difference red-edge index for rangeland biomass estimation. *Aircraft Engineering & Aerospace Technology, 95*(7), 1128–1136.

Sjöström, M., Ardö, J., Eklundh, L., El-Tahir, B. A., El-Khidir, H. A. M., Hellström, M., Pilesjö, P., & Seaquist, J. (2009). Evaluation of satellite based indices for gross primary production estimates in a sparse savanna in the Sudan. *Biogeosciences, 6*(1), 129–138.

Song, J., Gao, S., Zhu, Y., & Ma, C. (2019). A survey of remote sensing image classification based on CNNs. *Big Earth Data, 3*(3), 232–254. https://doi.org/10.1080/20964471.2019.1657720

Southworth, J., Zhu, L., Bunting, E., Ryan, S. J., Herrero, H., Waylen, P. R., & Hill, M. J. (2016). Changes in vegetation persistence across global savanna landscapes, 1982–2010. *Journal of Land Use Science, 11*, 7–32.

Southworth, J., Bunting, E., Zhu, L., Ryan, S. J., Herrero, H. V., Waylen, P., Muñoz-Carpena, R., Campo-Bescós, M. A., & Kaplan, D. (2018). Using a coupled dynamic factor – Random forest analysis (DFRFA) to reveal drivers of spatiotemporal heterogeneity in the semi-arid regions of Southern Africa. *PLoS One, 13*(12), Article e0208400. https://doi.org/10.1371/journal.pone.0208400

Southworth, J., Ryan, S. J., Herrero, H. V., Khatami, R., Bunting, E. L., Hassan, M., Muir, C. S., & Waylen, P. (2023). Latitudes and land use: Global biome shifts in vegetation persistence across three decades. *Frontiers in Remote Sensing, 4*. https://doi.org/10.3389/frsen.2023.1063188

Staver, A. C., Archibald, S., & Levin, S. A. (2011). Tree cover in Sub-Saharan Africa: Rainfall and fire constrain forest and savanna as alternative stable states. *Ecology, 92*(5), 1063–1072.

Storey, E. A., Stow, D. A., & O'Leary, J. F. (2016). Assessing postfire recovery of chamise chaparral using multi-temporal spectral vegetation index trajectories derived from landsat imagery. *Remote Sensing of Environment, 183*, 53–64.

Thoma, D. P., Munson, S. M., Irvine, K. M., Witwicki, D. L., & Bunting, E. L. (2016). Semi-arid vegetation response to antecedent climate and water balance windows. *Applied Vegetation Science, 19*(3), 413–429. https://doi.org/10.1111/avsc.12232

Thoma, D. P., Munson, S. M., & Witwicki, D. L. (2019). Landscape pivot points and responses to water balance in national parks of the southwest US. *Journal of Applied Ecology, 56*(1), 157–167. https://doi.org/10.1111/1365-2664.13250

Tian, Z., Fan, J., Yu, T., de Leon, N., Kaeppler, S. M., & Zhang, Z. (2025). Mitigating NDVI saturation in imagery of dense and healthy vegetation. *ISPRS Journal of Photogrammetry and Remote Sensing, 227*, 234–250.

Townshend, J. R. G., & Justice, C. O. (1986). Analysis of the dynamics of African vegetation using the normalized difference vegetation index. *International Journal of Remote Sensing, 7*(11), 1435–1445. https://doi.org/10.1080/01431168608948946

Tucker, C. J., Pinzon, J. E., Brown, M. E., Slayback, D. A., Pak, E. W., Mahoney, R., Vermote, E. F., & El Saleous, N. (2005). An extended AVHRR 8-km NDVI dataset compatible with MODIS and SPOT vegetation NDVI data. *International Journal of Remote Sensing, 26*(20), 4485–4498. https://doi.org/10.1080/01431160500168686

Tucker, C. J. (1979). Red and photographic infrared linear combinations for monitoring vegetation. *Remote Sensing of Environment, 8*(2), 127–150. https://doi.org/10.1016/0034-4257(79)90013-0

Tucker, C. J. (1980). Remote sensing of leaf water content in the near infrared. *Remote Sensing of Environment, 10*(1), 23–32. https://doi.org/10.1016/0034-4257(80)90096-6

Wang, J., Rich, P. M., Price, K. P., & Kettle, W. D. (2005). Relations between NDVI, grassland production, and crop yield in the central great plains. *Geocarto International, 20*, 5–11.

Wang, L., Yan, J., Mu, L., & Huang, L. (2020). Knowledge discovery from remote sensing images: A review. *WIREs Data Mining and Knowledge Discovery, 10*(5), Article e1371. https://doi.org/10.1002/widm.1371

Watts, L. M., & Laffan, S. W. (2014). Effectiveness of the BFAST algorithm for detecting vegetation response patterns in a semi-arid region. *Remote Sensing of Environment, 154*, 234–245.

Waylen, P., Southworth, J., Gibbes, C., & Tsai, H. (2014). Time series analysis of land cover change: Developing statistical tools to determine significance of land cover changes in persistence analyses. *Remote Sensing, 6*, 4473–4497.

Wilson, B. T., Knight, J. F., & McRoberts, R. E. (2018). Harmonic regression of landsat time series for modeling attributes from national forest inventory data. *ISPRS Journal of Photogrammetry and Remote Sensing, 137*, 29–46.

Wulder, M. A., White, J. C., Goward, S. N., Masek, J. G., Irons, J. R., Herold, M., Cohen, W. B., Loveland, T. R., & Woodcock, C. E. (2008). Landsat continuity: Issues and opportunities for land cover monitoring. *Remote Sensing of Environment, 112*(3), 955–969.

Wulder, M. A., Masek, J. G., Cohen, W. B., Loveland, T. R., & Woodcock, C. E. (2012). Opening the archive: How free data has enabled the science and monitoring promise of landsat. *Remote Sensing of Environment, 122*, 2–10.

Xie, Q., Dash, J., Huang, W., Peng, D., Qin, Q., Mortimer, H., Casa, R., Pignatti, S., Laneve, G., & Pascucci, S. (2018). Vegetation indices combining the red and red-edge spectral information for leaf area index retrieval. *IEEE Journal of Selected Topics in Applied Earth Observations and Remote Sensing, 11*(5), 1482–1493.

Xu, D., & Guo, X. (2014). Compare NDVI extracted from landsat 8 imagery with that from landsat 7 imagery. *American Journal of Remote Sensing, 2*(2), 10–14.

Xu, L., Zhang, S., He, Z., & Guo, Y. (2009). The comparative study of three methods of remote sensing image change detection. In *2009 17th international conference on geoinformatics* (pp. 1–4). https://ieeexplore.ieee.org/abstract/document/5293490/.

Xue, J., & Su, B. (2017). Significant remote sensing vegetation indices: A review of developments and applications. *Journal of Sensors, 2017*, 1–17. https://doi.org/10.1155/2017/1353691

Zhang, K., Kimball, J. S., Nemani, R. R., Running, S. W., Hong, Y., Gourley, J. J., & Yu, Z. (2015). Vegetation greening and climate change promote multidecadal rises of global land evapotranspiration. *Scientific Reports, 5*, Article 15956.

Zhao, H., Chang, C., Wang, Z., & Zhao, G. (2025). A large-scale agricultural land classification method based on synergistic integration of time series red-edge vegetation index and phenological features. *Sensors, 25*(2), 503.

Zheng, Z., Yu, Y., Yang, X., Yuan, X., & Hou, Z. (2025). A novel framework for long-term forest disturbance monitoring: Synergizing the LandTrendr algorithm with CNN in northeast China. *Remote Sensing, 17*(21), 3521.

Zhou, Q., Zhu, Z., Xian, G., & Li, C. (2022). A novel regression method for harmonic analysis of time series. *ISPRS Journal of Photogrammetry and Remote Sensing, 185*, 48–61.

Zhu, Z., & Woodcock, C. E. (2014). Continuous change detection and classification of land cover using all available landsat data. *Remote Sensing of Environment, 144*, 152–171. https://doi.org/10.1016/j.rse.2014.01.011

Zhu, Z., Piao, S., Myneni, R. B., Huang, M., Zeng, Z., Canadell, J. G., Ciais, P., Sitch, S., Friedlingstein, P., Arneth, A., et al. (2016). Greening of the Earth and its drivers. *Nature Climate Change, 6*, 791–795.

Zhu, X. X., Tuia, D., Mou, L., Xia, G.-S., Zhang, L., Xu, F., & Fraundorfer, F. (2017). Deep learning in remote sensing: A comprehensive review and list of resources. *IEEE Geoscience and Remote Sensing Magazine, 5*(4), 8–36.

GEOAI Methods for Land Use and Land Cover Classification: Machine Learning and Deep Learning Approaches

5

Jane Southworth [1], *Mohammad Safaei* [1], *Mashoukur Rahaman* [1], *Hannah Herrero* [2], *Audrey Culver Smith* [1], *Bewuket B. Tefera* [1], *Carly S. Muir* [3] *and Ali R. Alruzuq* [4]
[1]Department of Geography, University of Florida, Gainesville, FL, United States; [2]Department of Geography & Sustainability, University of Tennessee, Knoxville, TN, United States; [3]School of Life Sciences & Sustainability, Virginia Commonwealth University, Richmond, VA, United States; [4]Department of Geography and Geographic Information Systems, Imam University, Riyadh, Saudi Arabia

Key terms and definitions

Artificial Intelligence (AI): A broad class of computational techniques designed to perform tasks that typically require human intelligence, including pattern recognition, classification, and decision-making, widely applied in remote sensing for automated image analysis.
Autoencoders: Unsupervised neural network architectures that learn compact latent representations of high-dimensional data by minimizing reconstruction error, commonly used for feature extraction, dimensionality reduction, and anomaly detection in remote sensing imagery.
Classification: The proportion of correctly classified observations relative to the total number of observations, commonly used as a primary evaluation metric for land use and land cover (LULC) classification models.
Convolutional Neural Networks (CNNs): A class of deep learning models specifically designed for image analysis that use convolutional filters to learn spatial hierarchies of features, widely applied to land cover classification and segmentation tasks.
Deep Learning (DL): A subset of machine learning based on multi-layer neural networks that automatically learn hierarchical feature representations from raw data, particularly effective for high-dimensional and complex remote sensing datasets.
Decision Tree (DT): A supervised machine learning algorithm that classifies data through a series of rule-based splits, valued for interpretability and low computational requirements in land cover mapping applications.
Evaluation Metrics: Quantitative measures such as accuracy, precision, recall, F1-score, and confusion matrices used to assess and compare the performance of machine learning and deep learning models.

Remote Sensing, Big Data, and GeoAI. https://doi.org/10.1016/B978-0-443-26741-3.00011-8

Earth Observation (EO) Data: Data collected by satellite or airborne sensors that provide repeated measurements of the Earth's surface, forming the foundation for large-scale land system and land cover analyses.
F1-Score: A harmonic mean of precision and recall that provides a balanced assessment of classification performance, particularly useful for evaluating models with imbalanced land cover classes.
Generative Adversarial Networks (GANs): A deep learning framework consisting of a generator and discriminator trained adversarially, used in remote sensing for data augmentation, image enhancement, and addressing class imbalance.
Hyperspectral Imagery: Remote sensing data characterized by hundreds of narrow spectral bands, enabling detailed discrimination of land cover types but requiring advanced ML or DL methods due to high dimensionality.
Interpretability: The degree to which a model's internal logic and decision-making process can be understood by humans, often higher in traditional ML models than in complex DL architectures.
Land Change Science (LCS): An interdisciplinary field focused on understanding the drivers, patterns, and consequences of land use and land cover change, closely aligned with Land System Science.
Land System Science (LSS): A research framework that examines land systems as coupled human–environment systems, integrating social, ecological, and technological processes across spatial and temporal scales.
Land Use and Land Cover (LULC): A classification framework distinguishing between human land use activities (e.g., agriculture, urbanization) and biophysical land cover types (e.g., forest, grassland, water).
Land Use and Land Cover Change (LULCC): The dynamic transformation of land systems over time resulting from natural processes and human activities, a central focus of land system science research.
Machine Learning (ML): A subset of AI that enables algorithms to learn patterns from data without explicit programming, commonly applied to LULC classification using labeled training datasets.
Multispectral Imagery: Remote sensing data composed of a limited number of broad spectral bands, widely used in land cover mapping due to its availability and compatibility with traditional ML approaches.
Object-Based Image Analysis (OBIA): A classification approach that groups pixels into homogeneous objects prior to classification, contrasting with pixel-based methods and often applied in high-resolution imagery.
Overfitting: A modeling issue where a classifier performs well on training data but poorly on unseen data, commonly addressed through cross-validation, regularization, and careful feature selection.
Pixel-Based Classification: A classification method that assigns land cover labels to individual pixels based on their spectral and ancillary attributes, frequently used with medium-resolution satellite imagery.
Random Forest (RF): An ensemble machine learning algorithm that combines multiple decision trees to improve classification accuracy and robustness, widely used in remote sensing due to strong performance and interpretability.

Recurrent Neural Networks (RNNs): Deep learning architectures designed to process sequential data, including time-series remote sensing imagery, enabling modeling of temporal land cover dynamics.
Remote Sensing (RS): The science of acquiring information about the Earth's surface without direct contact, using satellite or airborne sensors to monitor land systems over space and time.
Synthetic Aperture Radar (SAR): An active remote sensing technology that uses microwave signals to collect data independent of cloud cover and illumination conditions, valuable for land cover and environmental monitoring.
Training Data: Labeled reference data used to train supervised ML and DL models, with availability and quality strongly influencing classification performance.
U-Net: A convolutional neural network architecture designed for image segmentation that enables pixel-level land cover classification through an encoder–decoder structure.
Vision Transformers (ViTs): Deep learning architectures based on self-attention mechanisms that model long-range spatial relationships in imagery, increasingly applied in high-resolution remote sensing tasks.

1. Background

The integration of Artificial Intelligence (AI), particularly Machine Learning (ML) and Deep Learning (DL), with remotely sensed imagery has reshaped the field of land system science (LSS) and land use and land cover classification (LULC). Historically, traditional classification methods such as Maximum Likelihood Classification and k-Nearest Neighbors dominated LULC tasks, relying on manual feature engineering and often struggling to process the complexity of remotely sensed data. The rise of AI in the 21st Century introduced new techniques that could manage large, diverse datasets with higher accuracy and efficiency (Karpatne et al., 2016; Lambin et al., 2001). The advancements in sensor technology, coupled with the growing availability of Earth Observation data, have further driven the shift toward AI-powered methods (Rindfuss et al., 2004; Verburg et al., 2013).

In today's world, the need for precise and scalable land classification is critical due to rapid environmental changes such as deforestation, urban sprawl, and agricultural expansion. Remote sensing provides vital insights into these changes, while AI helps process the increasing volume of data and improve classification accuracy (Turner et al., 2021). AI-powered methods, particularly ML and DL, enable researchers to handle large-scale datasets from satellites, aerial imagery, and ground sensors, making it possible to monitor land changes with greater detail and frequency than ever before (Hermosilla et al., 2022; Liu, 2015).

1.1 Objective

The goal of this chapter is to examine the application of AI, particularly ML and DL techniques, in the classification of remotely sensed (RS) data for LSS. It aims to provide an overview of how these technologies have enhanced the ability to classify land cover types and track land use changes over time. By focusing on both the technical challenges and practical implementations of AI in remote sensing, the chapter will outline the key issues related to data availability, computational resources, and model interpretability. This chapter addresses several geospatial challenges, including how to improve the accuracy of land cover classifications, manage the vast amounts of remotely sensed data, and ensure that AI models can adapt to various geospatial contexts. These challenges are increasingly important in a world where environmental changes must be monitored closely to support sustainable land management and policy-making efforts.

2. Theoretical foundations

2.1 Concept 1: Land system science (LSS)

LSS, also known as Land Change Science (LCS), is an interdisciplinary field that examines the complex interactions between human and environmental systems, with a focus on understanding how human activities transform terrestrial ecosystems. At its core, LSS investigates land use and land cover changes (LULCC), exploring how these changes affect both natural systems (such as forests, grasslands, and water bodies) and socio-economic systems (urban areas, agricultural land, and infrastructure) (Turner et al., 2007). LSS is closely tied to sustainability and global environmental change research, aiming to provide insights into how land system dynamics influence biodiversity, ecosystem services, and climate change (Verburg et al., 2013). LSS approaches land systems as complex social-ecological systems (SES), integrating social, ecological, and technological processes across spatial and temporal scales. These systems are analyzed across different spatial and temporal scales, acknowledging that changes in one component of the system, such as deforestation or urbanization, can lead to cascading effects on other components, including biodiversity loss, soil degradation, and alterations in the global carbon cycle (Geist et al., 2006; Lambin et al., 2001).

2.1.1 Key elements of land system science

1. **LULCC**: At the core of LSS is the study of LULCC, which encompasses the transformation of natural landscapes due to human activities like agriculture, forestry, and urban expansion (Turner et al., 2016). LULCC research investigates both the biophysical aspects of land cover (such as vegetation, water, and soil) and the human dimensions of land use (such as agricultural practices, settlement patterns, and resource management) (Verburg et al., 2013).
2. **Human-Environment Interactions**: LSS emphasizes the bidirectional relationship between humans and the environment. Human activities, such as deforestation or intensive agriculture, directly alter land cover, while changes in land systems, such as land degradation or biodiversity loss, in turn impact human livelihoods, food security, and water availability (Rindfuss et al., 2004). This feedback loop between human and natural systems is central to understanding the drivers and consequences of land change.

3. **Global Environmental Change**: LSS is situated within the broader context of global environmental change research, addressing critical issues such as climate change, desertification, and ecosystem degradation (Turner et al., 2007). Land system changes, driven by human activities, contribute significantly to global challenges, from greenhouse gas emissions due to deforestation to habitat fragmentation and loss of biodiversity (Verburg et al., 2016).

2.1.2 Significance

The significance of LSS lies in its capacity to provide a holistic view of land changes, integrating diverse, and often disparate, data sources and methodological approaches to analyze land system dynamics. Understanding LULCC is critical for addressing a wide range of environmental and socio-economic challenges, from mitigating climate change to promoting sustainable land use practices (Hibbard et al., 2010). LSS research has led to key insights into how land use decisions, driven by economic, political, and cultural factors, shape the Earth's surface and affect ecosystem functioning, for example through agricultural expansion and intensification in response to global markets, urban growth guided by planning and infrastructure investments, deforestation linked to commodity production and land tenure systems, and conservation actions such as protected area establishment and land-use zoning (Turner et al., 2021). One of the central objectives of LSS is to support sustainable development by providing actionable insights for land management and policymaking. For example, studies on deforestation in the Amazon have shown how unsustainable land use practices, such as illegal logging and slash-and-burn agriculture, contribute to biodiversity loss and carbon emissions, while alternative land management strategies can promote conservation and sustainable development (Rindfuss et al., 2004). In addition to its environmental importance, LSS also plays a key role in understanding socio-economic development. Urbanization is one of the most significant drivers of land use and land cover change (LULCC). It has far-reaching impacts on land systems, including increased demand for agricultural land and the transformation of infrastructure and transportation networks (Turner et al., 2021). LSS helps in predicting and managing the implications of urban expansion, ensuring that land use planning is informed by both ecological and social considerations.

2.1.3 Technological integration and remote sensing

RS is a cornerstone of LSS, offering a means to collect large-scale, high-resolution data on land cover and land use changes over time. The integration of RS with advanced technologies like Geographic Information Systems (GIS) and AI methods, particularly ML and DL, has enhanced the ability to monitor and analyze land systems (Gorelick et al., 2017; Tamiminia et al., 2020). The advent of satellite-based Earth Observation (EO) data has allowed researchers to track land changes in near real-time, providing critical insights for policymaking and land management (Verburg et al., 2016). With the growing availability of big data from remote sensing, AI techniques such as ML and DL have become essential tools in LSS. These techniques allow researchers to process vast amounts of imagery, identify patterns, and classify land cover types with unprecedented accuracy (Chen et al., 2017; Karpatne et al.,

2016). By leveraging AI, LSS can move beyond traditional methods of land classification, facilitating the analysis of complex datasets that span multiple temporal and spatial scales (Liu et al., 2017).

2.1.4 Challenges in land system science

While LSS has made significant progress in understanding land change dynamics, it faces several challenges. Data availability and quality are major issues, particularly in regions where satellite imagery is sparse or inaccessible (Verburg et al., 2016). Moreover, the complexity of land systems—marked by the interplay between social and ecological factors—requires interdisciplinary collaboration and innovative methodologies (Hibbard et al., 2010). A key challenge in LSS is the trade-off between data resolution and coverage. High-resolution imagery is ideal for detecting fine-scale land cover changes, but it is often expensive and available for limited geographical areas. On the other hand, lower-resolution data can provide global coverage but may lack the detail needed for local-level analysis (Chen et al., 2017). Addressing these trade-offs is crucial for advancing land system research. In addition, the application of AI techniques like ML and DL in LSS presents both opportunities and challenges. While AI models can handle large and complex datasets, they require significant computational resources and expertise, making them less accessible to researchers with limited technical capabilities (Karpatne et al., 2016). Moreover, the "black box" nature of some AI models, particularly DL, raises concerns about model interpretability and trustworthiness in decision-making (Shih et al., 2019).

2.2 Concept 2: Machine learning

ML is a subset of AI that involves the development of algorithms capable of learning from data without being explicitly programmed for every specific task. In the context of RS and LSS, ML is primarily used for LULC classification by recognizing patterns in large, complex datasets (e.g., multispectral satellite imagery, spectral indices such as NDVI, or long-term Landsat time-series observations). The foundation of ML lies in its ability to automatically identify patterns in data, making it ideal for classifying features across varying spatial and temporal scales (Karpatne et al., 2016). For example, algorithms such as Random Forest (RF), Decision Trees (DT) or Support Vector Machines (SVM) can be trained on labeled satellite imagery to distinguish between forest, cropland, and urban land cover types (Lary et al., 2016). In LULC, ML employs a bottom-up approach, which contrasts with traditional top-down models based on physical principles and expert knowledge (Karpatne et al., 2016). This enables ML algorithms to process large-scale Earth Observation datasets, such as satellite imagery and aerial photographs, identifying patterns that might be imperceptible to human analysis.

2.2.1 Types of machine learning algorithms

1. **Supervised Learning**: This approach involves training the algorithm on labeled data, where the outcome (land cover type) is known. The model learns to associate input data (e.g., spectral signatures from satellite imagery) with specific land cover classes, refining its

predictions through iterative adjustments (Chen et al., 2017). Supervised ML methods, such as RF and SVM, are widely used in remote sensing for their accuracy and ease of implementation. For example, RF has been extensively applied to classify agricultural, forest, and urban land cover types using multispectral satellite imagery, achieving high accuracy with relatively limited training data.

2. **Unsupervised Learning**: Unlike supervised learning, unsupervised learning works without labeled data. Instead, the algorithm groups data into clusters based on similarities, allowing it to classify pixels in satellite imagery without predefined categories (Liu et al., 2017). For example, K-means and ISODATA clustering have been used to identify broad land cover groupings in regions where reference data are unavailable, such as initial mapping of vegetation and surface types in data-scarce landscapes.
3. **Semi-Supervised Learning**: A hybrid approach where the model is trained on a combination of labeled and unlabeled data. This method is beneficial when large volumes of data are available, but labeling is resource-intensive (Karpatne et al., 2016). Semi-supervised learning has the potential to improve model accuracy while reducing the amount of labeled data required. For example, wrapper-based semi-supervised approaches have been used to improve LULC classification accuracy by iteratively expanding training datasets using confidently classified unlabeled pixels.
4. **Reinforcement Learning:** Reinforcement learning (RL) is a paradigm in which an agent learns optimal decision policies through interaction with its environment, receiving rewards or penalties based on performance. In remote sensing, RL has been applied to adaptive problems such as sensor scheduling, feature selection, and data acquisition. For example, Deep Q-Network (DQN) frameworks have been used to optimize multi-sensor sampling strategies, balancing data quality and energy consumption more effectively than fixed sampling approaches (Huang et al., 2025). Additional applications include adaptive path planning for marine observation platforms and hyperspectral band selection using reward-driven decision processes (Guo et al., 2025; Zhang et al., 2023).

2.2.2 Significance

ML models are particularly effective in applications where labeled training data can be obtained and where large volumes of heterogeneous geospatial data, such as multispectral and hyperspectral imagery, must be analyzed (Yu et al., 2017). The adaptability of ML algorithms makes them ideal for diverse geographic regions and varying environmental conditions, from urban areas to remote forests (Ahmed & Lin, 2021; Zhang et al., 2019). In LSS, ML plays a crucial role in monitoring and understanding land cover change by analyzing temporal sequences of images, allowing researchers to detect shifts in land use over time. For instance, RF has been effectively used to monitor deforestation patterns in tropical regions, leveraging spectral data from satellite imagery to identify changes in vegetation cover (Ghamisi et al., 2016). Similarly, SVM models have been employed to classify urban and agricultural land cover types with high precision, aiding in resource management and urban planning (Zhang, Harrison, et al., 2020).

While ML has been transformative in LULC classification, it is not without challenges. A key limitation to some types of ML is the need for extensive training data, which may not always be available or accurately labeled, particularly in remote or underdeveloped regions. Additionally, ML models are prone to overfitting, where the model performs well on training data but poorly on new, unseen data (Alhassan et al., 2020). To address this, researchers use techniques such as cross-validation, where the available dataset is repeatedly partitioned into training and validation subsets to test model performance on independent data, and regularization to improve model generalization (Lary et al., 2016). Another critical aspect of ML is its scalability. With the growing availability of big data from remote sensing platforms, ML offers scalable solutions for processing and analyzing vast datasets, making it a cornerstone technology for LULCC studies (Tamiminia et al., 2020).

2.3 Concept 3: Deep learning

DL is a specialized subset of machine learning that uses neural networks with multiple layers to analyze and classify data. In contrast to traditional ML algorithms, which rely on manual feature extraction, DL models automatically learn hierarchical features from the input data, making them particularly powerful for image-based tasks (Liu, 2015). Convolutional Neural Networks (CNNs) are the most commonly used deep learning architecture in remote sensing, renowned for their ability to capture spatial hierarchies in data, such as edges, textures, and shapes in satellite imagery (Zhang et al., 2019). DL models, such as CNNs, Recurrent Neural Networks (RNNs), and Long Short-Term Memory (LSTM) networks, are designed to handle high-dimensional data, making them ideal for analyzing complex and voluminous datasets in remote sensing (Magalhães et al., 2022). CNNs in particular excel in image classification tasks by learning spatial features, while RNNs are effective in processing time-series data, such as monitoring temporal changes in land cover (Campos-Taberner et al., 2023). Recently, Vision Transformers (ViTs) have emerged as a powerful alternative to CNNs, leveraging self-attention mechanisms to capture both local and global dependencies in remote sensing imagery (Dosovitskiy et al., 2021).

2.3.1 Types of deep learning architectures

1. **CNNs**: CNNs are specifically designed for image analysis, making them the preferred choice for remote sensing tasks such as land cover classification, object detection, and segmentation (Zhou, Xu, et al., 2023). CNNs process input imagery through convolutional and pooling layers to learn spatial features at multiple scales, enabling the detection of complex patterns in satellite imagery (Du et al., 2020). For example, CNNs have been widely used to classify heterogeneous land cover types: such as urban areas, croplands, and forest mosaics, from multispectral and hyperspectral satellite imagery with higher accuracy than traditional ML approaches.
2. **RNNs**: RNNs are well-suited for analyzing sequential data, such as time-series satellite imagery, due to their ability to retain information from previous inputs (Masolele et al., 2021). Long Short-Term Memory (LSTM) networks, a common RNN variant, are frequently used to model long-term temporal dependencies. For example, RNNs and LSTMs have been applied to multi-year satellite time series to track seasonal vegetation dynamics and detect land cover change trajectories over time.
3. **U-Net**: U-Net is a CNN architecture specifically designed for image segmentation, enabling pixel-level classification with high spatial precision. It has been particularly effective in land

cover mapping tasks where accurate boundary delineation is required (Zhou, Chen, et al., 2023). For example, U-Net models have been used to segment agricultural fields, urban infrastructure, and natural vegetation classes from high-resolution satellite imagery, supporting detailed land cover inventories and change detection.

4. **Generative Adversarial Networks (GANs)**: GANs consist of a generator and a discriminator network trained in an adversarial framework to produce realistic synthetic data. In RS, GANs have been applied to data augmentation, image super-resolution, and class balancing, helping to address data scarcity and imbalanced training samples (Tian et al., 2022). For example, GAN-based super-resolution approaches have been used to enhance coarse-resolution satellite imagery, improving downstream land cover classification performance in heterogeneous landscapes.
5. **Autoencoders**: Autoencoders are unsupervised neural networks that learn compact latent representations of high-dimensional data, enabling feature extraction, dimensionality reduction, and anomaly detection (Michelucci, 2022). In remote sensing, they are particularly useful for handling spectrally rich imagery where labeled data are limited (Li et al., 2023). For example, convolutional autoencoders have been used to denoise satellite imagery and extract informative features prior to land cover classification or change detection analyses (Bajaj et al., 2020).
6. **Transformers:** Transformer architectures leverage self-attention mechanisms to model long-range dependencies and global contextual relationships within data (Vaswani et al., 2017). Vision Transformers (ViTs) process imagery as sequences of image patches, capturing both local and global spatial structure (Lin et al., 2022). For example, ViT-based models have been applied to land cover classification and multi-temporal change detection using high-resolution satellite imagery, often achieving performance comparable to or exceeding CNN-based approaches (Uparkar et al., 2023; Wahid et al., 2023).

2.3.2 Significance

DL's significance in remote sensing stems from its ability to handle large, complex datasets with minimal manual intervention. By automatically extracting features from satellite imagery, DL models can achieve higher classification accuracy than traditional ML algorithms, especially in tasks involving high-dimensional data (Gorelick et al., 2017). The hierarchical nature of DL architectures allows them to learn both low-level and high-level features, making them particularly effective in identifying subtle differences in land cover types (Zhang, Han, et al., 2020). DL's application in LULC classification is transformative, particularly in challenging environments where traditional methods struggle. For example, CNNs have been used to classify open pit mining areas, outperforming traditional ML models in accuracy and robustness (Zhou, Xu, et al., 2023). Similarly, DL models can classify urban land cover in high-resolution satellite imagery, where fine details such as building outlines and road networks are crucial for accurate classification (Chen et al., 2017).

However, despite its advantages, DL poses significant challenges. DL models require large amounts of labeled training data to perform effectively, which can be difficult to obtain, particularly in regions with limited access to high-quality imagery (Alhassan et al., 2020). Additionally, DL models are computationally intensive, requiring specialized hardware such as Graphics Processing Units (GPUs) for training and inference, making them less accessible to researchers with limited computational resources (Chen & Ran, 2019). Another major challenge is model interpretability. While DL models can achieve superior performance, their decision-making processes are often considered a "black box," making it difficult to explain how they arrive at specific classifications (Shih et al., 2019). This lack of transparency is a significant drawback when applying DL to critical decision-making areas such as environmental monitoring and urban planning, where explainable models are preferred. Despite these challenges, DL continues to evolve, with new architectures and techniques being developed to address its limitations. Lightweight DL models, for instance, are being designed to reduce computational demands, making them more accessible for real-time applications (Choudhary et al., 2022). Moreover, techniques such as transfer learning, where models pre-trained on one dataset are adapted to another, are reducing the need for large, labeled datasets (Hong et al., 2021).

3. Decision making process

The remaining portions of this chapter reviews and synthesizes existing studies on the application of ML and DL techniques in LSS, specifically in the context of LULC classification. The goal was to examine how these methods have been applied across various studies and to compare their effectiveness, based on published findings. To guide researchers in selecting the most appropriate technique, we developed a decision tree framework, informed by an extensive review of the literature on the strengths and limitations of both ML and DL approaches (Southworth et al., 2024).

3.1 Review approach

Our methodology involved a systematic review of the literature on ML and DL applications in LULC classification. The review focused on studies that utilized these AI techniques in different remote sensing contexts, comparing performance across multiple datasets, regions, and types of imagery. We compiled findings from a wide range of peer-reviewed articles to build a comprehensive understanding of the conditions under which each method excels or faces challenges. The decision tree model was developed based on insights gained from this review, considering key factors such as data type, research objectives, computational resources, and model accuracy (Fig. 5.1). This methodology allows us to provide a clear framework for researchers in the field of LSS, helping them navigate the decision-making process between ML and DL techniques (Southworth et al., 2024).

3.2 Step 1: Data type and system studied

In the reviewed literature, ML and DL were applied to various data types, including multispectral, hyperspectral, and panchromatic imagery. The type of imagery used was a key determinant in the choice of classification technique, as some data types are better suited to certain models than others.

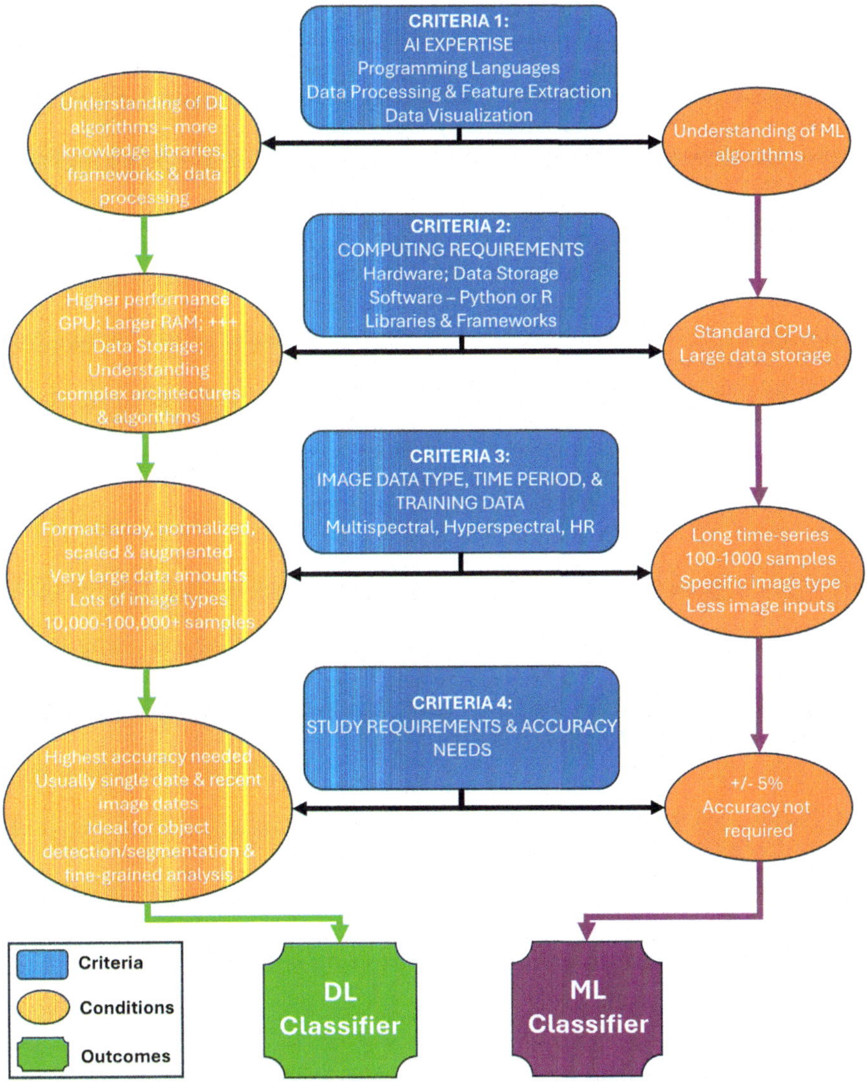

Figure 5.1 A decision tree exploring four criteria to select the appropriate classification approach between machine learning (ML) and deep learning (DL) for applications in land system science (LSS).

- **Multispectral Imagery**: Many studies that employed ML techniques, such as RF and SVM, worked with multispectral imagery due to its lower dimensionality and easier interpretability. For example, Zhang et al. (2019) found that RF was particularly effective for classifying urban areas and agricultural land in multispectral satellite data, with high classification accuracy and lower computational costs. In contrast, DL techniques such as CNNs have also been applied to multispectral imagery, with studies like Barbierato et al. (2020) highlighting CNNs' ability to outperform ML models in more complex landscapes, such as those with heterogeneous vegetation cover, like urban forest ecosystems.

- **Hyperspectral Imagery**: In studies using hyperspectral imagery, DL methods generally provided better results than ML due to the high-dimensional nature of hyperspectral data. CNNs excel at extracting spatial and spectral features from hyperspectral imagery, enabling more detailed land cover classifications. For example, Yu et al. (2017) demonstrated that CNNs achieved significantly higher accuracy than traditional ML methods when classifying vegetation types in hyperspectral imagery, due to the model's ability to automatically learn complex features.
- **LiDAR Data:** LiDAR data has long been leveraged for deriving three-dimensional metrics such as height percentiles and canopy cover, which traditional ML algorithms like RF and SVM can then use for tasks ranging from forest canopy classification to urban structure detection and biomass estimation (Dalponte et al., 2008). More recently, DL approaches have begun to exploit the raw 3D point clouds directly: PointNet employs shared MLPs and a global pooling operation to perform end-to-end classification and segmentation on unordered point sets (Qi et al., 2017), while 3D CNN-based architectures like VoxNet convert the point cloud into a voxel grid (voxelize) and apply 3D convolutional filters to capture local geometric features for object recognition in real time (Maturana & Scherer, 2015).
- **Radar Data:** Radar imagery, especially Synthetic Aperture Radar (SAR), provides cloud-penetrating, day-and-night observations that are invaluable for flood mapping and crop monitoring. Classical ML techniques such as SVM and RF have demonstrated strong performance in crop classification tasks using time-series polarimetric SAR data (Xie et al., 2021). On the DL front, convolutional LSTMs have been applied to radar time-series for precipitation nowcasting, effectively capturing both spatial and temporal dependencies in noisy radar sequences (Shi et al., 2015). Recent advancements have begun to leverage Vision Transformer (ViT) architectures to model long-range spatial relationships in SAR imagery, offering improved land-cover classification capabilities by addressing the locality limitations of CNNs (Dong et al., 2022).

The reviewed literature indicates that DL models, particularly CNNs, are more suitable for complex, high-dimensional datasets such as hyperspectral imagery, while ML models may suffice for simpler tasks using multispectral data (Chen et al., 2017; Du et al., 2020).

3.3 Step 2: Research objectives and complexity

The complexity of the research objective was another factor that influenced the choice between ML and DL in the studies we reviewed. Tasks ranged from basic LULC classifications to more intricate analyses of land cover dynamics over time.

- **Low-Complexity Tasks**: For relatively simple classification tasks, such as distinguishing between a few land cover types (e.g., urban vs. rural), ML methods such as RF and SVM were commonly used. Studies such as Chen et al. (2017) found that RF was sufficient for classifying urban expansion areas with high accuracy, using limited computational resources. In such cases, the simplicity of the task did not require the advanced feature extraction capabilities of DL.
- **High-Complexity Tasks**: In contrast, DL methods were preferred in studies that required high precision or involved complex land cover types. For example, Zhang, Harrison, et al. (2020) used CNNs to classify land cover in highly heterogeneous landscapes, where traditional ML models struggled to capture subtle variations in spectral signatures. The DL

model's ability to learn from large datasets and handle complex data made it more effective for these high-precision tasks.

- **Machine Learning for Moderate Complexity Tasks:** Traditional ML methods such as RF, SVM, and k-Nearest Neighbors (KNN) have demonstrated strong performance in relatively moderate or structured remote sensing classification problems, particularly where the number of land cover classes is limited, or where training data are scarce. These models rely on handcrafted features (e.g., vegetation indices, spectral ratios) and require fewer computational resources, making them attractive for rapid implementation in operational contexts (Belgiu & Drăguţ, 2016; Lary et al., 2016). Their interpretability allows practitioners to understand key drivers behind classification outcomes, aiding in policy-relevant applications such as basic change detection or feature importance analysis. However, as spatial resolution increases or class heterogeneity grows, the reliance on human-designed features may limit their scalability and generalization capacity.
- **DL for Complex Remote Sensing Problems:** In contrast, DL approaches, including CNNs, U-Net, DeepLabv3+, and ViTs, excel in complex remote sensing tasks that require learning from high-dimensional, nonlinear, or multi-temporal data. DL models are capable of automatically extracting hierarchical features from raw satellite imagery, capturing subtle spatial and spectral variations critical for fine-grained tasks such as crop-type classification, urban structure delineation, or spatiotemporal change mapping (Ma et al., 2019; Zhu et al., 2017). While they demand larger labeled datasets and robust computational infrastructure, their capacity to handle noisy or imbalanced data and model intricate patterns has led to state-of-the-art performance across a range of remote sensing benchmarks (Ball et al., 2017). As such, DL models are increasingly preferred for high-resolution, multi-class, or time-sensitive applications in agriculture, land monitoring, and disaster assessment.

Across the literature, DL models were shown to outperform ML models in high-complexity tasks, where intricate spatial patterns and high data dimensionality were involved (Magalhães et al., 2022). This distinction informed our decision tree, guiding researchers to choose DL for more challenging classification tasks while reserving ML for simpler objectives.

3.4 Step 3: Computational resources and infrastructure

The reviewed studies highlighted a difference in the computational requirements of ML and DL techniques. DL methods are computationally intensive, often requiring specialized hardware such as GPUs or access to cloud computing platforms.

- **ML Computational Requirements:** Traditional ML models such as RF, SVM, and gradient boosting frameworks like XGBoost and LightGBM are generally easier to implement, requiring fewer hyperparameters and simpler training processes (Lary et al., 2016; Liu et al., 2017; Rahaman et al., 2025; Shih et al., 2019). These models can be effectively trained on standard CPUs, making them suitable for environments with limited computational infrastructure (Alhassan et al., 2020; Lary et al., 2016). Additionally, ML models are more energy-efficient, have shorter training times, and are accessible through user-friendly libraries like scikit-learn, which facilitate rapid prototyping and experimentation. These characteristics make ML an attractive choice for moderate-scale applications, especially in resource-constrained research settings.

- **DL Computational Requirements:** DL models, such as CNNs, LSTM networks, and RNNs, are considerably more complex. They are highly sensitive to initializations, hyper-parameter settings, and data preprocessing (Chen & Ran, 2019; Du et al., 2020). Training these models typically demands substantial computational power, including access to GPUs or even Tensor Processing Units (TPUs), especially when working with large, high-resolution, or multi-temporal satellite imagery. DL training is often time-consuming and energy-intensive, which can be a barrier in many research environments. However, powerful open-source libraries such as TensorFlow, PyTorch, Keras, and HuggingFace provide scalable solutions for developing and deploying DL models across CPUs, GPUs, and TPUs, enabling researchers to tackle complex remote sensing tasks with high accuracy, provided the necessary infrastructure is available (Ma et al., 2019; Rahaman et al., 2025; Sarker, 2021; Zhu et al., 2017).

This aspect of computational resources is a critical factor in our decision tree, which helps researchers determine the feasibility of using DL based on their available infrastructure.

3.5 Step 4: Training data availability

The availability of labeled training data was a central issue in many of the reviewed studies. Both ML and DL models rely on high-quality training data, but the volume and quality of data required differ significantly between the two.

- **ML Training Data Requirements**: Many studies noted that ML models require smaller volumes of labeled data to perform effectively. Hermosilla et al. (2022) showed that RF could achieve good classification results with relatively limited training data, provided the data was well-curated and representative. However, these models often rely on manual feature engineering to enhance performance, which can introduce human bias.
- **DL Training Data Requirements**: In contrast, DL models demand large amounts of labeled data for training. Several studies, including Alhassan et al. (2020) and Zhou, Xu, et al. (2023), emphasized the need for substantial training datasets to prevent overfitting and ensure generalization in DL models. Data augmentation techniques, such as flipping and rotating images, were commonly used to artificially increase the size of training datasets, but these methods have limitations in terms of capturing real-world variation or in introducing human bias in this process.

Based on these findings, our final decision tree framework (Fig. 5.1) incorporates the volume and quality of available training data as a critical factor in selecting between ML and DL. In cases where large, labeled datasets are not available, ML may be the more practical option.

3.6 Step 5: Accuracy and model interpretability

Accuracy requirements and the interpretability of the models were key considerations in the reviewed studies.

- **ML Accuracy and Interpretability**: In many cases, ML models such as RF and SVM provided sufficient accuracy for LULC classification tasks. For instance, Hermosilla et al. (2022) found that RF achieved high accuracy in land cover classification while maintaining transparency in its decision-making process, which was particularly important for applications such as environmental management and policymaking. ML models generally offer higher interpretability compared to DL models, which is a significant advantage in fields where transparency is crucial.

- **DL Accuracy and Interpretability**: DL models often outperformed ML models in terms of classification accuracy, especially in studies involving complex land cover types. For example, Zhang, Han, et al. (2020) and Campos-Taberner et al. (2023) demonstrated that CNNs consistently achieved higher accuracy than RF and SVM in detecting subtle land cover changes in high-dimensional datasets. However, the "black box" nature of DL models raises concerns about their interpretability. Several studies, including Shih et al. (2019), pointed out that the lack of transparency in DL models can be a drawback when explaining the model's decision-making process is essential.

Our developed decision tree framework illustrated here in Fig. 5.1 reflects the trade-offs between accuracy and interpretability, helping researchers decide whether the improved performance of DL models justifies the additional complexity and opacity (Southworth et al., 2024).

3.7 Evaluation metrics

To synthesize the findings across the reviewed studies, we examined the evaluation metrics most commonly used to compare ML and DL models. These included:

- **Accuracy**: The percentage of correctly classified instances.
- **Precision and Recall**: Metrics that assess the model's ability to correctly classify positive examples (precision) and its ability to retrieve all relevant instances (recall).
- **F1-Score**: A balance between precision and recall, providing a more holistic view of model performance.
- **Confusion Matrix**: Used in multiple studies to illustrate the breakdown of model predictions, showing true positives, true negatives, false positives, and false negatives.

Across the reviewed literature, DL models generally achieved higher accuracy and F1-scores than ML models, especially in high-complexity tasks, while ML models maintained advantages in terms of interpretability and lower computational demands (Ghamisi et al., 2016; Southworth et al., 2024; Tamiminia et al., 2020).

4. Discussion

4.1 Comparative analysis of case studies

The comparative analysis of ML and DL techniques for LULC classification reveals clear differences in their effectiveness across various studies and applications (Fig. 6.1; Southworth et al., 2024). ML techniques, particularly RF and SVM, have historically been effective for simpler classification tasks with smaller datasets and lower computational requirements. For example, Chen et al. (2017) demonstrated that RF performed well in classifying urban and rural areas using multispectral satellite imagery, achieving high accuracy without requiring extensive computational infrastructure. These models were easier to interpret and required less data for training, making them accessible for resource-limited studies. However, as LULC tasks became more complex, DL models, particularly CNNs, began to outperform traditional ML techniques. For instance, the study by Zhang, Han, et al. (2020) showed that CNNs significantly outperformed RF in classifying complex land cover types such as urban green spaces and mixed agricultural areas.

A key finding across multiple case studies is that DL models, such as CNNs and LSTM networks, excel in environments where large datasets are available, and the land cover features are complex or heterogeneous (Campos-Taberner et al., 2023). For example, Yu et al. (2017) employed CNNs for hyperspectral data classification and found that DL achieved significantly better accuracy in detecting vegetation types in diverse ecosystems compared to SVM and RF. This success is attributed to CNNs' ability to learn from multi-scale features and extract spatial patterns that are otherwise difficult to capture using traditional ML models. While DL models tend to outperform ML models in terms of classification accuracy, their advantages diminish in scenarios where data availability or computational resources are limited. Studies like Hermosilla et al. (2022) highlighted the robustness of RF in such cases, where limited labeled data and computational power meant that the computational overhead of training a DL model was not justified. Moreover, DL models, despite their higher accuracy, often face issues of overfitting, especially when applied to smaller datasets (Shih et al., 2019).

4.2 Discussion on the results

DL's ability to automatically learn features from input data makes it particularly effective in handling high-dimensional and heterogeneous datasets (Zhang, Harrison, et al., 2020). However, this improved accuracy comes at the cost of higher computational demands and the need for extensive labeled training data (Alhassan et al., 2020). As demonstrated in the study by Du et al. (2020), CNNs significantly outperformed RF and SVM in classifying complex land systems but required access to GPUs and cloud-based platforms for effective training. On the other hand, ML models like RF and SVM remain viable options for researchers with limited access to computational resources or large datasets. The interpretability of these models also provides an important advantage in applications where understanding the decision-making process is critical, such as environmental management and policymaking (Sarker, 2021). The review highlights that ML models, while sometimes less accurate than DL models, are easier to deploy and maintain, especially in resource-constrained environments (Southworth et al., 2024).

The decision tree model framework (Fig. 5.1) synthesized from these findings helps guide researchers toward the appropriate AI technique based on the specific requirements of their LULC tasks. For simpler tasks with limited data, ML remains a practical and efficient solution. However, for high-complexity tasks, particularly those requiring detailed classification of heterogeneous landscapes, DL provides significant performance gains (Gorelick et al., 2017).

4.3 Implications

The increasing integration of AI techniques, particularly ML and DL, is poised to transform the field of LSS by enabling more accurate and efficient LULC classification. As LULCC represents the central analytical focus of LSS, advances in AI-based classification directly enhance the capacity of LSS to quantify how human decisions: such as urban expansion, agricultural intensification, and conservation planning, translate into observable changes in land cover over space and time. These applications can enhance the precision of monitoring land changes, which is essential for addressing global challenges such as deforestation, urban sprawl, and climate change (Verburg et al., 2013). DL's ability to analyze complex, high-dimensional data allows researchers to monitor land cover changes at a finer scale and with greater temporal frequency than ever before (Zhang, Han, et al., 2020). This can lead to improved decision-making in fields such as urban planning, conservation, and resource management. For instance, CNNs have already demonstrated their potential in identifying illegal deforestation activities in remote tropical areas, providing real-time monitoring that can inform rapid policy interventions (Ghamisi et al., 2016). Similarly, DL applications in urban land cover classification can support sustainable development by offering more detailed insights into the patterns of urban growth, helping planners optimize land use to balance environmental conservation and urban expansion (Turner et al., 2021).

The potential for these AI-driven methods to process vast amounts of remotely sensed data from multiple sources, including satellite imagery, aerial photography, and ground-based sensors, also opens up new opportunities for interdisciplinary collaboration. The integration of social and ecological data into LSS models, facilitated by AI, can provide a more holistic understanding of human-environment interactions, leading to more effective land management strategies (Rindfuss et al., 2004).

4.4 Future prospects

The prospects of AI in LSS are promising, with continued advancements in both ML and DL methods expected to further enhance land system monitoring and analysis. As DL architectures become more efficient and less reliant on large datasets, they will likely become even more accessible to researchers, allowing for broader adoption in LSS (Choudhary et al., 2022). The development of "lightweight" DL models, which require less computational power and memory, is already addressing some of the current limitations of DL applications in LULC (Hong et al., 2021). Moreover, the growing availability of open-source LULC datasets, such as those from NASA's Landsat program or ESA's Sentinel missions, will continue to fuel AI research in LSS. These datasets will help to overcome one of the major bottlenecks in DL, access to large volumes of labeled training data. As data-sharing platforms and collaborative efforts expand, the ability to apply DL models to new regions and ecosystems will become increasingly feasible (Zhou, Xu, et al., 2023).

Another promising direction for AI in LSS is the integration of unsupervised and semi-supervised learning methods, which can help reduce the reliance on labeled data. Techniques such as generative adversarial networks (GANs) and transfer learning are already showing promise in enabling DL models to perform well even with smaller datasets (Tamiminia et al., 2020). These approaches will likely drive significant advancements in how we analyze and interpret land cover changes, particularly in underrepresented regions where labeled data is scarce. The broader application of AI-driven LULC classification in decision support systems for land use planning and natural resource management also holds significant potential. AI models, integrated with geospatial decision-making tools, could offer real-time insights for policymakers, helping to mitigate the impacts of urbanization, agricultural expansion, and environmental degradation (Lambin et al., 2001). The future of AI in LSS thus not only promises advancements in research but also in practical applications that can directly influence policy and land management practices.

5. Conclusion

This chapter explores the comparative roles of ML and DL techniques in the field of LSS, with a particular focus on LULC classification tasks. Through a detailed review of existing studies, we have demonstrated the strengths and weaknesses of both approaches, providing a decision-making framework to guide researchers in selecting the most appropriate method for their specific tasks (Fig. 5.1).

The key takeaways from this chapter are as follows:

1. **ML in LULC**: Traditional ML techniques, such as RF and SVM, have been highly effective for simpler LULC classification tasks, particularly when computational resources and labeled data are limited. ML models are computationally efficient, easier to interpret, and well-suited to less complex land systems with moderate amounts of training data (Chen et al., 2017; Hermosilla et al., 2022).
2. **DL in LULC**: DL methods, particularly CNNs, excel in high-dimensional data contexts where the classification task is more complex. These models have shown superior performance in handling multispectral and hyperspectral imagery, enabling detailed land cover classifications in diverse and heterogeneous landscapes. However, DL methods come with higher computational costs, demand large datasets, and are less interpretable than traditional ML techniques (Yu et al., 2017; Zhang, Han, et al., 2020).
3. **Decision Tree Framework**: The decision tree model framework developed in this research (Fig. 5.1) provides researchers with a structured approach to selecting the appropriate AI method based on factors such as data type, computational resources, research objectives, and accuracy requirements. This framework highlights that while DL may offer higher accuracy in certain cases, ML remains a practical solution for many LULC tasks, especially when resources are limited (Ghamisi et al., 2016; Sarker, 2021).

5.1 Future directions

As the field of LSS continues to evolve, the integration of AI techniques such as ML and DL will play an increasingly critical role in advancing the study of LULCC, human–environment interactions, and broader processes of global environmental change. Future advancements in AI are expected to mitigate some of the current limitations of DL models, particularly their high demand for computational resources and large labeled datasets. The development of lightweight DL models and semi-supervised learning techniques could expand the accessibility of DL for researchers with fewer resources (Choudhary et al., 2022; Hong et al., 2021). At the same time, the growing availability of open-source LULC datasets from satellite missions such as Landsat and Sentinel will enable more consistent and comparable analyses of land system dynamics across regions and time, strengthening the empirical foundations of LSS. These advances support deeper investigation of how human decision-making, governance, and socio-economic drivers translate into observable land cover outcomes, reinforcing the coupled social–ecological perspective central to land system science. Looking forward, the integration of AI into decision support systems for land use planning and environmental management represents a particularly promising direction. Such systems can provide near–real-time insights for policymakers, supporting responses to deforestation, urbanization, agricultural expansion, and climate-related land change processes (Zhou, Xu, et al., 2023). Finally, the increasing incorporation of DL models into widely used geospatial software platforms will further lower technical barriers, facilitating broader interdisciplinary collaboration and expanding the practical application of AI-driven approaches within LSS.

5.2 Closing remarks

In conclusion, the comparison between ML and DL in LSS highlights that each approach offers distinct advantages depending on the scale, complexity, and objectives of a given study. While DL provides powerful tools for high-resolution and high-complexity LULC analyses, ML remains a robust and efficient option for many land system applications, particularly where data or computational resources are limited.

By situating these methodological choices within the broader framework of LSS, linking LULCC, human–environment interactions, and global environmental change, this chapter emphasizes that advances in AI are most impactful when aligned with clear theoretical and decision-making goals. The decision tree framework presented in this chapter (Fig. 5.1) offers practical guidance for selecting appropriate AI approaches based on data availability, research objectives, and interpretability needs. As AI technologies continue to advance, the combined use of ML and DL is poised to enhance our capacity to understand, monitor, and manage land systems, supporting more sustainable land use practices and informed decision-making in the face of ongoing environmental change.

Case studies

Case Study 1: Balancing precision and practicality: Classifying almond crops from space

Rahaman, M., Southworth, J., Wen, Y., & Keellings, D. (2025). Assessing model trade-offs in agricultural remote sensing: A review of machine learning and deep learning approaches using almond crop mapping. *Remote Sensing*, *17*(15), https://doi.org/10.3390/rs17152670.

Background

Advances in remote sensing have transformed agricultural monitoring, enabling the identification and tracking of crops over large areas with unprecedented speed and specificity (Rahaman et al., 2025). Accurate land cover classification is fundamental for many agricultural applications, such as crop management, yield forecasting, and resource allocation. In California's Central Valley, one of the world's most productive agricultural regions, precise mapping of almond orchards is particularly valuable for stakeholders ranging from farmers to policymakers (Case Study Figure 1.1). The ongoing expansion of remote sensing technologies has prompted the geospatial science community to explore and compare analytical approaches for classifying complex, multi-temporal image data. Two main paradigms dominate this landscape: traditional machine learning (ML) models, including Decision Tree (DT), K-Nearest Neighbor (KNN), and Random Forest (RF) algorithms; and contemporary deep learning (DL) architectures, such as Unet, MAnet, and DeepLabv3+. While DL has gained significant attention due to its impressive performance in various image processing tasks, it requires substantial labeled data and computational resources. In contrast, traditional ML models are widely regarded for their efficiency and lower resource requirements, but their upper limits of accuracy remain a point of ongoing research and debate.

Implementation

To objectively compare these two paradigms, the researchers conducted a systematic experiment for almond crop mapping in California's Central Valley (Rahaman et al., 2025). The evaluation involved the implementation of three state-of-the-art DL models: Unet, MAnet, and DeepLabv3+, as well as three conventional ML models: DT, KNN, and RF. A key difference in model development was the extent and nature of training data. The DL models were trained with an extensive, multi-year dataset, incorporating imagery from 2008 to 2022, representing a significant investment in data collection and preparation. By contrast, the ML models were trained using only a single year of data, reflecting a common constraint in many operational or resource-limited mapping scenarios. Model performance was assessed based on accuracy in correctly identifying almond orchards, and the researchers also carefully tracked the computational demands and practical feasibility of each approach.

Results

The comparative evaluation yielded important insights into the strengths and limitations of each modeling approach. All models achieved high levels of accuracy in almond crop classification. The DL models: Unet, MAnet, and DeepLabv3+, demonstrated the highest performance, with overall accuracy rates ranging from 97.3% to 97.5%. In close comparison, the traditional ML models: DT, KNN, and RF, achieved accuracies between 96.6% and 96.8% (Case Study Figure 1.2 and Case Study Table 1.1).

This less-than-one-percentage-point difference illustrates that, within the context of this study, conventional ML models can produce results nearly equivalent to those

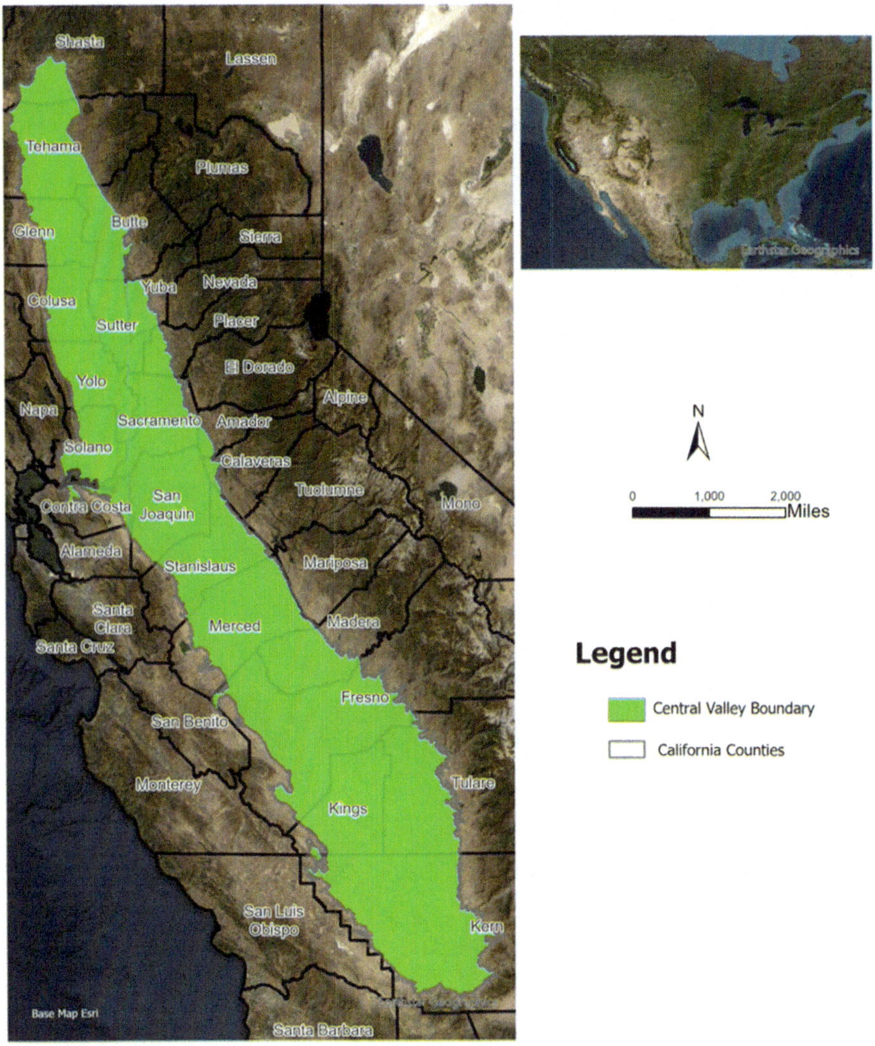

Case Study Figure 1.1 Study area map showing the central valley of California and its constituent counties.
Adapted from: Rahaman, M., Southworth, J., Wen, Y., & Keellings, D. (2025). Assessing model trade-offs in agricultural remote sensing: A review of machine learning and deep learning approaches using almond crop mapping. *Remote Sensing, 2025,* 17(15), 2670; https://doi.org/10.3390/rs17152670.

of modern DL techniques when applied to agricultural land cover mapping. Importantly, these marginal accuracy gains for DL models came with significant additional requirements: DL approaches necessitated multi-year labeled datasets for training and extensive computational resources, including powerful hardware and extended processing time (Rahaman et al., 2025).

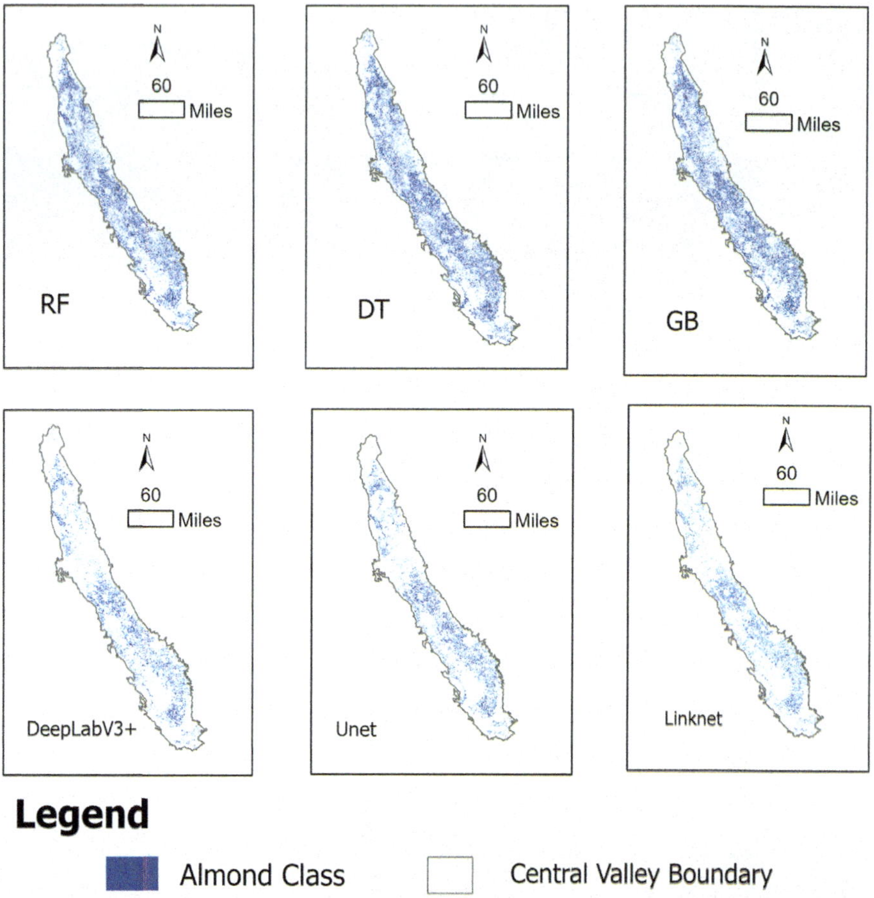

Case Study Figure 1.2 Spatial distribution of predicted almond class in California's central valley using Top three machine learning (RF, DT, GB) and deep learning (DeepLabV3+, Unet, linknet) models.
Adapted from: Rahaman, M., Southworth, J., Wen, Y., & Keellings, D. (2025). Assessing model trade-offs in agricultural remote sensing: A review of machine learning and deep learning approaches using almond crop mapping. *Remote Sensing,* 2025, 17(15), 2670; https://doi.org/10.3390/rs17152670.

In contrast, traditional ML models delivered robust and competitive classification results while using only a single year's data for training and requiring an order of magnitude less computational capacity (Case Study Table 1.1). These findings demonstrate that traditional ML methods remain highly effective and practical, especially in scenarios with constrained data availability or limited computational infrastructure. The study's results are consistent with prior research in remote sensing and land cover mapping, which has similarly found that DL techniques offer incremental, rather than transformative, improvements over mature ML approaches when applied to well-structured classification tasks (Southworth et al., 2024).

Case Study Table 1.1 Comparison of the top 3 ML and DL models based on precision, recall, F1-score, and overall accuracy.

Model	Precision	Recall	F1-score	Overall accuracy
Random forest (RF)	0.71	0.73	0.72	96.798
Decision tree (DT)	0.7	0.74	0.72	96.65
Gradient boosting (GB)	0.7	0.74	0.72	96.615
DeepLabV3+	0.8	0.66	0.7	97.502
Unet	0.79	0.66	0.7	97.465
Linknet	0.8	0.64	0.69	97.455

Impact

The practical implications of these findings are substantial for both the research community and practitioners in agricultural monitoring. The evidence suggests that traditional ML models are often sufficient for operational land cover mapping, supporting workflows that require rapid deployment, adaptability, and cost-effectiveness. This is particularly relevant for stakeholders with limited access to large, curated datasets or high-end computing environments. Where the highest precision and accuracy are imperative, such as in research settings, regulatory compliance, or precision agriculture initiatives with access to abundant training data and computational resources, DL models may provide additional value. However, the incremental gains in accuracy need to be weighed against the significantly greater demands in data preparation and computational investment.

This study encourages a balanced, context-sensitive approach to model selection in agricultural remote sensing. Rather than defaulting to the most complex or resource-intensive method, practitioners are advised to align their methodological choices with the specific goals, constraints, and available resources of their projects. Looking forward, the researchers recommend continued validation of these approaches across diverse agricultural environments and crop types, emphasizing the need to assess model adaptability under changing management practices and evolving climate conditions. Overall, the study provides a robust benchmark and a systematic framework for decision-making in land cover classification, reinforcing the importance of efficiency, practicality, and fitness-for-purpose in agricultural geospatial analytics (Rahaman et al., 2025).

Case Study 2: Blending technology and ecology: Monitoring land-cover transformations in Addo elephant National park

Safaei, M., Southworth, J., Gibbes, C., Herrero, H. V., Rahaman, M., Tefera, B. B., & Blackburn, J. K. (2025). Land-cover classification in Addo Elephant National Park: Analyzing the impact of variables, classifiers, and object-based approach. *Ecological Informatics*, Volume 90, https://doi.org/10.1016/j.ecoinf.2025.103279.

Background

Effective conservation management in protected areas like Addo Elephant National Park (AENP), South Africa (Case Study Figure 2.1), relies heavily on accurate and up-to-date mapping of land-cover types and understanding how these change over time (Safaei et al., 2025). AENP is a region of exceptional biodiversity and ecological complexity, shaped by both natural dynamics and anthropogenic influences. As land-cover patterns directly affect habitat availability, biodiversity, and ecosystem health, robust monitoring techniques are essential for detecting subtle or emerging changes that may have significant conservation implications. Traditional land-cover mapping, while useful, often lacks the granularity and adaptability needed to capture nuanced shifts in such dynamic landscapes.

Recent advances in satellite remote sensing have provided vast amounts of imagery at increasing temporal and spatial resolution. Alongside this, machine learning (ML) and deep learning (DL) techniques for image classification have shown promise in extracting meaningful information from complex, multidimensional data. However, the relative strengths of these approaches, especially in ecologically diverse,

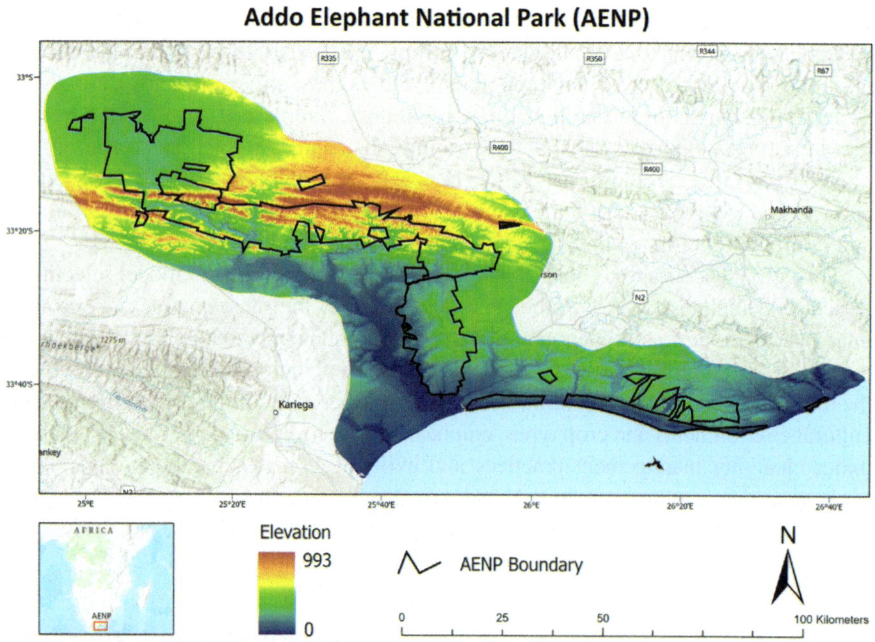

Case Study Figure 2.1 The location of Addo Elephant National park in South Africa, along with its surrounding landscape. The elevation gradient is depicted, ranging from sea level up to 993 m.

Adapted from: Safaei, M., Southworth, J., Gibbes, C., Herrero, H. V., Rahaman, M., Tefera, B. B., & Blackburn, J. K. (2025). Land-cover classification in Addo Elephant National Park: Analyzing the impact of variables, classifiers, and object-based approach. *Ecological Informatics*, Volume 90, https://doi.org/10.1016/j.ecoinf.2025.103279.

changeable regions like AENP, warrant systematic evaluation. This study sets out to compare ML and DL classification methods for land-cover mapping within and around AENP, aiming to both improve map accuracy and deepen understanding of the park's evolving land-cover dynamics over a 20-year period (Safaei et al., 2025).

Implementation

The researchers constructed a rigorous classification workflow centered on three key years: 2002, 2014, and 2022, using multi-temporal Landsat satellite imagery. Six major land-cover classes were defined, reflecting the principal vegetation types and human land-uses within this landscape. Image processing involved not only extracting spectral bands but incorporating a wide range of predictor variables: spectral indices such as the Normalized Difference Vegetation Index (NDVI), time-series composites, textural metrics to capture landscape heterogeneity, and terrain attributes from a digital elevation model (DEM), notably elevation and slope. Classification was carried out using two leading ML algorithms: Random Forest (RF) and Support Vector Machines (SVM) and a DL semantic segmentation model (UNet++). The team systematically evaluated both pixel-based and object-based approaches: the former involves classifying each individual pixel, while the latter segments the imagery into homogeneous "objects" first, then assigns classes based on aggregated information. This comparison allowed for a nuanced appraisal of methodology performance in the context of real-world park monitoring, where balance between accuracy and operational efficiency is critical.

The model training leveraged ground truth data from field surveys and very-high-resolution imagery, supporting a robust, supervised learning framework. Extensive hyperparameter optimization and feature selection were undertaken to ascertain the most informative predictors for each classifier. Performance was assessed using standard metrics, primarily overall accuracy and class-wise confusion matrices, against a reserved validation set.

Results

The analyses yielded several clear outcomes (Case Study Figure 2.2 and Case Study Table 2.1). First, across all years and all tested scenarios, the RF classifier consistently achieved the highest accuracy, with overall classification accuracy of 89.1% (2002), 91.2% (2014), and 91.9% (2022). SVM and the DL UNet++ model, while performing reasonably, did not surpass the well-tuned RF. Importantly, pixel-based classification methods outperformed object-based approaches in this context, suggesting that, at Landsat's spatial resolution, preserving granular spectral detail at the pixel level was more beneficial than aggregating information at coarser object scales.

Feature importance analyses revealed that terrain variables, particularly elevation and slope, provided substantial discriminatory power when distinguishing between vegetation types and separating natural cover from human land uses such as agriculture. Time-series composites of NDVI and similar metrics further enhanced class separability, capturing phenological trends and vegetation health over time. Texture

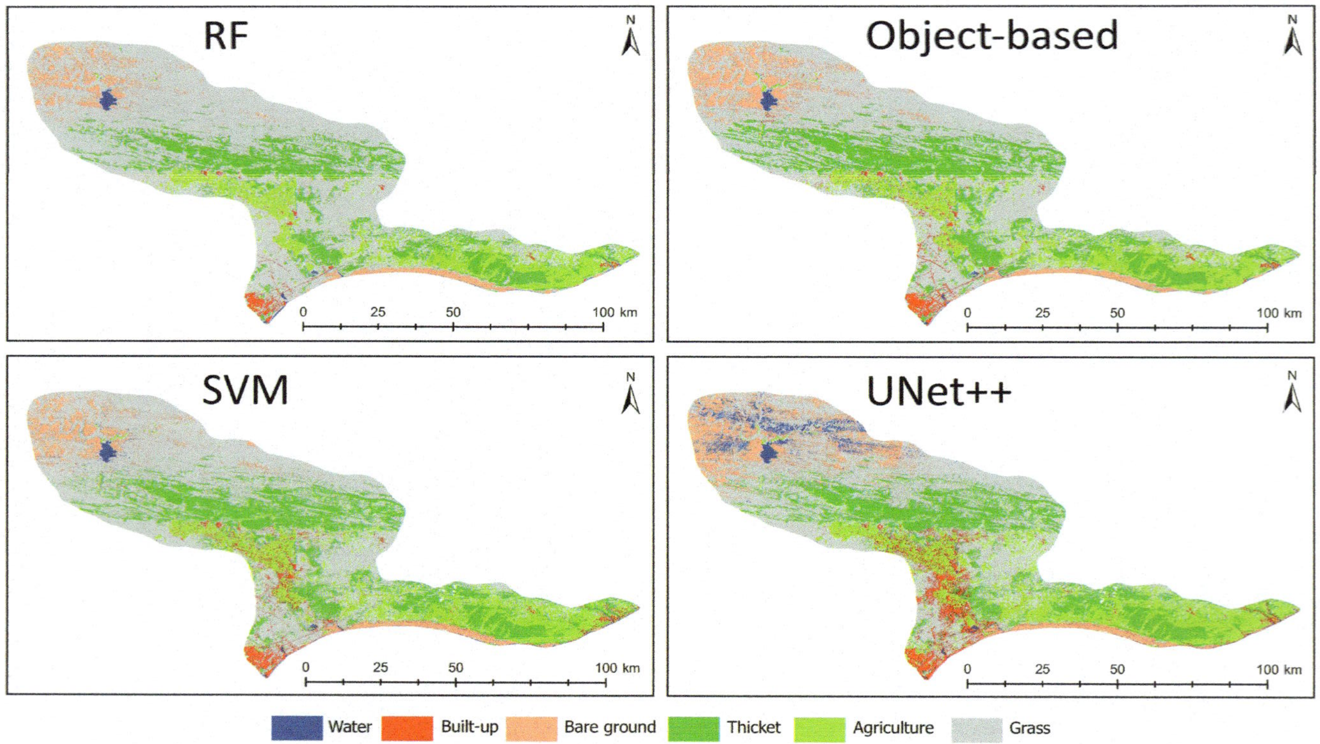

Case Study Figure 2.2 Land-cover maps of Addo Elephant National park and a 10 km buffer zone for the year 2022, generated using machine learning (ML) algorithms (Random forest (RF), support vector machines (SVM), object-based methods), and the deep learning (DL) algorithm UNet++.

Adapted from Safaei, M., Southworth, J., Gibbes, C., Herrero, H. V., Rahaman, M., Tefera, B. B., & Blackburn, J. K. (2025). Land-cover classification in Addo Elephant National Park: Analyzing the impact of variables, classifiers, and object-based approach. *Ecological Informatics*, Volume 90, https://doi.org/10.1016/j.ecoinf.2025.103279.

Case Study Table 2.1 Overall Accuracy (OA) and F1-score (%) of the machine learning (ML) and deep learning (DL) models.

Model		**Water**	**Built-up**	**Bare ground**	**Thicket**	**Agriculture**	**Grass**
	OA	**F1-score**					
RF	95.1	99.5	92.8	99.5	99.0	92.3	87.9
SVM	83.0	95.4	76.6	89.6	94.0	79.2	61.9
Object-based	91.8	98.7	85.9	82.6	96.2	93.5	91.5
UNet++	82.5	86.1	80.4	57.0	79.9	82.3	86.8

metrics contributed valuably as well, especially in differentiating heterogeneous woody thickets from more uniform grassland or cropland classes.

The finalized land-cover maps for AENP and its buffer zone depicted grassland as the predominant land-cover, both within and immediately outside park boundaries. Within the park, thicket was also a major component, whereas outside, agricultural land followed grass in extent. Despite the duration of the study interval (20 years), the analysis detected only minor overall changes in land-cover distribution: less than 3% total change was observed, indicating general landscape stability. Most transitions inside AENP occurred between thicket and grassland, while outside the park, grass–agriculture exchange was most common. These findings align with ecological expectations, highlighting AENP's role in maintaining natural vegetation, but also suggesting localized shifts that may warrant closer management attention (Safaei et al., 2025).

Impact

The significance of these findings is multi-faceted. Methodologically, the study demonstrates that high-accuracy, reliable land-cover maps for complex protected landscapes are best produced using advanced, carefully optimized ML models, with RF classifiers and a diverse suite of predictor variables proving most effective in this case. The relative underperformance of object-based methods and the UNet++ DL approach at 30-meter resolution underscores the importance of matching model and data to landscape heterogeneity and the spatial scale of analysis. The approach's success highlights the strategic value of integrating spectral, temporal, textural, and topographic information for landscape monitoring, resulting in classification outputs that can sensitively detect even small but ecologically relevant changes.

From a management and policy perspective, the results are reassuring but instructive. The modest rate of observed land-cover changes inside the park points to the ongoing effectiveness of AENP's management strategies in preserving natural and semi-natural habitat. However, the landscape is not static: the recurring transitions between thicket and grassland, potentially driven by factors such as herbivore pressure, fire regime, or climate variability, call for adaptive management. These subtle shifts, though numerically small, can have outsized effects on habitat quality, species composition, and broader ecosystem processes. Outside the park, the observed stability belies a dynamic interface with agriculture, implying the need for continued vigilance regarding land-use encroachment.

The broader impact of this work is to set a high standard for remote sensing-based land-cover monitoring in conservation contexts. By showing the importance of multi-source data integration and rigorous classifier selection, the study paves the way for future improvements, such as leveraging higher-resolution sensors, experimenting with new segmentation or active remote sensing (e.g., LiDAR, radar) inputs, and developing scenario models to forecast future land-use pressures. Ultimately, the approach offers a blueprint for protected area custodians and policymakers tasked with safeguarding biodiversity in the face of evolving threats. This study not only advances technical best practices in ecological remote sensing but directly informs conservation strategies in and around Addo Elephant National Park, enabling more precise, responsive, and evidence-based management for years to come (Safaei et al., 2025).

Further reading

Foundational Land System Science & LULCC

- Geist, H. J., & Lambin, E. F. (Eds.). (2006). Land-use and land-cover change: Local processes and global impacts. Springer.

  Why read this? This edited volume is a foundational reference for understanding how local land-use decisions scale to global environmental impacts and remains central to land change and land system science.

- Turner, B. L., Lambin, E. F., & Reenberg, A. (2007). The emergence of land change science for global environmental change and sustainability. *Proceedings of the National Academy of Sciences*, *104*(52), 20,666–20,671.

  Why read this? A seminal paper that formalizes land change science as a core discipline linking human–environment interactions to sustainability and global change research.

- Turner, B. L., Lambin, E. F., & Verburg, P. H. (2021). From land-use/land-cover to land system science. *Ambio*, *50*(7), 1291–1294.

  Why read this? This concise synthesis clarifies the conceptual shift from mapping LULC patterns to analyzing coupled land systems, directly aligning with the theoretical framing of this chapter.

Machine learning & deep learning in remote sensing

- Karpatne, A., Jiang, Z., Vatsavai, R. R., Shekhar, S., & Kumar, V. (2016). Monitoring land-cover changes: A machine-learning perspective. *IEEE Geoscience and Remote Sensing Magazine*, *4*(2), 8–21.

  Why read this? Provides a clear, early synthesis of how machine learning reshaped land cover change analysis, emphasizing scalability and pattern discovery.

- Liu, P., Choo, K. K. R., Wang, L., & Huang, F. (2017). SVM or deep learning? A comparative study on remote sensing image classification. *Soft Computing*, *21*, 7053–7065.

  Why read this? One of the earliest and most accessible studies directly comparing traditional ML and deep learning approaches for remote sensing classification.

- Ma, L., Liu, Y., Zhang, X., Ye, Y., Yin, G., & Johnson, B. A. (2019). Deep learning in remote sensing applications: A meta-analysis and review. *ISPRS Journal of Photogrammetry and Remote Sensing*, *152*, 166–177.

  Why read this? A comprehensive review of deep learning architectures, applications, and performance trends that situates DL within broader remote sensing workflows.

GeoAI, big data, and scaling land system science

- Gorelick, N., Hancher, M., Dixon, M., Ilyushchenko, S., Thau, D., & Moore, R. (2017). Google Earth Engine: Planetary-scale geospatial analysis for everyone. *Remote Sensing of Environment*, *202*, 18–27.

  Why read this? A landmark paper that fundamentally changed how large-scale land system and LULC analyses are conducted using cloud-based geospatial computing.

- Tamiminia, H., Salehi, B., Mahdianpari, M., Quackenbush, L., Adeli, S., & Brisco, B. (2020). Google Earth Engine for geo-big data applications: A meta-analysis and systematic review. *ISPRS Journal of Photogrammetry and Remote Sensing*, *164*, 152–170.

  Why read this? Synthesizes how cloud platforms enable scalable ML and DL applications for land cover mapping and land system analysis.

Interpretability, decision support, and policy relevance

- Shih, H., Stow, D. A., & Tsai, Y. H. (2019). Guidance on and comparison of machine learning classifiers for Landsat-based land cover and land use mapping. *International Journal of Remote Sensing*, *40*(4), 1248–1274.

  Why read this? Offers practical guidance on classifier choice, interpretability, and operational trade-offs—critical for policy-relevant land system applications.

- Southworth, J., Smith, A. C., Safaei, M., Rahaman, M., Alruzuq, A., Tefera, B. B., Muir, C. S., & Herrero, H. V. (2024). Machine learning versus deep learning in land system science: A decision-making framework for effective land classification. *Frontiers in Remote Sensing*, *5*, 1,374,862.

  Why read this? Directly extends the ML–DL decision framework developed in this chapter, providing a structured approach to method selection within LSS.

References

Ahmed, M. H., & Lin, L. S. (2021). Dissolved oxygen concentration predictions for running waters with different land use land cover using a quantile regression forest machine learning technique. *Journal of Hydrology, 597*, Article 126213.

Alhassan, V., Henry, C., Ramanna, S., & Storie, C. (2020). A deep learning framework for land-use/land-cover mapping and analysis using multispectral satellite imagery. *Neural Computing & Applications, 32*(12), 8529–8544. https://doi.org/10.1007/s00521-019-04349-9

Bajaj, K., Singh, D. K., & Ansari, M. A. (2020). Autoencoders based deep learner for image denoising. *Procedia Computer Science, 171*, 1535–1541. https://doi.org/10.1016/j.procs.2020.04.164

Ball, J. E., Anderson, D. T., & Chan, C. S. (2017). Comprehensive survey of deep learning in remote sensing: Theories, tools, and challenges for the community. *Journal of Applied Remote Sensing, 11*(4), Article 042609. https://doi.org/10.1117/1.JRS.11.042609

Barbierato, E., Bernetti, I., Capecchi, I., & Saragosa, C. (2020). Integrating remote sensing and street view images to quantify urban forest ecosystem services. *Remote Sensing, 12*(2), 329. https://doi.org/10.3390/rs12020329

Belgiu, M., & Drăguţ, L. (2016). Random forest in remote sensing: A review of applications and future directions. *ISPRS Journal of Photogrammetry and Remote Sensing, 114*, 24–31. https://doi.org/10.1016/j.isprsjprs.2016.01.011

Campos-Taberner, M., García-Haro, F. J., Martínez, B., Sánchez-Ruiz, S., Moreno-Martínez, Á., Camps-Valls, G., et al. (2023). Land use classification over smallholding areas in the European common agricultural policy framework. *ISPRS Journal of Photogrammetry and Remote Sensing, 197*, 320–334. https://doi.org/10.1016/j.isprsjprs.2023.02.005

Chen, B., Huang, B., & Xu, B. (2017). Multi-source remotely sensed data fusion for improving land cover classification. *ISPRS Journal of Photogrammetry and Remote Sensing, 124*, 27–39. https://doi.org/10.1016/j.isprsjprs.2016.12.008

Chen, J., & Ran, X. (2019). Deep learning with edge computing: A review. *Proceedings of the IEEE, 107*(8), 1655–1674.

Choudhary, K., DeCost, B., Chen, C., Jain, A., Tavazza, F., Cohn, R., et al. (2022). Recent advances and applications of deep learning methods in materials science. *Npj Computational Materials, 8*, 59. https://doi.org/10.1038/s41524-022-00734-6

Dalponte, M., Bruzzone, L., & Gianelle, D. (2008). Fusion of hyperspectral and LIDAR remote sensing data for classification of complex forest areas. *IEEE Transactions on Geoscience and Remote Sensing, 46*(5), 1416–1427.

Dong, H., Zhang, L., & Zou, B. (2022). Exploring vision transformers for polarimetric SAR image classification. *IEEE Transactions on Geoscience and Remote Sensing, 60*, 1–18. https://doi.org/10.1109/TGRS.2021.3137383

Dosovitskiy, A., Beyer, L., Kolesnikov, A., Weissenborn, D., Zhai, X., Unterthiner, T., ... Houlsby, N. (2021). An image is worth 16x16 words: Transformers for image recognition at scale. In *International conference on learning representations (ICLR)*. https://arxiv.org/abs/2010.11929.

Du, L., McCarty, G. W., Zhang, X., Lang, M. W., Vanderhoof, M. K., Li, X., et al. (2020). Mapping forested wetland inundation in the Delmarva Peninsula, USA using deep convolutional neural networks. *Remote Sensing, 12*(4), 644. https://doi.org/10.3390/rs12040644

Geist, H., McConnell, W., Lambin, E. F., Moran, E., Alves, D., & Rudel, T. (2006). Causes and trajectories of land-use/cover change. In *Land-use and land-cover change: Local processes and global impacts* (pp. 41–70). Berlin, Heidelberg: Springer Berlin Heidelberg.

Ghamisi, P., Chen, Y., & Zhu, X. X. (2016). A self-improving convolution neural network for the classification of hyperspectral data. *IEEE Geoscience Remote Sensing Letters, 13*(10), 1537–1541. https://doi.org/10.1109/lgrs.2016.2595108

Gorelick, N., Hancher, M., Dixon, M., Ilyushchenko, S., Thau, D., & Moore, R. (2017). Google Earth engine: Planetary-scale geospatial analysis for everyone. *Remote Sensing of Environment, 202*, 18–27. https://doi.org/10.1016/j.rse.2017.06.031

Guo, Y., Wang, Q., Hu, B., Qian, X., & Ye, H. (2025). Two-stage unsupervised hyperspectral band selection based on deep reinforcement learning. *Remote Sensing, 17*(4).

Hermosilla, T., Wulder, M. A., White, J. C., & Coops, N. C. (2022). Land cover classification in an era of big and open data: Optimizing localized implementation and training data selection to improve mapping outcomes. *Remote Sensing of Environment, 268*, Article 112780. https://doi.org/10.1016/j.rse.2021.112780

Hibbard, K., Janetos, A., van Vuuren, D. P., Pongtatz, J., Rose, S. K., Betts, R., … Feddema, J. J. (2010). *Research priorities in land use and land-cover change for the Earth system and integrated assessment modelling*.

Hong, D., Hu, J., Yao, J., Chanussot, J., & Zhu, X. X. (2021). Multimodal remote sensing benchmark datasets for land cover classification with a shared and specific feature learning model. *ISPRS Journal of Photogrammetry and Remote Sensing, 178*, 68–80. https://doi.org/10.1016/j.isprsjprs.2021.05.011

Huang, W., Zhan, J., Sun, Y., Han, X., An, T., & Jiang, N. (2025). *Context-aware adaptive sampling for intelligent data acquisition systems using DQN*. arXiv preprint arXiv.

Karpatne, A., Jiang, Z., Vatsavai, R. R., Shekhar, S., & Kumar, V. (2016). Monitoring land-cover changes: A machine-learning perspective. *IEEE Geoscience and Remote Sensing Magazine, 4*(2), 8–21.

Lambin, E. F., Turner, B. L., Geist, H. J., Agbola, S. B., Angelsen, A., Bruce, J. W., … Xu, J. (2001). The causes of land-use and land-cover change: Moving beyond the myths. *Global Environmental Change, 11*(4), 261–269.

Lary, D. J., Alavi, A. H., Gandomi, A. H., & Walker, A. L. (2016). Machine learning in geosciences and remote sensing. *Geoscience Frontiers, 7*(1), 3–10.

Li, Y., Bai, Y., & Zhang, H. (2023). Lightweight 3D dense autoencoder network for hyperspectral image classification with limited training samples. *Sensors, 23*(20), 8635. https://doi.org/10.3390/s23208635

Lin, T., Wang, Y., Liu, X., & Qiu, X. (2022). A survey of transformers. *AI Open, 3*, 111–132. https://doi.org/10.1016/j.aiopen.2022.10.001

Liu, P. (2015). A survey of remote-sensing big data. *Frontiers in Environmental Science, 3*, 45.

Liu, P., Choo, K. K. R., Wang, L., & Huang, F. (2017). SVM or deep learning? A comparative study on remote sensing image classification. *Soft Computing, 21*, 7053–7065.

Ma, L., Liu, Y., Zhang, X., Ye, Y., Yin, G., & Johnson, B. A. (2019). Deep learning in remote sensing applications: A meta-analysis and review. *ISPRS Journal of Photogrammetry and Remote Sensing, 152*, 166–177.

Magalhães, I. A. L., de Carvalho Júnior, O. A., de Carvalho, O. L. F., de Albuquerque, A. O., Hermuche, P. M., Merino, É. R., et al. (2022). Comparing machine and deep learning methods for the phenology-based classification of land cover types in the Amazon biome using sentinel-1 time series. *Remote Sensing, 14*(19), 4858. https://doi.org/10.3390/rs14194858

Masolele, R. N., De Sy, V., Herold, M., Marcos, D., Verbesselt, J., Gieseke, F., … Martius, C. (2021). Spatial and temporal deep learning methods for deriving land-use following deforestation: A pan-tropical case study using Landsat time series. *Remote Sensing of Environment, 264*, Article 112600.

Maturana, D., & Scherer, S. (September 2015). Voxnet: A 3d convolutional neural network for real-time object recognition. In *2015 IEEE/RSJ international conference on intelligent robots and systems (IROS)* (pp. 922–928). Ieee.

Michelucci, U. (2022). *An introduction to autoencoders*. arXiv. https://arxiv.org/abs/2201.03898.

Qi, C. R., Su, H., Mo, K., & Guibas, L. J. (2017). Pointnet: Deep learning on point sets for 3d classification and segmentation. In *Proceedings of the IEEE conference on computer vision and pattern recognition* (pp. 652–660).

Rahaman, M., Southworth, J., Wen, Y., & Keellings, D. (2025). Assessing model trade-offs in agricultural remote sensing: A review of machine learning and deep learning approaches using almond crop mapping. *Remote Sensing, 17*(15), 2670. https://doi.org/10.3390/rs17152670

Rindfuss, R. R., Walsh, S. J., Turner, B. L., Fox, J., & Mishra, V. (2004). Developing a science of land change: Challenges and methodological issues. *Proceedings of the National Academy of Sciences, 101*(39), 13976–13981. https://doi.org/10.1073/pnas.0401545101

Safaei, M., Southworth, J., Gibbes, C., Herrero, H. V., Rahaman, M., Tefera, B. B., & Blackburn, J. K. (2025). Land-cover classification in Addo Elephant National park: Analyzing the impact of variables, classifiers, and object-based approach. *Ecological Informatics, 90*. https://doi.org/10.1016/j.ecoinf.2025.103279

Sarker, I. H. (2021). Deep learning: A comprehensive overview on techniques, taxonomy, applications and research directions. *SN Computer Science, 2*(6), 420. https://doi.org/10.1007/s42979-021-00815-1

Shi, X., Chen, Z., Wang, H., Yeung, D. Y., Wong, W. K., & Woo, W. C. (2015). Convolutional LSTM network: A machine learning approach for precipitation nowcasting. *Advances in Neural Information Processing Systems, 28*.

Shih, H., Stow, D. A., & Tsai, Y. H. (2019). Guidance on and comparison of machine learning classifiers for Landsat-based land cover and land use mapping. *International Journal of Remote Sensing, 40*(4), 1248–1274. https://doi.org/10.1080/01431161.2018.1524179

Southworth, J., Smith, A. C., Safaei, M., Rahaman, M., Alruzuq, A., Tefera, B. B., Muir, C. S., & Herrero, H. V. (2024). Machine learning *versus* deep learning in land system science: A decision-making framework for effective land classification. *Frontiers in Remote Sensing, 5*, Article 1374862. https://doi.org/10.3389/frsen.2024.1374862

Tamiminia, H., Salehi, B., Mahdianpari, M., Quackenbush, L., Adeli, S., & Brisco, B. (2020). Google Earth engine for geo-big data applications: A meta-analysis and systematic review. *ISPRS Journal of Photogrammetry and Remote Sensing, 164*, 152–170.

Tian, C., Zhang, X., Lin, J. C. W., Zuo, W., Zhang, Y., & Lin, C. W. (2022). *Generative adversarial networks for image super-resolution: A survey*. arXiv preprint arXiv.

Turner, B. L., Geoghegan, J., Lawrence, D., Radel, C., Schmook, B., Vance, C., … Ogenva-Himmelberger, Y. (2016). Land system science and the social–environmental system: The case of Southern Yucatán Peninsular region (SYPR) project. *Current Opinion in Environmental Sustainability, 19*, 18–29.

Turner, B. L., Lambin, E. F., & Reenberg, A. (2007). The emergence of land change science for global environmental change and sustainability. *Proceedings of the National Academy of Sciences, 104*(52), 20666–20671. https://doi.org/10.1073/pnas.0704119104

Turner, B. L., Lambin, E. F., & Verburg, P. H. (2021). From land-use/land-cover to land system science: This article belongs to Ambio's 50th anniversary collection. Theme: Agricultural land use. *Ambio, 50*, 1291–1294. https://doi.org/10.1007/s13280-021-01510-4

Uparkar, O., Bharti, J., Pateriya, R. K., Gupta, R. K., & Sharma, A. (2023). Vision transformer outperforms deep convolutional neural network-based model in classifying X-ray images. *Procedia Computer Science, 218*, 2338–2349. https://doi.org/10.1016/j.procs.2023.01.209

Vaswani, A., Shazeer, N., Parmar, N., Uszkoreit, J., Jones, L., Gomez, A. N., Kaiser, Ł., & Polosukhin, I. (2017). Attention is all you need. *Advances in Neural Information Processing Systems, 30*, 5998–6008. https://arxiv.org/abs/1706.03762.

Verburg, P. H., Dearing, J. A., Dyke, J. G., Van Der Leeuw, S., Seitzinger, S., Steffen, W., & Syvitski, J. (2016). Methods and approaches to modelling the Anthropocene. *Global Environmental Change, 39*, 328–340.

Verburg, P. H., Erb, K. H., Mertz, O., & Espindola, G. (2013). Land system science: Between global challenges and local realities. *Current Opinion in Environmental Sustainability, 5* (5), 433–437. https://doi.org/10.1016/j.cosust.2013.08.001

Wahid, A., Yahya, M., Breslin, J. G., & Intizar, M. A. (2023). Self-attention transformer-based architecture for remaining useful life estimation of complex machines. *Procedia Computer Science, 217*, 456–464. https://doi.org/10.1016/j.procs.2022.12.241

Xie, Q., Lai, K., Wang, J., Lopez-Sanchez, J. M., Shang, J., Liao, C., Zhu, J., Fu, H., & Peng, X. (2021). Crop monitoring and classification using polarimetric RADARSAT-2 time-series data across growing season: A case study in Southwestern Ontario, Canada. *Remote Sensing, 13*(7), 1394. https://doi.org/10.3390/rs13071394

Yu, S., Jia, S., & Xu, C. (2017). Convolutional neural networks for hyperspectral image classification. *Neurocomputing, 219*, 88–98. https://doi.org/10.1016/j.neucom.2016.09.010

Zhang, X., Han, L., Han, L., & Zhu, L. (2020). How well do DL-based methods for land cover classification and object detection perform on high resolution remote sensing imagery? *Remote Sensing, 12*(3), 417. https://doi.org/10.3390/rs12030417

Zhang, C., Harrison, P. A., Pan, X., Li, H., Sargent, I., & Atkinson, P. M. (2020). Scale sequence joint deep learning (SS-JDL) for land use and land cover classification. *Remote Sensing of Environment, 237*, Article 111593. https://doi.org/10.1016/j.rse.2019.111593

Zhang, J., Liu, Y., & Zhou, W. (2023). Adaptive sampling path planning for a 3d marine observation platform based on evolutionary deep reinforcement learning. *Journal of Marine Science and Engineering, 11*(12), 2313.

Zhang, C., Sargent, I., Pan, X., Li, H., Gardiner, A., Hare, J., et al. (2019). Joint deep learning for land cover and land use classification. *Remote Sensing of Environment, 221*, 173–187. https://doi.org/10.1016/j.rse.2018.11.014

Zhou, G., Chen, W., Qin, X., Li, J., & Wang, L. (2023). Lithological unit classification based on geological knowledge-guided deep learning framework for optical stereo mapping satellite imagery. *IEEE Transactions on Geoscience and Remote Sensing, 61*, 1–16.

Zhou, G., Xu, J., Chen, W., Li, X., Li, J., & Wang, L. (2023). Deep feature enhancement method for land cover with irregular and sparse spatial distribution features: A case study on open-pit mining. *IEEE Transactions on Geoscience and Remote Sensing, 61*, 1–20. https://doi.org/10.1109/tgrs.2023.3241331

Zhu, X. X., Tuia, D., Mou, L., Xia, G. S., Zhang, L., Xu, F., & Fraundorfer, F. (2017). Deep learning in remote sensing: A comprehensive review and list of resources. *IEEE Geoscience and Remote Sensing Magazine, 5*(4), 8–36.

Integration of Remote Sensing, Big Data, and GeoAI

6

Yaqian He[1,2], *Di Yang*[3], *Brad Peter*[1], *Xiao Huang*[4] *and Faith Vire*[5]
[1]Department of Geosciences, University of Arkansas, Fayetteville, AR, United States; [2]Department of Geography, Indiana University Bloomington, Bloomington, IN, United States; [3]Department of Geography, University of Florida, Gainesville, FL, United States; [4]Department of Environmental Sciences, Emory University, Atlanta, GA, United States; [5]Department of Geography, University of Central Arkansas, Conway, AR, United States

1. Introduction

The field of remote sensing (RS) has undergone rapid development in recent years, driven by advancements in technology, data availability, and analytical techniques. Integrating RS with big data and GeoAI has opened new frontiers in Earth observation and environmental monitoring, offering unprecedented opportunities to understand our planet's ecosystems (Li et al., 2016) and provide feasible pathways for sustainable development.

RS, the practice of acquiring information about the Earth's surface without direct contact, has long been a basis of geospatial science. Traditionally, it has relied on satellite and airborne platforms to collect data across various spectral bands. However, RS has expanded dramatically with the advent of new sensors, platforms, and data sources. These include high-resolution satellite imagery (e.g., WorldView 3) (Li et al., 2015), light detection and ranging (lidar) (Reutebuch et al., 2005), unmanned aerial vehicles (UAVs) (Mohsan et al., 2022), Synthetic Aperture Radar (SAR) (Tsokas et al., 2022), and crowdsourced geospatial data (Basiri et al., 2019). The rapid expansion of these data sources has resulted in a massive increase in the volume, variety, and velocity of geospatial data, which are key characteristics of big data. This immense information presents both challenges and opportunities for the RS community (Yang et al., 2019). On one hand, it offers the potential for more detailed and frequent observations of Earth systems. On the other hand, it necessitates new approaches to data management, processing, and analysis (Gorelick et al., 2017).

GeoAI, an emerging field, utilizes machine learning, deep learning, and other AI approaches to extract meaningful information from complex geospatial datasets. This convergence of RS, big data, and GeoAI is transforming how we observe, analyze, and comprehend our planet (Janowicz et al., 2020). The synergy between RS, big data, and GeoAI is evident in numerous applications. In agriculture, this integration enables precision farming by providing timely information on crop health, soil moisture, and yield prediction (Liakos et al., 2018). In urban planning, it facilitates the mapping of land use changes, infrastructure monitoring, and the development of smart

Remote Sensing, Big Data, and GeoAI. https://doi.org/10.1016/B978-0-443-26741-3.00003-9

cities (Koumetio Tekouabou et al., 2022). In disaster management, it allows for rapid damage assessment and response coordination through near-real-time analysis of satellite and aerial imagery (Linardos et al., 2022; Voigt et al., 2016).

A critical component of the integration of RS, big data, and GeoAI is data fusion–the process of combining data from multiple sources to produce more consistent, accurate, and useful information than that provided by any individual source (Zhang, 2010a). In the context of RS, data fusion can merge data from different sensors, platforms, or even disparate data types (e.g., satellite imagery with ground-based observations) (He et al., 2016a,b). Data fusion can be implemented at different levels (i.e., pixel-level, feature-level, and decision-level fusion) (Huang et al., 2023). Pixel-level fusion combines raw data from multiple sources before any further processing. Feature-level fusion extracts features from different data sources and then combines these features. Decision-level fusion integrates the results after each data source has been processed independently (Pohl and van Genderen, 2014). Despite its great potential and flexibility, data fusion in RS remains a tremendous challenge, particularly when dealing with datasets that vary widely in spatial, temporal, and spectral resolutions. Integrating these multi-scale datasets requires careful consideration of the appropriate spatial and temporal scales for the intended application (Pohl and van Genderen, 2014). For example, when fusing high-resolution satellite data (e.g., WorldView, GeoEye) with lower-resolution weather or climate data (e.g., ERA5), the discrepancy in spatial scales can lead to misalignment and decreased accuracy in the final product. Techniques such as spatial downscaling or upscaling may be necessary to harmonize the data before fusion (Sun et al., 2024). Additionally, the diverse RS data, including optical imagery, radar, lidar, and other sensor types, introduces further complexity to the data fusion process. These data sources have different physical properties, measurement methods, strengths, and shortcomings. Effectively integrating these various data requires specific algorithms and thoughtful attention (Zhang, 2010b).

In this chapter, we will survey data fusion methods, examine scale issues in multi-source data, and outline the main steps in RS data integration. We will also draw upon real-world case studies to highlight both successful and failed data integration, underscoring the challenges and best practices in this domain.

2. Theoretical foundations

2.1 Data fusion

In RS, data fusion techniques generally attempt to improve the estimation of an object or area on Earth's surface through the combination of information from multiple sensors (Ghamisi et al., 2019; Schmitt and Zhu, 2016; Zhu et al., 2018). The output of these data fusion techniques is a set of synthesized images with the temporal frequency of one sensor and the spatial resolution of another sensor, or the harmonization of two sensors with similar spatial and temporal resolutions to create consistent observations (Zhu et al., 2018).

2.1.1 Spatiotemporal data fusion

Early image fusion in RS focuses on combining high-resolution panchromatic and low-resolution multispectral bands from a single sensor to create high-resolution multispectral images (Chen et al., 2015; Gharbia et al., 2014; Zhu et al., 2018). This image fusion is different from spatiotemporal data fusion in that it does not alter the temporal resolution of the imagery (Zhu et al., 2018). Spatiotemporal data fusion is increasingly popular in RS applications because remote sensors, such as space-borne satellites, usually have only one area of high resolution (Sisheber et al., 2022). For example, the Landsat family of sensors have high spatial resolution (e.g., 30 m) but coarse temporal resolution (e.g., 16-Day), while Moderate Resolution Imaging Spectroradiometer (MODIS) sensors have high temporal resolution (e.g., daily or 8-Day) and coarse spatial resolution (e.g., 250 m). Imagery with high spatial resolution is optimal because these images offer greater amounts of geographic information than images with a coarse spatial resolution (Sisheber et al., 2022). Imagery with a high temporal resolution, on the other hand, is beneficial because of its ability to capture rapid changes in Earth's surface and its enhanced likelihood of retrieving images without atmospheric disturbances (Gao et al., 2006). Spatiotemporal data fusion can create images with high resolutions of all types, which is impossible to accomplish with a single sensor. Data with enhanced resolution can be used to overcome the tradeoff between spatial and temporal resolutions. As such, the fusion of RS imagery from different sources can produce richer, more precise characterizations of Earth's surface.

Many different spatiotemporal data fusion techniques exist, and new methods are constantly being developed, largely because fusion performance varies across sensors, spatial and temporal scales, and land surface conditions (Zhu et al., 2018). Zhu et al. (2018) identified as many as 44 different spatiotemporal data fusion methods. They group fusion methods into five categories: unmixing-based, weight function-based, Bayesian-based, learning-based, and hybrid. *Unmixing-based fusion methods* are grounded in the concept of spectral mixture analysis. These techniques assume that each pixel in a coarse-resolution image is a mixture of different endmembers. By decomposing these mixed pixels and then reconstructing them at a finer resolution, these methods can produce higher-resolution fused images. A typical example is the Spatial and Temporal Reflectance Unmixing Model (STRUM) proposed by Gevaert and García-Haro (2015). STRUM uses high-resolution land cover maps to guide the unmixing process, allowing for the generation of daily, high-resolution surface reflectance data from coarser observations. *Weight function-based methods* assign different weights to input images based on their relevance to the desired output. These weights are then used to combine the input images into a fused product at a finer scale. The Spatial and Temporal Adaptive Reflectance Fusion Model (STARFM) by Gao et al. (2006) is a classic example of this approach. STARFM uses a weight function that takes into account spectral similarity, temporal proximity, and spatial distance to blend images (e.g., Landsat and MODIS data), producing synthetic images with higher spatial and temporal resolutions. STARFM has been extensively applied in various environmental monitoring research (Dhillon et al., 2023; Zhang et al.,

2014). Despite its wide application, STARFM assumes that temporal changes within each coarser pixel are consistent across all land cover classes, making it more suitable for homogeneous landscapes like large agricultural areas (Qin et al., 2022). To address these limitations, many improved methods were developed (Hilker et al., 2009). For instance, Zhu et al. (2010) introduced Enhanced STARFM (ESTARFM), an approach that enhances spatiotemporal fusion accuracy in regions with heterogeneous landscapes. *Bayesian-based fusion methods* use probabilistic models to combine different data sources, accounting for uncertainties in the input data and model parameters. These methods can provide not only fused estimates but also measures of uncertainty in the fusion results (Li et al., 2013). Zhu et al. (2024) applied the Bayesian maximum entropy (BME) method considering multi-spatial-scale uncertainties to generate fine-scale surface soil moisture data. With the advent of machine learning and deep learning, *learning-based fusion methods* have gained tremendous attention. These approaches use training data to learn the relationship between low-resolution and high-resolution images, which can then be applied to new data. Machine learning algorithms, such as random forest, can fuse data from multiple data sources (Ke et al., 2016). Although deep learning is still in its early stages for spatiotemporal data fusion, it has already been successfully applied to fuse RS imagery (Li and Hsu, 2022). For example, Wang and Li (2021) developed a new GeoAI method that enables deep learning from multi-source geospatial data (i.e., digital elevation model [DEM] and National Agriculture Imagery Program [NAIP]) for natural feature detection. *Hybrid methods* combine two or more of the above approaches to leverage their respective strengths. These methods often yield improved results by addressing the limitations of individual techniques. For instance, Zhu et al. (2016) introduced the flexible spatiotemporal data fusion (FSDAF) method, a hybrid approach that integrates concepts from unmixing-based methods with spatial interpolation and STARFM. FSDAF enhances spatiotemporal fusion accuracy in regions with heterogeneous landscapes and abrupt land cover changes.

Each of these spatiotemporal data fusion approaches has their strengths and limitations. For example, unmixing-based methods can handle complex spectral mixtures but often struggle with the temporal dynamics. Weight function-based methods are considered more computationally efficient but may not capture the fine spatial details. Bayesian methods provide uncertainty estimates, but this can make these approaches much more computationally intensive. Learning-based methods can capture complex relationships but to do so accurately requires large training datasets, which may not always be available. Hybrid methods aim to balance these trade-offs but as a result of this may be more complex to implement and tune (Moreno-Martínez et al., 2020). Therefore, there is no single best method to be recommended but rather, the choice of fusion method depends on the specific application, the characteristics of the input data, and the desired output. As RS technology continues to advance, we can only expect further innovations in these fusion methods, particularly in the GeoAI approaches, and so consideration of the best technique will only continue to be an important consideration for each researcher.

2.2 Scale issue

Within RS data fusion, beyond selecting an appropriate fusion approach, practitioners must also determine the spatial and temporal scales, at which the data should be integrated. These decisions are closely related to the scale issue. The issues of scale are foundational to the discipline of geography and RS and arguably present in any analysis that operates with a spatial or temporal component (Haggett, 1970; Harris, 2006; Meentemeyer, 1989; Thrift, 1977). While spatial and temporal selections must be made due to various constraints, as a matter of practicality, or simply as an inescapable reality, there is inevitably some degree of uncertainty introduced that can make results relative only to those bounds (Peter et al., 2024). Problematically, the effects of such selections can lead to differing conclusions drawn and errors propagated, particularly in multi-input models (Lunetta et al., 1991). These long-standing phenomena have been coined as the Modifiable Areal Unit Problem (MAUP) and the Modifiable Temporal Unit Problem (MTUP). The MAUP is the observation that spatial aggregations and analyses will vary depending on the spatial scale of observation and/or the delineations of the zones used, referred to as the scale problem and zone problem within MAUP, respectively (Openshaw, 1983; Wong, 2004). To manage problems such as MAUP, methods range from sensitivity analyses, multi-scale evaluations, and simply acknowledging them as limitations (Comber et al., 2022; Duque et al., 2018; Manley, 2014). Similarly, the MTUP is the observation that temporal aggregations and analyses will vary depending on the temporal scale of observation and/or temporal resolution used (Cheng & Adepeju, 2014; Çöltekin et al., 2011). Both MAUP and MTUP change how data are presented and perceived at the cartography, visualization, and quantification stages (Bird, 1989; Peter et al., 2024). Figs. 6.1 and 6.2 illustrate how different scales, i.e., spatial and temporal selections, can present different views of the same underlying data.

A subset of MAUP, the Ecological Fallacy Problem (EFP) is a fallacy in which assumptions are made about an individual based on data aggregated at a broader scale (Piantadosi et al., 1988). The EFP is commonly referenced in association with vector data, e.g., assuming that a home within a county will match the median home value of that county would be fallacious. Less discussed is how EFP manifests in raster-based analyses. An example in the raster context would be the use of coarse-resolution imagery to describe a locally complex landscape. To assume that any location within a classification will match the characteristics of the complete pixel would be fallacious (Hsieh et al., 2001). This is a challenge in smallholder agricultural landscapes where land use is mixed at local scales and one 500-meter spatial resolution pixel (e.g., MODIS land cover and land use [LCLU] (Friedl and Sulla-Menashe, 2022) may contain multiple LCLU types, such as trees, buildings, agriculture, and water bodies (Fig. 6.3). It is a reminder that RS scientists must be cognizant of the underlying people, environment, and wildlife within the pixels they manipulate and map.

With the continual launch of new platforms, the growing availability of satellite data, and advances in GeoAI and big data analytics, practitioners are able to link more and more datasets with varying spatial, temporal, and spectral characteristics for myriad purposes (Janowicz et al., 2020; Yang et al., 2022). Among Landsat,

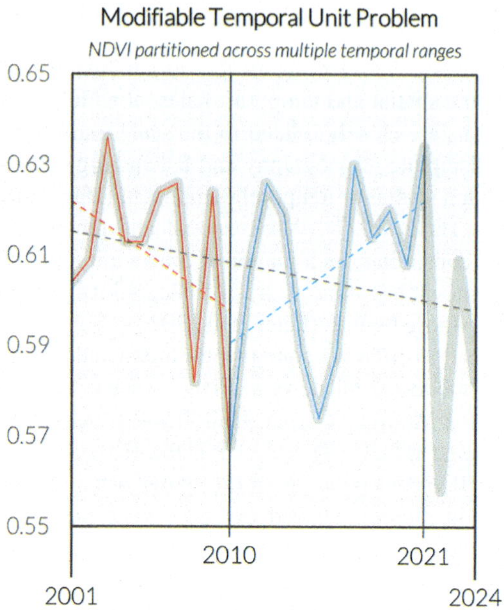

Figure 6.1 Variable time representations of agricultural productivity across the district of lilongwe in Central malawi. Normalized Difference Vegetation Index (NDVI) data were aggregated using the mean for the months between january and april from 2001 to 2024. Data Source: MODIS Terra Daily NDVI: MOD13Q1.061, 250-m spatial resolution and 16-day temporal resolution (Didan, 2021).

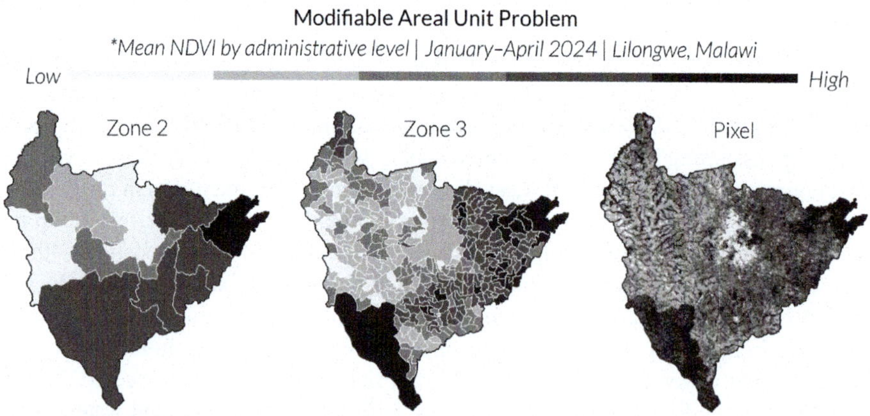

Figure 6.2 Variable geographic representations of vegetation productivity across the district of lilongwe in Central malawi. NDVI data were aggregated using the mean for the months between january and april for 2024.
Data Source: MODIS Terra Daily NDVI: MOD13Q1.061, 250-m spatial resolution and 16-Day temporal resolution (Didan, 2021). Zone boundaries collected from GADM Version 4.1.
*Minimum and maximum ranges vary per zonal selection and were symbolized using natural breaks to illustrate the spatial heterogeneity.

Sentinel-2, and MODIS, spatial resolutions range between 10 and 1000 m, temporal resolutions range from 5 to 16 days, and spectral resolutions are unique to each sensor. There is not a ubiquitous standard practice to merge datasets of variable spatial resolutions, and data loss can take place when resampling, particularly when resampling from finer to coarser spatial resolutions. Furthermore, users of raw Landsat and Sentinel-2 imagery will notice cross-tile discrepancies, which often appear as straight lines along mosaic edges. Recognizing these types of scale challenges, agencies such as NASA, USGS, and ESA have worked to deliver analysis-ready, harmonized data. One such example is the Harmonized Landsat and Sentinel-2 (HLS) product, which aims to create a seamless time-series of surface reflectance data across both systems (Claverie et al., 2018). As a platform, Google Earth Engine (GEE) uses a downsampling approach that allows multi-source data to be integrated without a need to resample or reproject to a common grid (Gorelick et al., 2017). As discussed previously, advanced technical solutions to problems of mixed spatial, temporal, and spectral resolutions include machine/deep learning to fuse data from multiple sources (Li et al., 2022).

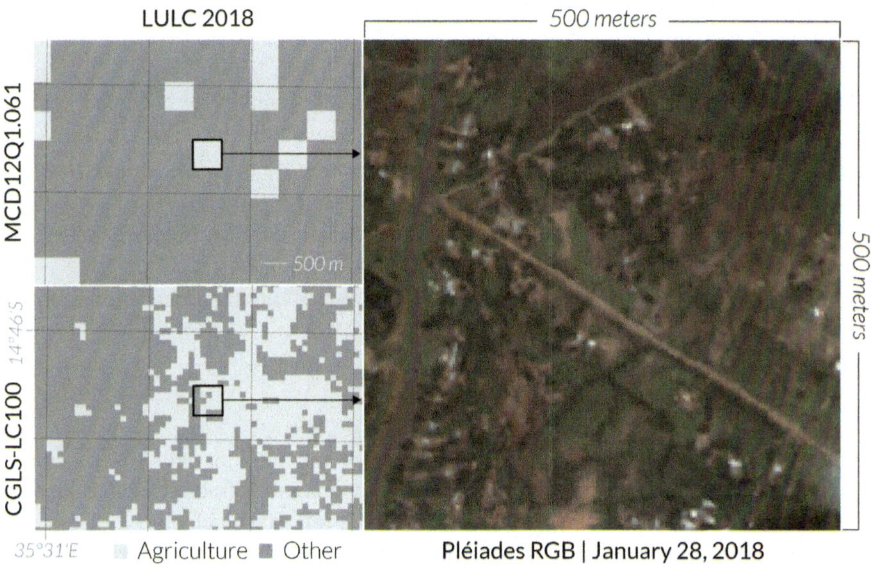

Figure 6.3 Elucidating the ecological fallacy problem in raster classifications. Spatial organization of a local-scale farming landscape in Central malawi related to global-scale agricultural land-use maps.
Data Sources: (1) MODIS Land Cover Type: MCD12Q1.061, 500-m spatial resolution, and annual temporal resolution (Friedl and Sulla-Menashe, 2022) and (2) Copernicus Global Land Cover Layers: CGLS-LC100, 100-m spatial resolution, and annual temporal resolution (Buchhorn et al., 2020).

3. Methodologies

In this section, we demonstrate several widely used fusion methods as examples for data integration that can remedy scale issues across different datasets. We provide detailed, step-by-step instructions on how to apply each of these methods.

3.1 *Improved spatial and temporal reflectance unmixing model (STRUM)*

Gevaert and García-Haro (2015) developed the STRUM method, an unmixing-based method that requires one pair of coarse-fine resolution images at the same time (t_0) and a coarse-resolution image at the prediction time (t_k) (Gevaert and García-Haro, 2015). The STRUM method can match spatial scales from different resolution images by linking endmembers, which is discussed below. The STRUM fusion includes four main steps (Fig. 6.4): (1) classify the fine-resolution image to obtain the endmembers of coarse resolution image; (2) calculate endmember component fraction within each coarse pixel; (3) unmix the coarse pixels based on the endmember fractions; and (4) assign reflectance values to the fine-resolution pixels.

Step 1. Classification- Various classification algorithms, such as K-nearest neighbors, support vector machines, and random forests, can be employed to generate classes or endmembers in fine-resolution image (He et al., 2017; Maxwell et al., 2018). However, if existing land cover data is available for the same area, the classification step may be unnecessary.

Step 2. Calculating the abundance- To calculate the fraction for each endmember, this step applies a $[n, n]$ moving window of coarse-resolution pixels to the classified map. It will generate an endmember fraction matrix A, a $[n^2, k]$ matrix with n^2 rows and k columns. Each row is corresponding to each pixel within the moving window and each column is for each endmember (Gevaert and García-Haro, 2015). The moving window traverses the entire coarse-resolution image.

Step 3. Unmixing- The unmixing is performed by solving the linear mixing model Eq. (6.1) (Gevaert and García-Haro, 2015).

$$diff = Ae + \varepsilon \tag{6.1}$$

where, $diff$ is the reflectance difference between the moving window in coarse-resolution image at the time t_0 and t_k; A is the endmember fraction matrix derived from Step 2; e is a column vector that contains the reflectance change for each endmember; and ε is the residual. The assumption is that the reflectance difference change from coarse-resolution images at different time steps are the same as the ones from the fine-resolution images. Therefore, if the e is solved for each endmember, it is able to predict the fine-resolution reflectance at the time t_k.

Traditionally, unmixing methods solve the e by minimizing the residuals ε of the linear model. However, it may produce unrealistic reference values (e.g., negative value) for the endmembers (Zurita-Milla et al., 2008). To remedy the issue, Gevaert

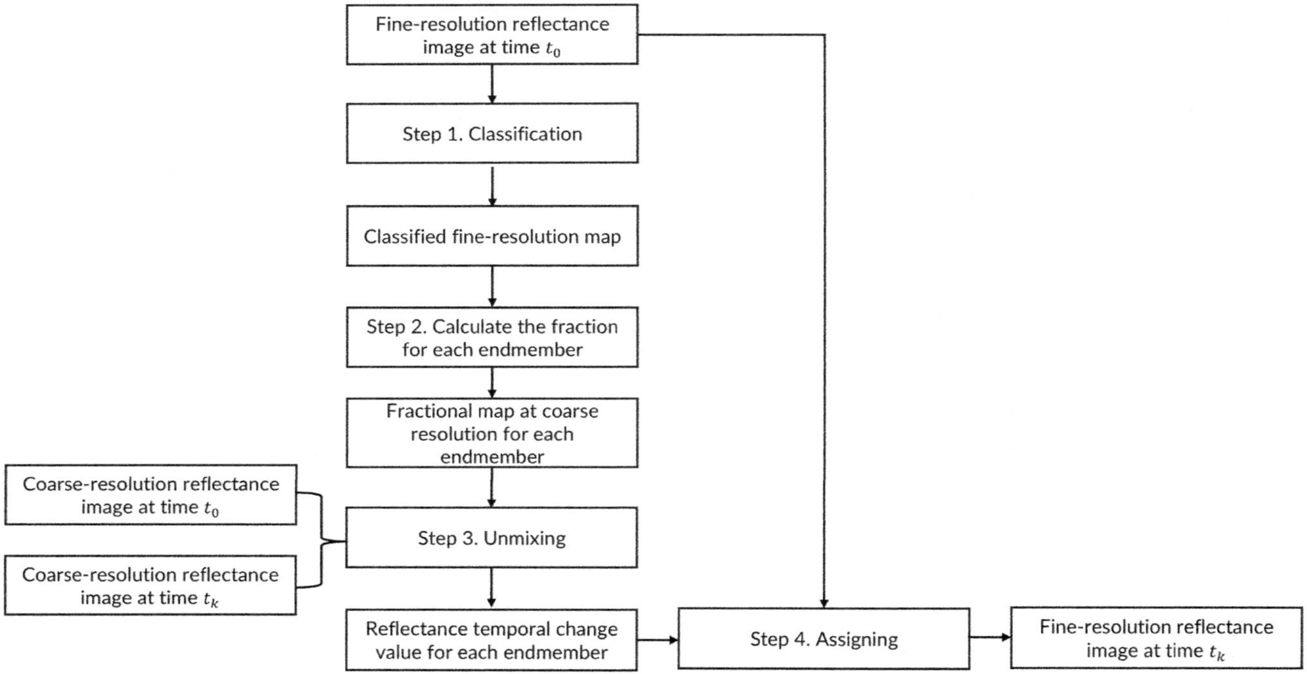

Figure 6.4 Flowchart of STRUM fusion method.

and García-Haro (2015) used a Bayesian method to constrain the estimation of the e by including the prior end member means e_0 from coarse-resolution pixels with high abundance levels for each class.

Step 4. Assigning- The final step predicts the fine-resolution reflectance values $L(t_k)$ in each moving window at the time t_k based on the e and the corresponding fine-resolution reflectance values $L(t_0)$ at the time t_0 Eq. (6.2).

$$L(t_k) = L(t_0) + e \tag{6.2}$$

Using unmixing-based fusion methods, one critical factor must be considered: the relationship between the number of clusters and the size of the moving window. Spectral unmixing can be particularly sensitive to co-linearity issues, where high correlations between endmembers may cause problems with matrix inversion (García-Haro et al., 2005). Although it is theoretically possible to use as many classes (k) as there are coarse-resolution pixels in a neighborhood (n^2), the presence of highly correlated data can significantly distort the estimates (Gevaert and García-Haro, 2015). Therefore, users can use n^2 that is substantially larger than the number of classes k or exclude some rare classes (e.g., the fractional value is less than 0.1) (Gevaert and García-Haro, 2015).

3.2 Spatial and temporal adaptive reflectance fusion model (STARFM)

The STARFM fusion method developed by Gao et al. (2006), is a weighted function-based approach. The STARFM method also requires at least one pair of coarse-fine resolution images from the same time (t_k) and a coarse-resolution image at the prediction time (t_0). The STARFM fusion method can address both spatial and temporal scale issues by using weights, which are discussed below. There are five main steps in the STARFM fusion method (Fig. 6.5), including (1) resampling coarse-resolution reflectance images at the time t_k and t_0 to fine-resolution; (2) searching pixels that are spectrally similar to the central pixel within a moving window; (3) filtering spectrally similar pixels to exclude poor-quality observations; (4) determining weights based on fine-resolution and resampled coarse-resolution reflectance images at the time t_k; and (5) assigning reflectance values to the fine-resolution pixels.

Step 1. Resampling- There are multiple resampling methods, such as nearest-neighbor method, bilinear interpolation method, cubic convolution method, amongst others (He et al., 2018). Given the interval nature of reflectance data, users may consider using bilinear interpolation or cubic convolution methods.

Step 2. Searching spectrally similar pixels- There are two ways to obtain spectrally similar pixels. An unsupervised classification method (e.g., Interactive Self-Organizing [ISO] in Gulhane et al. (2023)) can be applied to the fine-resolution image at the time t_k and group pixels into different classes. The pixels in the same class are considered spectrally similar. Another way is using thresholds in surface reflectance to find spectrally similar pixels in the fine-resolution image at the time t_k (Gao et al.,

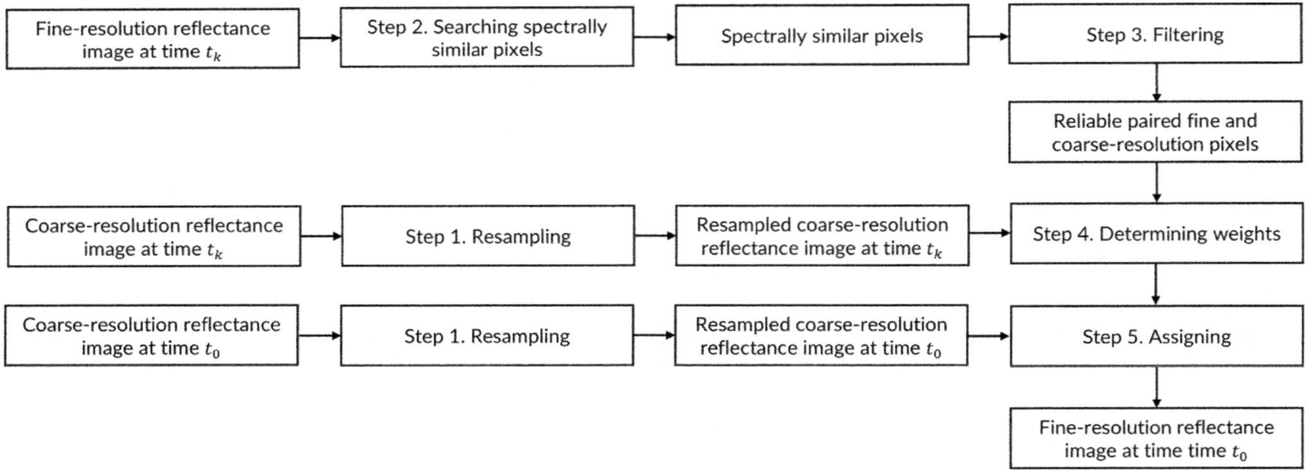

Figure 6.5 Flowchart of STARFM fusion method.

2006). Gao et al. (2006) used a method like the unsupervised classification but searched neighbor pixels within the local moving window that are spectrally similar to the central pixel.

Step 3. Filtering- The spectrally similar pixels derived from Step 2 may have poor-quality reflectance values, which need to be removed. These poor-quality data can be filtered out based on the quality layer provided by the data producers (Gao et al., 2006). Pixels are also removed if they cannot provide additional information. For example, good pixels should satisfy the conditions shown in Eqs. (6.3) and (6.4) (Gao et al., 2006).

$$S_{ijk} < \max\left(\left|L\left(x_{w/2}, y_{w/2}, t_k\right) - M\left(x_{w/2}, y_{w/2}, t_k\right)\right|\right) \tag{6.3}$$

$$T_{ijk} < \max\left(\left|M\left(x_{w/2}, y_{w/2}, t_k\right) - M\left(x_{w/2}, y_{w/2}, t_0\right)\right|\right) \tag{6.4}$$

where, $S_{ijk} = \left|L\left(x_i, y_j, t_k\right) - M\left(x_i, y_j, t_k\right)\right|$ is the reflectance difference between resampled coarse-resolution (M) and fine-resolution (L) images at a given location (x_i, y_j) at the time t_k; $T_{ijk} = \left|M\left(x_i, y_j, t_k\right) - M\left(x_i, y_j, t_0\right)\right|$ is the temporal difference between the coarse-resolution (M) images at the time t_k and t_0; w is the size of the moving window.

Step 4. Determining weights- The weights w_{ijk} determine the contributions of each neighboring pixel to the estimated reflectance of the central pixel in the moving window. The weights are based on the spectral difference S_{ijk}, temporal differences T_{ijk}, and location distance $d_{ijk} = \sqrt{\left(x_{w/2} - x_i\right)^2 + \left(y_{w/2} - y_j\right)^2}$ Eq. (6.5) (Gao et al., 2006).

$$W_{ijk} = \left(1 / C_{ijk}\right) / \sum_{i=1}^{w} \sum_{j=1}^{w} \sum_{k=1}^{n} \left(1 / C_{ijk}\right) \tag{6.5}$$

where, $C_{ijk} = S_{ijk} \times T_{ijk} \times d_{ijk}$; n is the number of pairs of coarse-fine resolution images.

Step 5. Assigning- Finally, the fine-scale reflectance at the time t_0: $L\left(x_{w/2}, y_{w/2}, t_0\right)$ is estimated by Eq. (6.6).

$$L\left(x_{\frac{w}{2}}, y_{\frac{w}{2}}, t_0\right) < \sum_{i=1}^{w} \sum_{j=1}^{w} \sum_{k=1}^{n} W_{ijk} \times \left(M\left(x_i, y_j, t_0\right) + L\left(x_i, y_j, t_k\right) - M\left(x_i, y_j, t_k\right)\right) \tag{6.6}$$

3.3 Bayesian maximum entropy (BME)

BME is a non-linear geostatistical method that can blend multi-source datasets. BME can address scale issues due to discrepancies in spatial resolution from different datasets by incorporating an error model related to covariance (Li et al., 2013). In addition, BME can integrate both hard data and soft data. The soft data are measurements with errors and are described in the form of probability functions or intervals of values, whereas the hard data are exact measurements with negligible errors (Christakos, 2002). In RS fusion, we consider fine-resolution data as hard data and coarse-resolution data as soft data. Furthermore, under the Bayesian paradigm, BME can provide an estimated error variance from the posterior probability distribution (Christakos et al., 2004). More importantly, BME does not have limitations on the number of pairs of coarse and fine-resolution data. The main steps in BME fusion are shown in Fig. 6.6.

Step 1. Spatiotemporal trend removal- To satisfy the assumption of second-order stationarity (constant mean and variance) over the entire spatiotemporal domain, which is required by BME, the removal of the spatiotemporal trend is necessary Eq. (6.7).

$$x(s,t) = \overline{x(s,t)} + \varepsilon(s,t) \tag{6.7}$$

where s represents space and t represents time. $\overline{x(s,t)}$ is the space-time mean trend of the reflectance value, which is estimated by Gaussian kernel smoothing (Yu et al., 2007). The residual $\varepsilon(s,t)$ is homogenous in space and stationary in time.

Step 2. Error model- The scaling transformation is the basic procedure for unifying the multi-resolution data into one scale, but arbitrary resampling from coarse resolution to fine resolution may induce new uncertainty. BME utilizes the error model (Lee et al., 2009) to establish the relationship between coarse and fine-resolution data. The procedure is described below.

Let $X(s)$ be defined as a variable at a fine scale centered at s, and $Z(s')$ is defined as the average of $X(s)$ over a 2-D spatial domain A centered on s':

$$Z(s') = \int d\mu X(\mu) \,/\, \|As'\| \tag{6.8}$$

To establish the relationship between the fine-scale $X(s)$ and the coarse-scale $Z(s')$, a new spatial random field $Y(s,s')$ is defined:

$$Y(s,s') = X(s) - Z(s') \tag{6.9}$$

When the spatial random field is homogeneous and the mean trend is removed, the expected value of $Y(s,s')$, i.e., $E(Y(s,s'))$ is zero. The variance describing the uncertainty associated with $Y(s,s')$ can be calculated as:

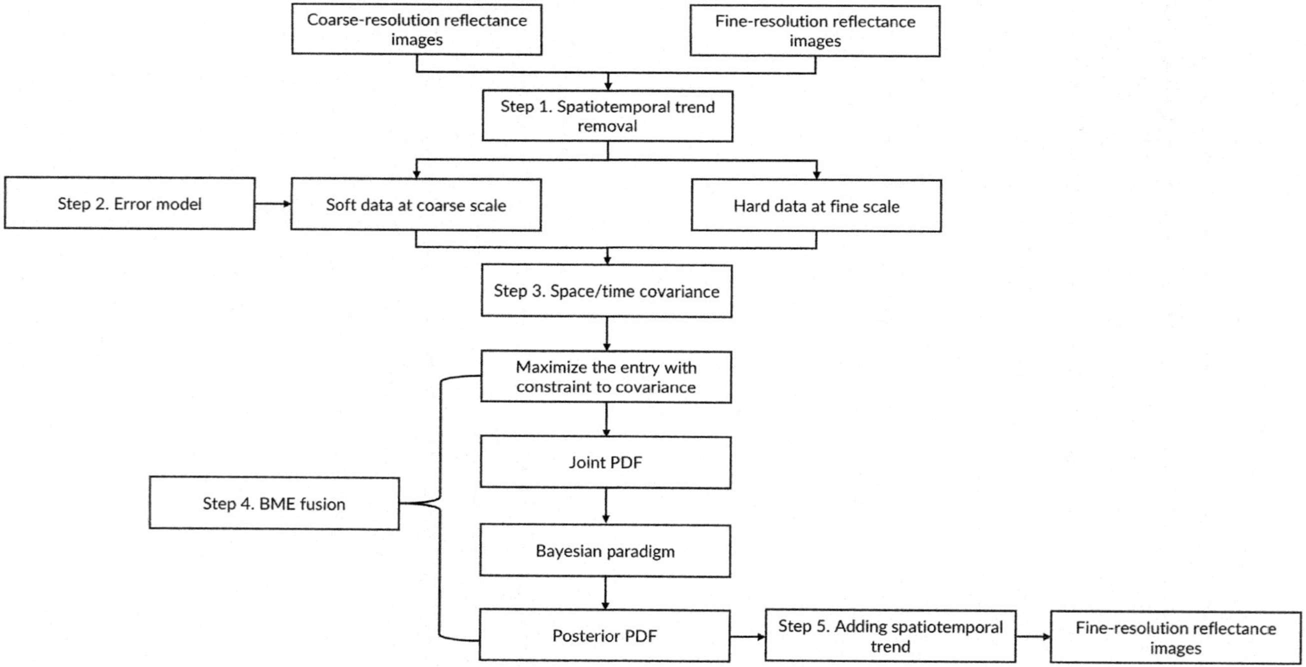

Figure 6.6 Flowchart of the BME fusion method.

$$\sigma_Y^2 = \sigma_X^2 - 2\|As'\|^{-1}\int d\mu C_X(|s' - \mu|) + \|As'\|^{-2}\int d\mu \int d\mu' C_X(|\mu - \mu'|) \tag{6.10}$$

Where σ_X^2 is the variance of the $X(s)$ and C_X is the covariance between $X(s)$ and $Z(s')$. Therefore, we can obtain a probabilistic soft datum χ_s for the fine-scale X at point s given a coarse-scale observed value z at point s':

$$f_S(\chi_s|z) = \phi(\chi_s; E(Y(s,s')) + z, \sigma_Y^2)) \Rightarrow N(E(Y(s,s')) + z, \sigma_Y^2) \tag{6.11}$$

The high-resolution data is considered to be hard data. The coarse-resolution data are transformed to corresponding soft data with a Gaussian probability distribution using the error model described above.

Step 3. Space/time covariance model- In geostatistics, the covariance $c(\gamma, \tau)$ is a function of space and time, where γ is the spatial lag and τ is the temporal lag. It decreases with increasing distance and time (Cressie, 1993). There are a number of covariance functions, such as exponential function, spherical function, Matérn covariance function, and so on (Li et al., 2013).

Step 4. BME fusion- With hard and soft data as well as covariance function, it is ready to conduct BME fusion. Assuming that various adjacent observations from RS data are available with irregular spatial and temporal gaps, the nonlinear mean estimation $\overline{X_k}$ at the location (x, y) at time t, can be estimated from Eq. (6.12) (Christakos et al., 2004).

$$\overline{x_k} = \int x_k f(x_k|x_{soft}, x_{hard})dx_k \tag{6.12}$$

where $f(x_k|x_{soft}, x_{hard})$ is a posterior probability density function (PDF) under the adjacent pixel observations, x_{soft} and x_{hard} denote the soft and hard values of the adjacent grids, respectively.

The posterior PDF at the estimation point updates from the prior PDF in the Bayesian rule when hard and soft data are involved. According to Bayesian rule, $f(x_k|x_{soft}, x_{hard})$ can be expressed as Eq. (6.13):

$$f(x_k|x_{soft}, x_{hard}) = \frac{f(x_{soft}, x_{hard}, x_k)}{f(x_{soft}, x_{hard})} = \frac{f_G(x_{map})}{f(x_{soft}, x_{hard})} \tag{6.13}$$

where $f(x_{soft}, x_{hard})$ is the prior PDF of soft and hard data at the spatiotemporal adjacent pixels, $f_G(x_{map})$ is a joint PDF and achieved by maximizing the entropy under the constraint of the general knowledge G. The general knowledge can be logical principles, scientific theories, physical laws, and summary statistics (Li et al., 2013). In RS, the general knowledge can be the covariance of the reflectance or vegetation index field and is expressed as $\varphi_\alpha(x_{map})$.

For a continuous variable, the entropy H is defined as (Shannon, 1948):

$$H = -\int dx_{map} f_G(x_{map}) \log f_G(x_{map}) \quad (6.14)$$

Introducing the Lagrange multiplier λ_α, maximizing the entropy is equivalent to maximizing the relation:

$$\begin{aligned} \mathrm{L}\left[f_G(x_{map})\right] = & -\int dx_{map} f_G(x_{map}) \log f_G(x_{map}) \\ & -\sum_{\alpha=0}^{N} \lambda_\alpha \left[\int \varphi_\alpha(x_{map}) f_G(x_{map}) dx_{map} - \overline{\varphi_\alpha(x_{map})}\right] \end{aligned} \quad (6.15)$$

where is $\overline{\varphi_\alpha(x_{map})}$ is the expected value of $\varphi_\alpha(x_{map})$. Setting the partial derivatives to zero and solving the system of equations with respect to the λ_α yields the maximum entropy solution to joint PDF $f_G(x_{map})$ (Christakos et al., 2004):

$$f_G(x_{map}) = \frac{\exp\left[\sum_{\alpha=0}^{N} \lambda_\alpha \varphi_\alpha(x_{map})\right]}{\int \exp\left[\sum_{\alpha=0}^{N} \lambda_\alpha \varphi_\alpha(x_{map})\right] dx_{map}} \quad (6.16)$$

The joint PDF $f_G(x_{map})$ is substituted in Eqs. (6.12) and (6.13) to solve the nonlinear mean estimation of $\overline{x_k}$ and the corresponding variance is calculated.

Step 5. Adding trend- The spatiotemporal trend was added back to the estimated mean to obtain fused values.

3.4 Neural network-based fusion methods

The previous three fusion methods are commonly applied at the pixel level, as they typically use reflectance images or derived vegetation index images as their data sources (Li et al., 2022). GeoAI-based fusion methods are widely adopted to integrate data with different characteristics (e.g., lidar-optical and SAR-optical) at the feature and decision levels. The GeoAI-based methods can match spatial scales from data sources by extracting feature maps, which are discussed below. There are many different GeoAI-based fusion methods, such as the deep generative fusion method and the deep convolutional spatiotemporal network (Bouabid et al., 2020; Tan et al., 2019). Here, we will introduce a fusion framework developed by Chen et al. (2017), who utilized convolutional neural networks (CNNs) and fully connected deep neural networks (DNNs) to fuse lidar and optical images. There are three main steps which are shown in Fig. 6.7.

Step 1. Feature extraction- This step uses two convolution layers and two pooling layers to extract features from optical reflectance images (e.g., Landsat) and lidar images, respectively (Chen et al., 2017). Users can adjust the number of convolution and

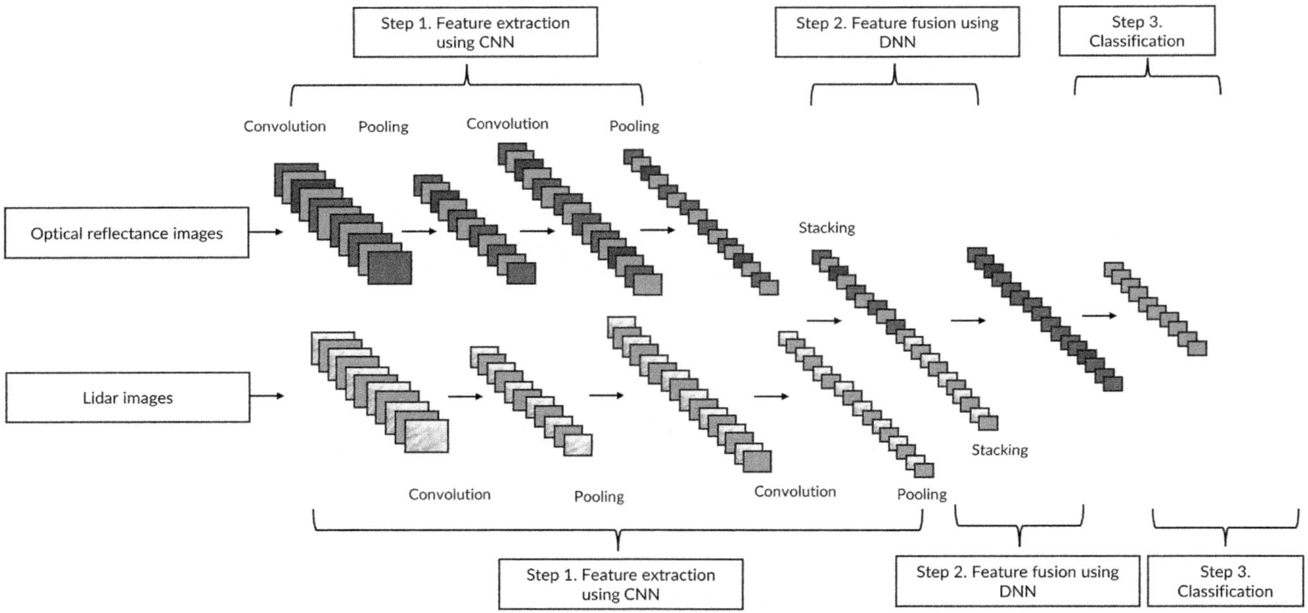

Figure 6.7 Flowchart of neural network-based methods.

pooling layers depending on the specific applications. A convolutional layer with nonlinear operation is shown in Eq. (6.17) (Chen et al., 2017).

$$x_j^l = f\left(\sum_{i=1}^{M} x_i^{l-1} * k_{ij}^l + b_j^l\right) \tag{6.17}$$

where x_i^{l-1} is the *ith* feature map of the previous $(l-1)th$ layer; x_j^l is the *jth* feature map of the current *lth* layer, *M* is the number of input feature maps; k_{ij}^l *and* b_j^l are randomly initialized and set to zero, respectively, which are subsequently fine-tuned through backpropagation; $f()$ is a nonlinear function known as the activation function, which accounts for the nonlinear relationships between different layers. The rectified linear unit (ReLU) activation function is the most commonly used (Agarap, 2026); $*$ is the convolution operation.

The pooling operation is used to reduce the resolution of the feature maps. It involves combining neighboring cells into a single, larger cell, which is filled with a statistic derived from the original cells. The most common pooling operation is the max pooling, in which the maximum cell value is populated into the large cell.

Step 2. Feature fusion- Once features from lidar and optical images are extracted, they are fused using DNNs (Chen et al., 2017). DNNs include many fully connected layers (or hidden layers [i.e., additional layers between the input layer and the output layer]). The number of hidden layers depends on specific applications. DNNs also use the ReLU activation function to consider the nonlinear relationships between different hidden layers. To minimize the overfitting issue, users can use techniques, such as dropouts, where some neurons and their associated connections are turned off (Srivastava et al., 2014), or batch normalization, which forces the activations of each layer to have zero means and unit variants (Ioffe and Szegedy, 2015).

Step 3. Classification- The final step is directly related to the certain application. Here, Chen et al. (2017) used logistic regression to produce the final classified map. The labeled training samples optimize the CNNs and DNNs through backpropagation. It is worth noting that users can use different methods in this step to stratify their specific requirements.

Case studies

In this section, we provide two case studies that show both successful and unsuccessful attempts at integrating multi-source RS data. These case studies not only offer valuable lessons on best practices in data fusion but also shed light on the importance of proper procedures in ensuring data integration success.

Case study 1: A successful integration project in land cover and land use classification

(Land cover mapping via crowdsourced multi-directional views: The more directional views, the better, *International Journal of Applied Earth Observation and Geoinformation*, 2023).

Crowdsourced land cover datasets have emerged as important geospatial data sources because of their great potential to provide human-centric ground observations. This study developed an end-to-end classification framework to classify land cover data collected by citizen scientists using the NASA Global Learning and Observations to Benefit the Environment (GLOBE) Observer tool. Specifically, we explored classification performance with different levels of viewing involvement under different image fusion strategies using various popular deep-learning algorithms.

Each of the land cover images taken by a NASA GLOBE Observer at the same time has six-directional views, including north, south, east, west, and up and down directions. Because up and down directions offer less information for inferring land cover categories, they are discarded in the study. Therefore, there are only four cardinal directions (i.e., north, south, east, and west). To assess the impacts of multi-directional views on classification accuracy, this study tested several scenarios, including single-directional, double-directional, triple-directional, and all four directional views at early fusion, later fusion, and score fusion strategies, respectively (Case Study Figure 1.1). The early fusion is a pixel-level fusion, the late fusion is a feature-level fusion, and the score fusion is a decision-level fusion.

To comprehensively explore the effects of fusion strategies and the involvement of the number of views on classification accuracy, this study used four popular deep learning architectures, including VGG (Simonyan and Zisserman, 2015), ResNet (He et al., 2016a,b), MobileNet (Howard et al., 2017), and EfficientNet (Tan and Le, 2021). We chose these four algorithms, as their overall designs are divergent from each other, presenting the current mainstream development of CNNs (Huang et al., 2021).

The results show that the late fusion (i.e., feature-level fusion) consistently achieves the highest weighted F1 score, regardless of the number of views or the deep learning algorithms applied (Case Study Figure 1.2). This may be attributed to the fact that late fusion can extract distinct features, enabling the encapsulation of more intricate and high-level representations of the data. Our results also show that the classification scenario with EfficientNet, the involvement of quadruple views, and the late fusion strategy leads to the highest weighted F1 score, indicating the adoption of multi-directional views for crowdsourced land cover datasets is necessary.

Overall, MobileNet and EfficientNet consistently deliver the strongest performance across all conditions, regardless of the number of views incorporated or the fusion stage applied. Across all architectures, model accuracy improves as additional views are included, indicating that multi-view information provides complementary features that enhance discrimination. Among the fusion strategies, feature-level fusion achieves the highest performance, suggesting that integrating learned representations before classification allows the network to capitalize on cross-view relationships more effectively than other fusion stages.

Case study 2: A failed integration project in Agriculture.

Optical imagery is a powerful tool for monitoring agricultural production. Simple indices such as NDVI are used to inform farmers and policymakers of both inter- and intra-farm plant variability that can facilitate famine prediction, large-scale targeted adaptation, and local-scale precision agriculture, i.e., spatially varied yield

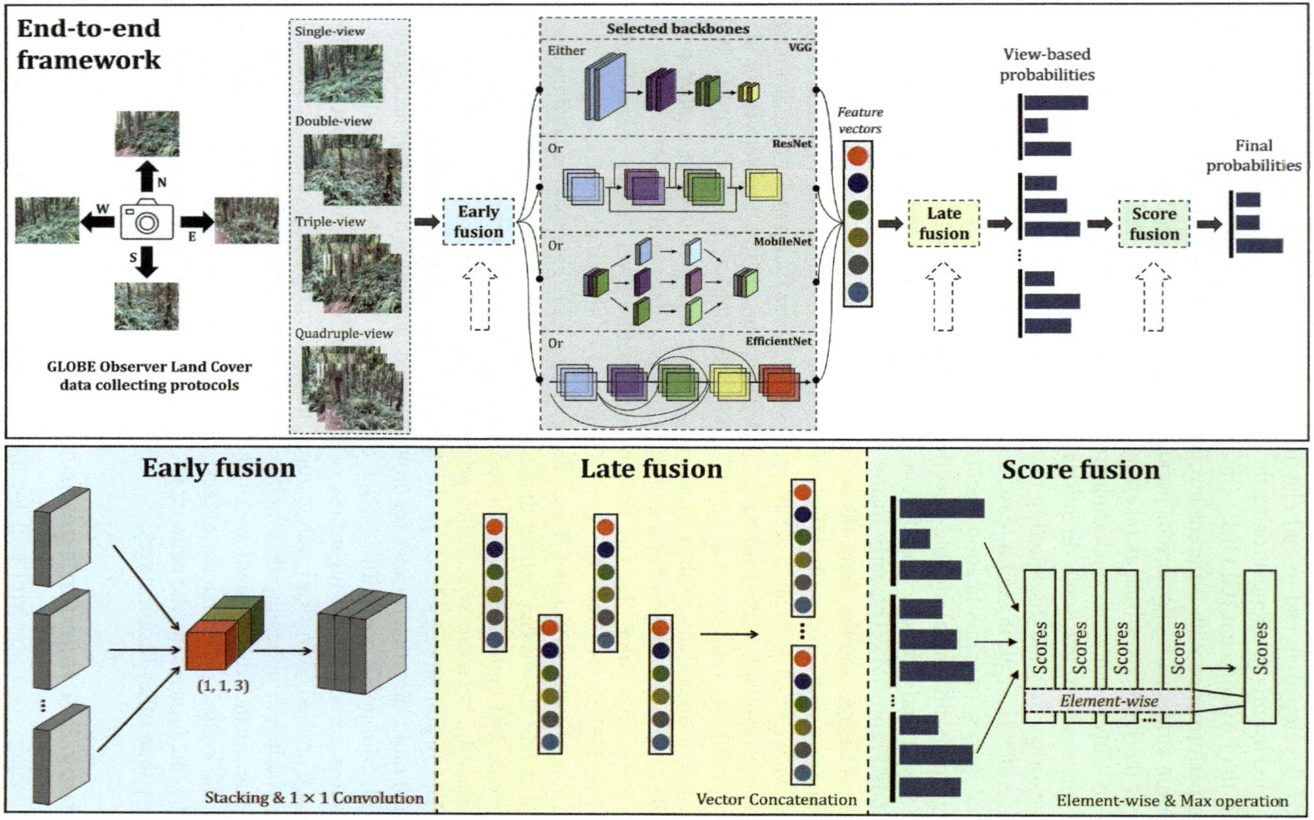

Case Study Figure 1.1 Deep learning fusion of image data with multi-directional views for land cover classification.
This figure is from Huang, X., Yang, D., He, Y., Nelson, P., Low, R., McBride, S., Mitchell, J., & Guarraia, M., 2023. Land cover mapping via crowdsourced multi-directional views: The more directional views, the better. *International Journal of Applied Earth Observation and Geoinformation*, *122*, 103382. https://doi.org/10.1016/j.jag.2023.103382).

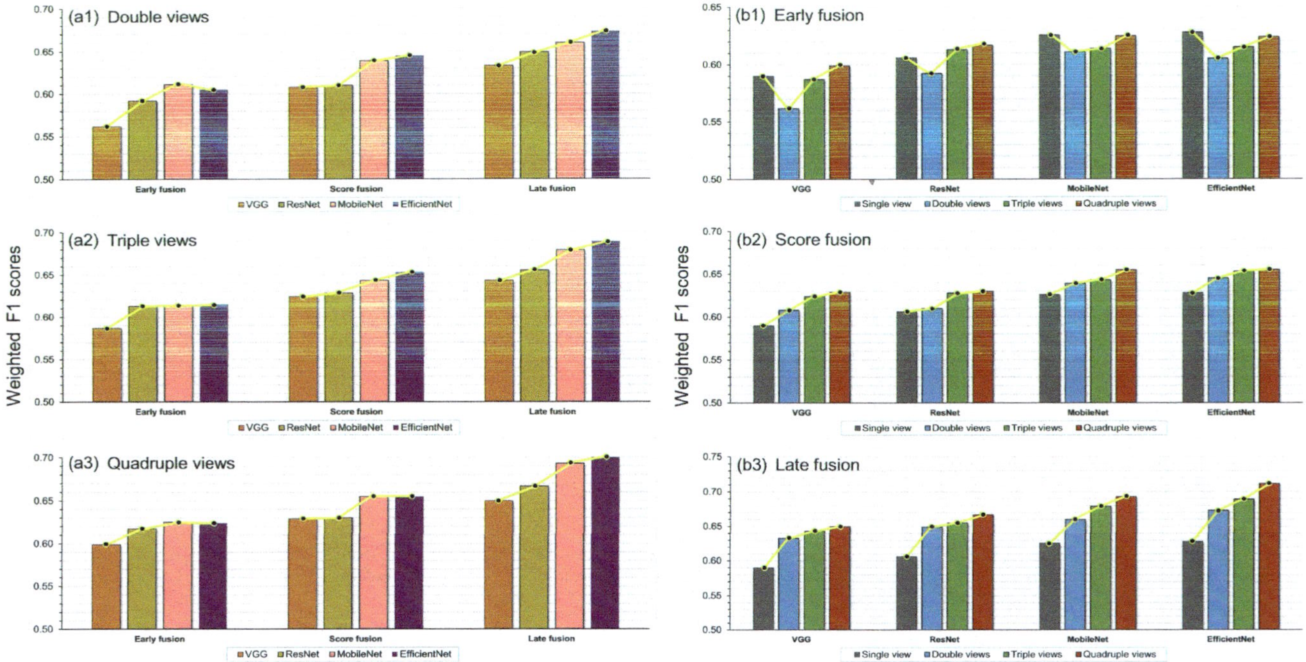

Case Study Figure 1.2 Classification accuracy using different deep learning algorithms with different levels of directional view involvement under different fusion strategies.

This figure is from Huang, X., Yang, D., He, Y., Nelson, P., Low, R., McBride, S., Mitchell, J., & Guarraia, M., 2023. Land cover mapping via crowdsourced multi-directional views: The more directional views, the better. *International Journal of Applied Earth Observation and Geoinformation*, *122*, 103382. https://doi.org/10.1016/j.jag.2023.103382).

management strategies (Nakalembe et al., 2021; Sishodia et al., 2020). A major problem in optical imaging is reliance on ground-level visibility, which is often limited during crop growth periods where rainfall and clouds are abundant (Case Study Figure 2.1). Problems of clouds were apparent in a study by Peter et al. (2020), which attempted to relate farm-scale productivity to sensors from the drone to the satellite scale to test a full range of spatial resolutions. Optical imagery of a smallholder farm in Central Malawi was collected from Sentinel-2, Planet Labs, Pléiades, and a senseFly eBee AG at spatial resolutions of 10 m and 20 m, 3 m, 2 m, and 7 cm, 14 cm, and 27 cm, respectively. The in situ data were collected using a handheld SPAD, which uses a reflectance ratio to estimate plant nitrogen (Bullock and Anderson, 1998). While the in situ data and sUAS images were collected on the same day (20 February 2018), there was a failure to acquire images from many platforms on or near that date, negatively impacting the integrability or feasibility of relating the imagery. The most cloud-free image available from Pléiades was available on 28 January 2018 and no cloud-free images were available from SPOT 6 and WorldView-2; however, images were available from Planet Labs and Sentinel-2 on 17 February 2018 and 20 February 2018, respectively. Some results of the study were unsurprising, particularly that the drone-scale was more effective at measuring intra-farm variability than the satellite platforms, though there was a major limitation of weak date alignment.

The failure example highlights the limitations of relying solely on optical systems in cloud-prone regions, suggesting a need for operational strategies that fuse multi-source RS data, such as radar imagery. Moreover, failed attempts can illuminate new research opportunities, such as designing more robust fusion frameworks that explicitly model temporal uncertainty or fill gaps due to cloud contaminations.

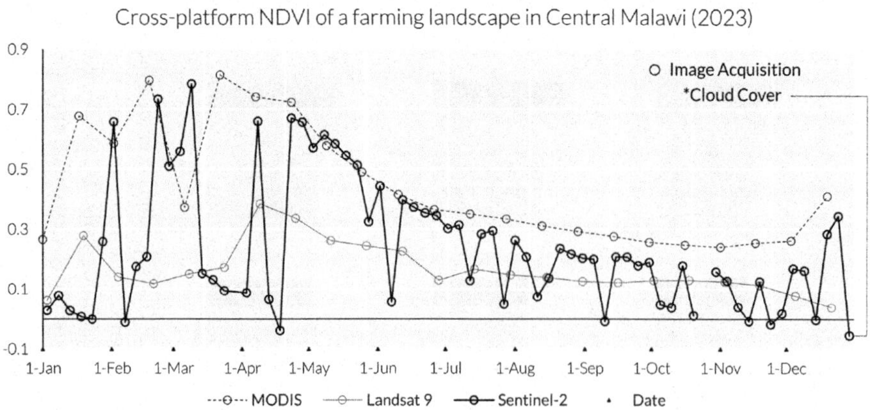

Case Study Figure 2.1 Problems of clouds and satellite revisit rates: A time series of Sentinel-2, landsat 9, and MODIS NDVI. Data was collected from a de-identified point in Central malawi.

Data sources: (1) Harmonized Sentinel-2 MSI: MultiSpectral Instrument, Level-2A, (2) USGS Landsat 9 Level 2, Collection 2, Tier 1, and (3) MODIS Terra Daily NDVI.

4. Discussion

In the successful data fusion case study, a key factor is the careful selection of images to be fused, which ensures that the data are compatible in terms of temporal attributes. Typically, these images are taken simultaneously, but from different directional views to capture complementary information. Temporal misalignment of images, if unresolved, can lead to a failure in data fusion, as discussed in case study 2. To avoid this, it is critical to use images that are temporally aligned or captured within proximity of each other.

Another important consideration is the choice of fusion algorithms and the fusion stage itself. Even when images are temporally aligned, the selection of the appropriate fusion algorithm can tremendously influence integration results. Different algorithms may emphasize different data characteristics, such as spatial resolution, spectral fidelity, or feature extraction (Li et al., 2022). Furthermore, the stage at which the fusion occurs—whether at the pixel, feature, or decision level—can also affect the results, as discussed in case study 1. Since there is no universal algorithm that is suitable for all data sources, it is recommended to test different fusion algorithms at various fusion stages to determine the best combination for a specific application. This highlights the need for conducting sensitivity analyses in the selection of fusion techniques, especially when working with diverse datasets.

Correct data integration can enhance the spatial and temporal coverage of datasets, while also improving the resolution of existing data. For instance, the application of the Highly Scalable Temporal Adaptive Reflectance Fusion Model (HISTARFM) by Moreno-Martínez et al. (2020) enabled the fusion of Landsat and MODIS reflectance data to generate Landsat-like data with a 30 m spatial resolution and a monthly temporal resolution from 2009 to 2021, covering the entire United States. This dataset with spatial and temporal completeness can improve various downstream applications, such as land surface classification, change detection, and natural hazard monitoring. Tavus et al. (2020) combined Sentinel-1 SAR and Sentinel-2 optical data to enhance flood extent mapping accuracy in Turkey's Ordu Province, demonstrating how effective data fusion can improve the detection and analysis of critical environmental events.

In contrast, incorrect data integration can lead to flawed or unreliable results. For example, Dai and Khorram (1998) demonstrated that even sub-pixel misregistration between multi-temporal images could lead to significant errors in change detection analysis. They found that a misregistration of 0.2 pixels could induce a 10% error in change detection.

5. Conclusion

This chapter explored the integration of RS, big data, and GeoAI, highlighting both challenges and opportunities in data fusion across various spatial and temporal scales. We examined several key methods for data fusion, including STRUM, STARFM,

BME, and neural network-based fusion methods. Through case studies, we illustrated both successful and unsuccessful examples of integrating multi-source data and provided best practices for data fusion. This chapter highlights:

- Data fusion can greatly enhance the spatial and temporal resolutions and coverage of datasets.
- The choice of fusion algorithm and the stage at which fusion occurs (pixel, feature, or decision level) can greatly influence the results of data integration.
- Accurate spatial and temporal alignment of images is essential for effective data fusion.
- Careful preprocessing, including proper alignment and registration of images, is essential to avoid errors in change detection and other analyses.
- While integrating data from multiple sources (e.g., optical and SAR imagery) can overcome the limitations of individual data types, it also introduces new challenges that must be carefully managed.

Along with the development of RS, big data, and GeoAI, in the future, we will:

- Increase the use of machine learning and deep learning algorithms, as they can help address the challenges from these increasingly complex and diverse datasets.
- Integrate non-traditional data sources, such as social media data, Internet of Things (IoT) sensor networks, and citizen science observations, as they are likely to become more prevalent.
- Fuse data in real-time with advances in computing power and algorithm efficiency, which is crucial for applications such as disaster response and urban management.
- Develop cloud-based fusion platforms, as the volume of Earth observation data continues to increase.

The integration of RS, big data, and GeoAI presents great opportunities for advancing our understanding and management of Earth systems, paving the path for sustainable development. It is crucial for researchers to continue developing robust data fusion methods that can handle the increasing volume, variety, and velocity of geospatial data. Moreover, as we develop data integration methods, we must remain mindful of the potential pitfalls to ensure the accuracy and reliability of our results. This requires ongoing collaboration across disciplines, from RS specialists and data scientists to domain experts in various fields of application.

Acronyms

CNNs	convolutional neural networks
DNNs	deep neural networks
EFP	Ecological Fallacy Problem
FSDAF	flexible spatiotemporal data fusion
GEE	Google Earth Engine
GLOBE	Global Learning and Observations to Benefit the Environment
HISTARFM	Highly Scalable Temporal Adaptive Reflectance Fusion Model
HLS	Harmonized Landsat and Sentinel-2
LCLU	land cover and land use

MAUP	Modifiable Areal Unit Problem
MTUP	Modifiable Temporal Unit Problem
NAIP	National Agriculture Imagery Program
NDVI	Normalized Difference Vegetation Index
PDF	posterior probability density function
RS	Remote Sensing

References

Agarap, A. F. (2026). Deep learning using rectified linear units. *arXiv*. https://doi.org/10.48550/arXiv.1803.08375

Basiri, A., Haklay, M., Foody, G., & Mooney, P. (2019). Crowdsourced geospatial data quality: Challenges and future directions. *International Journal of Geographical Information Science, 33*, 1588–1593. https://doi.org/10.1080/13658816.2019.1593422

Bird, H. J. (1989). *The changing worlds of geography*. Oxford: Clarendon Press.

Bouabid, S., Chernetskiy, M., Rischard, M., & Gamper, J. (2020). Predicting landsat reflectance with deep generative fusion. *arXiv*. https://doi.org/10.48550/arXiv.2011.04762

Buchhorn, M., Lesiv, M., Tsendbazar, N.-E., Herold, M., Bertels, L., & Smets, B. (2020). Copernicus global land cover layers—collection 2. *Remote Sens (Basel), 12*, 1044. https://doi.org/10.3390/rs12061044

Bullock, D. G., & Anderson, D. S. (1998). Evaluation of the minolta SPAD-502 chlorophyll meter for nitrogen management in corn. *Journal of Plant Nutrition, 21*, 741–755. https://doi.org/10.1080/01904169809365439

Chen, B., Huang, B., & Xu, B. (2015). Comparison of spatiotemporal fusion models: A review. *Remote Sensing (Basel), 7*, 1798–1835. https://doi.org/10.3390/rs70201798

Chen, Y., Li, C., Ghamisi, P., Jia, X., & Gu, Y. (2017). Deep fusion of remote sensing data for accurate classification. *IEEE Geoscience and Remote Sensing Letters, 14*, 1253–1257. https://doi.org/10.1109/LGRS.2017.2704625

Cheng, T., & Adepeju, M. (2014). Modifiable temporal unit problem (MTUP) and its effect on space-time cluster detection. *PLoS One, 9*. https://doi.org/10.1371/journal.pone.0100465

Christakos, G. (2002). On the assimilation of uncertain physical knowledge bases: Bayesian and non-Bayesian techniques. *Advances in Water Resources, 25*, 1257–1274. https://doi.org/10.1016/S0309-1708(02)00064-7

Christakos, G., Kolovos, A., Serre, M. L., & Vukovich, F. (2004). Total ozone mapping by integrating databases from remote sensing instruments and empirical models. *IEEE Transactions on Geoscience and Remote Sensing, 42*, 991–1008. https://doi.org/10.1109/TGRS.2003.822751

Claverie, M., Ju, J., Masek, J. G., Dungan, J. L., Vermote, E. F., Roger, J.-C., Skakun, S. V., & Justice, C. (2018). The Harmonized Landsat and Sentinel-2 surface reflectance data set. *Remote Sensing of Environment, 219*, 145–161. https://doi.org/10.1016/j.rse.2018.09.002

Çöltekin, A., De Sabbata, S., Willi, C.,V., Pfister, S., Kuhn, M., & Lacayo, M. (2011). Modifiable temporal unit problem. In *ICC2011 workshop "Persistent Problems in Geographic Visualization." Paris, France.*

Comber, A., Harris, P., Bratkova, K., Phe, H. H., Kieu, M., Bui, Q. T., Nguyen, T. T. H., Wanjau, E., & Malleson, N. (2022). Handling the MAUP: Methods for identifying appropriate scales of aggregation based on measures on spatial and non-spatial variance. *AGILE: GIScience Series, 3*, 1–5. https://doi.org/10.5194/agile-giss-3-30-2022

Cressie, N. A. C. (1993). *Statistics for spatial data*. Wiley. https://doi.org/10.1002/9781119115151

Dai, X., & Khorram, S. (1998). The effects of image misregistration on the accuracy of remotely sensed change detection. *IEEE Transactions on Geoscience and Remote Sensing, 36*, 1566–1577. https://doi.org/10.1109/36.718860

Dhillon, M. S., Dahms, T., Kübert-Flock, C., Liepa, A., Rummler, T., Arnault, J., Steffan-Dewenter, I., & Ullmann, T. (2023). Impact of STARFM on Crop Yield Predictions: Fusing MODIS with Landsat 5, 7, and 8 NDVIs in bavaria germany. *Remote Sensing (Basel), 15*, 1651. https://doi.org/10.3390/rs15061651

Didan, K. (2021). *MODIS/Terra Vegetation Indices 16-Day L3 Global 250m SIN Grid V061*. NASA EOSDIS Land Processes Distributed Active Archive Cente. https://doi.org/10.5067/MODIS/MOD13Q1.061 (Accessed 9.6.24).

Duque, J. C., Laniado, H., & Polo, A. (2018). S-maup: Statistical test to measure the sensitivity to the modifiable areal unit problem. *PLoS One, 13*. https://doi.org/10.1371/journal.pone.0207377

Friedl, M., & Sulla-Menashe, D. (2022). *MODIS/Terra+Aqua Land Cover Type Yearly L3 Global 500m SIN Grid V061*. NASA EOSDIS Land Processes Distributed Active Archive Center. https://doi.org/10.5067/MODIS/MCD12Q1.061 (Accessed 9.6.24).

Gao, F., Masek, J., Schwaller, M., & Hall, F. (2006). On the blending of the landsat and MODIS surface reflectance: Predicting daily landsat surface reflectance. *IEEE Transactions on Geoscience and Remote Sensing, 44*, 2207–2218. https://doi.org/10.1109/TGRS.2006.872081

García-Haro, F. J., Sommer, S., & Kemper, T. (2005). A new tool for variable multiple end-member spectral mixture analysis (VMESMA). *International Journal of Remote Sensing, 26*, 2135–2162. https://doi.org/10.1080/01431160512331337817

Gevaert, C. M., & García-Haro, F. J. (2015). A comparison of STARFM and an unmixing-based algorithm for landsat and MODIS data fusion. *Remote Sensing of Environment, 156*, 34–44. https://doi.org/10.1016/j.rse.2014.09.012

Ghamisi, P., Gloaguen, R., Atkinson, P. M., Benediktsson, J. A., Rasti, B., Yokoya, N., Wang, Q., Hofle, B., Bruzzone, L., Bovolo, F., Chi, M., & Anders, K. (2019). Multisource and multitemporal data fusion in remote sensing: A comprehensive review of the state of the art. *IEEE Geoscience and Remote Sensing Magazine, 7*, 6–39. https://doi.org/10.1109/MGRS.2018.2890023

Gharbia, R., Azar, A. T., Baz, A. E., & Hassanien, A. E. (2014). Image fusion techniques in remote sensing. *arXiv*. https://doi.org/10.48550/arXiv.1403.5473

Gorelick, N., Hancher, M., Dixon, M., Ilyushchenko, S., Thau, D., & Moore, R. (2017). Google earth engine: Planetary-scale geospatial analysis for everyone. *Remote Sensing of Environment, 202*, 18–27. https://doi.org/10.1016/j.rse.2017.06.031

Gulhane, V. A., Rode, S. V., & Pande, C. B. (2023). Correlation analysis of soil nutrients and prediction model through ISO cluster unsupervised classification with multispectral data. *Multimedia Tools and Applications, 82*, 2165–2184. https://doi.org/10.1007/s11042-022-13276-2

Haggett, P. (1970). Scale components in geographical problems. In P. Haggett (Ed.), *Frontiers in geographical teaching*. London: Routledge.

Harris, T. M. (2006). Scale as artifact: GIS, ecological fallacy, and archaeological analysis. In *Confronting scale in archaeology* (pp. 39–53). Springer US. https://doi.org/10.1007/0-387-32773-8_4

He, Y., Bo, Y., Chai, L., Liu, X., & Li, A. (2016). Linking in situ LAI and fine resolution remote sensing data to map reference LAI over cropland and grassland using geostatistical regression method. *International Journal of Applied Earth Observation and Geoinformation, 50*, 26–38. https://doi.org/10.1016/j.jag.2016.02.010

He, Y., Lee, E., & Warner, T. A. (2017). A time series of annual land use and land cover maps of china from 1982 to 2013 generated using AVHRR GIMMS NDVI3g data. *Remote Sensing of Environment, 199*, 201–217. https://doi.org/10.1016/j.rse.2017.07.010

He, Y., Warner, T. A., McNeil, B. E., & Lee, E. (2018). Reducing uncertainties in applying remotely sensed land use and land cover maps in land-atmosphere interaction: Identifying change in space and time. *Remote Sensing (Basel), 10*. https://doi.org/10.3390/rs10040506

He, K., Zhang, X., Ren, S., & Sun, J. (2016). Deep residual learning for image recognition. In *2016 IEEE conference on computer vision and pattern recognition (CVPR)* (pp. 770–778). IEEE. https://doi.org/10.1109/CVPR.2016.90

Hilker, T., Wulder, M. A., Coops, N. C., Linke, J., McDermid, G., Masek, J. G., Gao, F., & White, J. C. (2009). A new data fusion model for high spatial-and temporal-resolution mapping of forest disturbance based on landsat and MODIS. *Remote Sensing of Environment, 113*, 1613–1627. https://doi.org/10.1016/j.rse.2009.03.007

Howard, A. G., Zhu, M., Chen, B., Kalenichenko, D., Wang, W., Weyand, T., Andreetto, M., & Adam, H. (2017). *MobileNets: Efficient convolutional neural networks for Mobile vision applications*. https://doi.org/10.48550/arXiv.1704.04861.

Hsieh, P.-F., Lee, L. C., & Chen, N.-Y. (2001). Effect of spatial resolution on classification errors of pure and mixed pixels in remote sensing. *IEEE Transactions on Geoscience and Remote Sensing, 39*, 2657–2663. https://doi.org/10.1109/36.975000

Huang, X., Yang, D., He, Y., Nelson, P., Low, R., McBride, S., Mitchell, J., & Guarraia, M. (2023). Land cover mapping via crowdsourced multi-directional views: The more directional views, the better. *International Journal of Applied Earth Observation and Geoinformation, 122*. https://doi.org/10.1016/j.jag.2023.103382

Huang, X., Zhu, D., Zhang, F., Liu, T., Li, X., & Zou, L. (2021). Sensing population distribution from satellite imagery via deep learning:model selection, neighboring effects, and systematic biases. *IEEE Journal of Selected Topics in Applied Earth Observations and Remote Sensing, 14*, 5137–5151. https://doi.org/10.1109/JSTARS.2021.3076630

Ioffe, S., & Szegedy, C. (2015). *Batch normalization: Accelerating deep network training by reducing internal covariate shift*. MachineLearning.

Janowicz, K., Gao, S., McKenzie, G., Hu, Y., & Bhaduri, B. (2020). GeoAI: Spatially explicit artificial intelligence techniques for geographic knowledge discovery and beyond. *International Journal of Geographical Information Science, 34*, 625–636. https://doi.org/10.1080/13658816.2019.1684500

Ke, Y., Im, J., Park, S., & Gong, H. (2016). Downscaling of MODIS one kilometer evapotranspiration using Landsat-8 data and machine learning approaches. *Remote Sensing (Basel), 8*, 215. https://doi.org/10.3390/rs8030215

Koumetio Tekouabou, S. C., Diop, E. B., Azmi, R., Jaligot, R., & Chenal, J. (2022). Reviewing the application of machine learning methods to model urban form indicators in planning decision support systems: Potential, issues and challenges. *Journal of King Saud University - Computer and Information Sciences, 34*, 5943–5967. https://doi.org/10.1016/j.jksuci.2021.08.007

Lee, S.-J., Yeatts, K. B., & Serre, M. L. (2009). A bayesian maximum entropy approach to address the change of support problem in the spatial analysis of childhood asthma prevalence across north carolina. *Spatial and Spatio-temporal Epidemiology, 1*, 49–60. https://doi.org/10.1016/j.sste.2009.07.005

Li, A., Bo, Y., Zhu, Y., Guo, P., Bi, J., & He, Y. (2013). Blending multi-resolution satellite sea surface temperature (SST) products using bayesian maximum entropy method. *Remote Sensing of Environment, 135*, 52–63. https://doi.org/10.1016/j.rse.2013.03.021

Li, S., Dragicevic, S., Castro, F. A., Sester, M., Winter, S., Coltekin, A., Pettit, C., Jiang, B., Haworth, J., Stein, A., & Cheng, T. (2016). Geospatial big data handling theory and methods: A review and research challenges. *ISPRS Journal of Photogrammetry and Remote Sensing, 115*, 119–133. https://doi.org/10.1016/j.isprsjprs.2015.10.012

Li, J., Hong, D., Gao, L., Yao, J., Zheng, K., Zhang, B., & Chanussot, J. (2022). Deep learning in multimodal remote sensing data fusion: A comprehensive review. *International Journal of Applied Earth Observation and Geoinformation, 112*. https://doi.org/10.1016/j.jag.2022.102926

Li, W., & Hsu, C.-Y. (2022). GeoAI for large-scale image analysis and machine vision: Recent progress of artificial intelligence in geography. *ISPRS International Journal of Geo-Information, 11*, 385. https://doi.org/10.3390/ijgi11070385

Li, D., Ke, Y., Gong, H., & Li, X. (2015). Object-based urban tree species classification using Bi-Temporal WorldView-2 and WorldView-3 images. *Remote Sensing (Basel), 7*, 16917–16937. https://doi.org/10.3390/rs71215861

Liakos, K., Busato, P., Moshou, D., Pearson, S., & Bochtis, D. (2018). Machine learning in agriculture: A review. *Sensors, 18*, 2674. https://doi.org/10.3390/s18082674

Linardos, V., Drakaki, M., Tzionas, P., & Karnavas, Y. (2022). Machine learning in disaster management: Recent developments in methods and applications. *Machine Learning and Knowledge Extraction, 4*, 446–473. https://doi.org/10.3390/make4020020

Lunetta, R., Congalton, R., Fenstermaker, L., Jensen, J., Mcgwire, K., & Tinney, L. R. (1991). Remote sensing and geographic information system data integration- error sources and research issues. *Photogrammetric Engineering and Remote Sensing, 57*, 677–687.

Manley, D. (2014). Scale, aggregation, and the modifiable areal unit problem. In *Handbook of regional science* (pp. 1157–1171). Berlin, Heidelberg: Springer Berlin Heidelberg. https://doi.org/10.1007/978-3-642-23430-9_69

Maxwell, A. E., Warner, T. A., & Fang, F. (2018). Implementation of machine-learning classification in remote sensing: An applied review. *International Journal of Remote Sensing*. https://doi.org/10.1080/01431161.2018.1433343

Meentemeyer, V. (1989). Geographical perspectives of space, time, and scale. *Landscape Ecology, 3*, 163–173. https://doi.org/10.1007/BF00131535

Mohsan, S. A. H., Khan, M. A., Noor, F., Ullah, I., & Alsharif, M. H. (2022). Towards the unmanned aerial vehicles (UAVs): A comprehensive review. *Drones, 6*, 147. https://doi.org/10.3390/drones6060147

Moreno-Martínez, Á., Izquierdo-Verdiguier, E., Maneta, M. P., Camps-Valls, G., Robinson, N., Muñoz-Marí, J., Sedano, F., Clinton, N., & Running, S. W. (2020). Multispectral high resolution sensor fusion for smoothing and gap-filling in the cloud. *Remote Sensing of Environment, 247*. https://doi.org/10.1016/j.rse.2020.111901

Nakalembe, C., Becker-Reshef, I., Bonifacio, R., Hu, G., Humber, M. L., Justice, C. J., Keniston, J., Mwangi, K., Rembold, F., Shukla, S., Urbano, F., Whitcraft, A. K., Li, Y., Zappacosta, M., Jarvis, I., & Sanchez, A. (2021). A review of satellite-based global agricultural monitoring systems available for africa. *Global Food Security, 29*. https://doi.org/10.1016/j.gfs.2021.100543

Openshaw, S. (1983). The modifiable areal unit problem. In S. Openshaw (Ed.), *Concepts and techniques in modern geography* (pp. 1–41). Norwich, UK: GeoBooks.

Peter, B. G., Messina, J. P., Breeze, V., Fung, C. Y., Kapoor, A., & Fan, P. (2024). Perspectives on modifiable spatiotemporal unit problems in remote sensing of agriculture: Evaluating rice production in vietnam and tools for analysis. *Frontiers in Remote Sensing, 5*. https://doi.org/10.3389/frsen.2024.1042624

Peter, B. G., Messina, J. P., Carroll, J. W., Zhi, J., Chimonyo, V., Lin, S., & Snapp, S. S. (2020). Multi-spatial resolution satellite and sUAS imagery for precision agriculture on smallholder farms in Malawi. *Photogrammetric Engineering & Remote Sensing, 86*, 107–119. https://doi.org/10.14358/PERS.86.2.107

Piantadosi, S., Byar, D. P., & Green, S. B. (1988). The ecological fallacy. *American Journal of Epidemiology, 127*, 893–904. https://doi.org/10.1093/oxfordjournals.aje.a114892

Pohl, C., & van Genderen, J. (2014). Remote sensing image fusion: An update in the context of digital Earth. *International Journal of Digital Earth, 7*, 158–172. https://doi.org/10.1080/17538947.2013.869266

Qin, P., Huang, H., Tang, H., Wang, J., & Liu, C. (2022). MUSTFN: A spatiotemporal fusion method for multi-scale and multi-sensor remote sensing images based on a convolutional neural network. *International Journal of Applied Earth Observation and Geoinformation, 115*. https://doi.org/10.1016/j.jag.2022.103113

Reutebuch, S. E., Andersen, H.-E., & McGaughey, R. J. (2005). Light detection and ranging (LIDAR): An emerging tool for multiple resource inventory. *Journal of Forestry, 103*, 286–292. https://doi.org/10.1093/jof/103.6.286

Schmitt, M., & Zhu, X. X. (2016). Data fusion and remote sensing: An ever-growing relationship. *IEEE Geoscience and Remote Sensing Magazine, 4*, 6–23. https://doi.org/10.1109/MGRS.2016.2561021

Shannon, C. E. (1948). A mathematical theory of communication. *TheBellSystemTechnicalJournal, 27*, 379–423.

Simonyan, K., & Zisserman, A. (2015). Very deep convolutional networks for large-scale image recognition. *arXiv*. https://doi.org/10.48550/arXiv.1409.1556

Sisheber, B., Marshall, M., Mengistu, D., & Nelson, A. (2022). Tracking crop phenology in a highly dynamic landscape with knowledge-based Landsat–MODIS data fusion. *International Journal of Applied Earth Observation and Geoinformation, 106*. https://doi.org/10.1016/j.jag.2021.102670

Sishodia, R. P., Ray, R. L., & Singh, S. K. (2020). Applications of remote sensing in precision agriculture: A review. *Remote Sensing (Basel), 12*, 3136. https://doi.org/10.3390/rs12193136

Srivastava, N., Hinton, G., Krizhevsky, A., Sutskever, I., & Salakhutdinov, R. (2014). Dropout: A simple way to prevent neural networks from overfitting. *Journal of Machine Learning Research*, 1929–1958.

Sun, T., Yan, N., Zhu, W., & Zhuang, Q. (2024). Assessing a machine learning-based downscaling framework for obtaining 1km daily precipitation from GPM data. *Heliyon, 10*.

Tan, Z., Di, L., Zhang, M., Guo, L., & Gao, M. (2019). An enhanced deep convolutional model for spatiotemporal image fusion. *Remote Sensing (Basel), 11*, 2898. https://doi.org/10.3390/rs11242898

Tan, M., & Le, Q. (2021). EfficientNetV2: Smaller models and faster training. In M. Meila, & T. Zhang (Eds.), *Proceedings of the 38th international conference on machine learning, proceedings of machine learning research* (pp. 10096–10106). PMLR.

Tavus, B., Kocaman, S., Nefeslioglu, H. A., & Gokceoglu, C. (2020). A fusion approach for flood mapping using Sentinel-1 and Sentinel-2 datasets. *The International Archives of the Photogrammetry, Remote Sensing and Spatial Information Sciences, XLIII-B3*, 641–648. https://doi.org/10.5194/isprs-archives-XLIII-B3-2020-641-2020

Thrift, N. (1977). *An introduction to time-geography (Concepts and techniques in modern geography; no. 13)*. Geo Abstracts Ltd.

Tsokas, A., Rysz, M., Pardalos, P. M., & Dipple, K. (2022). SAR data applications in earth observation: An overview. *Expert Systems with Applications, 205*. https://doi.org/10.1016/j.eswa.2022.117342

Voigt, S., Giulio-Tonolo, F., Lyons, J., Kučera, J., Jones, B., Schneiderhan, T., Platzeck, G., Kaku, K., Hazarika, M. K., Czaran, L., Li, S., Pedersen, W., James, G. K., Proy, C., Muthike, D. M., Bequignon, J., & Guha-Sapir, D. (2016). Global trends in satellite-based emergency mapping. *Science, 353*, 247–252. https://doi.org/10.1126/science.aad8728, 1979.

Wang, S., & Li, W. (2021). GeoAI in terrain analysis: Enabling multi-source deep learning and data fusion for natural feature detection. *Computers, Environment and Urban Systems, 90*. https://doi.org/10.1016/j.compenvurbsys.2021.101715

Wong, D. W. S. (2004). The modifiable areal unit problem (MAUP). In *WorldMinds: Geographical perspectives on 100 problems* (pp. 571–575). Netherlands, Dordrecht: Springer. https://doi.org/10.1007/978-1-4020-2352-1_93

Yang, L., Driscol, J., Sarigai, S., Wu, Q., Chen, H., & Lippitt, C. D. (2022). Google earth engine and artificial intelligence (AI): A comprehensive review. *Remote Sensing (Basel), 14*, 3253. https://doi.org/10.3390/rs14143253

Yang, C., Yu, M., Li, Y., Hu, F., Jiang, Y., Liu, Q., Sha, D., Xu, M., & Gu, J. (2019). Big earth data analytics: A survey. *Big Earth Data, 3*, 83–107. https://doi.org/10.1080/20964471.2019.1611175

Yu, H.-L., Kolovos, A., Christakos, G., Chen, J.-C., Warmerdam, S., & Dev, B. (2007). Interactive spatiotemporal modelling of health systems: The SEKS–GUI framework. *Stochastic Environmental Research and Risk Assessment, 21*, 647. https://doi.org/10.1007/s00477-007-0172-8

Zhang, J. (2010a). Multi-source remote sensing data fusion: Status and trends. *International Journal of Image and Data Fusion, 1*, 5–24. https://doi.org/10.1080/19479830903561035

Zhang, J. (2010b). Multi-source remote sensing data fusion: Status and trends. *International Journal of Image and Data Fusion, 1*, 5–24. https://doi.org/10.1080/19479830903561035

Zhang, F., Zhu, X., & Liu, D. (2014). Blending MODIS and landsat images for urban flood mapping. *International Journal of Remote Sensing, 35*, 3237–3253. https://doi.org/10.1080/01431161.2014.903351

Zhu, Z., Bo, Y., Sun, T., Zhang, X., Sun, M., Shen, A., Zhang, Y., Tang, J., Cao, M., & Wang, C. (2024). A downscaling-and-fusion framework for generating spatio-temporally complete and fine resolution remotely sensed surface soil moisture. *Agricultural and Forest Meteorology, 352*. https://doi.org/10.1016/j.agrformet.2024.110044

Zhu, X., Cai, F., Tian, J., & Williams, T. (2018). Spatiotemporal fusion of multisource remote sensing data: Literature survey, taxonomy, principles, applications, and future directions. *Remote Sensing (Basel), 10*, 527. https://doi.org/10.3390/rs10040527

Zhu, X., Chen, J., Gao, F., Chen, X., & Masek, J. G. (2010). An enhanced spatial and temporal adaptive reflectance fusion model for complex heterogeneous regions. *Remote Sensing of Environment, 114*, 2610–2623. https://doi.org/10.1016/j.rse.2010.05.032

Zhu, X., Helmer, E. H., Gao, F., Liu, D., Chen, J., & Lefsky, M. A. (2016). A flexible spatiotemporal method for fusing satellite images with different resolutions. *Remote Sensing of Environment, 172*, 165–177. https://doi.org/10.1016/j.rse.2015.11.016

Zurita-Milla, R., Clevers, J., & Schaepman, M. E. (2008). Unmixing-based landsat TM and MERIS FR data fusion. *IEEE Geoscience and Remote Sensing Letters, 5*, 453–457. https://doi.org/10.1109/LGRS.2008.919685

7 Geovisualization and Remote Sensing: Bridging Complexity and Interpretation

Cerian Gibbes[1], *Suiyuan Wang*[2], *Fuzhen Yin*[2], *Thi Hong Diep Dao*[1], *Brandon Vogt*[1], *Rachel Bump*[1], *Zachary Davison*[1], *Kyle Owens*[1] and *Evan Pickard*[1]
[1]Department of Geography & Environmental Studies, University of Colorado Colorado Springs, Colorado Springs, CO, United States; [2]Department of Geography, State University of New York at Buffalo, Buffalo, NY, United States

Key terms and definitions

Cartography: The art, science, and technology of making maps. Also, the study of maps as scientific documents and works of art.
Visual representation: The transformation of raw data into visual formats, including images, diagrams, charts, maps, or other visual elements, to support the identification of patterns and relationships.
Visual thinking: The use of visual representations of data to cognitively process ideas and promote thought about a concept or problem.
Visual communication: The use of visual representations to convey information or ideas.
LiDAR: An active remote sensing technology that uses laser pulses to detect objects.
Data sonification: The process of translating data into sound.
Apophenia: The tendency to see connections or patterns between unrelated or random things.

1. Introduction

Society is inundated with and increasingly driven by immense amounts of geospatial data. The widespread use of personal computers, mobile devices, intelligent autonomous sensors (such as those used in precision agriculture), and extensive connectivity enables continuous collection of data and increases human interactions in online environments and information spaces (Batty & Miller, 2000). Virtual spaces, such as Google Earth, have seamlessly combined terabytes of satellite imagery, aerial photography, geographic datasets, and user-contributed content. This allows individuals–expert and nonexpert–to easily engage with geographic data running from a

Remote Sensing, Big Data, and GeoAI. https://doi.org/10.1016/B978-0-443-26741-3.00005-2

local street-level scale to global scales. Combined with rapid advancements in remote sensing and the proliferation of big data (link to chapters 1-3), these virtual spaces have fundamentally transformed how geographic information is collected, analyzed, interpreted, and shared. Central to this transformation is geovisualization. Geovisualization offers a dynamic approach to spatial data visualization that supports the exploration, interpretation, and communication of geographic phenomena.

Geovisualization is the interactive representation of spatial information and analysis. It encompasses both the visual outputs, such as maps, graphs, or hybrid displays, and the methodologies, ranging from exploratory to explanatory, used to create them. Importantly, geovisualization centers on a thematic focus. As such, data and representations are tailored to a specific topic of interest for users, whether applied or theoretical. Through the emphasis on a specific topic, geovisualization moves raw data toward meaningful and contextualized representations of environmental and social phenomena. Focusing on data cleaning, analysis, and presentation on a theme or topic of interest enables knowledge synthesis by aligning distinct datasets with the chosen theme, making complex information accessible and actionable. This focus allows users to quickly understand and draw knowledge from the relevant features of datasets. Although the foundations of geovisualization are found in cartography, the field today is highly interdisciplinary. This disciplinary-crossing engagement with geovisualization is because of the recognition that data visualization, visual thinking, and visual communication are central to the scientific process (DiBiase, 1990, pp. 13–18). That is, visualization of spatial data is valuable not only to share results but is also valuable when used at multiple stages of analysis because it supports the exploration of datasets and the development of new ideas or questions, and developing new ideas or research questions. Spatial data visualization refers to the process of representing spatial datasets in a visual format, often through maps, 3D models, or interactive platforms. This form of visualization is instrumental in understanding patterns, relationships, and processes that are otherwise difficult to discern from raw data. DiBiase (1990, pp. 13–18) and MacEachren (1994) present the most widely accepted models of visual thinking and communication. They define these as processes that range from the private to the public and that move from exploration of data to synthesis of knowledge.

The emergence of interactive maps and digital tools for analysis in the 1980s and 1990s marked a shift in how the use of spatial data was conceptualized. Maps have been used for the sharing of scientific knowledge and in everyday life. Geographic data was presented through maps in a static two-dimensional form. Maps have guided decision-making based on the spatial information they present. Map-supported decision-making ranged from identifying efficient routes from one place to another to sharing weather forecasts. Maps provided, and continue to provide, valuable insights into our environment by simplifying the complexity of reality into clear visual representations. The ability to use spatial information and relationships presented in maps is supported by cartographic design principles. Cartographic design principles offer guidelines to ensure that maps are tailored to the intended audience, effectively communicate information, and are visually appealing. For example, when visualizing vegetation health using satellite-derived normalized difference vegetation index

(NDVI) data, an appropriate color scheme is one that supports the sequential ordering of NDVI values. These values range from −1 to one where lower values indicate nonvegetated surfaces (e.g., water, bare soil) and higher values represent healthy vegetation. As such, the use of a gradient that transitions from brown (nonvegetation) to yellow (low vegetation) to dark green (healthy vegetation) offers the audience a clear representation of the meaning of the values (Fig. 7.1). Furthermore, avoidance of red-green combinations enhances accessibility, considering individuals with color vision limitations. Through the application of these color use principles, the presentation of satellite imagery becomes more intuitive and supports accurate interpretation of the satellite data.

Prior to the transformation, in the 1980s and 1990s, of the way spatial data was understood and utilized, maps were the primary form for using and sharing spatial data. However, as technology has advanced, interactive and dynamic forms of data representation are increasingly the norm. The convergence of technological advancement with growth in the quantity and complexity of data has fostered perfect conditions for the maturation of geovisualization. Big data is defined by volume, velocity, variety, and complexity, and demands new approaches to visualization (Kitchin & McArdle, 2016). The importance of visualization is further acknowledged through the expanded definition of big data, which itself includes visualization. For spatial data, visualization approaches will need to support the identification of patterns and relationships across space and time and to meet the increasingly varied applications of spatial data. Geovisualization has emerged as a critical tool able to unite advanced computational techniques with visual representation, and in doing so, bridge the gap between complex spatial datasets and actionable information.

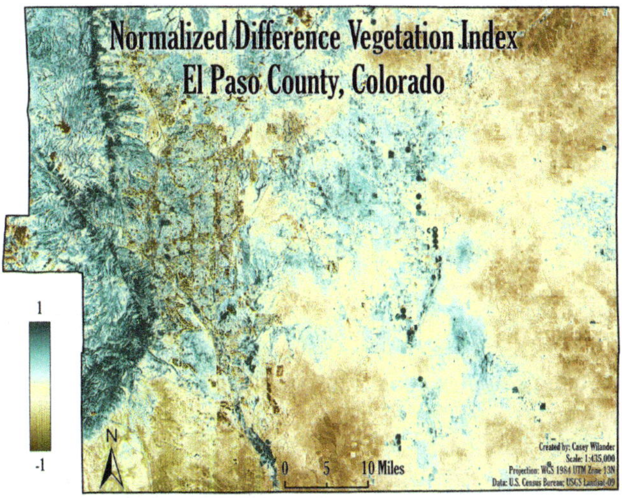

Figure 7.1 Normalized difference vegetation index (NDVI) data shown using a gradient that transitions from brown (nonvegetation) to yellow (low vegetation) to dark green (healthy vegetation).

The emergence of big data has redefined the scope of spatial analysis, presenting both challenges and opportunities for geovisualization. Satellites collect large amounts of remotely sensed (RS) data, contributing to the emergence of big data for studies of humans and their environments. For example, the Copernicus program highlights the role of daily production of image products to support land monitoring (Copernicus, 2026). RS data are a valuable form of spatial data for multiple reasons. First, it offers wide spatial coverage. Satellites can capture data across extensive geographic areas, providing a global perspective that is difficult to achieve through in situ observations. This capability is essential for monitoring large-scale phenomena such as deforestation or climate change. Remote sensing also offers standardization and repeatability. Earth observations are recorded over the same locations at regular intervals, supporting the development of datasets suitable for time-series analysis. Repeatability is combined with near real-time capabilities for some satellite systems, which is invaluable for disaster management, as in the case of wildfires, floods, or hurricanes. These data are multispectral and increasingly have high spatial resolution, e.g., Maxar constellation, which provides detailed observations that enable precise analysis, such as determining crop yield or the identification of buildings (Ji et al., 2018; Zhao et al., 2020).

The accessibility of remote-sensing datasets democratizes spatial data and makes it a powerful tool not only for research but for policymakers and the public. Beyond these characteristics, the combined use of satellite imagery and other spatial datasets, such as ground-based measurements or social data, creates comprehensive representations of the physical and social worlds and their interactions. This integration enables a richer understanding of complex processes and their impacts, such as urbanization or climate change. Remote sensing is an important source of spatial data that can be integrated with other spatial data to improve data visualization, visual thinking, and visual communication.

The current inundation of spatial data from satellite imagery to GPS data to social media geotags, combined with the value of data visualization, offers opportunities to study Earth's systems at scales and resolutions previously unexamined. However, this abundance also introduces complexities, such as data heterogeneity, high dimensionality, and computational demands.

2. Diversity of applications

Geovisualization addresses a wide range of global challenges and has emerged in the past decade as a key tool helping to address some of the world's most pressing concerns. Transforming complex geographic data into interactive, intuitive visuals has empowered researchers, policymakers, and communities to understand spatial relationships and make informed decisions. Geovisualization applications span a wide range of issues, from responding to natural disasters to addressing public health needs, managing urban growth, and safeguarding biodiversity (for example, see Aslam et al., 2023; Phang et al., 2023; Shennan et al., 2023; Willyono et al., 2024). The ability of geovisualization to make complex data practical and useful has proven essential to the current global challenges. For example, in the wake of the 2015 Nepal earthquake, geospatial tools such as OpenStreetMap played a critical role. Volunteers worldwide worked to overlay high-resolution satellite images with real-time data from affected areas, creating detailed maps of destroyed buildings, blocked roads, and isolated communities. This supported humanitarian rescue efforts.

The insights gained from geovisualization techniques applied to disasters are also evident for slow-moving disasters such as climate change. Geovisualization helps monitor and mitigate the effects of climate change. Geovisualization applications such as Global Forest Watch enable near real-time tracking of forest loss through the visualization of satellite data. Singapore's Smart Nation initiative also uses remote-sensing data and geovisualization to link the needs of the population with the management and planning of urban amenities.

Urban settings offer fertile ground for geovisualization applications. This is in part due to the ability of these applications to spatialize multiple forms of information. In addition to improving the management and planning of cities, geovisualization helps with public health logistics, as seen during the COVID-19 pandemic. The Johns Hopkins University COVID-19 Dashboard, though primarily vector-based and not remote sensing-focused, became a globally recognized tool, allowing users to track the spread of the virus (Lan et al., 2021; Suroweic, 2021). The aggregation and visualization of data from multiple sources in an interactive map and dashboard not only informed public health officials but also empowered individuals to understand the spread and changes in the pandemic. This geovisualization supported health policy decisions and served as an effective form of communication with the public. While this example is not remotely sensing focused, it represents a recent and readily understood example of how visualization of data related to complex challenges, such as a pandemic, enhances information dissemination during times of high uncertainty, when the nature of knowledge is particularly dynamic.

Geovisualization is also central to managing biodiversity and conservation globally. For example, tracking wildlife migration is important as habitat ranges decrease. Movebank brings together wildlife movements and hundreds of environmental parameters from global remote-sensing datasets. Such tools have informed the design of wildlife corridors, connections for fragmented habitats, and reducing risks for both animals and local communities (Bohrer et al., 2014).

Global food security remains a pressing challenge that is acute in regions where agriculture is vulnerable to shifting climate patterns. In parts of Africa, many farmers depend on rain-fed agriculture. Tools like those offered by Sentinel Hub provide valuable insights into crop health, soil moisture, and pest monitoring. Through the geovisualization, farmers can adjust farming practices, and food aid organizations are better able to anticipate shortages, which in turn helps build food security and resilience in communities that face high levels of environmental variation. Even in cases of human displacement, geovisualization supports aid efforts. Satellite imagery combined with other forms of geospatial data has been used by the United Nations High Commissioner for Refugees (UNHCR) to locate resources in refugee camps (Mooney et al., 2017).

An example of an application that exemplifies core geovisualization principles is Google Timelapse. Google Timelapse offers access to large unprocessed satellite datasets and interactive representations of Earth's changing landscapes. One of its key contributions is its ability to illustrate the interconnected nature of global challenges, demonstrating how environmental and anthropogenic changes are spatially and temporally linked. Climate change, deforestation, urbanization, and glacial retreat are not isolated events but part of larger global processes. Google Timelapse enables users to observe these phenomena over time, reinforcing the necessity of spatiotemporal analysis. Google Timelapse enables spatiotemporal analysis for multiple audiences. Researchers can analyze spatial patterns and trends, while educators and the broader public can access simplified geovisualizations. The versatility of geovisualizations through technological capabilities and their ability to accommodate big datasets with multiple dimensions for a wide range of audiences across disciplines has made geovisualization central to addressing current and future global challenges.

3. Interactive and multidimensional data visualization

3.1 Visualizing volume, velocity, and variety in remote-sensing data

Spatially explicit big data often exceeds the capacities of traditional visualization techniques. Static maps, for instance, fall short of representing the dynamic, multidimensional nature of spatial datasets, such as satellite images, that span more than one time, space, and thematic dimension. RS big data exhibits volume, velocity, and variety. These are known as the three "Vs" and are optimized through visualization. Remote-sensing data are being generated and stored at immense scales. For example, NASA's clearinghouse for open data offers tens of thousands of datasets, ranging from terabytes to petabytes of data freely available to the public (see for a visual representation https://data.nasa.gov/data_visualizations.html). Remote sensing data are now available from many different sources, ranging from private to public, and the diversity of sensors is great, including lasers, radar, and optical sensors. In addition, temporal and spatial resolution vary significantly, making these datasets useful for an increasing range of applications.

3.2 Ideation as a foundation for geovisualization

The continuously expanding remote sensing data environment necessitates rapid image processing approaches, including geovisualization processes, for societal benefit. To ensure these processes are effective, it is essential to understand how ideation and visual paradigms shape the transformation of raw spatial data into meaningful visual representations. Ideation is a foundational step in geovisualization, where researchers, designers, and analysts explore and refine ideas for transforming spatial data into meaningful visual representations. This creative process bridges the gap between raw geographic information and the communicative potential of visual storytelling. It requires careful consideration of the visualization's theme or topic, purpose, and audience. Beyond esthetic considerations, ideation is deeply tied to analytical goals, spatial reasoning, and pattern interpretation. It shapes the narrative structure of visual representations, influencing decisions about scale, symbology, interactivity, and data integration.

3.3 Visual paradigms and interactivity in spatial knowledge construction

Central to this step in geovisualization are visual paradigms. Visual paradigms are the conceptual frameworks through which spatial data are encoded, represented, and understood. Visual paradigms determine how information is emphasized and how users engage with the visualization. Whether drawing on traditional cartographic accuracy, thematic mapping, or immersive experience, the visual paradigm significantly affects the accessibility and clarity of spatial knowledge provided through the visual representation.

Building upon the role of visual paradigms, the process of producing effective visual representations must consider how the visualization would best accommodate the defining features of big data—the three "Vs". The three "Vs" applied to RS data combine to form rich and multidimensional datasets that, when used with other forms of spatial data, create complex data landscapes. From this complexity, the development of innovative visualization methods arises. Geovisualization techniques center on the use of interactivity, multidimensionality, data integration, and customization. Interactivity gives users the ability to manipulate the data and explore the data from varied perspectives. Multidimensionality supports space-time explorations of data and can include the use of multidimensional datasets as well as the integration of multiple low-dimensional datasets. Customization creates visualizations that are audience specific and thus appropriate for prespecified applications. The geovisualization approaches that bring together interactivity, multidimensionality, data integration, and customization include dashboards, 3D visualizations, and immersive technologies. These approaches integrate across scales and accommodate the diverse nature of RS data.

Interactive dashboards provide user-friendly options for geovisualization by providing dynamic interfaces for the exploration and analysis of spatial data. The combined use of charts, graphs, and maps in these dashboards enables users to gain deeper insights into spatial phenomena and patterns. Interactive dashboards are widely used across disciplines for a range of earth observation purposes. For example, the Coastal Inundation Dashboard maintained by the National Oceanic and Atmospheric Administration (NOAA) provides real-time water level information across the U.S. and its territories (Fig. 7.2). It combines this information with satellite imagery of cloud cover to help users understand flood risks from climatic events such as hurricanes and long-term trends such as rising sea levels (https://tidesandcurrents.noaa.gov/inundationdb/). The design of the Coastal Inundation Dashboard supports flood mitigation and responses at varied temporal scales. Interactive dashboards facilitate zooming, panning, filtering, and querying, thus giving users a wide range of data exploration options. Widgets such as dropdown menus, sliders, and time-series playbacks provide additional control, making it easier to tailor the visualization to user

Figure 7.2 Coastal Inundation Dashboard showing real-time water level conditions–Minor flood level inundation in red and sea level rise above mean higher high water (MHHW) in blue–Across the southeastern United States and its territories.

needs. Dashboards tend to combine multiple datasets, such as satellite imagery and other socioenvironmental records. This cohesive use of multiple spatial datasets offers a more holistic perspective of geographic phenomena, and the interactive features present effective and accessible communication of spatial relationships.

3.4 Advancing spatial understanding through multidimensional visualization

The role of 3D, 4D, and multidimensional visualizations in the analysis and interpretation of spatial data is increasingly important as data complexity and availability grow. These advanced visualization techniques provide powerful tools for understanding spatial phenomena by representing data in forms that closely mirror real-world structures and environments. Moving beyond traditional 2D representations, 3D, 4D, and multidimensional visualizations enhance our ability to represent, analyze, interpret, and communicate the complex realities present in RS datasets. Multidimensional visualizations are now enhanced through interactive features. Interactivity is considered one of the most defining features of geovisualization (Çöltekin et al., 2017). As such, interactivity has become embedded in many of the geovisualization techniques used to represent Earth's systems. NASA's WorldWind is an example of multidimensionality in geovisualization, and programs such as the 3D Elevation Program (3DEP), by USGS, continue to respond to the need for multidimensional representations of the environment to support decision-making in a wide range of fields, from natural resource management to infrastructure and safety.

Multidimensional visualizations allow data users to present the world to an audience in a more realistic fashion than 2D representations. Landscapes, infrastructure, and natural phenomena take on a real-world representation that is more easily understood by nonexperts. For example, digital elevation models (DEMs) and multidimensional terrain models generated from LiDAR data can illustrate surface features in ways that flat maps cannot. LiDAR technology exemplifies the intersection of big data and geovisualization, capturing extensive x, y, and z point coordinates relative to sensor positions and optionally recording additional attributes such as laser intensity (i) and digital images in red, green, and blue (r, g, b). These attributes, when integrated into a LiDAR point cloud, create visually rich datasets that support intuitive visual thinking and communication (MacEachren et al., 2004; Mathewson, 1999). The ability to render elevation ranges into color or grayscale classes further enhances interpretation and analytical insights (Buckley et al., 2004; Schloss et al., 2018). Unlike raster-based DEMs, LiDAR data inherently captures 3D spatial relationships. The compatibility of LiDAR data with both vector and raster analytical tools ensures versatile applications across multiple geospatial environments. In Schindling & Gibbes (2014), LiDAR capabilities enhanced the detection of anthropogenic activity and identification of archaeological sites. Visualizations that retain the multidimensionality of datasets are valuable for understanding elevation gradients, watershed boundaries, and slope stability in applications such as urban planning, disaster management, and hydrological modeling.

Different forms of multidimensionality are defined in distinctive ways. For example, Verdonck et al. (2020) describe the difference between 3D and 4D representations as centered on the definition of what is represented, emphasizing that 3D representations focus on individual objects and prioritize spatial aspects while 4D representations integrate both space and time. Because of the inclusion of time as a dimension, 4D visualizations are particularly useful for understanding dynamic processes. Processes that vary over both space and time, for example, changes in land cover, are well represented using 4D visualization. In addition, 4D representations integrate additional variables into the space-time framework. This is typically facilitated using color gradients and textures. For instance, Ruols et al. (2024) use a 4D visualization of a near-terminus glacier collapse using drone-based ground-penetrating radar (GPR) to explore the rapid evolution of subsurface processes. Similarly, Zhang et al. (2019) demonstrate the use of a 4D model of flooding to support and visualize effective response planning. 4D models support interactions with the audience that can enhance understanding of dynamic processes, such as the freezing of time frames and the use of experimentation to explore time and space.

Interactive data exploration is a hallmark of 3D, 4D, and multidimensional visualizations. Advanced platforms allow users to navigate through datasets, zoom into specific areas of interest, and manipulate variables in real time. This interactivity not only improves understanding but also aids collaboration among stakeholders by making data accessible to both experts and nonexperts. These visualization techniques are essential for addressing complex, interdisciplinary challenges, and to make informed decisions. As technology continues to advance, these visualizations are likely to become even more immersive and realistic.

3.5 Virtual and mixed reality for spatial data exploration

Immersive technologies, including mixed reality (MR) and virtual reality (VR), are advancing how we visualize and interact with spatial data including remote sensing data. Immersive environments allow users to "step into" data visualizations. VR offers a fully immersive experience by placing users in a computer-generated environment where they can explore remote sensing data in multiple dimensions. This form of visualization makes it easier for a user to engage multiple senses as they identify data extremes and/anomalies, assess spatial variation, and consider the context within which spatial relationships exist. Game development software has facilitated the use of VR, although it does not support the application of remote sensing methods. However, the increasing familiarity in the broader public with these environments makes the use of VR suitable for a wide range of users. For example, Rienow et al. (2020) demonstrate the suitability of VR for geographic education in K-12. Immersion in VR environments supports an intuitive understanding of geographic characteristics such as scale and elevation, which can be challenging to grasp with 2D visualizations.

MR blends digital and physical worlds and, in doing so, takes this a step further by enabling users to interact with digital representations of remote sensing data while still engaging with their physical surroundings. MR can enhance our understanding of human-environment interactions while protecting the environment under study and reducing costs. For example, Brito et al. (2019) describe that an MR framework can improve the workflow for building inspectors, while ensuring safety and reducing costs.

One advantage of using immersive technologies for geovisualization is their ability to handle large and complex datasets, such as those generated by remote sensing platforms. Many applications of MR and VR use LiDAR scans, which are rich in information (Fig. 7.3). Immersive environments make it easier to explore these complex datasets by providing a sense of depth and perspective. LiDAR is often used to measure complex structures, such as tree canopies or urban infrastructure, because it captures detail that is important to understand how those complex systems work. Through dynamic sensor positioning or mobility, LiDAR data can represent all perspectives of objects, including complex interiors and overhangs, offering unparalleled detail. This capability enables the collection of multiple z (elevation) values for identical x, y coordinates, pushing the boundaries of geovisualization. Furthermore, LiDAR data, collected as irregularly spaced vector points, can be transformed into surfaces via triangular irregular networks (TINs) or interpolated into raster formats using methods like inverse distance weighting (IDW), natural neighbor, or kriging. These inherent characteristics of LiDAR contribute to the collection of complex datasets, which in turn necessitate the use of visualization techniques appropriate for high levels of complexity.

The integration of LiDAR data into big data workflows underscores its versatility across numerous applications. In infrastructure, it supports real-time monitoring of powerline sag and assesses the structural integrity of bridges and overpasses. Industry applications optimize logistics, guiding the maneuvering of large machinery within complex building layouts. Geoscience leverages LiDAR to develop early warning

Figure 7.3 LiDAR scan showing a rich depth and perspective of data representing XYZ.

systems for unstable slopes, fault slip rates, and geothermal deformations, while mining employs it to estimate excavation volumes through comparative pre- and postmining scans. In advertising, LiDAR enables projection mapping by deriving precise distance values to create custom lenses for complex facades, and in entertainment, it integrates real-world 3D landscapes, such as racetracks, into video games for enhanced realism. Public safety benefits from rapid documentation of accident scenes to expedite roadway clearance while preserving critical details, and bathymetric applications utilize green lasers to measure water depths in aquatic environments. LiDAR also aids with cultural heritage preservation through digital replication, such as the 2019 Notre-Dame Cathedral restoration.

In addition to MR and VR, data sonification and distributed visualization are redefining the applications of geovisualization. By providing multisensory experiences, these techniques allow users to interact with spatial datasets in intuitive ways. Data sonification, the process of translating data into sound, offers a novel approach to representing spatial information, particularly for patterns that are difficult to perceive visually. Distributed visualization leverages cloud computing to process and render large-scale spatial datasets, enabling real-time analysis and visualization across geographically dispersed users.

4. Geovisualization environments

The environments within which geovisualization occurs are varied; however, they incorporate the above-described features of geovisualization: interactivity and space-time multidimensionality. Geovisualization environments also integrate varied data sources and facilitate customization. These environments blend concepts from geographic information science (GIScience), cartography, and data visualization to represent spatial phenomena.

Widely used geovisualization environments include ArcGIS tools from Esri, QGIS, Cesium, Leaflet, Tableau, Google Earth, and Google Earth Engine. ArcGIS, including features available in ArcGIS Pro, ArcGIS Online, and mobile applications, has robust geovisualization capabilities that include both 2D and 3D visualization. Open-source platforms offer alternative environments and increase accessibility, thereby opening geovisualization capabilities to broader populations of users. QGIS provides a wide range of features, including data visualization and geospatial analysis, and has a large global user community. Cesium, with origins in the aerospace and defense industries, is an open-source platform designed for 3D geospatial visualization. Leaflet and other web-based platforms for interactive mapping have an appeal for developers with limited experience with mapping software. With Leaflet, a user can display map tiles stored on public servers and customize the map; for example, by adding markers and interactive pop-ups. Environments such as Tableau, while not open source, are increasingly used across many disciplines and in business settings. Tableau's increasing popularity stems from advances in the use of Visual Query Language (VizQL), which linked drag-and-drop features to database queries and visualizations. Environments particularly popular for remote sensing data include Google Earth and Google Earth Engine. Google Earth is a user-friendly geovisualization tool, while Google Earth Engine is an advanced environment that facilitates cloud-based large-scale geospatial data analysis.

Geovisualization environments have undergone significant evolution, transforming from two-dimensional representations to dynamic, interactive, and immersive representations. The advent of Geographic Information Systems (GIS) in the 1960s marks the start of decades of development of digital tools to handle, analyze, and visualize spatial data. Early GIS environments focused on data management and basic visualization, such as layering of datasets. While revolutionary at the time, these systems were primarily used by experts, and the outputs were static and limited in dimension. As computational power grew, so too did interactive geovisualization environments. Many of these environments are now commonly used by nonexperts. Web-based mapping services, such as Google Earth and ArcGIS Online, make spatial data and its visualization accessible to a broad audience. The increase in breadth of users holds potential to bring the science based on analysis of spatial data closer to the public. Thus geovisualization serves as a bridge between data and decision-making in part through its ability to increase citizen engagement in the process of translating science to policy.

The integration of artificial intelligence (AI) is shaping geovisualization through the automation of data analysis. GeoAI combines machine learning, spatial statistics, and geoinformatics to analyze and interpret large-scale spatial datasets. Coalescing with cloud computing and big data, geovisualization environments provide real-time insights on global scales, accommodating increasingly complex spatial questions. In remote sensing, AI-integrated models allow for the automatic identification of features, detection of patterns, and prediction making using imagery datasets. Nikçi et al. (2024) demonstrate that the use of AI and machine learning improved the accuracy with which buildings were detected and differentiated in 3D models. Deep learning methods offer fast and accurate approaches for mapping land use. Rane et al. (2024) present an interesting study of how the combined use of deep learning algorithms with AI language models creates AI-powered spatial analysis systems that are easy to use and simplify the results from complex spatial analysis and thus ensure data-supported decision-making. In addition, they describe the AI potential to address inconsistent observations and large amounts of geospatial information available in satellite imagery.

AI advances in geovisualization environments reflect a continued trajectory of evolution, whereby geovisualization draws from cutting-edge technologies and growing availability of datasets to enable multidisciplinary spaces for scientific exploration and practical problem-solving. AI-integrated environments adapt analysis to user queries and evolve continuously based on data inputs. The integration of models based on generative, recurrent, and convolutional neural networks within visualization environments, automates the identification of objects, detection of trends, anomalies, and patterns in remote sensing and other spatial data. The nexus of GeoAI, geovisualization, and big remote sensing data represents a next frontier in geovisualization.

5. Challenges and future directions

The convergence of GeoAI, geovisualization, and remote sensing data holds great potential to advance our understanding of Earth's systems, yet issues of data quality, accessibility, computational requirements, and ethical considerations remain central to successful advancement. Data quality refers to how accurate, reliable, and suitable spatial datasets are for specific predefined uses. Effective geovisualization depends on the use of high-quality data. Data quality issues include errors and completeness. Data errors generally fall into three main categories: systematic errors, gross errors, and random errors. Systematic errors happen because of issues with the tools or instruments used to collect or create the data. These errors tend to be consistent throughout the dataset. Gross errors, or "observational blunders" (El-Hakim & Ziemann, 1984), are errors caused by human mistakes during the process of data entry, analysis, or visualization. Random errors are often considered noise in a dataset and are not caused intentionally. Error, regardless of form, reduces the accuracy of geovisualization. The completeness of datasets and metadata can also present challenges for accurate representations of geographic phenomena. Incomplete data can occur for a range of reasons, including technical issues, as occurred when Landsat seven Enhanced Thematic Mapper Plus (ETM+) developed a scan line corrector (SLC) failure in 2003, or, as a result of naturally occurring variation, such as cloud cover. For example, during the rainy season in tropical zones, ever-present cloud cover obscures the Earth's surface from satellite sensors, which in turn leads to gaps in earth observations.

The currentness of spatial datasets, how up to date they are, has been identified as a data quality issue (Morrison, 1995). The temporal aspect of spatial data should be considered with regard to how well a dataset represents the geographic phenomenon. However, many uses of remote sensing data seek to provide multitemporal or retrospective representations of geographic phenomena. For example, an image from 20 years ago provides a representation of the Earth's surface from 20 years ago, which might be relevant for certain applications. Although technological advancements have improved image quality and resolutions, an older image often contains valuable observations that hold value or may become more valuable over time. For example, Rizayeva et al. (2023) used historical Corona imagery to create a baseline representation of land cover, with particular interest in agriculture and forest, in the Caucasus Mountains. Historical imagery holds value in the digital humanities for "deep mapping" visualization, a method that involves creating multifaceted representations of a place (McAuliffe et al., 2017).

Accessibility remains a concern for expanded use of geovisualization. Despite the value and enthusiasm around VR and MR for geovisualization, these approaches rely on expensive hardware, advanced computational resources, and the development of user-friendly software tailored to nonexperts. Augmented reality (AR) could provide another interactive approach to geovisualization. AR overlays of spatial data onto the physical world could unlock the full potential of geovisualization for big data studies of Earth's systems. This would serve to improve our ability to perceive valuable information and use this information in decision-making processes. However, AR technology is limited by accessibility and remains in development. Increased accessibility of technology will facilitate the expanded integration of immersive technologies into geovisualization workflows. As data quantities grow, the computational resources required for processing and visualization become increasingly demanding.

Cloud-based solutions and distributed computing offer pathways to address this challenge, but they simultaneously raise questions about data security.

The increased interconnectedness of GeoAI and geovisualization raises important concerns about model interpretability and bias. Machine learning applications for the analysis of satellite imagery are common. Yet, data limitations, for example, what data was available to train the model, may yield biases that could reproduce existing inequities (Janowicz et al., 2022). Thus the importance of data quality and model evaluation remains important as GeoAI and geovisualization continue to coevolve.

In addition to model-based or algorithmic bias, ethical questions around privacy, transparency, and misinformation are pivotal to the future development of GeoAI and geovisualization. As such, Degbelo (2022) proposes using the FAIR geovisualizations. FAIR stands for Findable, Accessible, Interoperable, and Reusable. These principles were initially created with reference to Open Science (Wilkinson, 2016); however, Degbelo (2022) proposed an adaptation suitable for geovisualization and uses backcasting to demonstrate the value of applying the FAIR principles. Visual representations, including representations based on remote sensing datasets such as land cover classifications, are not impartial. Rather, there are multiple ways in which the process of data visualization can impart bias. For example, Wickham et al. (2010) ask, "How do we know what we see is really there?" In posing this question, and through multiple visualizations of data, the authors seek to encourage geovisualization creators and users to avoid apophenia. Fundamentally, FAIR geovisualizations are those that explicitly consider the multiple ways visual representations can reduce the impartiality of geovisualizations, are easily reproduced, and enhance fairness.

The future of geovisualization lies in its ability to integrate seamlessly with other technological advancements, such as the Internet of Things (IoT) and advanced simulation models. These advancements will offer holistic analyses of geographic phenomena. Future developments in geovisualization will, ideally, bridge the gap between data collection, interpretation, and sharing actionable information. The continued democratization of spatial data, including remote sensing data, will likely be driven by the continued expansion of open-access policies and collaborative platforms. For example, as of September 2024, freely available LiDAR data are available or are in production for 98.3% of the United States. These highly accurate, high-resolution data are provided at no cost by the US Geological Survey (USGS) through LidarExplorer and 3D Elevation Program (3DEP). The former allows users to visualize LiDAR datasets in web-based viewers, and the latter supports the download of raw point cloud files in LiDAR Aerial Survey (.las) format or in a rasterized lossless TIFF format. This increased accessibility of both data visualizations and raw data will empower a wider range of users to engage with geovisualizations or RS data.

Case studies

The following case studies highlight the capabilities of geovisualization to address environmental and societal challenges through the synthesis of knowledge from multiple datasets, including RS data. The first case study analyses climate change impacts on the hydrological cycle, using remote sensing data to visualize changes in the surface area of water and ice in lakes and glaciers. Through the use of geovisualization techniques, this case study focuses on local, small-scale feature recognition in the presentation of complex environmental phenomena in a comprehensible and visually engaging manner, supporting visual inference and knowledge sharing. The second case study introduces a framework for producing high-resolution, annual population maps at the census block level, combining geospatial data such as very-high-resolution (VHR) satellite imagery, GIS tax parcels, and US Census data with statistical and machine learning models. This integration demonstrates how geovisualization connects diverse datasets to extract meaningful patterns, trends, and predictions relevant to a specific topic. Both case studies exemplify thematically oriented geovisualizations. While remote sensing provides measurements of the physical landscape, the creation of effective geovisualizations requires data cleaning and extraction processes of RS data in concert with complementary datasets to align and synthesize relevant information for users. These examples showcase how geovisualization not only enhances understanding of specific environmental and social dynamics but also serves as a critical tool for bridging technical research with actionable insights.

Case study 1: Visualizing changes in the surface area of water and ice in lakes and glaciers through image vectorization

This case study aims to utilize remote sensing image vectorization and cartographic designs as an effective geoprocessing method to quantify and visualize changes in the surface area of lakes and glaciers. By adopting this approach, we seek to enhance our understanding of the dynamic nature of these water bodies and their responses to environmental changes over time. Our specific focus is on analyzing the changes in water surface area for three prominent lakes in the United States: Lake Mead in Nevada, Lake Powell in Utah and Arizona, and the Great Salt Lake in Utah. In addition, we will examine the changes in the ice surface area of the Nisqually Glacier in Washington State.

Existing studies indicate that satellite imagery time series can effectively monitor changes in the water surface areas of lakes and the ice surface areas of glaciers. These studies have identified both long-term and short-term variations in surface area, particularly at shoreline regions. They utilize a range of techniques, including advanced imagery classification, Google Earth Engine, and machine learning, or straightforward imagery comparison with dynamic time slice visualization or sliding maps (e.g., Murray & Khaki, 2021; Bonnema et al., 2022; Bazilova & Kääb, 2022). While sophisticated imagery classification and machine learning techniques generally improve estimation accuracy, their application is often constrained by high computational costs, long processing times, and thus a small study area. On the other hand, visualization techniques that solely employ satellite imagery with time slices or sliding maps provide effective visual representations of changes without requiring extensive processing. Sliding maps typically compare surfaces between two timestamps and require user interaction to slide the maps and reveal differences. In contrast, dynamic change displays with time slices or multiple small map layouts can eliminate the need for user interaction with sliding maps. Nevertheless, visualization methods that rely solely on imagery have limitations in quantifying changes or conducting systematic analyses of numerous lakes and glaciers.

This study integrates valuable satellite imagery with effective geoprocessing tools to quantify surface changes in the three lakes mentioned and the Nisqually Glacier. Our approach focuses on minimizing computational costs while applying cartographic design principles to enhance the base map's stunning imagery, ensuring clear communication of these changes. This work establishes a foundation to further establish a solid research methodology for automating surface area change analysis of lakes and glaciers globally.

The data used in this study consists of cloud-free satellite imagery from the Sentinel-2 Land Cover Explorer collected over each lake and glacier. This imagery is then analyzed using a geoprocessing technique called vectorization, which converts the satellite images into vector format. Vectorization of raster data is a vital and widely used function in geographic information analysis. The process involves transforming integer raster data, which is composed of a grid of pixels with a range of integer attribute values, into vector graphics in the form of polygons, following ESRI 's raster-to-polygon conversion algorithm (Esri, 2025; Zhang & Pan, 2001). Essentially, the algorithm operates on an integer raster dataset and groups continuous cells with the same values into polygons. It then creates arcs from bordering cells to define the polygon boundaries. Notably, raster cells with a NoData value do not produce polygons in the output.

For our analysis, we focus on the polygons that represent surface water or ice. Polygons indicating the water or ice surface of the lakes and glaciers under investigation are manually identified and saved for further mapping. We repeat the vectorization process for images taken at different time intervals for each study area, generating distinct polygons that represent the surface area at a specific timestamp. This method effectively estimates the surface area of each lake or glacier at each timestamp, enhancing our ability to quantify these areas and analyze their changes over time. In addition, it allows us to apply cartographic design techniques to the vector data, effectively illustrating these changes through static maps.

Our analysis and cartographic work produce static maps that effectively illustrate the surface changes occurring in lakes and glaciers, as demonstrated in Case Study Figures 1.1, 1.2, and 1.3. For instance, Case Study Figure 1.1 depicts the water surface changes in the northern area of Lake Mead, Nevada, from 2000 to 2023. We utilized imagery taken in 2000, 2002, 2007, 2014, 2022, and 2023 to facilitate vectorization and quantify these changes. Notably, our findings reveal a significant retreat of the shoreline, highlighting the dynamic shifts in the lake's landscape over time.

Case Study Figure 1.2 illustrates the changes in water surface area for Lake Powell during the period from 2017 to 2022, as well as for the Great Salt Lake from 2017 to 2023. Our analysis shows that the total area of Lake Powell in 2017 was 192.36 km^2. By 2022, this had decreased to 116.06 km^2. Over the 5-year period, Lake Powell lost

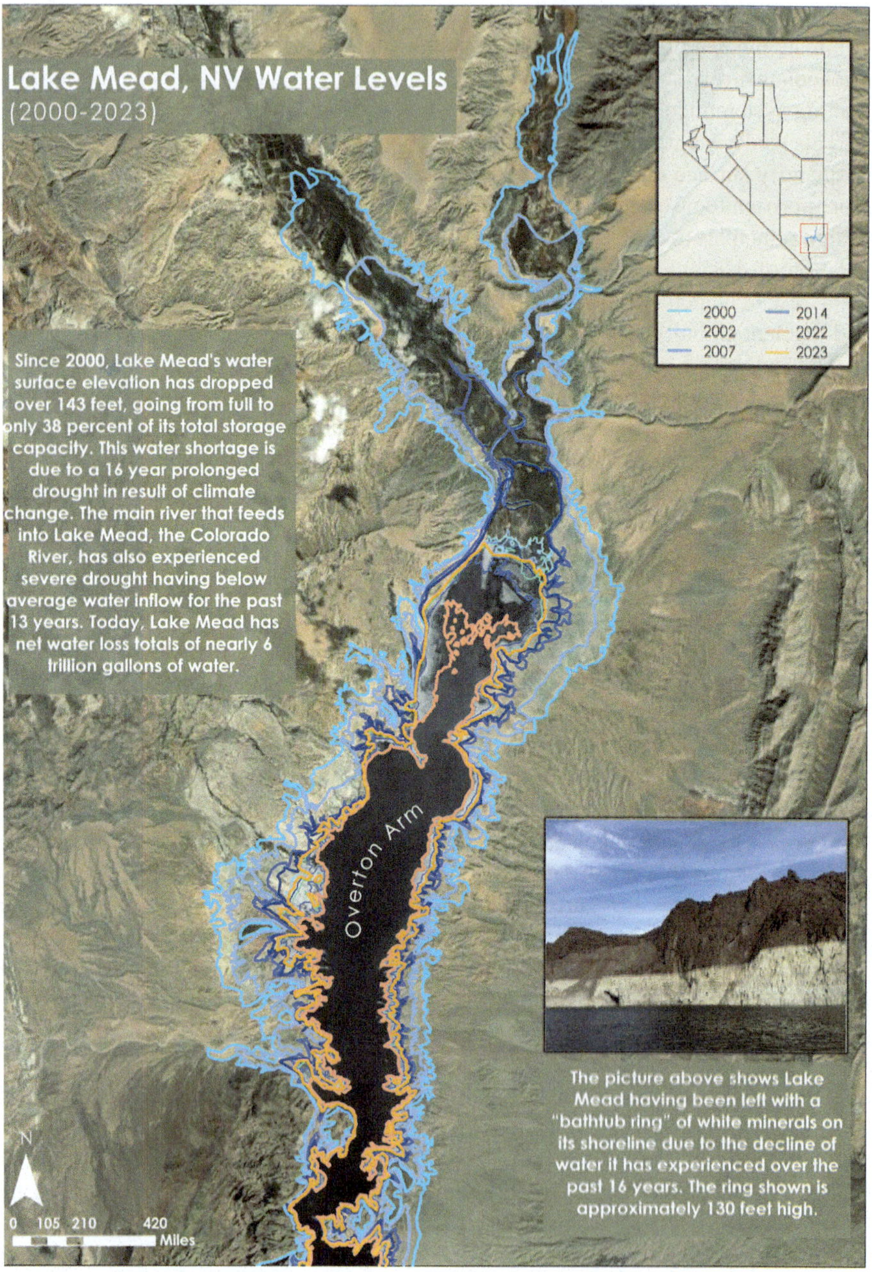

Case Study Figure 1.1 The change in water surface area of Lake Mead, Nevada, USA, during 2000–2023.

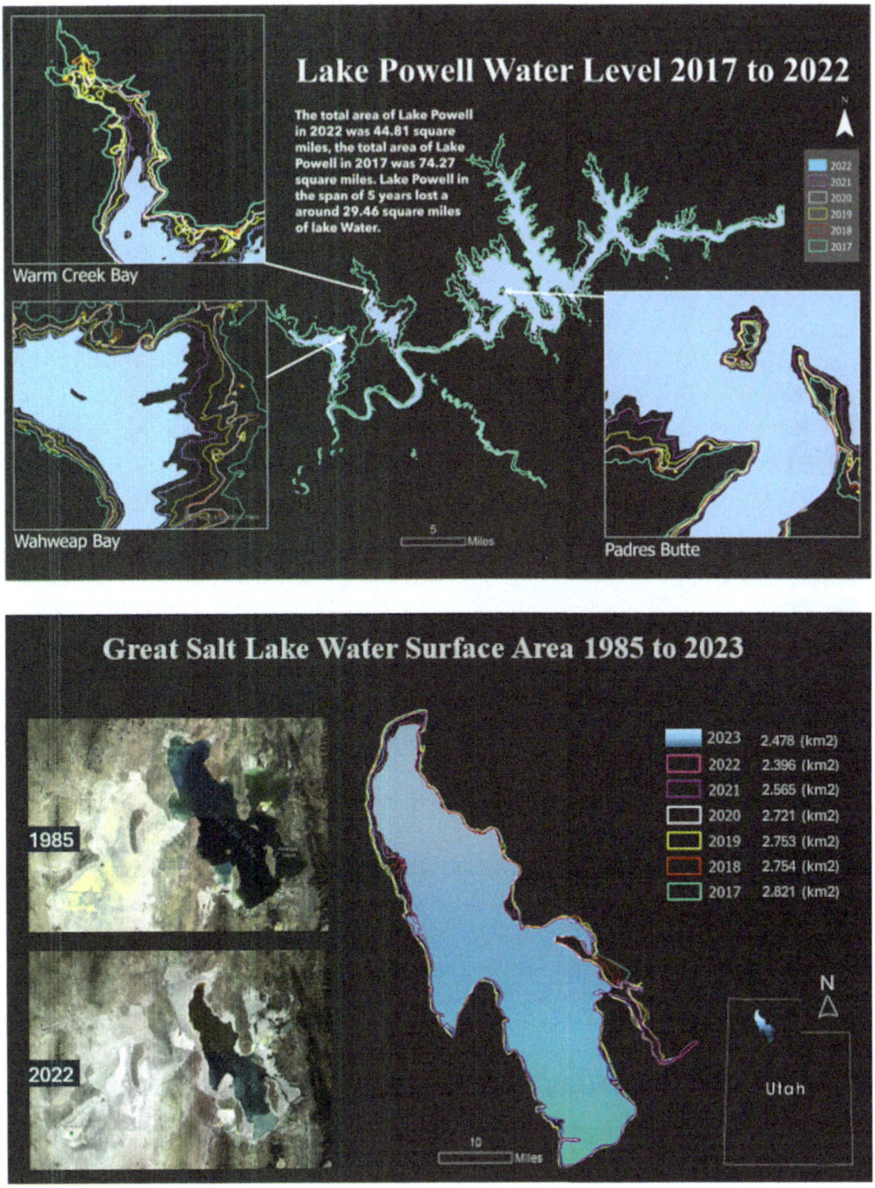

Case Study Figure 1.2 Change in water surface area of Lake Powell, Utah and Arizona, USA, during 2017–2022 (*top*) and Great Salt Lake, Utah, USA, during 1985–2023 (*bottom*).

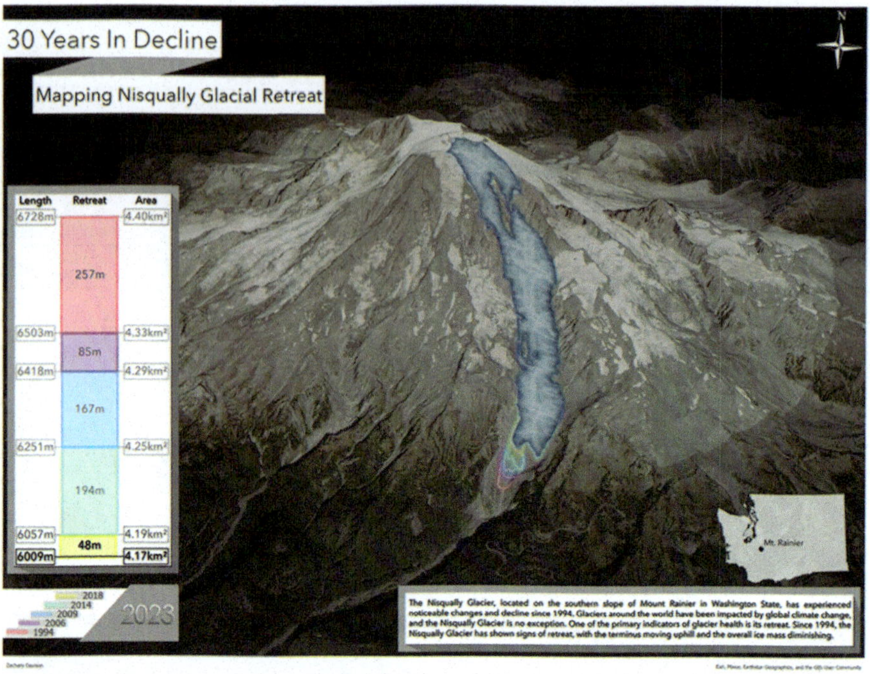

Case Study Figure 1.3 Depicting 3 decades of decline in ice surface area for the nisqually Glacier, Washington, USA.

approximately 75 km^2 of water surface area. In the case of the Great Salt Lake, the water surface area lost approximately 0.34 km^2 over the last 6 years, although satellite imagery indicates a significant reduction in surface area compared to measurements in 1985.

In addition, our analysis of the ice surface area of the Nisqually Glacier from 1994 to 2023 indicates a notable retreat, with the surface area decreasing from 4.40 km^2 in 1994 to 4.17 km^2 in 2018, as illustrated in Case Study Figure 1.3.

In summary, this study explores an effective method that combines valuable satellite imagery with robust geoprocessing tools to quantify surface changes in three lakes and the Nisqually Glacier. The results we presented offer clear visualization and important analyses for understanding environmental impacts and supporting effective management strategies.

Case study 2: Remote sensing applications for annual population estimation at the fine spatial scale: A case study of Colorado Springs

In the imagery format, remote sensing data contains spatial information about the earth that helps us understand vegetation, soils, and land use types. These datasets are standardized, nearly real time, with high spatial resolution and coverage. While remote sensing data alone is already a great asset to the field of geography and geovisualization, it has significant potential to be combined with other data formats (e.g., Gibbes et al., 2017) to mitigate the data shortage issues that have been witnessed in fields that mainly rely on surveys or censuses to collect data.

This case study demonstrates the potential of integrating GeoAI, remote sensing, and other forms of spatial data to produce fine-resolution geovisualization of population mapping. Population data at a fine spatial scale (e.g., census block) has been widely used in spatial analysis and modeling of various human behaviors, including residential segregation, disease spread, and other mobility and opinion dynamics (Dao et al., 2024; Jiang et al., 2024; Yin et al., 2024). However, in the United States, the finest population data at the census block level is only released every 10 years through decennial censuses, which makes the data difficult to meet research needs. In this case study, we present a research framework that combines remote sensing data with multiple forms of vector data and GeoAI to generate fine-resolution population mapping that can be updated annually.

This example uses the city of Colorado Springs (Colorado) as a case study. Colorado Springs, a midsize city with a population of 488,670 in 2023, has witnessed a steady population growth, a 16.4% increase since 2010 (U.S. Census Bureau, 2023). Dynamic population mapping at a fine resolution (e.g., census blocks) allows researchers to understand where the growth is happening and inform planning policies related to housing, zoning, and infrastructure.

For population estimation at the fine spatial scale, we apply a novel framework that combines dasymetric mapping, statistical modeling, and GeoAI (S. Wang & Wang, 2024). Specifically, the dasymetric mapping is a common and effective interpolation technique to split the population of a large geography unit into a finer spatial scale, considering local characteristics such as land use and land cover and building types. Statistical modeling can capture the relationship between population and local characteristics, while GeoAI allows us to extract local characteristics by combining remote sensing data with other GIS data (e.g., tax parcel). Based on other researchers' works (Georganos et al., 2021; Wang et al., 2018), the spatial distribution of population is primarily affected by building variables such as building area, count, volume, type, and the number of housing units. In this study, based on two building variables, including 1) building area information (provided by building footprints extracted from remote sensing data) and 2) detailed residential building classifications (provided by tax parcels), our method can provide a good prediction of population at the census block level. Specifically, we start with applying a Mask Region-based Convolutional Neural Network (Mask R–CNN) to detect building footprints as polygons from VHR satellite imagery from the National Agriculture Imagery Program (NAIP). The Mask R–CNN model has been pretrained using the publicly available dataset, Berlin building footprints provided by Esri (2021), with a high accuracy score and good performance on residential building detection. After detecting building footprints, the variable of building area can be obtained. Next, we have overlaid tax parcels to identify residential

Case Study Table 2.1 Datasets used for Annual Population Estimation in Colorado Springs.

Dataset	Spatial unit	Year	Variable
Decennial census	Census block (CB)	2020	Population at CB level
ACS-5 year estimation	Census block group (CBG)	2020, 2022, 2023	Population at CBG level
VHR NAIP satellite imagery	0.6 m (2019), 0.3 m (2023), raster	2019, 2023	Building area
Colorado Springs tax parcel	Parcel polygons	2024	Residential building classification
TIGER/Line shapefiles	CBG, CB	2020	

buildings and their classifications. The tax parcel data indicate that residential zone districts have 26 building types, including condo, single-family, two-family, three-family, multifamily (4–8 units), multifamily (over 9 units), etc. We kept all 26 types and assigned them to building footprints. Case Study Table 2.1 summarizes the dataset used in this case study, including their spatial unit, year ranges and the extracted variables.

After data collection and processing to obtain the necessary variables (see Case Study Table 2.1), we built a statistical model using ordinary least squares (OLS) regressions to capture the relationship between population and building characteristics, including building areas and residential building classification at the census block (CB) level. Eq. (7.1) defines the dependent and independent variables in a linear OLS:

$$P_i = \beta_0 + \sum_{k=1}^{26} \beta_k BA_{i,k} + \varepsilon_i \tag{7.1}$$

where P_i is the number of populations at the i-th CB, β_0 is the intercept, $BA_{i,k}$ represents the total building area of the k-th classification in the i-th CB, and the scalar ε_i accounts for the unobserved errors of the i-th CB.

After building the model using Eq. (7.1), we applied the workflow illustrated in Case Study Figure 2.1 for model fitting, assessment, and prediction. Specifically, we have fit the OLS model using 2020s data, including the population at the CB level from the decennial census, building areas extracted from VHR NAIP imagery, and building classification from tax parcel data (see Case Study Table 2.1). Then, we used error metrics including R-squared (R^2) and absolute error to assess the model's performance. The final model with an R^2 of 0.74 suggests a good fit. The fitted OLS model was used to predict population distribution for the year 2023. The outcomes from the OLS model served as a weighting layer of the dasymetric method to split the census block group population into the CB level.

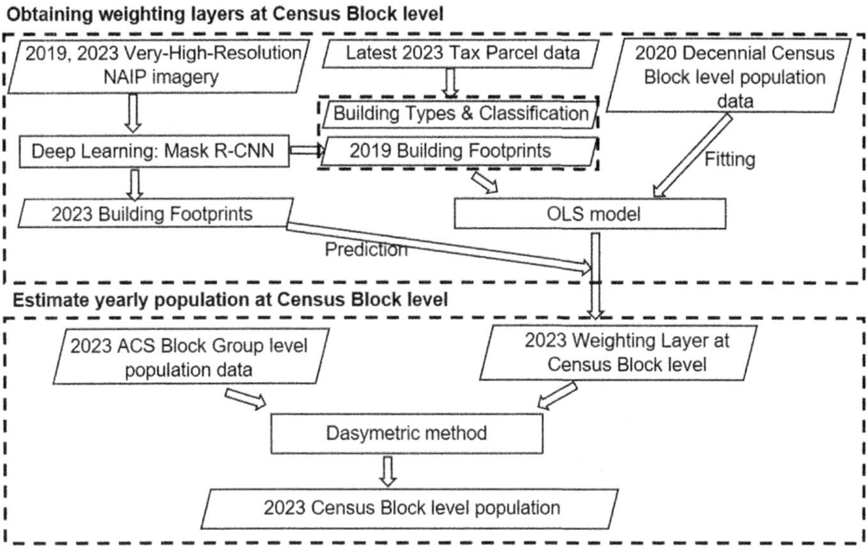

Case Study Figure 2.1 Analysis workflow of combining remote sensing imagery with GIS and census data to predict yearly population at a fine spatial scale (i.e., census block). This method incorporates a novel framework that integrates GeoAI (i.e., mask RCNN) with statistical modeling and the dasymetric method.

Case Study Figure 2.2 presents the result of the predicted population for the city of Colorado Springs at the census block (CB) level. A comparison of the population mapped using the census data (Case Study Figure 2.2A) with those mapped using our proposed framework (Case Study Figure 2.2B) suggests that our method is particularly good at capturing the spatial heterogeneity of population at a fine spatial scale, i.e., within block groups. Case Study Figure 2.2C compares the census data with our results by zooming into the Rustic Hills neighborhood in the east of the city that has a mix of land uses. While the census data indicates that the southern part of the neighborhood has a high population, our method more precisely locates the specific block where people are concentrated, with consideration given to the existence of apartment properties. The combined use of VHR remote sensing data with other forms of data (e.g., census and GIS data) has important potential to resolve the above-discussed data limitation issues and improve population mapping at fine spatial scales. In addition, this case study demonstrates that the interconnected use of multiple models and multiple visualizations of data is particularly important when geovisualizations likely serve as the basis for action through policy or planning, as is often done with population mapping. This case study highlights the essential analytical steps, from raw data to extraction, modeling, and synthesis, needed to generate meaningful insights. As demonstrated here, geovisualization serves as a critical tool to bridge technical research with practical understanding and informed decision-making.

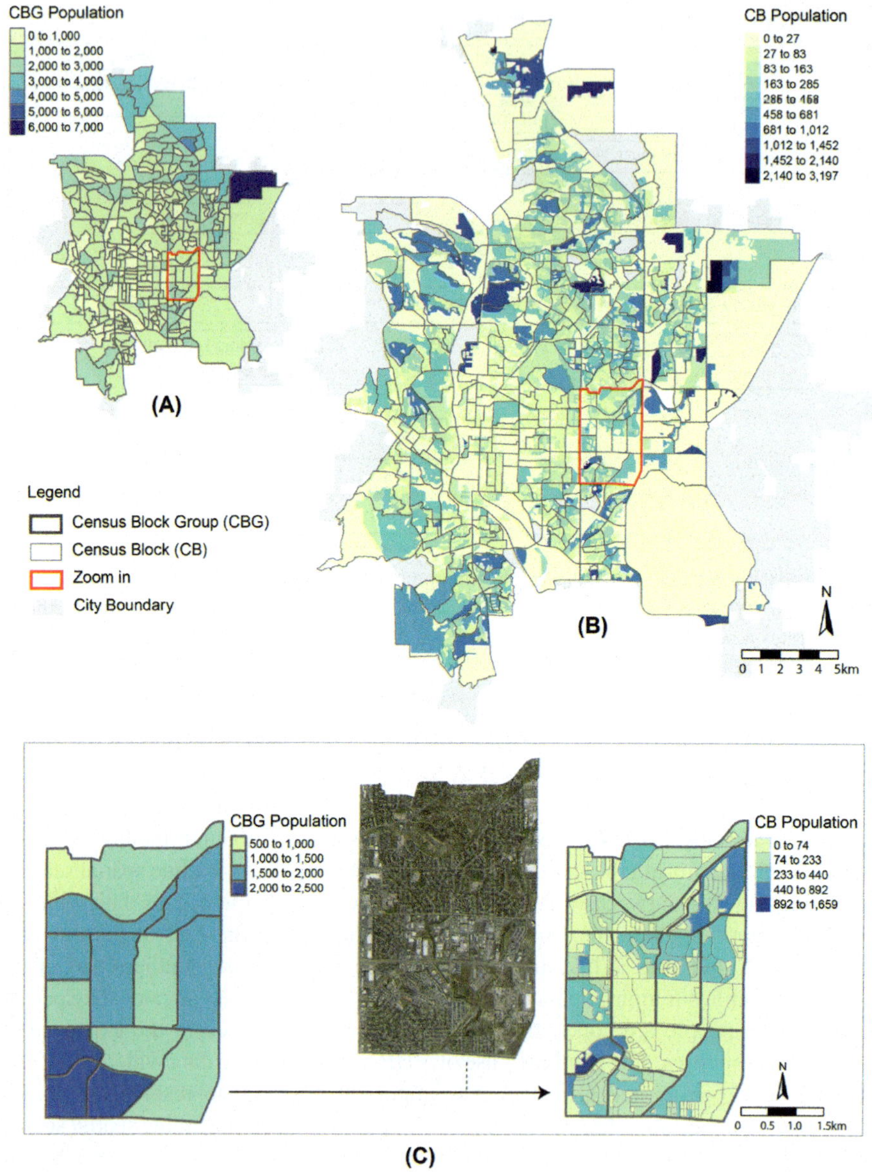

Case Study Figure 2.2 Population maps for the city of Colorado Springs, CO, in 2023. A: Population map at the census block group (CBG) level using ACS 5-year estimates. B: Map of population at the census block (CB) by incorporating remote sensing data. C: Zoom-in maps to explain the role of remote sensing data for predicting population at the fine spatial scale.

6. Conclusion

The case studies illustrate local, small-scale feature recognition and vectorization, the value of utilizing multiple geovisualizations, and the role of remote sensing in addressing data limitations. The remote sensing presented in the case studies captures real-world patterns, which, through geovisualization, are organized along with other forms of geospatial data in theme-driven ways that serve to connect data analysis to applied use. Geovisualization, used at the intersection of remote sensing, big data, and GeoAI, has become an indispensable tool for understanding the complexity of Earth's systems. Advancements in visualization techniques and addressing the challenges of data quality, scalability, and ethical implications, geovisualization will likely continue to transform spatial analysis and associated real-world decision-making. The relevance of spatial data interpretation is as significant as ever. Earth's systems are inherently multifaceted and characterized by interrelated processes that span multiple temporal and spatial scales. Understanding these systems demands not only robust data collection and methodologies, as is found in remote sensing, but also advanced tools to distill insights from the sheer volume and complexity of spatial data. In a data ubiquitous world, geovisualization will play a central role in addressing the global challenges of the 21st century, empowering researchers, policymakers, and communities to engage with spatial information in effective ways and to make data-driven decisions in a highly interconnected world.

References

Aslam, R. W., Shu, H., & Yaseen, A. (2023). Monitoring the population change and urban growth of four major Pakistan cities through spatial analysis of open-source data. *Annals of GIS, 29*(3), 355–367.

Batty, M., & Miller, H. J. (2000). Representing and visualizing physical, virtual and hybrid information spaces. In *Information, place, and cyberspace: Issues in accessibility* (pp. 133–146). Berlin, Heidelberg: Springer Berlin Heidelberg.

Bazilova, V., & Kääb, A. (2022). Mapping area changes of glacial Lakes using stacks of optical satellite images. *Remote Sensing, 14*(2022), 5973.

Bohrer, G., Beck, P. S., Ngene, S. M., Skidmore, A. K., & Douglas-Hamilton, I. (2014). Elephant movement closely tracks precipitation-driven vegetation dynamics in a kenyan forest-savanna landscape. *Movement Ecology, 2*, 1–12.

Bonnema, M., David, C. H., Frasson, R. P. de M., Oaida, C. M., & Yun, S.-H. (2022). The global surface area variations of Lakes and reservoirs as seen from satellite remote sensing. *Geophysical Research Letters, 49*(2022).

Brito, C., Alves, N., Magalhães, L., & Guevara, M. (2019). Mixed reality tool for the inspection of heritage buildings. *ISPRS Ann. Photogramm. Remote Sens. Spat. Inf. Sci., 4*, 25–29.

Buckley, A., Hurni, L., Kriz, K., Patterson, T., & Olsenholler, J. (2004). Cartography and visualization in mountain geomorphology. *Geographic Information Science and Mountain Geomorphology*, 1–27.

Çöltekin, A., Bleisch, S., Andrienko, G., & Dykes, J. (2017). Persistent challenges in geovisualization: A community perspective. *International Journal of Cartography, 3*, 115–139.

Copernicus (n.d.) Copernicus land monitoring service. European Union. https://www.copernicus.eu/en/copernicus-services/land. (Accessed 2/5/2026).

Dao, T. H. D., Dao, K. A., & Barboza-Salerno, G. (2024). Uncovering spatial patterns of residential settlements, segregation, and vulnerability of urban seniors using geospatial analytics and modeling techniques. *Urban Science, 8*(3), 81. https://doi.org/10.3390/urbansci8030081

Degbelo, A. (2022). FAIR geovisualizations: Definitions, challenges, and the road ahead. *International Journal of Geographical Information Science, 36*(6), 1059–1099.

DiBiase, D. (1990). *Visualization in the Earth sciences. Earth and mineral sciences, bulletin of the college of Earth and mineral sciences* (Vol 59). Pennsylvania State University.

El-Hakim, S., & Ziemann, H. (1984). A step-by-step strategy for gross-error detection. *Photogrammetric Engineering & Remote Sensing, 50*(6), 713–718.

Esri. (2021). *Mask RCNN trainning data Berlin (No. 7d3f633a325f4dcf962c82284098ce9d).* Arcgis [Dataset] https://ivt.maps.arcgis.com/home/item.html?id=7d3f633a325f4dcf962c82284098ce9d.

Esri. (2025). *Raster to Polygon.* Available at: https://pro.arcgis.com/en/pro-app/3.2/tool-reference/conversion/raster-to-polygon.htm. (Accessed 10 January 2025).

Georganos, S., Grippa, T., Niang Gadiaga, A., Linard, C., Lennert, M., Vanhuysse, S., Mboga, N., Wolff, E., & Kalogirou, S. (2021). Geographical random forests: A spatial extension of the random forest algorithm to address spatial heterogeneity in remote sensing and population modelling. *Geocarto International, 36*(2), 121–136. https://doi.org/10.1080/10106049.2019.1595177

Gibbes, C., Havlick, D. G., & Robb, J. R. (2017). Land use and land cover in a transitioning militarized landscape. *Journal of Land Use Science, 12*(2–3), 182–196. https://doi.org/10.1080/1747423X.2017.1313325

Janowicz, K., Sieber, R., & Crampton, J. (2022). GeoAI, counter-AI, and human geography: A conversation. *Dialogues in Human Geography, 12*(3), 446–458.

Ji, S., Wei, S., & Lu, M. (2018). Fully convolutional networks for multisource building extraction from an open aerial and satellite imagery data set. *IEEE Transactions on Geoscience and Remote Sensing, 57*(1), 574–586.

Jiang, N., Crooks, A., Yin, F., & Wang, B. (2024). Geographically-Explicit synthetic populations for agent-based models: A gallery of applications. In Z. Yang, & C. Krejci (Eds.), *Proceedings of the 2023 international conference of the computational social Science Society of the Americas* (pp. 158–172). Nature Switzerland: Springer. https://doi.org/10.1007/978-3-031-64193-0_10

Kitchin, R., & McArdle, G. (2016). What makes Big Data, Big Data? Exploring the ontological characteristics of 26 datasets. *Big Data & Society, 3*(1). https://doi.org/10.1177/2053951716631130

Lan, Y., Desjardins, M. R., Hohl, A., & Delmelle, E. (2021). Geovisualization of COVID-19: State of the art and opportunities. *Cartographica: The International Journal for Geographic Information and Geovisualization, 56*(1), 2–13.

MacEachren, A. M. (1994). Visualization in modern cartography: Setting the agenda. In A. M. MacEachren, & D. R. F. Taylor (Eds.), *Visualization in modern cartography* (pp. 1–12). Oxford, England: Pergamon.

MacEachren, A. M., Gahegan, M., Pike, W., Brewer, I., Cai, G., Lengerich, E., & Hardistry, F. (2004). Geovisualization for knowledge construction and decision support. *IEEE Computer Graphics and Applications, 24*(1), 13–17.

Mathewson, J. H. (1999). Visual-spatial thinking: An aspect of science overlooked by educators. *Science Education, 83*(1), 33–54.

McAuliffe, C. P., Lage, K., & Mattke, R. (2017). Access to online historical aerial photography collections: Past practice, present state, and future opportunities. *Journal of Map & Geography Libraries, 13*(2), 198–221.

Mooney, M. D., Tomaszewski, B., & French, M. B. (2017). Development of a block level geovisualization tool of Zaatari refugee camp community asset data. In *2017 IEEE Global Humanitarian Technology Conference (GHTC)* (pp. 1–8). IEEE.

Morrison, J. L. (1995). Spatial data quality. *Elements of spatial data quality, 161*, 1–12.

Murray, H., & Khaki, M. (2021). Analysis of Surface Water Areal changes using Remote Sensing Data. *Advanced Environmental Engingeering Research, 2*(3), 019. https://doi.org/10.21926/aeer.2103019

Nikçi, R., Župan, R., & Racetin, I. (2024). Geovisualization of buildings: AI vs. procedural modeling. *Applied Sciences, 14*(18), 2076-3417.

Phang, P., Yap, M. Y., Zakaria, S., & Labadin, J. (2023). Geo-visualization of Sarawak COVID-19 publicly available data employing open-source geospatial software. *Universal Journal of Public Health, 11*, 34–49.

Rane, J., Kaya, O., Mallick, S. K., & Rane, N. L. (2024). Artificial intelligence-powered spatial analysis and ChatGPT-driven interpretation of remote sensing and GIS data. *Generative Artificial Intelligence in Agriculture, Education, and Business*, 162–217.

Rienow, A., Lindner, C., Dedring, T., Hodam, H., Ortwein, A., Schultz, J., … Jürgens, C. (2020). Augmented reality and virtual reality applications based on satellite-borne and ISS-borne remote sensing data for school lessons. Journal of photogrammetry. *Remote Sensing and Geoinformation Science, 88*, 187–198.

Rizayeva, A., Nita, M. D., & Radeloff, V. C. (2023). Large-area, 1964 land cover classifications of Corona spy satellite imagery for the Caucasus Mountains. *Remote Sensing of Environment, 284*.

Ruols, B., Klahold, J., Farinotti, D., & Irving, J. (2024). 4D imaging of a near-terminus glacier collapse feature through high-density GPR acquisitions. *EGUsphere, 2024*, 1–26.

Schindling, J., & Gibbes, C. (2014). LiDAR as a tool for archaeological research: a case study. *Archaeological and Anthropological Sciences, 6*(4), 411–423.

Schloss, K. B., Gramazio, C. C., Silverman, A. T., Parker, M. L., & Wang, A. S. (2018). Mapping color to meaning in colormap data visualizations. *IEEE Transactions on Visualization and Computer Graphics, 25*(1), 810–819.

Shennan, K., Stow, D. A., Nara, A., Schag, G. M., & Riggan, P. (2023). Geovisualization and analysis of landscape-level wildfire behavior using repeat pass airborne thermal infrared imagery. *Fire, 6*(6), 240.

Suroweic, J. (2021). *'We had to get this right': How Johns Hopkins built the coronavirus tracking global dashboard: An oral history*. https://www.jhuapl.edu/news/news-releases/210426-JHU-COVID-dashboard-oral-history.

U.S. Census Bureau. (2023). *Age and Sex, American Community Survey 5-Year estimates subject tables*. U.S. Census Bureau. Table S0101 https://data.census.gov/table/ACSST1Y2023.S0101?q=population&g=160XX00US0816000. (Accessed 5 January 2025).

Verdonck, M., Gailly, F., & de Cesare, S. (2020). Comprehending 3D and 4D ontology-driven conceptual models: An empirical study. *Information Systems, 93*.

Wang, S., & Wang, L. (2024). A novel framework for mapping updated fine-resolution populations with remote sensing and Mobile phone data. *Journal of Remote Sensing, 4*, 0227. https://doi.org/10.34133/remotesensing.0227

Wang, L., Wang, S., Zhou, Y., Liu, W., Hou, Y., Zhu, J., & Wang, F. (2018). Mapping population density in China between 1990 and 2010 using remote sensing. *Remote Sensing of Environment, 210*, 269–281. https://doi.org/10.1016/j.rse.2018.03.007

Wickham, H., Cook, D., Hofmann, H., & Buja, A. (2010). Graphical inference for infovis. *IEEE Transactions on Visualization and Computer Graphics, 16*(6), 973–979.

Wilkinson, M. D. (2016). *The FAIR guiding principle for scientific data management and stewardship: Comment.*

Willyono, E., Bone, C., & Newell, R. (2024). Geovisualisation for effective management of invasive species: Bridging the knowing–doing gap. *Parks, 30*, 25–36.

Yin, F., Crooks, A., & Yin, L. (2024). How information propagation in hybrid spaces affects decision-making: using ABM to simulate Covid-19 vaccine uptake. *International Journal of Geographical Information Science*, 1–27. https://doi.org/10.1080/13658816.2024.2333930

Zhang, X., Li, N., & Pan, B. (2019). 4D Visualization of Flood Disaster Process for Emergency Response and Planning. *ISPRS International Journal of Geo-Information Conference Proceeding.*

Zhang, X., & Pan, Y. H. (2001). Vectorization technique for GIS grid data based on "grid technique". *Journal of Computer-Aided Design & Computer Graphics, 13*, 895–900.

Zhao, Y., Potgieter, A. B., Zhang, M., Wu, B., & Hammer, G. L. (2020). Predicting wheat yield at the field scale by combining high-resolution Sentinel-2 satellite imagery and crop modelling. *Remote Sensing, 12*(6), 1024.

Further reading

Kitchin, R., & McArdle, G. (2016). What makes big data, dig data? Exploring the ontological characteristics of 26 datasets. *Big Data & Society, 3*(1).

Kraak, M. (2020). In*Nature of geovisualization. International encyclopedia of human geography* (2nd ed., Vol. 6). https://doi.org/10.1016/B978-0-08-102295-5.10552-9

Zhang, X., Zhang, M., Jiang, L., & Yue, P. (2020). An interactive 4D spatio-temporal visualization system for hydrometeorological data in natural disasters. *International Journal of Digital Earth, 13*(11), 1258–1278.

Remote Sensing and Big Data Applications in Climate and Health

8

David Keellings and Mostafa Rezaali
Department of Geography, University of Florida, Gainesville, FL, United States

Key terms and definitions

Climate extremes: extreme events such as heat waves, hurricanes, drought, or degraded air quality that are by definition rare but have an outsized impact on society.
Health impacts: human health impacts associated with the occurrence of a climate extreme.
Analog forecasting: a method to compare current atmospheric patterns to those from the past in order to match patterns and make forecasts.
Big Data: ever growing large and complex datasets that are difficult to analyze with traditional methods.
Explainable AI: methods aimed at increasing human comprehension of how AI models work and increased confidence in their output.

1. Introduction

The climate has always been linked to human health. Indeed, humanity has been able to flourish on Earth due, at least partly, to a period of remarkably stable global climate. However, anthropogenic climate change is now rapidly impacting many physical systems on our planet. It is driving sea level rise and hydrological changes that are affecting food and water insecurity as well as degrading air quality, and, as you will see in a later chapter, is also linked to the increased incidence of infectious disease (Calvin et al., 2023). According to the Lancet and the World Health Organization, anthropogenic climate change is the greatest global health threat facing the world in the 21st Century (Atwoli et al., 2021). Perhaps one of the clearest and most direct linkages between climate and human health is the occurrence of extreme climatic events such as heat waves.

Heat waves are occurring more frequently across the globe, and under a warming climate are projected to increase in frequency, intensity, and duration (Calvin et al., 2023; Perkins-Kirkpatrick & Lewis, 2020; Raymond et al., 2020; Russo et al., 2017). Heat waves cause many adverse ecosystem and societal impacts, including diminishing the productivity of agriculture and vegetation (Ciais et al., 2005; Mishra & Cherkauer, 2010), degrading air quality (Shaposhnikov et al., 2014; Vautard

Remote Sensing, Big Data, and GeoAI. https://doi.org/10.1016/B978-0-443-26741-3.00009-X

et al., 2005), inducing infrastructure failures (Garcia-Herrera et al., 2010; McEvoy et al., 2012), and increasing energy use for cooling (Miller et al., 2008). Heat waves have also been linked to the enlargement of the areal extent of wildfires, resulting in severe vegetation and economic impacts (Zamuda et al., 2013). Extreme heat is detrimental to human health as it can cause acute heat stress, but also exacerbates preexisting medical conditions and causes mortality rates to increase (Ahmadalipour et al., 2019; Ahmadalipour & Moradkhani, 2018; Curriero et al., 2002; Ebi, 2008; Gasparrini et al., 2015). Recent work suggests that around half of the world's population will be annually exposed to deadly heat waves by 2100 and that even in a technologically developed and wealthy country like the U.S., there will likely be a four-to-six-fold increase in exposure to heat waves by 2070 (Jones et al., 2015; Mora et al., 2017). Despite the pervasive use of cooling technology in American homes, extreme heat continues to cause more deaths annually than any other form of weather extreme in the U.S (Davis & Gertler, 2015; NWS, 2016).

Heat waves are typically defined as a sequence of days with temperatures exceeding a high threshold of daily maximum or mean temperatures (Bunting et al., 2024). The heat wave threshold is often between the 90th and 99th percentiles of the local temperature distribution, and it may or may not include some measure of humidity. It is important to note that definitions of heat waves vary geographically. A temperature of 78°F (25.5°C) is a comfortable and not unusual spring day in Atlanta, Georgia, but is an extreme high temperature in Anchorage, Alaska. To constitute a heat wave, the abnormally high temperatures must also persist for several days. People can usually adapt to one hot day, but when there are multiple hot days in a row, the temperature starts to take a toll on people's health. Three consecutive hot days are generally considered to be the minimum duration for a heat wave. Numerous heat wave metrics have been developed (Anderson & Bell, 2009; Hajat et al., 2006; Keellings & Waylen, 2014; Meehl & Tebaldi, 2004; Peng et al., 2011; Photiadou et al., 2014; Tan et al., 2007), yet there is still much debate as to how best to define heat waves, and no definition or metric has been widely accepted (Perkins, 2016). The U.S. National Weather Service (NWS) issues public excessive heat watches and warnings based on elevated heat index temperatures (a combination of temperature and relative humidity). However, the NWS-defined heat index temperature threshold for excessive heat varies across forecast offices and is based on a collaboration between local NWS offices and partners. Unlike other climate extremes, such as hurricanes and tornadoes, heat waves have no universally accepted definition (Bunting et al., 2024). This complicates both the comparisons between heat wave events and the analysis of their impacts. Heat waves are generally associated with hot summertime conditions where an area of high pressure (often associated with no cloud cover, no rainfall, and light winds) remains stationary over a location for an extended length of time (Photiadou et al., 2014). Under these conditions, heat quickly builds up, similar to being stuck in a hot room with no ventilation. In addition, when heat waves happen during droughts, the heat wave gets intensified by the dry conditions, as there is little to no water to evaporate.

As seen in previous chapters, there are increasingly large volumes of remotely sensed data being collected at global scales and at finer spatial and temporal resolutions. Climate data are no exception to this rapid upward trend in availability. Indeed, we find ourselves surrounded by climate data and increasingly complex numerical weather prediction (NWP) models that aim to make sense of our atmosphere and predict its behavior. However, despite the increasing availability of data and ever more complex models, we fail to understand the atmospheric dynamics of extreme events or accurately predict/forecast their occurrence beyond a week to 10 days in advance. For example, we know that heat waves occurring in the summertime over North America are linked to slow modes of ocean-atmosphere teleconnections (such as the oscillations of El Niño and La Niña) and are preconditioned by land surface feedback from springtime soil moisture (Hoerling et al., 2013; Keellings & Waylen, 2014; Cowan et al., 2017). Yet, despite this evidence for predictability at seemingly seasonal scales, these heat wave events are currently not accurately forecasted beyond 2 weeks (Alley et al., 2019). AI is well-suited to working with large and complex datasets with its incredible ability to reduce complexity and learn patterns that other approaches are incapable of. Therefore, there is huge potential for GeoAI to transform our understanding of climate extremes and increase our ability to predict climate extremes such as heat waves.

As an example, in 2023, Google introduced its revolutionary GraphCast model (Lam, et al., 2022), an application of machine learning to climate forecasting. GraphCast was developed to predict weather and climate patterns by employing graph neural networks—an advanced computing method capable of processing complex relationships between spatial and temporal data sets. This model was trained on large datasets generated from climate simulations and real-world atmospheric measurements, giving it the capability of providing accurate forecasts worldwide. By applying advanced AI techniques, GraphCast holds promise to extend the predictability of extreme climate events like heat waves beyond what can be accomplished using traditional numerical weather prediction models. Not only can this innovation enhance our ability to forecast extreme weather events accurately, but it may also mitigate health and societal effects associated with climate change.

This chapter will present state-of-the-art GeoAI methods that are being developed and applied to remotely sensed datasets to better understand and predict extreme climate events with impacts on health.

2. Theoretical foundations

2.1 The curse of dimensionality

As more and more data become available, it becomes increasingly difficult to sort through it and find meaning. Big Data refers to the large volume of data that is growing exponentially over time. These data are so large that traditional analysis methods and computational systems may struggle to make sense of them. The sheer size of Big Data is not the only thing that makes it challenging to work with, as these data are also likely to be more complex as Big Data often has more dimensions. Dimensions are variables or attributes of data; for example, a dataset of cars might include many dimensions such as make, model, color, horsepower, safety features, etc. As data increases in dimensions, it becomes harder to work with, requiring more computational resources and processing time. With more dimensions comes more complexity, and it becomes increasingly difficult to spot patterns and trends. High dimensionality causes all sorts of problems when analyzing data, including making it difficult to visualize and decreasing the performance of models that try to represent or simulate these complex data. This can result in overcomplicated models that can't be generalized to new data.

The antidote to the curse of dimensionality is called dimensionality reduction. This is a process that reduces the number of variables in the dataset to a more manageable set of principal variables. Such methods make it possible to discard the least important variables and retain the most important ones. One such method is Principal Component Analysis (PCA), whereby a dataset comprised of many variables is reduced to a smaller set of uncorrelated variables termed principal components (Preisendorfer, 1988). These principal components are linear combinations of the original variables that capture as much information from the original dataset as possible. The number of variables is thereby reduced, and the new principal components make patterns easier to visualize and analyze. Autoencoders are another nonlinear method of dimensionality reduction. Autoencoders are a form of artificial neural network that begin by compressing input data into a low-dimensional representation called latent space. The autoencoder then reconstructs the original input from the compressed latent space. By working through this compression and reconstruction of the data, the autoencoder effectively learns how to reduce the high-dimensional data into a lower-dimensional representation without losing the essential features of the data. AI is very good at dimensionality reduction, and this makes it ideally suited to sorting through climate datasets that include many physical variables such as temperature, precipitation, pressure, wind, soil moisture, etc.

2.2 GeoAI analysis of extreme climate events

To study extreme events, they first must be identified within climatological datasets. As climate data grows larger in spatial coverage and detail, it has become necessary to develop methods to delineate extreme events from the matrix or background of data. Climate science has borrowed and adapted techniques from the fields of landscape ecology and remote sensing to spatially define extremes such as heat waves and hurricanes. Shape metrics have been used extensively to examine, for example, patches of forest on the landscape occurring within gridded remotely sensed imagery (Keellings et al., 2018). These metrics rely on simple geometric concepts of convex hull, perimeter, area, clusters, and solid area (see Table 8.1) to quantify size and shape. For example, Zick and Matyas have applied shape metrics to hurricane precipitation to improve our understanding of hurricane dynamics and structure as the storms evolve and make landfall (Zick & Matyas, 2016).

Recent progress in deep learning AI has made it possible to accurately detect, analyze, and predict complex patterns in multidimensional data. Two everyday

Table 8.1 Geometric parameters for the calculation of shape metrics.

Geometric parameter	Definition
Convex hull	Smallest convex polygon that contains the event
Perimeter	Distance around the boundary of the event
Area	Area of the event's convex hull
Clusters	Isolated clusters of the climate extreme within the encompassing convex hull
Solid area	Area of event (non-zero cells) within the encompassing convex hull

examples are facial recognition or the push toward self-driving automobiles that make use of AI to identify patterns in images that correspond to a person's facial features or road signage. In the field of climatology, the recent utilization of AI has produced promising results through borrowing these same pattern recognition methods. For example, AI has been applied to visually identify extreme events, including hurricanes, from multi-dimensional climate data and to forecast land surface temperatures from learning patterns in antecedent conditions (Chattopadhyay et al., 2020; Liu et al., 2016).

There is potential for AI to transform our understanding of complex atmospheric processes and greatly increase our ability to predict extreme events like heat waves. Prior to the development of NWP models, weather forecasting was strongly dependent on data-driven analog forecasting (Lorenz, 1969; Van den Dool, 2007). Following this method, experts would forecast future weather by comparing current weather patterns to previously observed patterns (or analogs) found in historical weather maps. Despite its long history and widespread use, there are limitations to analog forecasting. It is challenging to obtain similar enough analogs that will result in accurate forecasts, and time is required for experts to look through weather maps by hand. AI has been able to overcome these limitations through computational power, since deep learning discovers complicated patterns that human specialists might overlook, and AI has the capacity to work at high speeds across vast datasets.

Early work in the application of deep learning AI to climate extremes was undertaken by Liu et al. (2016) using neural networks to classify climate patterns and detect extreme events. They developed deep learning Convolutional Neural Networks (CNN) that could classify hurricanes, atmospheric rivers, and weather fronts within multivariate data with an 89%–99% accuracy (Liu et al., 2016). More recently, Chattopadhyay et al. (2020) took deep learning techniques a step further toward analog forecasting of extreme surface air temperatures over North America. They accurately (up to 80%) predicted temperatures one to 5 days ahead from pattern recognition of air pressure measurements (Chattopadhyay et al., 2020).

2.3 Explainable AI

AI methods are difficult to interpret, and although they may produce highly accurate predictions, it is often not clear how those predictions have come about. Therefore, there has been criticism of AI, particularly from researchers and scientists concerned with understanding physical mechanisms. AI learns patterns from input data, but it is notoriously difficult to understand what patterns have been learned and why. This has led to many referring to AI as a "black box". While much of the effort in developing new AI techniques has been focused on improving accuracy, it is of great importance to explore how AI has made predictions and how it has identified physically meaningful patterns within data. Explainable or interpretable AI aims to uncover these physically meaningful patterns by examining the reasoning behind AI predictions. Explainable AI is now at the forefront of deep learning and is only just beginning to be investigated in climate research. One of the few examples is Toms et al. (2020) use of AI neural networks to identify patterns in climate data (Toms et al., 2020). These techniques were demonstrated with two example cases. The first case was able to characterize the El Niño/La Niña phase just by analyzing sea surface temperature patterns across the Pacific Ocean with perfect (100%) accuracy. The second case was able to use northern Pacific sea surface temperatures to predict the sign (+ or −) of average land surface temperature anomalies along the west coast of North America up to 180 days in advance with a 67% accuracy. These authors then demonstrated the ability to interpret the developed AI networks through reverse mapping of the output onto the input. They utilized Optimal Input and Layerwise Relevance Propagation (LRP) to interpret the patterns learned by the neural network. Optimal Input works by feeding a null input into the previously trained and now frozen neural network. Backpropagation of loss (error of output vs. desired output) then updates the input, and this is repeated until the loss is minimized. Thus, patterns within the input that are most important for determining the output are identified. The optimized input can be thought of as a composite of patterns that lead to the output, in the same units/dimension of the original input. However, the real world is not a composite, so LRP considers one input and output sample at a time on a case-by-case basis and produces a heat map in the same dimensions as the input showing the most important patterns for that particular output.

These techniques allow us to project the AI learned patterns back onto the original dimensions of the input data. This aids our understanding of meaningful physical relationships, which in the examples above are the spatial structures of sea surface temperature anomalies directly linked to the El Niño/La Niña phase and the west coast land surface temperature anomalies. This allowed the authors to see what the AI network had learned about patterns of sea surface temperatures and identify the region of the Pacific that is most important for determining climate impacts. These interpretation techniques allow us to see inside the AI and reveal what has been learned about the underlying physical mechanisms.

3. Methodologies

3.1 Spatial event tracking

Extreme events are, by definition, rare, and data on these events have been lacking. What data did exist was often confined to weather station point locations or sporadic measurements by ships or weather balloons. Events such as heat waves and hurricanes were analyzed within these sparse datasets in terms of individual variables such as their peak temperature or winds. In the case of heat waves, this would generally take the form of time series analysis where observed daily temperatures were converted to exceedances above a high temperature threshold, and the event magnitude, duration, and frequency of threshold crossings could be calculated and modeled. Not until recently have extreme climate events been conceptualized as spatial events. These events are occurring in the atmosphere, and they have a size and shape that were not captured by analyzing time series data alone.

As there are now remotely sensed and spatially interpolated climate datasets (grids of values such as temperature) readily available, it is possible to examine events more holistically across space. This is where the spatial metrics, discussed previously, have come into play to quantify the size and shape of extreme events. However, to apply these spatial metrics, the event must first be delineated from the matrix or background of the underlying data. In the case of heat waves, the grid cells exceeding a heat wave threshold must be highlighted, and then AI methods can be utilized to identify the entire individual event. These AI methods fall under the category of cluster analysis, where the goal is to group similar observations into spatially continuous regions. Spatial clustering works by applying algorithms that aim to cluster spatial observations into coherent regions that are internally similar and spatially connected (Keellings & Moradkhani, 2020). For example, in marketing, the K-Means clustering algorithm could be applied to a spatial dataset of customer locations and purchase history in order to identify regions where customers buy similar products to inform the marketing strategy for a particular retail store. K-Means works by grouping data points or features into clusters of similar features while maximizing the differences between clusters. The algorithm begins by randomly picking k starting points that are centroids in the data space. Each data point is assigned to the closest centroid based on location and/or other attributes. Each centroid is then recalculated/moved based on the average of all data points assigned to it. Data points are then reassigned to the nearest cluster/centroid, and the process repeats until data points stabilize, i.e., don't change clusters. Through this repetitive process, the algorithm works toward ideal cluster locations with maximum similarity within clusters and maximum dissimilarity between clusters.

Climate data are generally represented in grids with each cell containing climate variables for a location across the land surface. When exploring extreme climate events, the clustering analysis tries to group all the grid cells together that have been identified as containing the extreme event. For example with heat waves, this process could involve the following steps working with a gridded temperature dataset: (1) highlight all grid cells where maximum temperature exceeds the 95th percentile of summertime maximum temperature and are, therefore, considered to be in heat wave, (2) perform clustering of the highlighted heat wave cells to identify all cells that are part of the same event based on distance between cells, (3) a convex hull (polygon) is then drawn to encompass all cells identified as being part of the same event, (4) the convex hull is given a unique identifier and then tracked based on spatial overlap from 1 day to the next. Then, spatial metrics such as areal extent can be calculated for each spatially defined event.

3.2 AI based prediction of extreme climate events

The application of AI in climatology is a very recent endeavor, but like applications of AI across all disciplines, it has the potential to transform our understanding and predictive ability. Much of this early work has relied on neural networks to learn patterns in data. Neural networks try to mimic the human brain as they consist of a network of layered interconnected nodes (Fan et al., 2020). When a neural network is supplied with input data, the first layer of nodes processes the input and then passes on the output to the next layer of nodes. As data flows through the network, the connections between nodes are strengthened or weakened to learn patterns and improve the accuracy of the prediction passed out of the final output layer. Each node is set up to perform simple calculations on the data it receives and to pass output to the next layer through the network. The connections between the nodes are weighted, and thus some connections can be strong and others weak. The network learns by adjusting these weights in a way that reduces the error in the final output.

Most work applying AI within the field of climatology has taken the form of a CNN architecture. CNN is a special form of neural network that is ideally suited to pattern recognition tasks like those climatologists face when trying to identify patterns in large atmospheric datasets (Chattopadhyay et al., 2020; Liu et al., 2016). CNNs have been used in facial recognition, image classification, search engines, and recommender systems. For example, Spotify uses CNNs to identify the types of music you like and may want to hear more of, not just by genre or artist (that's natural language processing), but also by patterns in the audio waveform like beats per minute, loudness, major/minor key, etc. A CNN takes advantage of the fact that an image is composed of smaller features by analyzing each in isolation. In this way, a CNN operates similarly to our eyes. In our visual cortex, individual neurons respond to visual stimuli in a receptive field, which is a restricted region, and all regions partially overlap. The CNN operates in a similar way by using square patches (kernels) that are a grid of weights that slide over the image and take the dot product of the underlying image to create convolutional layers (H. Lee & Song, 2019; Milosevic, 2020). This way, features are detected and learned by continually modifying/learning the weights. The use of these kernels also reduces the complexity of the image and thus greatly reduces the number of weights that need to be learned versus a traditional neural network. This greatly helps reduce computational time when applied to large and complex climate datasets. CNNs are, therefore, ideally suited to visual data processing/interpretation and are a foundational technology within the field of computer vision, and have been embraced in the application of AI in climate research.

GeoAI approaches to modeling climate extremes, like CNNs, require large datasets on which to train AI. This training is essential as AI must learn what weighting between nodes produces the most accurate output. The training data must also be labeled, i.e., these data represent a known heat wave or a hurricane, so that AI knows what the desired output should be. Once training is complete, AI can then be given new data, not used in training, to test its performance. A conceptual example of this process is a CNN trained to identify images of cats versus dogs. The AI is first presented with labeled images of cats and dogs to learn how to identify one versus the other, and then given new images to see if the training has been successful. Unfortunately, labeled training data is not readily available for climate extremes, and so the development of spatial event tracking, outlined previously, was necessary. That algorithmic approach using clustering allows for the creation of labeled datasets of climate extremes.

4. Discussion

4.1 Importance of GeoAI for climate and health

The dynamics of weather systems are some of the most complicated of any system that makes up our world. Therefore, modern weather forecasting is one of the great examples of science being brought to bear to understand and predict the behavior of our natural world. NWP is the principal method by which forecasters attempt to predict weather, and this is accomplished by solving physical equations using supercomputers. NWP has incrementally improved over the last few decades by providing increasingly accurate forecasts of weather variables like temperature and precipitation or hurricane tracks (Alley et al., 2019; Bauer et al., 2015). However, NWP-based forecasts are still limited to around 14 days into the future, and accuracy can only be improved through slow improvement to both the models and the approximations made by experts (Alley et al., 2019; Bauer et al., 2015). AI has the potential to improve forecasting by capturing and learning patterns within increasingly available big weather data that experts may have overlooked, or that would be difficult to represent in NWP equations. Subsequently, increased forecast accuracy with longer lead times will be highly beneficial in planning for and reducing the health impacts of climate extremes such as heat waves, hurricanes, and dangerous air quality.

4.2 Core methods

Climate data are inherently spatial, and much of the application of GeoAI to the field has focused on cluster analysis and visual learning of patterns. The case studies discussed in the chapter (section 5) show how GeoAI has been developed and applied to real-world climate problems. The studies focused on very different phenomena, heat waves and ground-level ozone, and applied different techniques, but they both demonstrate how AI has been used to overcome traditional limitations in how we examine and model these climate phenomena. The heat wave study has shown how GeoAI can be applied to more fully represent extreme climate events in a holistic manner by redefining these events as truly spatial in nature rather than as blips in a time series of temperature. The spatial extent of heat waves has been delineated using cluster analysis and the basic geographical principles of minimum convex hull and spatial overlap. This has allowed for a new understanding of heat waves and captured significant upward trends in their occurrence that had been missed by traditional analyses. The work on ground-level ozone has shown how deep learning AI can be applied to climate data to more accurately predict dangerous reductions in air quality. Particularly of note is how AI has worked with widely sampled and readily available data (e.g., temperature, humidity, population density) across Iran rather than with antecedent pollution data that is often not available across all locations. This study represents a clear example of how GeoAI can outperform traditional models by providing populations with higher resolution and more accurate knowledge about dangerous climatic events.

4.3 Future directions

This chapter has presented the great potential for GeoAI to overcome limitations in our ability to understand and predict in the area of climate-health interactions stemming from Big Data and the sheer complexity of our climate systems. The future of GeoAI in climate studies will likely involve the application of increasingly powerful visual deep learners to ever larger volumes of climate data. Future work in this direction should focus on the integration of additional geophysical variables as predictors and methods for dealing with subsequently large multidimensional datasets. The development of methods to increase prediction lead times toward seasonal time steps is also of great importance in the fight to adapt to ever more extreme weather events resulting from anthropogenic climate change. GeoAI may be a panacea able to simplify and look for patterns in increasingly large and complex datasets, as well as produce more and more accurate predictions at longer and longer lead times. However, it is crucial that these modeling efforts be interpretable and that they increase our physical understanding of the systems they represent. The explainable AI methods presented in this chapter (Section 2.3), as well as others, will become increasingly important as an essential companion to AI-based modeling approaches. We must move beyond the quest for ever higher accuracy toward the goal of increasing our knowledge of the physical processes that underpin the AI predictions.

5. Conclusion

In this chapter, we've presented how GeoAI can assist in climate health research and given two examples, one of tracking heatwaves across the conterminous United States, and a second one of predicting ground-level ozone in Iran. In one instance, GeoAI completed an analysis that wouldn't be possible with traditional methods; in the other, GeoAI performed better than traditional methods.

GeoAI represents a new way of understanding and predicting our climate. It is currently underutilized in climate science, and we have yet to realize its full potential to transform our understanding of our physical world. There has likely not been such potential for transformation since the advent of computers. It is perhaps fitting that this new technology, with its promise of great potential, harkens back to where weather forecasting began, with experts looking for patterns in data and issuing analog forecasts. GeoAI tries to mimic our own brains in how we see data and maps, and how we look for visual patterns. In a real sense, it learns and is able to discern patterns and relationships within Big Data that we cannot with traditional methods.

Case studies

The following case studies highlight applications of GeoAI to quantify and predict climate extremes from within large datasets. The first presents the development of a spatial tracking algorithm and spatial heat wave severity index across the conterminous United States of America. The second discusses how AI has been used to accurately forecast air quality across Iran.

Case study 1: Spatiotemporal evolution of heat wave severity and coverage across the United States

Keellings, D., & Moradkhani, H. (2020). Spatiotemporal evolution of heat wave severity and coverage across the United States. *Geophysical Research Letters*, *47* (9), e2020GL087097.

In this case study, we track heat waves across the conterminous United States. The study was the first to apply GeoAI techniques to heat waves and the first to conceptualize heat waves as spatial events. As discussed in this chapter, heat waves have an extremely detrimental impact on human health, and they are known to kill more people than any other form of hazardous weather in the United States. It has been observed that average temperatures are increasing across the United States due to anthropogenic climate change. However, an upward trend in extreme heat has not been as apparent. Traditionally, analysis of heat waves has taken the form of examining temperature data recorded at weather stations or found within individual grid cells of raster climate datasets. These analyses have explored heat waves as events and made an effort to characterize them as such in terms of frequency (how often they occur), magnitude (how hot they get), and duration (how long they last). Analyzing heat waves in this manner has often failed to detect any trend. For example, the frequency of heat waves exhibits no upward trend across the conterminous United States (Case Study Figure 1.1).

Despite increasing temperatures due to anthropogenic climate change, there appear to be no obvious increases in heat waves. The goal of this study was to conceptualize heat waves as spatial events that not only have the characteristics (frequency, magnitude, and duration) examined previously but also have spatial extent, like patches or blobs of warm air, hovering over the landscape. To get a better and more complete picture of the behavior of heat waves, GeoAI was used to spatially track observed heat waves and develop a new spatial severity metric to quantify heat waves within a gridded climate dataset between the years 1981 and 2018.

A hierarchical clustering algorithm was employed to delimit individual events. The clustering algorithm works by grouping all grid cells that are in a heat wave (all cells that have a maximum temperature above the 95th percentile of all temperatures for that summer month) based on the distance between cells. The distance at which heat wave cells are grouped is determined by a process called

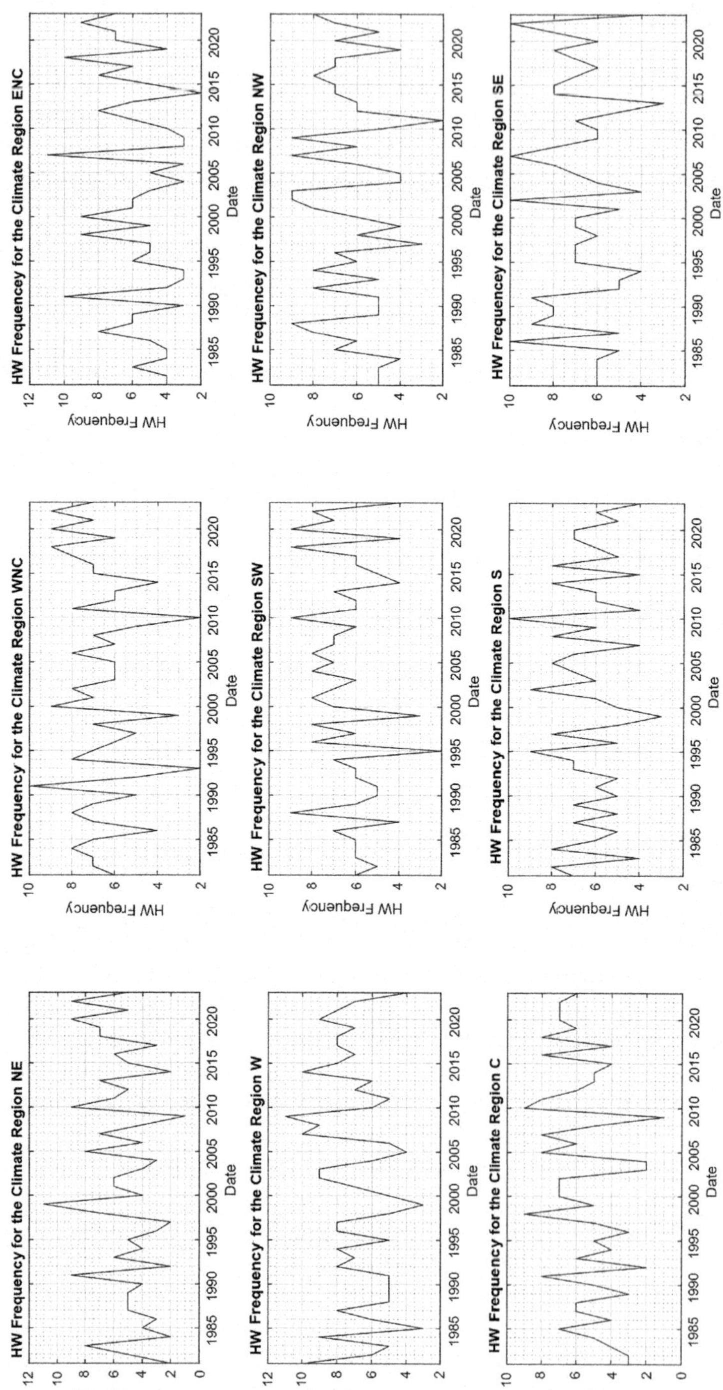

Case Study Figure 1.1 Time series of annual heat wave frequency in each of the nine U.S. climate regions (as defined by the national oceanic and atmospheric administration) from 1981 to 2018. For map locations of each climate region, please see Case Study Figure 1.3.

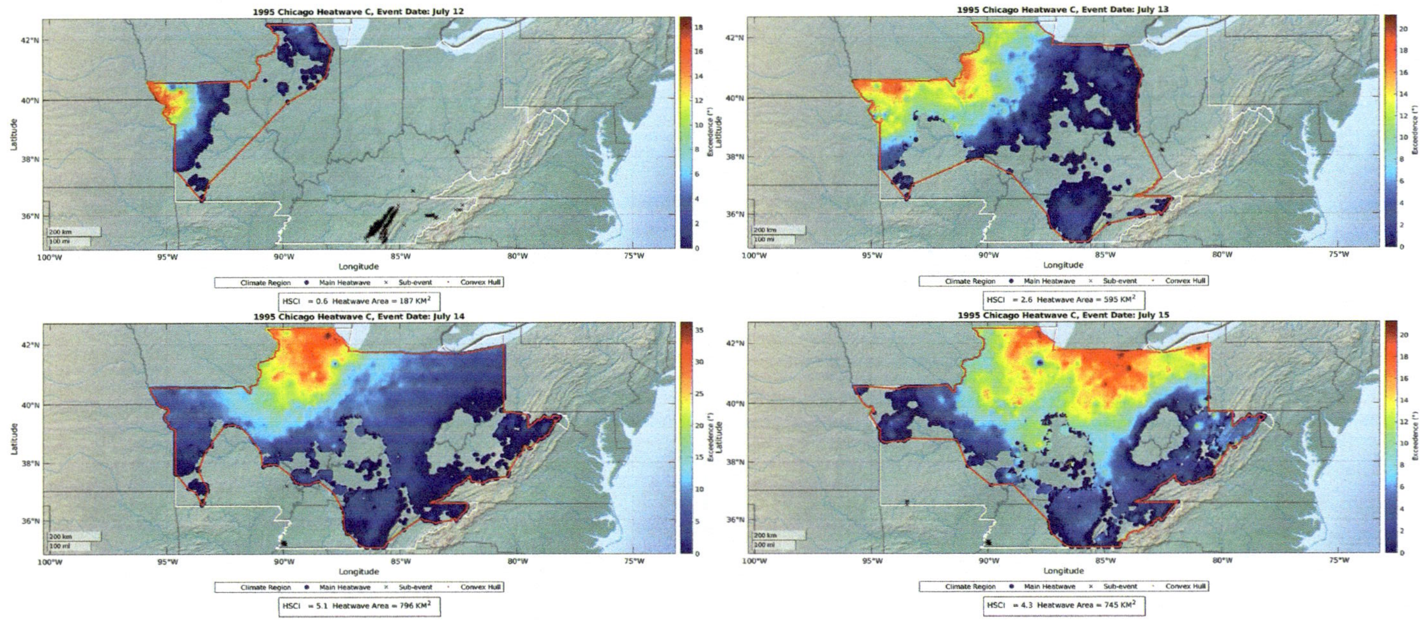

Case Study Figure 1.2 Conceptual example of the spatial heat wave tracking algorithm. Showing tracking of individual heat waves in the southern climate region, as well as the calculated area, average magnitude, and HSCI.

indicator kriging, whereby the distance at which there is no longer a strong statistical relationship between cells is calculated. Once the event is identified by this clustering, the spatial extent can be delineated by drawing a bounding polygon around all heat wave cells. The heat wave can then be tracked day-to-day based on spatial overlap between these bounding polygons (Case Study Figure 1.2).

All heat wave characteristics were combined in the new spatial Heat Severity and Coverage Index (HSCI) so that each event can be quantified, and the index can be summed to give annual scores. The HSCI is calculated by summing the product of event magnitude (the average exceedance of all cells in the event above their heat threshold) and areal extent (the size of the event). A conceptual example of the tracking process and calculation of the HSCI is illustrated in Case Study Figure 1.2.

When heat waves are examined using the new tracking methodology and quantified using the holistic HSCI, it can be seen that there are significant upward trends in most regions of the conterminous United States (Case Study Figure 1.3). Taking the conterminous United States as a whole, it becomes apparent that there has been a substantial increase in the total burden of heat waves during the study period (Case Study Figure 1.4). The study did not examine the causes of the observed increase in heat waves, but in the context of anthropogenic climate change, the findings indicate there has been a significant change in heat waves across the United States and that this trend may continue or intensify moving forward. These results demonstrate the importance of considering heat waves as spatial events, as the upward trend cannot be captured by examining heat wave characteristics individually.

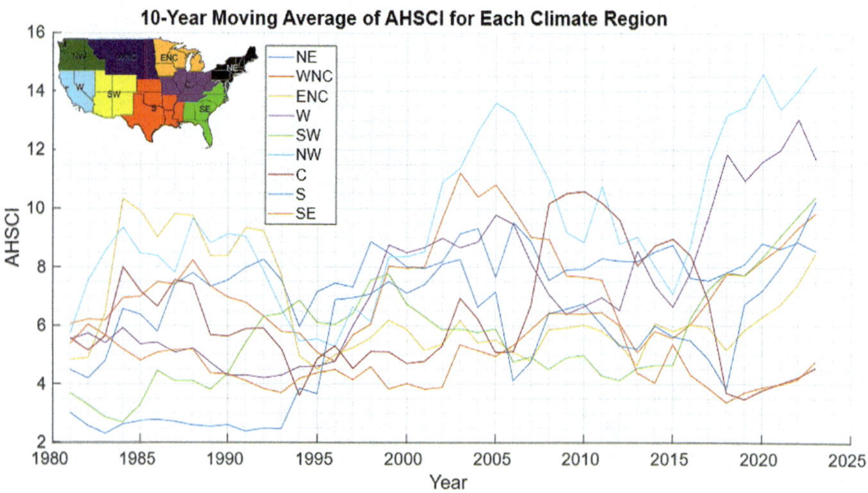

Case Study Figure 1.3 Annual HSCI accumulated for each climate region from 1981–2018. Regions are color-coded as indicated in the inset map.

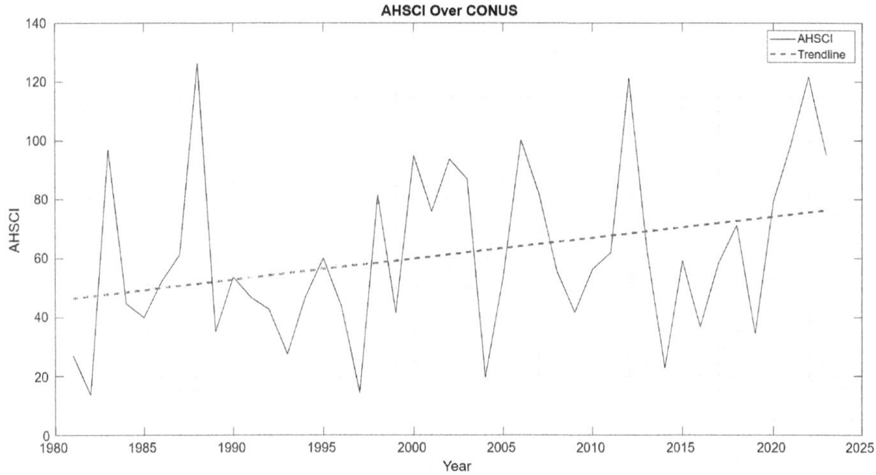

Case Study Figure 1.4 Annual accumulated HSCI for the entire conterminous United States. The red line shows a statistically significant linear trend ($\alpha < 0.05$).

The study also exemplifies the power of GeoAI as a tool to examine an extreme climate event in a manner that would not be possible without these machine learning techniques.

Case study 2: An ensemble deep learning approach to spatiotemporal tropospheric ozone forecasting: A case study of Tehran, Iran

Rezaali, M., Jahangir, M. S., Fouladi-Fard, R., & Keellings, D. (2024). An ensemble deep learning approach to spatiotemporal tropospheric ozone forecasting: A case study of Tehran, Iran. *Urban Climate*, *55*, 101,950.

AI has the potential to predict environmental and climate variability. In this study, we used deep learning to predict hourly ground-level ozone pollution in Tehran, Iran. Ground-level Ozone (GOC) is a major air pollutant in urban areas with negative effects on human health and the environment. GOC typically forms on hot days when volatile organic compounds and nitrogen oxides are released as pollutants from sources such as vehicles and power plants, and then react in the presence of sunlight. This reaction creates ozone (O_3), a secondary air pollutant that irritates the respiratory system and is also known as smog (Haines & Ebi, 2019). The traditional approaches to forecasting GOC often struggle due to the complex atmospheric chemistry, as well as the interaction of different meteorological factors such as solar radiation, temperature, and wind patterns that influence ozone.

Our approach uses advanced deep learning models, specifically Long Short-Term Memory (LSTM), Bidirectional LSTM (BiLSTM), and Gated Recurrent Units (GRU) to model spatiotemporal variations of ozone levels in Tehran. These are types of artificial neural networks that are specialized in remembering information in sequence. While other AI-based models might understand the meaning of a sentence, an LSTM also considers the sentences that came before, so it gets the context of the whole story. These models were chosen as they have the ability to capture the long-term dependencies within climate variables, i.e., if it's hot today, it was likely hot yesterday. This is important for accurately forecasting ozone levels, which are affected by daily cycles as well as seasonal variations. We further improved the performance of the models by incorporating wavelet transformations. This allowed us to detect both short-term fluctuations and long-term trends of ozone pollution.

Deep learning models were developed using data from the Advanced Research Weather Research Forecast model (WRF-ARW), which included inputs like surface temperature, solar radiation, and humidity. These inputs were carefully chosen as they are known to have a major impact on ozone formation. We used a K-Nearest Neighbor algorithm based on environmental and geographic characteristics such as population density and proximity to green space. This enabled a more accurate mapping of ozone levels across the city.

This study has made a major contribution by using a deep learning wavelet-based approach. In particular, the Wavelet Data-Driven Forecast Framework enabled models to predict GOCs without having to rely on data from previous ozone levels. This is an improvement on traditional models, which often rely on pollutant data that may be incomplete or not available in certain regions. Our approach can also be used in regions with limited data availability.

The study results demonstrate that wavelet-enhanced model predictions are highly accurate as assessed by a suite of performance metrics, including the Willmott Index of Agreement (WIA >0.95). Case Study Figure 2.1 shows the hourly predictions made by the wavelet-enhanced BiLSTM (WBiLSTM) for different months. This illustrates how well the model can capture seasonal variations of ozone levels. The model accurately predicted high episodes of ozone during summer months when solar radiation is at its peak. In the winter, the model was able to accurately predict lower concentrations.

Case Study Figure 2.2 shows the spatial distribution of predicted GOC over Tehran. It provides detailed insight into areas where ozone levels are higher, particularly in the central and southern regions, which have higher traffic and population density. The spatial distribution shows how the model can be used for pinpointing pollution hotspots and help develop more targeted interventions. These maps demonstrate the WBiLSTM capability in real-world urban air quality management applications.

The integration of AI and deep learning models within GeoAI is a powerful tool to advance environmental forecasting. This study showed the effectiveness of AI in predicting ground-level ozone levels across Tehran. It provided precise spatiotemporal insight that traditional models fail to capture. WBiLSTM and other wavelet-enhanced models of deep learning demonstrate how AI can be used to tackle complex atmospheric patterns and improve accuracy and reliability in air quality forecasts.

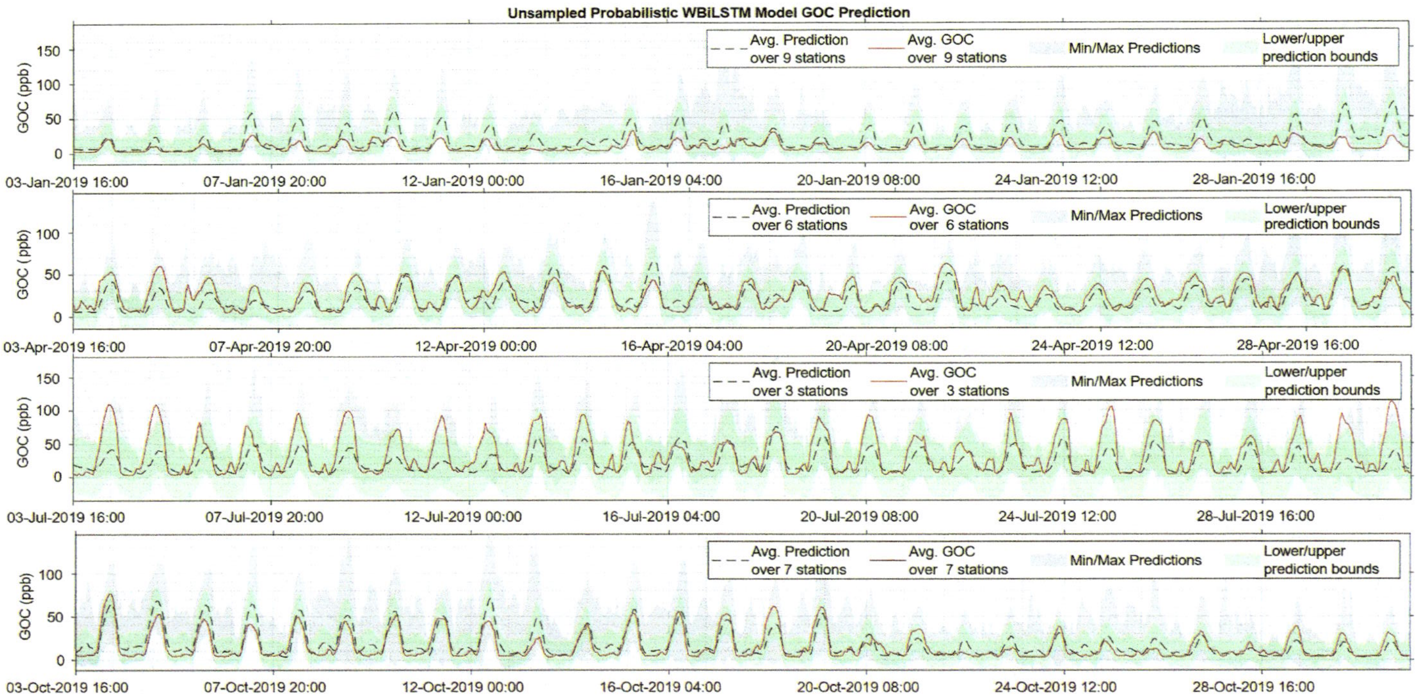

Case Study Figure 2.1 The hourly predictions of the WBiLSTM model for different months showing close relationship between the AI-based model predictions and the observed GOC.

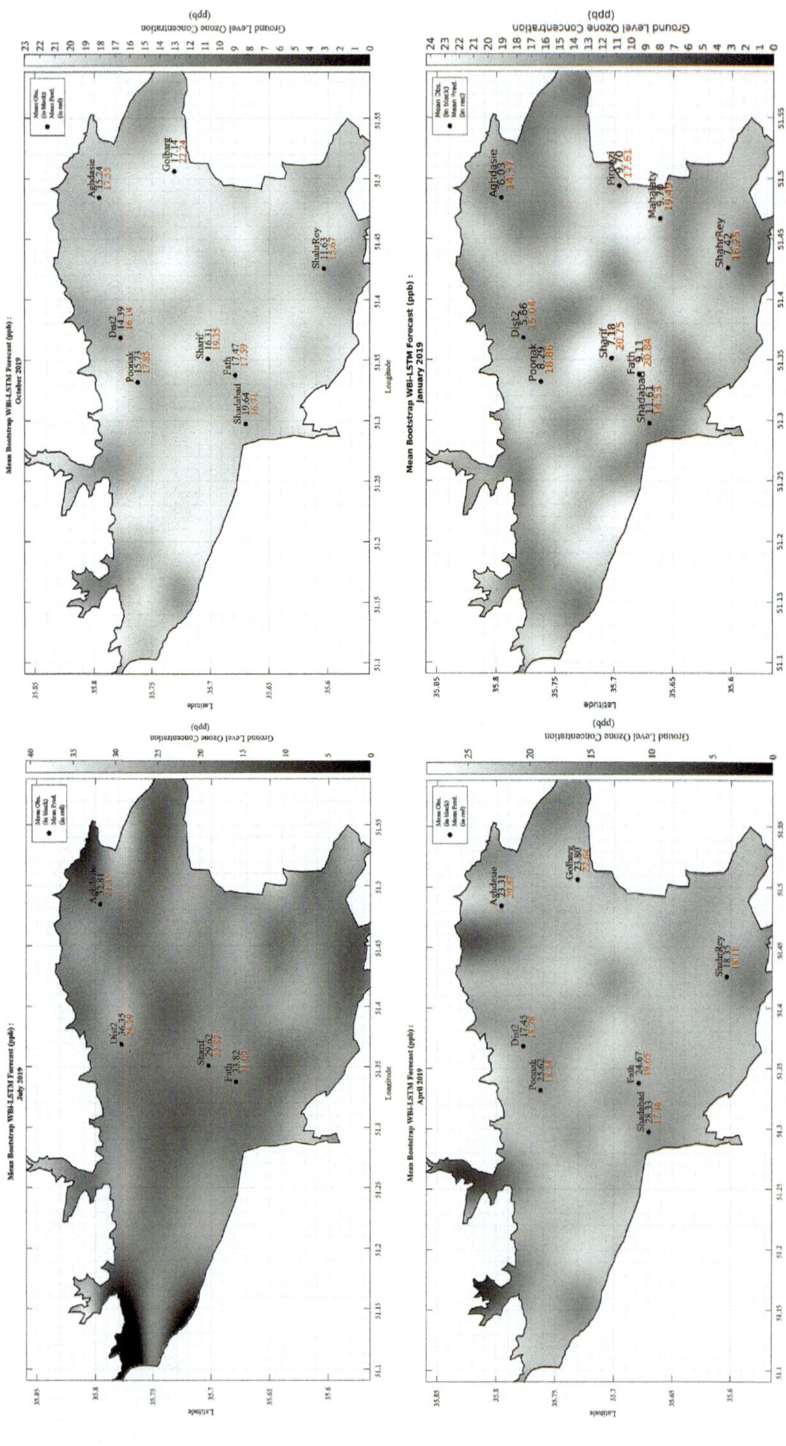

Case Study Figure 2.2 The forecasted GOC concentration map over Tehran.

References

Ahmadalipour, A., & Moradkhani, H. (2018). Escalating heat-stress mortality risk due to global warming in the Middle East and North Africa (MENA). *Environment International, 117*, 215–225. https://doi.org/10.1016/j.envint.2018.05.014

Ahmadalipour, A., Moradkhani, H., & Kumar, M. (2019). Mortality risk from heat stress expected to hit poorest nations the hardest. *Climatic Change, 152*(3–4), 569–579. https://doi.org/10.1007/s10584-018-2348-2

Alley, R. B., Emanuel, K. A., & Zhang, F. (2019). Advances in weather prediction. *Science, 363* (6425), 342–344.

Anderson, B. G., & Bell, M. L. (2009). Weather-related mortality. *Epidemiology, 20*(2), 205–213. https://doi.org/10.1097/EDE.0b013e318190ee08

Atwoli, L., Baqui, A. H., Benfield, T., Bosurgi, R., Godlee, F., Hancocks, S., … Vazquez, D. (2021). Call for emergency action to limit global temperature increases, restore biodiversity, and protect health. *The Lancet, 398*(10304), 939–941.

Bauer, P., Thorpe, A., & Brunet, G. (2015). The quiet revolution of numerical weather prediction. *Nature, 525*(7567), 47–55.

Bunting, E. L., Tolmanov, V., & Keellings, D. (2024). What is a heat wave: A survey and literature synthesis of heat wave definitions across the United States. *PLOS Climate, 3*(9).

Calvin, K., Dasgupta, D., Krinner, G., Mukherji, A., Thorne, P. W., Trisos, C., et al. (2023). IPCC, 2023: Climate change 2023: Synthesis Report. In H. Lee, & J. Romero (Eds.), *Contribution of working groups I, II and III to the sixth assessment report of the intergovernmental Panel on climate change [core writing team*. Geneva, Switzerland: IPCC. https://doi.org/10.59327/IPCC/AR6-9789291691647

Chattopadhyay, A., Nabizadeh, E., & Hassanzadeh, P. (2020). Analog forecasting of extreme-causing weather patterns using deep learning. *Journal of Advances in Modeling Earth Systems, 12*(2). https://doi.org/10.1029/2019MS001958

Ciais, P., Reichstein, M., Viovy, N., Granier, A., Ogée, J., Allard, V., et al. (2005). Europe-wide reduction in primary productivity caused by the heat and drought in 2003. *Nature, 437* (7058), 529–533. https://doi.org/10.1038/nature03972

Cowan, T., Hegerl, G. C., Colfescu, I., Bollasina, M., Purich, A., & Boschat, G. (2017). Factors contributing to record-breaking heat waves over the Great Plains during the 1930s dust bowl. *Journal of Climate, 30*(7). https://doi.org/10.1175/JCLI-D-16-0436.1

Curriero, F. C., Heiner, K. S., Samet, J. M., Zeger, S. L., Strug, L., & Patz, J. A. (2002). Temperature and mortality in 11 cities of the eastern United States. *American Journal of Epidemiology, 155*(1), 80–87. https://doi.org/10.1093/aje/155.1.80

Davis, L., & Gertler, P. (2015). Contribution of air conditioning adoption to future energy use under global warming. *Proceedings of the National Academy of Sciences of the United States of America, 112*(19), 5962–5967. https://doi.org/10.1073/pnas.1423558112

Ebi, K. L. (2008). Using health models to prepare for and cope with climate change: Guest editorial. *Climatic Change, 88*, 1–3. https://doi.org/10.1007/s10584-008-9402-4

Fan, J., Fang, L., Wu, J., Guo, Y., & Dai, Q. (2020). From brain science to artificial intelligence. *Engineering, 6*(3), 248–252.

Garcia-Herrera, R., Díaz, J., Trigo, R. M., Luterbacher, J., & Fischer, E. M. (2010). A review of the european summer heat wave of 2003. *Critical Reviews in Environmental Science and Technology*. https://doi.org/10.1080/10643380802238137

Gasparrini, A., Guo, Y., Hashizume, M., Lavigne, E., Zanobetti, A., Schwartz, J., et al. (2015). Mortality risk attributable to high and low ambient temperature: A multicountry observational study. *The Lancet, 386*(9991), 369–375. https://doi.org/10.1016/S0140-6736(14)62114-0

Haines, A., & Ebi, K. (2019). The imperative for climate action to protect health. *New England Journal of Medicine, 380*(3), 263–273.

Hajat, S., Armstrong, B., Baccini, M., Biggeri, A., Bisanti, L., Russo, A., Paldy, A., Menne, B., & Kosatsky, T. (2006). Impact of high temperatures on mortality: Is there an added heat wave effect? *Epidemiology, 17*(6), 632–638. https://doi.org/10.1097/01.ede.0000239688.70829.63

Hoerling, M., Kumar, A., Dole, R., Nielsen-Gammon, J. W., Eischeid, J., Perlwitz, J., et al. (2013). Anatomy of an extreme event. *Journal of Climate, 26*(9). https://doi.org/10.1175/JCLI-D-12-00270.1

Jones, B., O'Neill, B. C., Mcdaniel, L., Mcginnis, S., Mearns, L. O., & Tebaldi, C. (2015). Future population exposure to US heat extremes. *Nature Climate Change, 5*(7), 652–655. https://doi.org/10.1038/nclimate2631

Keellings, D., Bunting, E., & Engström, J. (2018). Spatiotemporal changes in the size and shape of heat waves over North America. *Climatic Change, 147*(1), 165–178.

Keellings, D., & Moradkhani, H. (2020). Spatiotemporal evolution of heat wave severity and coverage across the United States. *Geophysical Research Letters, 47*(9). https://doi.org/10.1029/2020GL087097

Keellings, D., & Waylen, P. (2014). Investigating teleconnection drivers of bivariate heat waves in Florida using extreme value analysis. *Climate Dynamics, 44*(11–12), 3383–3391. https://doi.org/10.1007/s00382-014-2345-8

Lam, R., Sanchez-Gonzalez, A., Willson, M., Wirnsberger, P., Fortunato, M., Alet, F., Ravuri, S., Ewalds, T., Eaton-Rosen, Z., Hu, W., & Merose, A. (2022). *GraphCast: Learning skillful medium-range global weather forecasting*. arXiv preprint arXiv: 2212.12794.

Lee, H., & Song, J. (2019). Introduction to convolutional neural network using Keras; an understanding from a statistician. *Communications for Statistical Applications and Methods, 26*(6). https://doi.org/10.29220/CSAM.2019.26.6.591

Liu, Y., Racah, E., Prabhat, Correa, J., Khosrowshahi, A., Lavers, D., et al. (2016). *Application of deep convolutional neural networks for detecting extreme weather in climate datasets*.

Lorenz, E. N. (1969). Atmospheric predictability as revealed by naturally occurring analogues. *Journal of the Atmospheric Sciences, 26*(4), 636–646.

McEvoy, D., Ahmed, I., & Mullett, J. (2012). The impact of the 2009 heat wave on Melbourne's critical infrastructure. *Local Environment, 17*(8), 783–796. https://doi.org/10.1080/13549839.2012.678320

Meehl, G. A., & Tebaldi, C. (2004). More intense, more frequent, and longer lasting heat waves in the 21st century. *Science, 305*(5686), 994–997. https://doi.org/10.1126/science.1098704

Miller, N. L., Hayhoe, K., Jin, J., & Auffhammer, M. (2008). Climate, extreme heat, and electricity demand in California. *Journal of Applied Meteorology and Climatology, 47*(6), 1834–1844. https://doi.org/10.1175/2007JAMC1480.1

Milosevic, N. (2020). Introduction to convolutional neural networks. *Introduction to Convolutional Neural Networks*. https://doi.org/10.1007/978-1-4842-5648-0

Mishra, V., & Cherkauer, K. A. (2010). Retrospective droughts in the crop growing season: Implications to corn and soybean yield in the Midwestern United States. *Agricultural and Forest Meteorology, 150*(7–8), 1030–1045. https://doi.org/10.1016/j.agrformet.2010.04.002

Mora, C., Dousset, B., Caldwell, I. R., Powell, F. E., Geronimo, R. C., Bielecki, C. R., et al. (2017). Global risk of deadly heat. *Nature Climate Change, 7*(7), 501–506. https://doi.org/10.1038/nclimate3322

NWS. (2016). *76-year list of severe weather fatalities*. Retrieved from http://www.nws.noaa.gov/om/hazstats/resources/%0Aweather_fatalities.pdf.

Peng, R. D., Bobb, J. F., Tebaldi, C., McDaniel, L., Bell, M. L., & Dominici, F. (2011). Toward a quantitative estimate of future heat wave mortality under global climate change. *Environmental Health Perspectives, 119*(5), 701–706. https://doi.org/10.1289/ehp.1002430

Perkins, S. E. (2016). *A review on the scientific understanding of heatwaves-their measurement, driving mechanisms, and changes at the global scale. Atmospheric Research.* Elsevier Ltd. https://doi.org/10.1016/j.atmosres.2015.05.014

Perkins-Kirkpatrick, S. E., & Lewis, S. C. (2020). Increasing trends in regional heatwaves. *Nature Communications, 11*(1). https://doi.org/10.1038/s41467-020-16970-7

Photiadou, C., Jones, M. R., Keellings, D., & Dewes, C. F. (2014). Modeling European hot spells using extreme value analysis. *Climate Research, 58*(3), 193–207. https://doi.org/10.3354/cr01191

Preisendorfer, R.Ä. N.Ü. (1988). *Principal component analysis in meteorology and oceanography* (Vol 17, p. 425). Elsevier Sci. Publ..

Raymond, C., Matthews, T., & Horton, R. M. (2020). The emergence of heat and humidity too severe for human tolerance. *Science Advances, 6*, 19. https://doi.org/10.1126/sciadv.aaw1838

Russo, S., Sillmann, J., & Sterl, A. (2017). Humid heat waves at different warming levels. *Scientific Reports, 7*(1). https://doi.org/10.1038/s41598-017-07536-7

Shaposhnikov, D., Revich, B., Bellander, T., Bedada, G. B., Bottai, M., Kharkova, T., et al. (2014). Mortality related to air pollution with the Moscow heat wave and wildfire of 2010. *Epidemiology, 25*(3), 359–364. https://doi.org/10.1097/EDE.0000000000000090

Tan, J., Zheng, Y., Song, G., Kalkstein, L. S., Kalkstein, A. J., & Tang, X. (2007). Heat wave impacts on mortality in Shanghai, 1998 and 2003. *International Journal of Biometeorology, 51*(3), 193–200. https://doi.org/10.1007/s00484-006-0058-3

Toms, B. A., Barnes, E. A., & Ebert-Uphoff, I. (2020). Physically interpretable neural networks for the geosciences: Applications to Earth System variability. *Journal of Advances in Modeling Earth Systems, 12*(9). https://doi.org/10.1029/2019MS002002

Van den Dool, H. M. (2007). *Empirical methods in short-term climate prediction*. Oxford University Press.

Vautard, R., Honore, C., Beekmann, M., & Rouil, L. (2005). Simulation of ozone during the August 2003 heat wave and emission control scenarios. *Atmospheric Environment, 39*(16), 2957–2967.

Zamuda, C., Mignone, B., Bilello, D., Hallett, K. C., Lee, C., Macknick, J., … Steinberg, D. (2013). *US energy sector vulnerabilities to climate change and extreme weather*. US Department of Energy. https://energy.gov/sites/prod/files/2013/07/f2/20130716-Energy%20Sector%20Vulnerabilities%20Report.pdf.

Zick, S. E., & Matyas, C. J. (2016). A shape metric methodology for studying the evolving geometries of synoptic-scale precipitation patterns in tropical cyclones. *Annals of the American Association of Geographers, 106*, 1217–1235. https://doi.org/10.1080/24694452.2016.1206460

Further reading

(A list of books, articles, and websites for readers interested in exploring the topic in depth)

https://www.climatechange.ai/

https://www.weforum.org/stories/2024/02/ai-combat-climate-change/

https://deepmind.google/discover/blog/graphcast-ai-model-for-faster-and-moreaccurate-global-weather-forecasting/

https://www.nationalacademies.org/news/2024/08/exploring-the-implications-ofai-for-climate

https://news.un.org/en/story/2023/11/1143187

https://www.npr.org/2024/01/02/1218677963/ai-climate-change-solutions-fireslithium-methane.

Remote Sensing and Big Data Applications in Coastal Environments

9

Phillipe A. Wernette[1] *and Katherine A. Serafin*[2]
[1]Department of Geography, Environment, and Spatial Sciences, Michigan State University, East Lansing, MI, United States; [2]Department of Geography, University of Florida, Gainesville, FL, United States

Key terms and definitions

Aeolian: relating to or arising from the action of wind.
Boundary conditions: describe the behavior or state of variables at the edges of a model domain, allowing for equations to be solved within a defined spatial and temporal framework.
Compound flooding: flooding that occurs when two or more flood drivers happen at the same time or in close succession and interact to produce more severe flooding than any single driver would alone. It usually highlights interactions between different flood drivers, such as combined coastal, riverine, and rainfall-driven flooding.
Deposition: the settling of sediment onto a beach or seabed when wind, waves, or currents lose enough energy to drop the material, resulting in new landforms such as beaches, dunes, or bars.
Erosion: the process by which wind, waves, currents, or tidal action removes material (e.g., sand) from a shoreline or seabed and transports it elsewhere, reshaping the coast over time.
Satellite-derived shoreline (SDS): the process of identifying shoreline (land-water interface) location from satellite imagery or other data through automated or semiautomated models.
Sea level anomaly: the difference between the sea surface height at a given time and location and its long-term average or expected sea level.
Shoreline: the transition between land and water, often described as the land-water interface.
Substrate: the uppermost layer of geologic (bedrock, sediment, soil, etc.) or biogenic (whole or crushed shells, woody debris, etc.) material that shapes landforms and influences processes.
Surrogate models: data-driven models that approximate the response of physics-based models, reducing the need for running full simulations of complex, physics-based models. They are also known as meta-models or emulators.

Remote Sensing, Big Data, and GeoAI. https://doi.org/10.1016/B978-0-443-26741-3.00007-6

Swash: the time-varying vertical excursion of waves around the mean water level. It relates to the water that uprushes onto the beach after a wave has broken.

Wave runup: the maximum elevation reached by waves, relative to the shoreline position in the absence of waves. It is the sum of the wave setup and swash. While the focus in this chapter is on wind-generated wave runup, wave runup can also be generated by tsunamis, seiches, or ship waves.

Wave setup: the elevation of the time-averaged mean water level at the shoreline caused by breaking waves.

1. Introduction

Coastal zones are transition zones extending from the ocean onto land (Fig. 9.1). The environment is shaped by climatological and atmospheric drivers of wind and weather patterns extending above the land and water surface, which, in turn, drag across the water surface and drive waves and currents that interact with the land. The coastline is one of the most dynamic environments on Earth and is home to an ever-growing number of people and other organisms.

At the coast, waves, tides, and sea levels interact with and move sediment, and, at the same time, waves, tides, and sea levels are influenced by the sediment in place. The *nearshore zone* is the location where waves and currents interact with the seabed or lakebed to move sediment. The *shoreline* marks the land-water interface, where waves transport sediment between the *beach* and *dunes* through complex feedback mechanisms, driving patterns in erosion and deposition across both the surface and subsurface. Consequently, the coastal zone includes not only the seabed and terrestrial surface we see today but extends below the modern surface to include a host of past landscapes (i.e., antecedent or relict surfaces) formed by different past conditions (Browder & McNinch, 2006; Gal et al., 2021; McNinch, 2004; Wernette & Houser, 2022; Wernette et al., 2018a, 2018b).

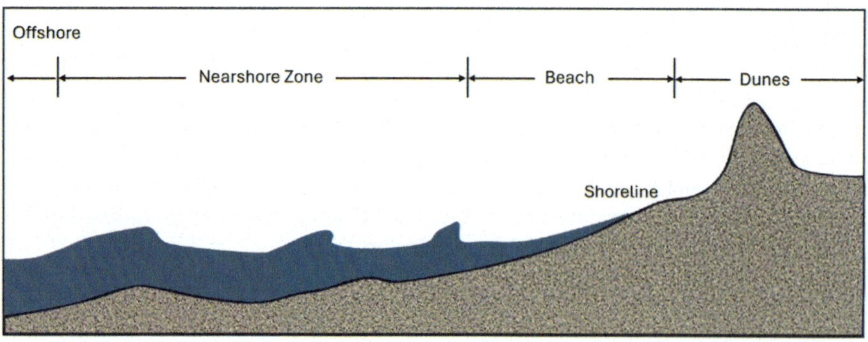

Figure 9.1 Cross-shore profile showing the coastal transition from nearshore waters across the beach and dunes to the land, where erosion and deposition shape the landscape.

The proximity of the ocean to land makes the coast a desirable place to live, work, and play, and coastlines are some of the most developed places on earth. Throughout history, humans living in coastal environments have been able to adapt to periodic flooding or erosion, changing the coastline. However, increasing development and population pressures and changes to the mean sea level and storminess have made the coupled human-natural system more vulnerable. This necessitates an understanding of how coastlines are changing and will continue to evolve.

Our current understanding of coastal systems is largely based on a combination of field and lab studies, which have been extended and expanded through the use of computational modeling. Field-based/in situ data collection is highly accurate, but long-term monitoring programs (e.g., >30 years) are sparse and only on developed coastlines (e.g., Duck, North Carolina, United States (US); Pacific Northwestern US; California, US; Japan, Australia). While models enable us to explore the complex interactions of different environmental variables, they are only as good as the completeness of their input data. We often assume that our input data capture the complete ensemble of processes interacting to form the landscape, as is common in many numerical, statistical, and empirical models used to predict benthic substrate and habitats (Diesing & Stephens, 2015; Reif et al., 2021), shoreline change (Calcraft et al., 2025; Simmons & Splinter, 2025; Vitousek et al., 2024; Wernette et al., 2017), beach and dune responses to storms (Roelvink et al., 2009; Sallenger, 2000; Stockdon et al., 2007), wave runup (Roelvink et al., 2009; Stockdon et al., 2006) and total water level (Ruggiero et al., 2001; Serafin & Ruggiero, 2014). Traditional sampling methods and in situ measurements, however, may not capture spatial or temporal information with appropriate density, and the dynamic and sometimes harsh nature of the coastline and its nearshore waters makes the domain difficult to sample (Holman & Haller, 2013).

Remote sensing (RS) in coastal science allowed for sensors to be placed at a distance from the dynamic coastal environment and survey a larger spatial extent at a lower cost (Holman & Haller, 2013). Initially, RS was used for broad-scale monitoring of coastal environments, but over time, it has expanded to include a range of passive (e.g., optical) and active (e.g., radar) remote sensors for more specialized techniques. Technological improvements in sensors, platforms, and computers have resulted in increasingly large and complex datasets. Satellite data continues to improve the frequency and resolution (spatial and spectral) of data used to advance our understanding of coastal processes and environments (e.g., Vitousek, Buscombe, et al., 2023; Vos et al., 2020). Crowd-sourced data further expands datasets by engaging communities to monitor environments in ways that would otherwise be impossible with stand-alone research experiments (e.g., Harley & Kinsela, 2022; Pucino et al., 2021; Theuerkauf et al., 2022). Underwater environments remain poorly mapped and understood relative to their terrestrial counterparts, although autonomous surface vessels, autonomous underwater vessels, remotely operated vessels, and other technology are enabling researchers and managers to map these challenging environments with greater frequency, accuracy, and efficiency. While traditional data has been limited in size and complexity, analyzing the abundance of new, larger data necessitates exploring increasingly sophisticated Artificial Intelligence (AI) and machine learning (ML) models.

A vast array of ML techniques has been developed and refined in recent decades, owing partly to improvements in computer components (i.e., graphics processing units—GPUs and central processing units—CPUs), which represent an exciting opportunity to distill very large data into valuable information more efficiently and accurately than ever before. Developing future generations of coastal researchers and managers necessitates understanding how we can leverage emerging technologies with past and future data to monitor and address coastal change in a continuously changing world. This chapter will highlight some of the challenges that traditional coastal research has faced, how RS and ML algorithms are revolutionizing coastal science, and likely future directions in an era of big data and AI. Three case studies will be presented to demonstrate the diverse and innovative ways that ML is being used to change the research and management landscape.

2. Theoretical foundations

Important theoretical foundations and considerations persist throughout coastal research focused on geographic transferability, field validation, scalability, and big data analytics. Although not unique to coastal environments, these concepts are especially pronounced in historical and modern coastal research in a variety of subdisciplines and may be addressable through ML and AI.

2.1 Geographic transferability or generalizability

Transferability of data, information, and knowledge from one location to another (i.e., geographic transferability) and through time (i.e., temporal transferability) remains a challenge in coastal science and management. The amount of data and complexity of coastal environments instill inherent limitations when attempting to transfer localized or data-limited knowledge to other locations, even adjacent to the location being studied. While highly localized knowledge is essential for management of resources at a fine spatial scale, it is often expected to be at least partially applicable to other geographic locations. Limiting the spatial scope of any field-based research is important because it enables us to apply boundary conditions to the observations, thereby making the problem more tractable than a completely unbounded region or time. However, these imposed boundary conditions may limit the extent to which insights gained at one site can be generalized to others. Within coastal science, two well-documented systems highlight challenges of geographic transferability: (1) the Field Research Facility (FRF) in Duck, North Carolina (NC), US, and (2) the coastal system at Greenwich Beach, Prince Edward Island (PEI), Canada.

The United States Army Corps of Engineers (USACE) has collaborated with and hosted researchers from all over the globe for decades with the goal of improving our understanding of the suite of coastal processes at the FRF. Located in Duck, NC, the USACE has been collecting and archiving coastal processes data since 1977, and this long-term data archive can provide valuable insight into coastal processes, as demonstrated by the large number of field experiments and corresponding research (e.g., Anderson et al., 2023; Birkemeier et al., 1996; Cohn et al., 2021, 2022a, 2022b; Kim et al., 2023; Lee et al., 1998; McCormack et al., 2025; O'Dea et al., 2021; Straub et al., 2023). Findings from research conducted at the FRF are frequently extrapolated to other sites, under the assumption that processes at FRF approximate those at the new location. However, because FRF has a specific coastal and nearshore slope, sediment characteristics (size and composition distribution), wind and wave regime, tidal range, vegetation composition, and climate, questions remain about how directly transferable some of this knowledge is to regions with different conditions.

Issues of geographic transferability can occur even over regional scales. Coastal geomorphology and processes have been extensively studied at Greenwich Beach, PEI, Canada since the early 2000s (Bauer et al., 2009, 2012, 2015; Davidson-Arnott et al., 2005; Delgado-Fernandez & Davidson-Arnott, 2009; Mathew et al., 2010; Ollerhead, et al., 2003, 2013), yet the geographic transferability of such findings to even nearby sites along PEI remains unclear. Geographic variability along the northern PEI coast is significant, owing to the many sandstone outcrops that anchor the coast in spots and allow the formation of pocket beaches and alongshore spits. This alongshore geologic variability, coupled with variations in sediment size, coastal orientation, and migrating inlets, produces significant geomorphic variability, as highlighted by George et al. (2021, 2024) and Lunardi et al. (2023). The importance of understanding alongshore geologic variability and geographic transferability extends beyond the modern geomorphology and processes but is one that directly affects beachgoers. For example, Wernette and Houser (2022) highlight the important role that 3D variations in geologic setting play in controlling rip current location through time at a site along the northern PEI coast by demonstrating their geospatial relationship to coastal water bodies behind the modern beach and dune system. This finding further emphasizes the importance of understanding issues of geographic transferability.

Another example of the assumed geographic transferability of research findings relates to wave runup, the maximum elevation reached by waves due to wave breaking relative to the shoreline in the absence of waves. Wave runup consists of both the wave setup, the time-averaged mean water level at the shoreline caused by breaking waves (Longuet-Higgins & Stewart, 1964), and swash, the time-varying, vertical excursion of waves around the mean water level (Stockdon et al., 2006). Wave runup is important for characterizing the vulnerability of coastal dunes and infrastructure to wave action; however, observations of wave runup have historically been limited to a few sites geographically and/or field experiments with short-term deployment (e.g., Holman & Guza, 1984). Thus, wave runup is most commonly estimated using empirical formulas related to easily obtainable parameters based on laboratory and limited field experiments (Power et al., 2019; da Silva et al., 2020). For example, the Stockdon et al., (2006) model is one of the most commonly used empirical formulations for estimating wave runup, and is based on 10 field experiments, in which 91% of the data comes from the USACE FRF (Stockdon et al., 2006), representing an intermediate to reflective beach as defined by Short and Wright (1983). Other empirical models have been developed to better reflect local field hydrodynamic and morphologic characteristics (e.g., Poate et al., 2016 for gravel beaches; Conlin et al., (2025) for composite beaches). Yet individual models, including the Stockdon et al. (2006) model, are still applied to locations where no field data originates, even though a single empirical model may not be the most accurate option over time as morphological conditions vary (Itzkin et al., 2025). It is thus necessary to understand that variation in external forcing factors (e.g., tidal range, wind regime, wave regime, sea level change) combined with different internal conditions (e.g., grain size, vegetation community, salinity, shoreline orientation) could produce different coastal trajectories.

2.1.1 Corroboration of numerical models and lab-based experiments with field observations

Another key consideration with transferability is the ability to or limitations encountered when corroborating lab-based experiments or numerical models with field-based observations. Lab-based experiments and numerical models can provide very valuable insight into complex interactions between one or more variables. By controlling boundary conditions and other internal variables, these experiments and models reduce the environmental complexity sufficiently so that researchers can use the experimental data to refine many coefficients and mathematical relationships in the governing equations underlying environmental change. However, in order for these equations to be applied in a management setting, it is necessary that the experiment findings be corroborated or tuned by field-based observations with various instrumentation.

Numerical modeling of coastal systems has increased substantially since the early 2000s, building on foundational work from the mid-1980s (Wright & Short, 1984) and leveraging technological improvements in computing speed. While foundational work, such as the beach state model (Wright & Short, 1984) sought to simply characterize coastal systems into a modal state by their morphology and wave regime, more recent numerical models include a much broader range of coastal processes (Ciarletta et al., 2019, 2020; Shawler et al., 2021). Corroborating subannual and decadal modeling efforts may be feasible through historical air photos, maps, and satellite data, although caution should be taken when extrapolating such models to very-long space or time scales.

Numerical modeling of coastal erosion can be done with numerous software and algorithms, ranging from more empirically driven or statistical models to physics-based models. One of the more popular sediment transport modeling frameworks is XBeach (Roelvink et al., 2009). This software models 2D waves, sediment transport, and morphological changes across the nearshore, beach, dune, and backbarrier. However, XBeach does not account for intrawave processes and nonlinearity in the water surface as waves shoal (Elsayed & Oumeraci, 2017). The cumulative effect of this short timeframe variability can result in overestimation of sediment erosion and morphological change, as noted by Elsayed and Oumeraci (2017). Discrepancies between the modeled sediment transport and morphological change and field observations underscore the importance of understanding the limitations of numerical, statistical, and physics-based models. More recent work by Sherman (2020) directly highlights the importance and challenge of corroborating lab-based research with field studies and vice-versa within aeolian sediment transport research.

2.2 Scaling-up or scaling-down

Scaling-up or scaling-down our understanding of the coast is challenging, given the dynamic and complex nature of the environment. Coastal systems reflect the interplay of geomorphologic, limnologic, atmospheric, and anthropogenic forces across an even wider range of spatial and temporal scales. Thus, linking fundamental processes that operate at small spatial and short temporal scales with those that drive large-scale, long-term coastal landform evolution remains a challenge for predicting geomorphic change.

One challenge with scaling research lies in the fundamental spatial scales at which different processes dominate (Fig. 9.2, Vitousek, Buscombe, et al., 2023). Lazarus et al. (2011) illustrate this challenge at two sites along the US Atlantic Coast, where shoreline changes over small spatial scales do not represent the main shoreline change signal, which is instead dominated by larger-scale changes. Wavelets, which are functions that separate data into different frequency components, are used to categorize shoreline change into different space and time groups. Comparing the shoreline changes across small and large spatial scales with coastal processes operating at different time scales reveals that wave run-up feedbacks operate at the finest spatial and temporal scales, followed by surf-zone currents, wave propagation over complicated bathymetry, and wave-driven alongshore-sediment-transport gradients. Intuitively, we can visually see these distinct spatial and temporal scales, when we consider distinguishing wave run-up processes and feedbacks (i.e., swash, backwash, percolation, exfiltration), which affect the coastal morphology on the order of subminute time scales and across very small distances (submeter to a couple of meters), from surf-zone currents, where variations in wind and currents can affect nearshore wave energy and shoaling to redistribute sediment over the entire nearshore (10s of meters) area across longer time scales (minutes to hours). However, this work largely examined space and time scales across a bidirectional gradient of process regimes.

Work examining process-form relationships along the Texas coast by Wernette et al. (2018a, 2018b; Fig. 9.3) and Weymer et al. (2018) further complicates the concept of spatial scaling. Using signal decomposition (continuous wavelet decomposition and cross-wavelet coherence), fluid turbulence analyses (bicoherence), and adapted economic forecasting models (autoregressive fractionally integrated moving average), this work suggested that scaling research up or down necessitates considering nested process regimes and that these process regimes interact nonlinearly. Instead of each process dominating at a particular set of space and time scales, it is more likely that modern processes and forms are the result of an interaction of finer-scale processes superimposed on broader-scale processes and variation. For example, instead of the underlying geology driving coastal evolution at broad space and time scales and controlling beach and dune morphology at those scales, it is more likely that process signals, such as wave runup and surf-zone currents, vary alongshore and interact with these finer- and broader-scale processes. This interaction

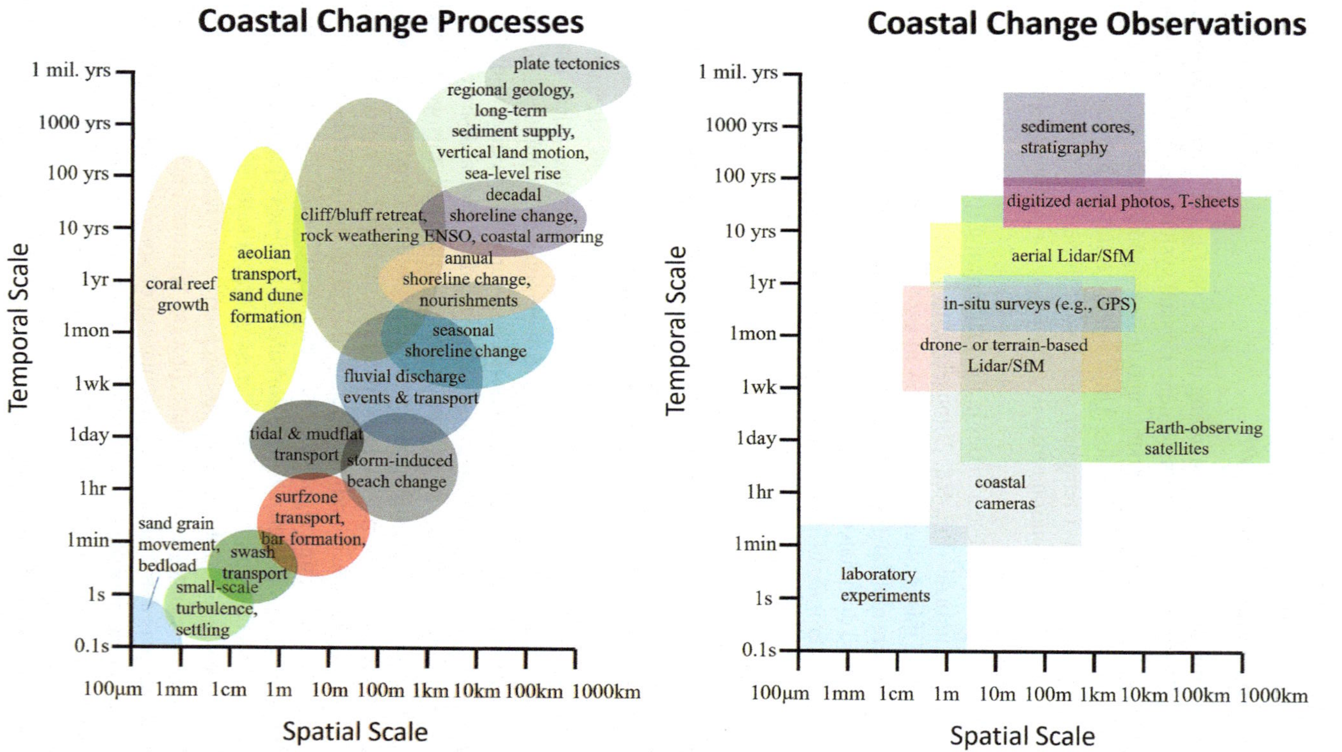

Figure 9.2 Coastal processes dominate over distinct spatial and temporal scales, although there may be nonlinear interactions that enhance or minimize the effects of a set of processes.
From Vitousek, S., Buscombe, D., Vos, K., Barnard, P. L., Ritchie, A. C., & Warrick, J. A. (2023). The future of coastal monitoring through satellite remote sensing. *Cambridge Prisms: Coastal Futures*, *1*, e10, CC BY 4.0 https://creativecommons.org/licenses/by/4.0/

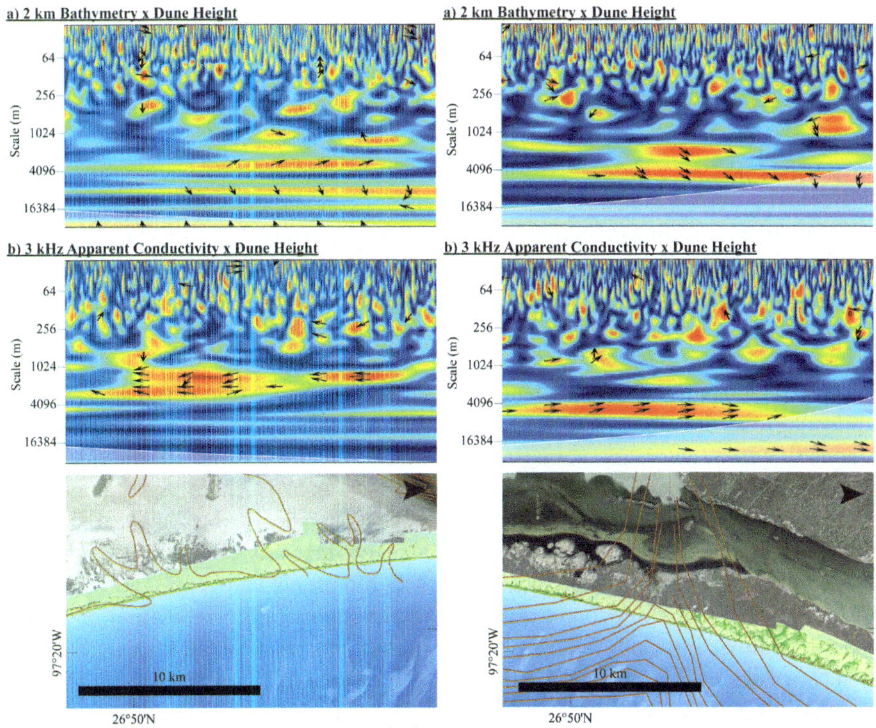

Figure 9.3 Wavelet coherence plots of different coastal process-form relationships along the Texas coast highlight the nonlinearity of offshore bathymetry and dune height, suggesting fine-scale and broad-scale processes actually interact in different constructive and destructive ways to affect coastal morphology. Plots (a) and (b) highlight the interaction of a 2 km offshore bathymetry profile with the modern dune height, and plots (c) and (d) highlight the interaction between the dune height and a proxy for subsurface framework geology. Warmer colors in a-d are more significant, with statistically significant areas outlined in back arrows. The phase relationship is shown by the arrow direction, with left-pointing arrows indicating antiphase relationships and right-pointing arrows indicating in-phase relationships. Modified from Wernette, P. A., Houser, C., Weymer, B. A., Bishop, M. P., Everett, M. E., & Reece, R. (2018a). Influence of a spatially complex framework geology on island geomorphology. Marine Geology 398, 151–162. doi: 10.1016/j.margeo.2018.01.011.

can either be constructive or destructive, depending on the location alongshore and through time. In areas where the processes are both aggradational (i.e., sediment is accumulating and morphology is growing) and constructive, these process regimes can combine to form significantly larger coastal landforms. Areas where process regimes are erosional and constructive are more likely to exhibit accelerated or prolonged erosion. However, if the process regimes are destructive, then the trajectory of these areas in space and time can be much more dynamic, sometimes exhibiting erosion and sometimes exhibiting aggradation.

Regardless of the site and subdiscipline, continued work on understanding these complex spatial scaling relationships is needed to better predict coastal change, especially in environments where human development affects the coastal processes by beach nourishment, seawall and jetty construction, and shoreline armoring. Accounting for nonlinear process interactions introduces added complexity in understanding how to effectively and accurately scale coastal research.

2.3 Big data analytics

With evolving and improving models, sensors, and platforms, data volume continues to increase very rapidly. Challenges associated with big data and analytics are ubiquitous across all subdisciplines in coastal research and are highlighted by examples below focused on benthic mapping and classification and flood forecasting/prediction efforts.

The U.S. Geological Survey (USGS) Great Lakes Science Center (GLSC) has been collecting over 10 TB annually of lakebed (i.e., benthic) images and point clouds since 2019 from autonomous underwater vessels (AUVs) throughout the Laurentian Great Lakes in a coordinated effort with the National Oceanic and Atmospheric Administration (NOAA) and other U.S. federal agencies and partners to ground-truth validated, high-fidelity benthic maps of the Great Lakes. While managing the data volume is a challenge, analyzing this data represents an even bigger challenge. Traditional methods for ground-truthing benthic maps use a combination of photos, videos, and existing maps and rely extensively on manual annotation and interpretation (Mabrouk et al., 2022). These methods can be valuable for relatively small areas, but they are not scalable when dealing with big data being collected in recent years.

While lakebed imagery provides detailed, site-specific insight into submerged landscapes, satellite observations offer a complementary perspective that enables analysis across broader temporal and spatial scales. Satellite observations provide critical global data on sea surface height, temperature, and other oceanographic parameters essential for understanding regional sea level dynamics, and global sea level model (e.g., hindcasts and/or reanalysis) data help to fill in temporal details for understanding extreme storm events. The large amount of data produced, however, means ML techniques are necessary for efficiently processing, analyzing, and identifying patterns and trends that are not readily apparent through traditional analytical methods. This is particularly important in the context of climate change, where understanding the interplay between sea level rise and other environmental variables is crucial for modeling complex sea level patterns necessary for effective coastal management and adaptation. For example, sea level anomalies represent a deviation from the average sea level and can occur due to storm events or large-scale climate patterns such as the El Niño Southern Oscillation (ENSO), which operates over interannual time scales. The many years of satellite altimetry observations of sea surface height can be used to produce an average sea level surface, and the difference between the average sea level and the sea surface measurement for a particular date is called the sea surface height anomaly.

Over seasonal to interannual timescales, accurately forecasting sea level anomalies can be difficult because they are influenced by different ocean and atmosphere processes such as wind patterns, atmospheric pressure changes, and ocean currents

(Ezer, 2015; Sheridan et al., 2017; Woodworth et al., 2019). However, strengths in ML techniques, such as handling complex dependency structures and their ability to model nonlinear relationships, highlight their broad applicability for being utilized for predicting sea level anomalies. The prediction of sea level anomalies on subseasonal to seasonal to multidecadal scales has been a priority in a joint effort between the National Aeronautics and Space Administration (NASA) and NOAA in order to produce annual integrated flood outlooks necessary for merging the risks associated with weather to climate timescales (e.g., NOAA, 2022).

Similarly, the multiple processes that drive extreme flooding events from inland areas to the coastline (e.g., tides, storm surges, streamflow, waves, rainfall) have complex interactions that are not fully understood and are hard to simulate (Green et al., 2025; Jafarzadegan et al., 2023). Models need to operate over various space scales, from local watersheds to regional coastlines, and the drivers of flooding operate over different time scales. Achieving high-resolution results while maintaining computational efficiency is a challenge. Modeling compound flooding—from the statistics of cooccurrence of individual processes to physics-based interactions remains a current academic and management priority.

3. Methodologies

ML and AI have the potential to transform coastal science research by enhancing the transferability of models and findings across different regions, handling issues of scale, and extracting meaningful information from big data. While a number of approaches exist, some of the most commonly used methodologies fall under two categories: supervised and unsupervised ML approaches. Supervised learning approaches learn from data to make predictions based on established relationships. For example, a model is trained on a dataset, where both inputs and outputs are known, so the model can map the inputs to the outputs. Examples of these techniques include radial basis functions (RBF), support vector machines (SVMs), support vector regression (SVR), Gaussian process regression (GPR), artificial neural networks (ANNs), decision trees (DTs), random forest (RF) regression, multilayer perceptrons (MLP), and deep learning networks. Unsupervised learning approaches uncover hidden patterns and structures without relying on target variables or predefined outputs. Unsupervised learning techniques include clustering (e.g., self-organizing maps (SOM), K-means, etc.) and dimensionality reduction (such as Principal Component Analysis). Below, we describe examples of specific ways the above techniques are used in some of the coastal science challenges outlined in Section 2.

3.1 Enhancing transferability

ML can enhance transferability of models developed from data in one location to another through feature extraction/selection and ensemble methods. For example, the transferability of wave runup formulations, which are used in place of measurements of wave runup, is an active research area due to its importance in predicting coastal change. A number of empirical formulations for wave runup exist and have been compiled into easily accessible libraries for general use (e.g., Leaman et al., 2020). To appropriately utilize a specific model, however, it is often necessary to have prior knowledge about the study site, because the empirical formulations are based on field conditions from a specific geography (Itzkin et al., 2025). At the same time, hydrodynamic and morphologic conditions may be changing rapidly over the course of a storm event, such that at the same location, one wave runup formulation may represent wave runup better from one time to another.

DTs, a supervised learning technique where a series of true or false decisions classify data into categories or numerical values, are particularly useful for this type of challenge. Labeled variable inputs of a number of predictors utilized in wave runup formulations (such as wave height, wave period, and beach slope, or other variables related to physical processes such as wave steepness) are related to the expected output, which is the empirical runup model that best predicts observed runup. The DT "branches" operations based on minimizing selection criteria during training, such as the Gini index or Entropy. Probabilities of the most accurate prediction corresponding to each wave runup formulation can be output from the DT model and then used to create a weighted average of wave runup from all empirical runup formulations (Itzkin et al., 2025). Itzkin et al. (2025) show that while a DT model only had an accuracy score of 0.31 for predicting wave runup in an example in coastal North Carolina, US, it was still an improvement from randomly or subjectively choosing an individual wave runup model to predict wave runup.

When applied to wave runup, ML methods such as DTs allow for the development of weighted ensembles that perform comparably or better than an individual formulation, allowing for more confidence in results, regardless of knowing the hydrodynamic and morphologic characteristics of a location, and thus transferability to other locations. This is evidenced in Itzkin et al. (2025), which showed the DT model trained on measurements from one location could be efficiently deployed to new locations with similar forcing and morphodynamic conditions without needing to be retrained. Models could easily be retrained, though, if more data becomes available. As with most transferability issues, the results are only as good as the input data, and it should be noted that often wave runup observations do not capture extreme events with very large wave conditions.

Remotely sensed Earth observations from satellites are also improving the transferability of coastal science by providing data at high resolution across large areas where data have not previously existed from traditional field collection techniques (e.g., satellite-derived shorelines, see Case study 2: Extracting shorelines globally from satellite data and Section 3.2 Handling scale issues). When coupled with ML models, such as neural networks for image classification, information from observations can be extracted efficiently and used to evaluate historical changes and drive future coastal change models.

3.2 Handling scale issues

Similar to the tools being used to address issues of geographic transferability and generalization, satellite data and ML models represent a significant potential to aid in addressing scaling issues. ML models can be used to segment a wide variety of data, including true-color imagery, hyperspectral imagery, and radar data.

Satellite-derived shorelines (SDSs) represent a significant advancement in our ability to remotely discern land-water interface positions across large areas with good accuracy. Although the concept of identifying the shoreline position from imagery or other remotely sensed data is not new to coastal research and has been done for decades (Moore, 2000; Wernette et al., 2017), leveraging unsupervised approaches with past and future satellite and aerial imagery to accurately characterize its position across a range of data is relatively new. In this context, SDS is referring to the automated extraction of a shoreline using machine learning or other models. Satellite imagery is ideal for geographical scaling and transferability because there is a wealth of data beginning in the 1970s and varying in its spatial and spectral resolutions (Fig. 9.4), enabling researchers, communities, and managers to look at past and present trends in coastal morphology.

Vos, Splinter, et al. (2019) created an open-source program called CoastSat that introduced ML into SDS research (Fig. 9.5; see Case study 2). This approach uses band differencing with pan-sharpened satellite imagery and a multilayer perceptron (MLP) model to classify pixels into four classes: (1) sand, (2) water, (3) whitewater, and (4) other land features. Otsu's thresholding and a marching squares algorithm are then used to identify the shoreline position from the imagery. Integrating the MLP model into shoreline detection is a computationally efficient approach to extracting pixel-level information from satellite imagery without supervision, which

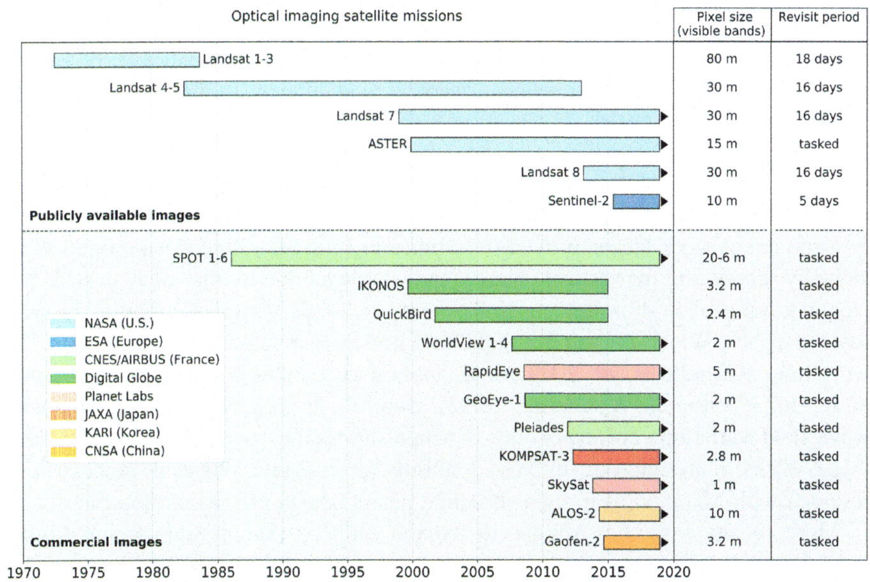

Figure 9.4 Imagery is available publicly and commercially beginning in the 1970s and continuing today, although the spatial and spectral resolutions vary from sensor to sensor (Vos, Harley, et al., 2019).

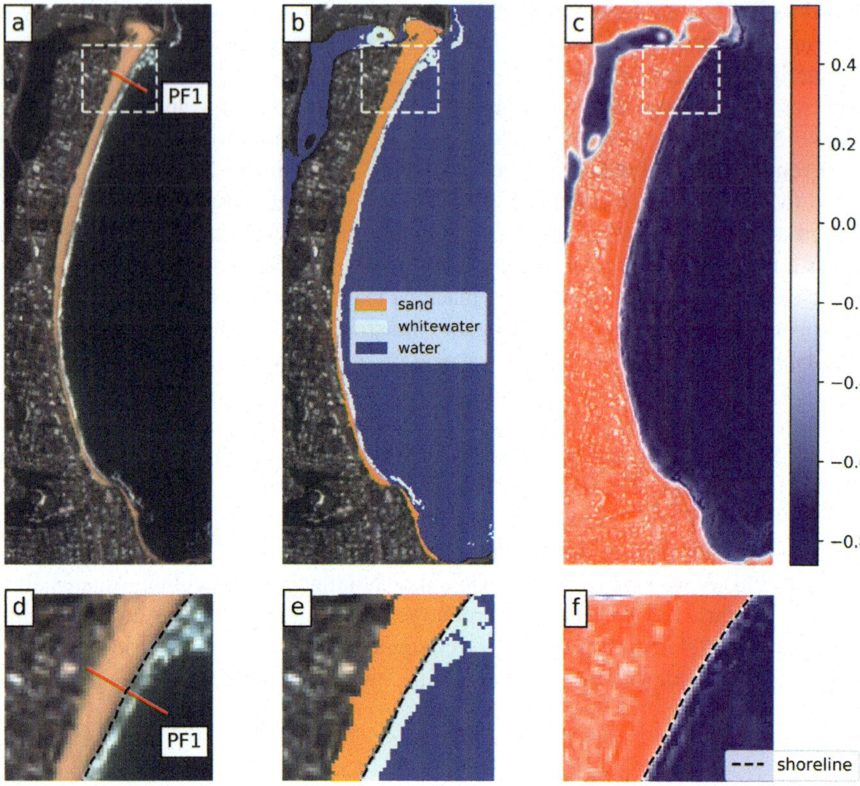

Figure 9.5 Vos, Splinter, et al. (2019) demonstrate the ability to efficiently and accurately extract shoreline position from satellite imagery by band-differencing, pan-sharpening, and application of an MLP model.

significantly improved the objectivity and efficiency of deriving geomorphic features from satellite imagery.

Subsequent work has improved our ability to infer shoreline position and beach slope by accounting for vertical changes in water level due to tidal stage and the horizontal translation in shoreline position (Vos et al., 2020), thereby facilitating broader-scale analyses. Wave setup and runup correction have also been integrated into SDS workflows (Castelle et al., 2021; Konstantinou et al., 2023; Vitousek, Buscombe, et al., 2023; Vitousek, Vos, et al., 2023), enabling us to account for instantaneous wave field variations corresponding to when the images were acquired. Together, these works represent our improved ability to integrate physical process-based research with ML models to map shoreline position with increasing accuracy.

More recent work has sought to expand on CoastSat by integrating higher-resolution (across space and time) data sources and applying semantic segmentation models. This new approach, called CoastSeg (Fitzpatrick et al., 2024), is primarily browser-based and has an improved workflow for deriving the shoreline position, making the program more user-friendly than CoastSat. By leveraging semantic segmentation ML models, CoastSeg is better able to utilize adjacent information in pixel classification, instead of the computationally intensive pixel-based approach used in CoastSat.

The evolution of SDS work to improve model performance, usability, efficiency, and accuracy is significant for coastal research, as we are now able to glean valuable multi-scale insight into spatial and temporal variability in shoreline position with tidal corrections at the local, regional, and even global scales with very minimal supervision. SDSs are being used to explore changes in coastal morphology along the coasts at a range of space and time scales (Dow et al., 2024; Janda et al., 2025; Vos et al., 2019a, 2023; Warrick et al., 2025). More broadly, Vos et al. (2023) leveraged SDSs to develop a shoreline dataset for the entire Pacific Rim over 40 years. The dataset produced by large-scale SDS research (e.g., Vos, 2025) has significant potential to aid our ability to scale-up localized research and scale-down global to regional research.

3.3 Extracting meaningful information from big data

With the abundance of data, fast and accessible techniques are necessary for processing, analyzing, and understanding patterns and relationships in data over time and space that are not easily visible to the eye. For this reason, RS and ML have become integral tools in forecasting, predicting, and validating sea level variability and flood events, over the ocean (e.g., Nieves et al., 2021; Wang et al., 2022), at the coast (e.g., Lee et al., 2021), and within estuaries subject to compounding ocean and inland processes (Jane et al., 2022; Moradian et al., 2024). Neural networks, a supervised learning approach that mimics neurons and the connectivity of the human brain, are one of the most commonly used methods for approaching the challenges surrounding big data analytics due to their ability to learn nonlinear and complex relationships between input and output data with limited human assistance (Geisz et al., 2024; Rivas, 2020). Neural networks are composed of layers, where data is introduced through an input layer, which is connected through transfer functions to intermediate layers and the output layer (Juan & Valdecantos, 2022). There is a range of transfer functions, including linear, sigmoidal, step like, and radial basis functions, to name a few (Duch & Jankowski, 1999). Neural networks require minimal data preprocessing. Some examples are: Deep neural network (DNN), convolutional neural network (CNN), artificial neural network (ANN), generative adversarial neural network (GANN), etc. Common neural network applications include water level prediction (ANN; Huang et al., 2003), sea level anomaly prediction (ANN; Lee et al., 2023), and wave runup/total water level (CNN; Kang et al., 2024; Buckley et al., 2024) and shoreline (Multilayer Perceptron (MLP); Vos, Splinter, et al., 2019) classification.

One example of the utility of neural networks coupled with the self-organizing map (SOM) clustering algorithm comes from recent advancements in understanding, predicting, and forecasting sea level anomalies using ML methods. While many studies have explored relationships between drivers of sea level anomalies such as ocean currents (Ezer et al., 2013; Park & Sweet, 2015), thermal expansion (Widlansky et al., 2020), atmospheric pressure due to the inverse barometer effect

(Piecuch & Ponte, 2015), and climate oscillations such as the El Niño-Southern Oscillation (Merrifield et al., 2012), skillful forecasts using numerical models over long lead times has been a challenge due to the multidimensional nature of sea level over time and space. ML techniques that leverage known climatological relationships and combine ANN, PCA, and SOM show their utility for understanding atmospheric drivers of sea level changes (Sheridan et al., 2017) and along the US West Coast for improving anomaly forecasting lead times up to 9 months (Lee et al., 2023).

SOMs utilize a neural network algorithm to determine the distribution of multidimensional data (Sheridan & Lee, 2011). The algorithm creates a two-dimensional matrix of nodes, or cluster centroids. They allow for intuitive visualization of classification/clustering of patterns. The number of nodes is selected by the user, and maps are initialized with either random values or eigenvectors of the data set (Sheridan & Lee, 2011). As each input vector is added, it is compared to the nodes to find the best match. A learning rate parameter helps gradually adjust the nodes and neighboring node positions to be more like the input data. This process is repeated many times, gradually organizing the nodes to reflect the structure of the input data. The result is that similar data points are grouped together, making it easier to identify patterns and relationships within the data (Gutiérrez et al., 2005). This allows the nodes to self-organize into cohesive patterns on a continuum, with similar nodes close to each other, and dissimilar nodes farther away (Sheridan & Lee, 2011).

Known relationships with atmospheric variables such as monthly average sea level pressure, wind direction, and speed at 10 m height, and monthly teleconnection and oscillation data can be used as predictors of sea level anomalies. For example, PCA can first be used to reduce data dimensionality with each environmental data set, and all retained principal components can be used as inputs to SOM-based classification. Once environmental data is organized into different nodes or clusters, anomalous sea levels can be evaluated during the days and matched to a SOM classification to associate high and low sea level anomalies as geographic responses to a driver (Sheridan et al., 2017). All in all, this approach begins to disentangle the drivers of sea level anomalies across broad geographic scales.

4. Discussion

The unique applications presented as case studies show the utility of ML and GeoAI for tackling complex challenges like efficient data extraction, classification, and modeling across a breadth of coastal subjects over the land-sea interface. These applications only provide a snapshot of the work being done across the coastal community. The presented case studies explore both simple and more complex uses of RS and ML. One of the simpler approaches to using ML is presented in Case study 3 to make computing compound flood levels more efficient for extracting the 100-yr event. It is not entirely surprising that scattered interpolation and RBF performed better than multiple linear regression because there may be nonlinear interactions in flooding zones, and multiple linear regression cannot account for those dynamics. While ML techniques with increasing complexity could be used to develop surrogate models in the case study presented herein, it may not be necessary for model inputs of only two variables (river discharge and ocean water level)—the accuracy was already fairly high for the scattered interpolation (RMSE between 1–2 inches). Another two-input (river discharge and ocean water level) application using a two-dimensional modeling approach tested 7 ML models (SVR, SVM, RBF, linear regression (LR), GPR, DT, and ANN) for producing flood forecasts of water level at different grid locations (Moradian et al., 2024) and found that RBF was the most accurate ML approach for predicting flooding. However, predicting water level or flood depth from more than 2 variables, for example, accounting for lateral inflow from tributaries, rainfall-runoff, groundwater processes, and/or wave setup, would not necessarily be easily handled by simple model choices, warranting more complex ML approaches for surrogate model development.

Similarly, once flood levels are expanded beyond the channel across the floodplain/estuary (2D), more advanced ML techniques might be needed to include spatial information, either into the floodplain or within large estuaries where local wind forces or currents may be important for water height, or time-variable (unsteady) processes. One example is Bass and Bedient (2018) who tested kriging models (also called Gaussian process regression) and ANN for developing surrogate models to rapidly predict joint storm surge and river flooding for a 2D hydrodynamic model coupled with a 1D hydraulic model and Parker et al. (2019), who developed a skillful surrogate model for water level within the Greys Harbor, Washington, US estuary using Gaussian process regression and 16 input parameters: wind speed, direction, sea surface pressure, wave height, wave period, mean wave direction, streamflow, base water level, and 8 tidal components. Continued testing of the best tools for surrogate models, which are also tractable, is an area of continuing research. Many studies only choose one model and justify that it does well, rather than testing a number of models with high predictive power.

While the combination of RS, GeoAI, and ML has washed over the coastal research landscape and changed it completely, there are still a number of continuing challenges and research directions related to an abundance of data. Perhaps the most significant challenge with GeoAI and ML in coastal research revolves around training data availability and development. The volume of data required to train many high-accuracy ML models can be cost-prohibitive because developing such data requires extensive resources for labeling and QA/QC. More powerful data-driven models will require more specialized labeling (Vitousek, Vos, et al., 2023). "Coast Train", a recent advance by Buscombe et al. (2023), is a labeled training dataset of coastal environments for pixel-level discrete classification from geospatial imagery to help coastal scientists train models for automatic identification and mapping. Not only does this prevent repetitive, time-consuming work, but it also provides a vetted classification dataset for transparent and reproducible outcomes. Although this remains the case, improvement in unsupervised and intelligent graphical modeling can help alleviate many of these issues by using existing very large models and data, such as Segment Anything (Kirillov et al., 2023), DINO (Caron et al., 2021), and Florence (Yuan et al., 2021). Using these foundational models can reduce the resources required to develop quality training data of sufficient volume for training new ML models for a variety of tasks.

As the number of sensors and data increases and processing techniques advance, another challenge that remains is the validation and verification of approaches. Continued application of ML algorithms across different, and at times, dissimilar, sites for testing their accuracy will remain important. For example, we do not necessarily know the best combination of thresholds and image indices for identifying the shoreline across varying coastal environments and tidal reference elevations (Vitousek, Buscombe, et al., 2023).

While deep learning transfer learning techniques are improving flood modeling approaches in data sparse regions based on models trained in locations with similar characteristics (e.g., Seleem et al., 2023; Xu et al., 2023), perhaps instead of only assuming transferability of knowledge from one geography to another, we should seek to understand systems similarities in form and trajectories, and the relationships between locations based on these data. In other words, what is the historical and current state of the system in its beach, dune, nearshore, and island (if applicable) morphology, and how does this compare to the historical and current morphology of other systems? Work by Lehner et al. (2026) suggests that instead of applying knowledge of one system to another system with the assumption that nearby locations are similar or that visibly similar sites have a similar trajectory, we can leverage machine learning techniques to understand the "fingerprint" of a system using its combined historical morphology and trajectory.

By developing a unique fingerprint of past and present morphology for a system based on historical observations, we can then relate this system to other systems that exhibit a similar trajectory. In this way, it is possible to relate two or more systems via processes and evolution instead of relying on their modern morphological similarity or dissimilarity with other systems. Even if the morphology of the two systems are different today, the recognition that they share a similar lineage/trajectory is valuable information about how they may evolve in the future.

5. Conclusion

5.1 Summary

ML models have improved significantly over the past decade and have revolutionized coastal research on many fronts, helping address once insurmountable or intractable challenges. ML models can be used to interrogate very large and complex data with greater efficiency and accuracy than ever before. They are being used to address historical challenges, such as geographic transferability or generalizability, scaling-up localized research to accurately capture geospatial variation in these processes at regional or global scales (Dow et al., 2024; Fitzpatrick et al., 2024; Janda et al., 2025; Vos et al., 2019a, 2019b, 2020, 2023; Warrick et al., 2025), efficiently and accurately mapping at very high resolution over broad geographic areas (Geisz et al., 2024; Wernette et al., 2022), and evaluating very large data including complex nonlinear interactions in order to predict future environmental change.

As computing resources continue to get faster, smaller, and more precise, our ability to efficiently and accurately measure and model coastal evolution will steadily improve. UAS, AUV, ROV, and other autonomous or semiautonomous systems have already been developed with onboard CPU and GPU processors, which now enable edge processing in near-real-time. Trained models can be uploaded to these small but powerful systems and used to process incoming data as it is streamed. Continued integration of these and other systems will only further expand our ability to efficiently capture, process, and link different complex environmental data, a task that was once impossible.

5.2 The path ahead

Instead of only considering coastal morphology in its current state and seeking similarities in the modern morphology between different geographic locations, future work could leverage machine learning techniques to quantitatively relate different geographies based on their evolutionary trajectory. These trajectories are the result of similar past and present process interactions occurring across different spaces and may not be visible by the eye or traditional quantitative technqiues. In the context of continued climatic uncertainty and variability, it is more important than ever to understand not only how processes have shaped and continue to reshape modern landscapes, but also how these processes are changing in order to anticipate future change. For example, Lehner et al. (2026), highlight the importance of understanding and linking geographies by trajectory instead of only modern form and is an important first step in this direction. In light of climate uncertainty, a trajectory-based approach is more adaptable to shifts in process regimes by recognizing that systems may transition between different evolutionary pathways/trajectories, informed by existing knowledge rather than starting from scratch.

5.3 Closing thoughts

As computer and sensor hardware continue to improve, we are likely to see an even greater push for the integration of ML models into existing and future workflows. Models that were once inaccessible due to limited computing hardware will become increasingly accessible and portable as the emphasis on developing user-friendly interfaces continues to grow. Yet, ML model application in coastal research continues to suffer from some limitations.

Current ML models struggle to predict water levels and coastal changes over the suite of space and time scales in response to nonlinear interaction of different process regimes across these different scales. Future work should seek to better understand and quantify these nonlinear process interactions across short- and long-time scales and fine- and broad-geographic scales. The specific AI/ML models being used to address the outlined theoretical foundations have evolved with the improvements in computer science and engineering.

For example, CNN models were initially developed in the 1980 and 90s to recognize handwritten digits (LeCun et al., 1989, 1998). It wasn't until the 2000s that these models became more feasible because of significant advancements in GPU hardware and software. Accelerated improvements in GPUs have further enabled us to adapt relatively simple CNN models to identify different objects for a wide range of applications. Today, a wide range of CNN-based models are used in coastal research to segment data and develop geographically transferable models. More recently, vision transformer models have gained traction in computer vision literature and applications because these models are capable of handling multi-scale input data with greater generalizability than other models. The different use cases and models emphasize the constantly changing landscape of AI/ML models and their range of applications. The methods used to interrogate growing coastal datasets will continue to evolve to include new AI/ML models as they are developed, thereby enabling us to better understand the suite of complex nonlinear interactions that drive coastal change across a range of space and time scales.

Case studies

AI and ML are increasingly used to address complex research questions in a range of subenvironments. The following case studies are designed to highlight both simple and more complex applications of ML and GeoAI spanning the land - water interface to 1) generate high-resolution and widespread geologic substrate maps from a combination of AUV, LIDAR, and multibeam echosounder (MBES), 2) quantify shoreline variability in beach morphology along the coast using satellite data, and 3) make physics-based flooding models more efficient through the use of surrogate models. Overall, these case studies display the variety of different questions being asked of ML and GeoAI in coastal environments.

Case study 1: Benthic habitat mapping

Publication: Geisz, J. K., Wernette, P. A., and Esselman, P. C. (2024). Classification of lakebed geologic substrate in autonomously collected benthic imagery using machine learning. *Remote Sensing*, *16*(7), 1264. https://doi.org/10.3390/rs16071264.

Research by Geisz et al. (2024) seeks to address the significant data process challenges associated with the TB of ground-truth data collected by USGS, NOAA, and others throughout the Great Lakes. This work leverages ML models to address issues of processing big data, scalability, and geographic transferability in mapping benthic habitats and ground-truthing these maps. While large areas of the lakebed can be mapped with airborne LIDAR and MBES surveys, ground-truthing these pointwise or interpolated data necessitates observing areas of the lakebed visually. Traditional methods for ground-truthing benthic maps are resource-intensive and dependent largely on remotely operated vessel (ROV), drop camera, and diver surveys in combination with physical sediment samples (Mabrouk et al., 2022).

Improvements in AUV vessels and payloads in recent decades have facilitated acquiring imagery and other data over larger areas with increasing efficiency. Since 2019, the USGS Great Lakes Science Center (GLSC) has collected over 10 TB of imagery and stereo point clouds of the lakebed in each of the five Laurentian Great Lakes on an annual basis, resulting in close to a petabyte (PB) of valuable lakebed data. These data serve several valuable tasks, including mapping and monitoring invasive plants and organisms, mapping valuable freshwater reef habitats, and ground-truthing NOAA's collaborative effort to improve our existing Great Lakes lakebed maps. By coupling AUV with MBES and LIDAR data, we can expand our high-fidelity maps of the Great Lakes faster and more accurately than ever before. However, one challenge with the massive amount of AUV imagery and point cloud data is the interpretation of these raw data into useful and actionable information and knowledge.

Mapping the benthic habitat within the framework of the Coastal and Marine Ecological Mapping Standard (CMECS; United States Federal Geographic Data Committee, 2012) includes classifying four components: water column, biotic, substrate, and geoform. Of these components, substrates, specifically geologic substrates, are easiest to visually identify and distinguish. Rocks generally have clearly defined edges, making them easy to detect and distinguish from the surrounding sediment.

Random forest (RF) and deep convolutional neural network (DCNN) models were trained independently for classifying a single whole AUV image into 9-class, 6-class, and 2-class geologic substrate classes (Case Study Table 1.1) based on the size of

Case Study Table 1.1 CMECS-based 9-class, 6-class, and 2-class schemas were used to train the RF and DCNN models.

<table>
<tr><th>9-Class</th><th>6-Class</th><th>2-Class</th></tr>
<tr><td>Bedrock (Be) (46)</td><td>Consolidated (Con) (46)</td><td rowspan="7">Coarse[a] (C) (5330)</td></tr>
<tr><td>Boulder (Bo) (1488)</td><td rowspan="2">Very coarse (VC) (2621)</td></tr>
<tr><td>Cobble (Co) (1133)</td></tr>
<tr><td>Pebble (Pe) (136)</td><td rowspan="2">Moderately coarse (MoC) (149)</td></tr>
<tr><td>Granule (Gran) (13)</td></tr>
<tr><td>Gravel mix (GM) (500)</td><td rowspan="2">Mixed coarse (MiC) (624)</td></tr>
<tr><td>Gravelly (Gr) (124)</td></tr>
<tr><td>Slightly Gravelly (SGr) (138)</td><td>Mixed (M) (138)</td><td rowspan="2">Fine (F) (1906)</td></tr>
<tr><td>Fine (F) (1768)</td><td>Fine (F) (1768)</td></tr>
<tr><td colspan="3">Coarse algae (CA) (1936)</td></tr>
</table>

[a]The 2-class scheme (coarse-fine) also included coarse algae (CA) images in the coarse class.
Modified from Geisz, J. K., Wernette, P. A., & Esselman, P. C. (2024). Classification of lakebed geologic substrate in autonomously collected benthic imagery using machine learning. *Remote Sensing*, *16*(7), 1264. doi: 10.3390/rs16071264,

rocks and sediments visible in the image. It was important that these classes nest into each other, where one or more classes from the 9-class schema would combine into a class in the 6-class schema, which would further combine into a class in the 2-class schema. These classification schemas were modified from CMECS to be more suitable for mapping geologic substrates from the USGS AUV images (based on image pixel size). CMECS was originally developed based on particle sizes using the Wentworth (1922) scale from physical grab samples. However, without physical samples, the classification schema was adapted to use the proportion of area each particle fraction covers in the image. With the USGS AUV images, it is not possible to resolve particles smaller than granules; therefore, any classes of granule or finer were aggregated into a single "fine" class of substrates.

Results demonstrate that RF and DCNN models had similar model accuracy for a 9-class, 6-class, and 2- class schema (Case Study Figure 1.1). The 2-class DCNN model predicted both coarse and fine class substrates with 96.1% accuracy, and the 2-class RF model predicted coarse and fine substrates with 97.4% and 93.4% accuracy, respectively. Classification accuracy was inversely related to the number of classes in the schema, where models with the 9-class schema were the poorest performing and models with the 2-class schema were significantly more accurate.

The most significant difference between the RF and DCNN models is the preprocessing time required for each. RF models rely on a series of feature vectors as inputs, meaning you must calculate each of these feature vector values for every point in the data prior to training or inferring with the model. Given the large data volume collected by groups such as the USGS and NOAA, the resources required to compute these values can be cost-prohibitive. Instead, the DCNN model simply requires the whole image as a model input and has a significantly faster inference speed. Because of these resource requirements, DCNN models were generally preferable to the RF models for classifying geologic substrate from benthic imagery.

One notable limitation of this approach is its inability to see through dense vegetation that can obscure the rock edges. In these cases, the image classification or object detection models may mistake smaller rocks for larger ones or fail to recognize the rock edges completely. The inverse is also a challenge for the RF and DCNN approaches, where particles granule or smaller are too small to detect and distinguish from each other. It is important to note that the exact minimum mappable particle size is a function of the image ground sample size (i.e., pixel size), which is, itself, a combination of the sensor hardware, image acquisition distance, vessel speed, and environmental characteristics, such as turbidity and water clarity. If the particles are too small relative to the image resolution, then it is only possible to classify the geologic substrate to the Fine Unconsolidated Substrate Subclass with this approach.

Future work should seek to extend our ability to characterize the geologic substrate and other objects by moving from whole-image classification to instance segmentation of rock clasts within the images. *SegmentEveryGrain* is an instance segmentation model that seeks to identify and outline every rock in an image (Sylvester, 2024). An in-development tool called *RockFinder* uses a similar approach for detecting, segmenting, and sizing rocks in benthic imagery and video, although it is different in that training data is developed by leveraging *GroundingDINO* (Liu et al., 2023) to

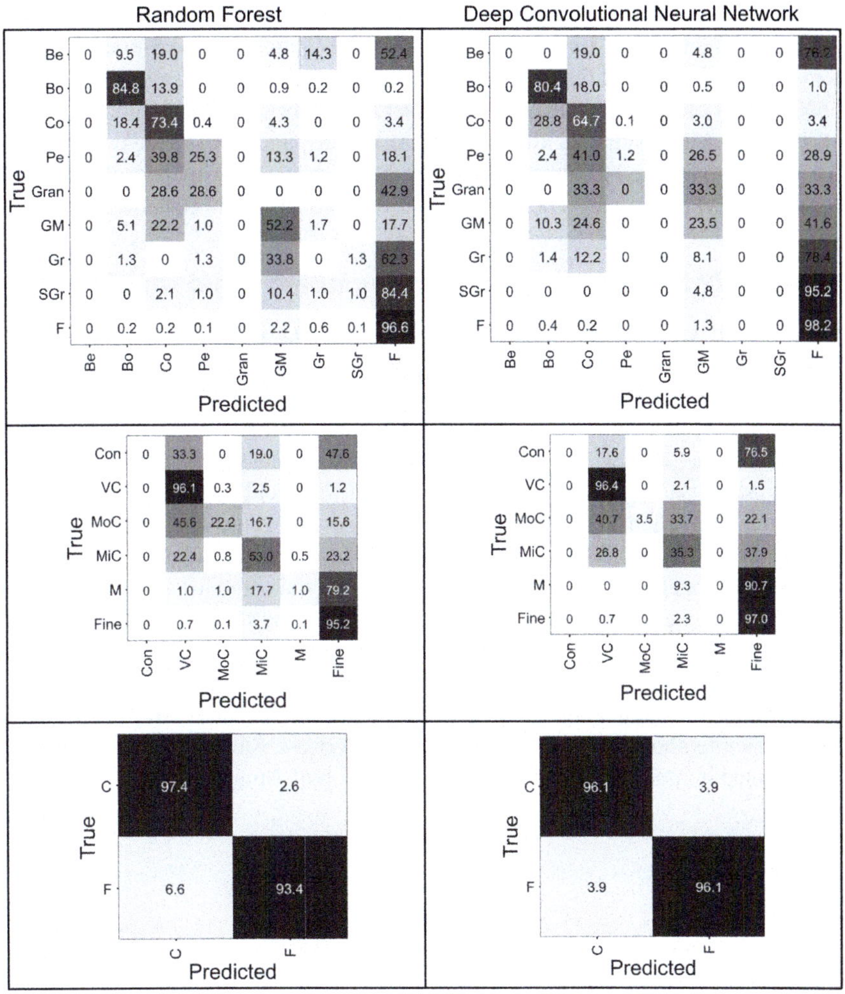

Case Study Figure 1.1 Classification accuracy percentages for the RF and DCNN models for the 9-class, 6- class, and 2-class schemas.
Modified from Geisz, J. K., Wernette, P. A., & Esselman, P. C. (2024). Classification of lakebed geologic substrate in autonomously collected benthic imagery using machine learning. *Remote Sensing*, *16*(7), 1264. doi: 10.3390/rs16071264, CC BY 4.0 https://creativecommons.org/licenses/by/4.0/

identify and generate the rock objects. Regardless of the specific approach, future work should seek to (1) improve our ability to efficiently and accurately detect and segment various objects from imagery and point cloud data, (2) scale these across broader geographic areas, and (3) improve our ability to accurately classify geologic substrate in classification schemas more complex than two classes.

A complementary series of computer vision programs and workflows is being developed by the USGS and NOAA to automate the identification and classification of invasive *Dreissena* mussels (Morris et al., 2024), submerged aquatic vegetation (Moradi et al., 2025), and fish (Moradi, Esselman, Geisz, & Roussi, 2024; Moradi et al., 2024) for mapping and natural resource assessment. First steps at extending these AUV image observations of geologic substrate to predict nearshore substrate composition through all of Lake Michigan with RF models are already underway. However, scaling-up the pointwise, highly localized ground-truth observations to map geologic substrate across a DEM of the nearshore zone for all of Lake Michigan remains a challenge.

Case study 2: Extracting shorelines globally from satellite data

Publication: Vos, K., Harley, M. D., Splinter, K. D., Simmons, J. A., and Turner, I. L. (2019). Sub-annual to multi-decadal shoreline variability from publicly available satellite imagery. *Coastal Engineering*, *150*, 160–174.

Monitoring coastlines is crucial for understanding and managing environmental changes such as coastal erosion, sea-level rise, and habitat loss. Yet very few locations around the world have been regularly monitored with field surveys. When field data is collected, it is often before and after a storm event; thus, long records of coastal change are limited. This case study explores the automated extraction of shoreline positions using satellite data, which combines RS, big data, and ML. The focus is on understanding the accuracy of the horizontal position of the shoreline over different time scales for use in understanding how coastlines are changing.

The open-source *CoastSat* toolkit written in Python, can be used to output mapped shorelines from satellite imagery. The workflow involves accessing, preprocessing, and extracting shorelines from satellite imagery (Case Study Figure 2.1). Each step is explained in more detail below. First, Tier 1 Top of Atmosphere (TOA) reflectance

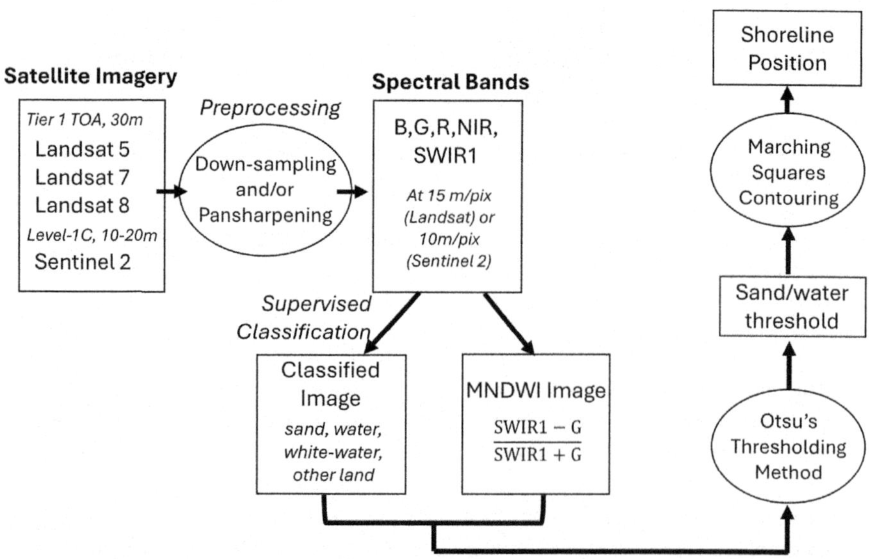

Case Study Figure 2.1 Flowchart (modified from Vos, Harley, et al., 2019) showing the procedure for extracting shorelines from publicly available satellite imagery retrieved from Google Earth Engine. Squares represent inputs/outputs and circles represent models/methods.

images from Landsat 5,7, and 8 and Sentinel—2 Level - 1C scenes were retrieved with Google Earth Engine. Images from Landsat 1 – 3 were excluded due to the coarser pixel size/resolution (80 m), and images from Landsat 4 were excluded due to limited data availability at sites of interest. The relevant spectral bands to download are the visible bands (Blue, Green, and Red), the near-infrared bands (NIR), the short-wave infrared band (SWIR1), and the quality assessment (QA) band for the cloud mask. Landsat 7 and 8 also include a panchromatic band, which combines multiple spectral bands. The panchromatic band was also downloaded to increase the spatial resolution from 30 to 15 m of the Blue, Red, and Green bands in Landsat 8 and the Red, Green, and NIR bands in Landsat 7, following the approach of (Liu et al. (2017). All other bands (NIR, SWIR1—Landsat 8; Blue, SWIR1—Landsat 7; all bands from Landsat 5) were down-sampled to 15 m using bilinear interpolation. All the images with more than 10% cloud cover over the whole study site or georeferencing accuracy larger than 10 m were not used.

To automatically extract the shoreline at a subpixel resolution, the first step is to find or "classify" all image pixels that correspond to sand or water. On sandy beaches, wave breaking makes this boundary more difficult to define, due to the presence of white-water from breaking waves and wave runup, which has a unique spectral signature due to its aerated nature. A supervised image classification labeled each pixel into four classes instead of the usual 2: "sand", "water", "white-water", and "other", which contained the remaining land features like buildings or vegetation. A training set with 1500 pixels in each class was manually digitized from images at one beach system (Narrabeen-Collaroy) to train a MLP—a Neural Network classifier algorithm commonly used for generating landcover maps. The classifier was trained with 70% of the data, withholding 30% for testing.

The final step in shoreline extraction is finding the land-water border using a Modified Normalized Difference Water Index (MNDWI), which is calculated as MNDWI = (SWIR1—Green)/(SWIR1 + Green), for each pixel where SWIR1 and Green are pixel intensity in the short-wave infrared band. The MNDWI is sorted by classification type, and a threshold separating sand and water (rather than a global threshold) is selected. The Marching Squares contouring algorithm (Cipolletti et al., 2012) is used to determine the final shoreline. Finally, a tidal correction is applied to translate the shoreline elevation to a single tidal reference, because each image is taken at different tidal heights.

Shorelines were extracted from satellite data using the methods described above at cross-shore transects of interest at a set of coastal locations, which included Narrabeen-Collaroy, Australia, Duck, North Carolina, US, Truc Vert, France, Moruya beach in Australia, and Tairua beach, New Zealand. All of the beaches are microtidal, with mean tidal ranges less than 2 m, except for Tairua beach, which is classified as meso-macrotidal, with a mean tidal range of 3.7 m. Each of these study sites was chosen due to their long-term histories of monthly shoreline data obtained from field surveys or onsite video monitoring to be used for validation and comparison. Of the five sites, the earliest field survey began in 1976 for Narrabeen-Collaroy beach, and the latest survey began in 2007 at Moruya beach. Because early Landsat (1–3) imagery was not used, shoreline observations extracted from the satellite imagery ranged from 12 to just over 30 years long.

At all beaches, hundreds of shorelines could be extracted. For example, at Narrabeen-Collaroy, a total of 502 satellite-derived shorelines were extracted at five cross-shore transects between 1987–2018, which is approximately 100 shorelines per transect. The Root Mean Square Error (RMSE) of the satellite-extracted shorelines compared to field-defined shorelines ranged from 7–13 m, with an average of approximately 10 m. For Narrabeen-Collaroy beach, 90% of the errors fell within 14 m. There was no significant difference found between data obtained from the different Landsat 5 (30 m resolution), Landsat 6–8 (10 m resolution), and Sentinel-2 (10 m resolution). An example of shorelines extracted and shoreline change rates from the 1980s to 2020 at Duck, NC are shown in Case Study Figure 2.2 for two example transects that are erosive (b) and accretive (c).

How a beach varies over time can only be quantified based on how often measurements are made. For example, if a beach is surveyed only once a year, there's no knowledge of how that beach might change seasonally. If a beach is only surveyed when there is an extreme event, then there's no understanding of the year-to-year changes of the beach. Thus, measurements dictate how much is known about how the shoreline varies. This work also tested whether satellites could be used to evaluate the variability within a year and year to year, by quantifying the magnitude of shoreline variability at timescales of 1 month, 6 months, 1 year, and 1 decade. It was found that satellite imagery publicly available from the 1980–2020s can capture shoreline variability over timescales longer than 6 months, as long as the change in the shoreline behavior exceeds the shoreline accuracy (which is ~10 m). This means that if a shoreline only changes a little year to year (like that of Duck, North Carolina, US, which on average varied from 4–7 m between a month and a year), satellite imagery could not be used to confidently distinguish coastal change. This is also why at most locations satellite imagery could not be used to confidently distinguish month-to month shoreline changes; the magnitude of change ranged from 4–8 m at all beaches. However, rapid shoreline changes due to extreme storm events that rework the coastline by meters or nourishment where sand is added to the beach also could be captured from the shorelines extracted by satellite imagery (e.g., Case Study Figure 2.2b).

The results of this study build off the work of Luijendijk et al., 2018, which was the first big-data application of identifying yearly shorelines from satellite imagery on a global scale, and found 24% of the world's sandy beaches were eroding at rates greater than 0.5 m/yr from 1984–2016. The open-source *CoastSat* python toolbox (Vos, Splinter, et al., 2019) developed for this application derives shorelines from individual rather than yearly composite satellite images, allowing daily to weekly shoreline monitoring as well as long-term trends. Satellites provide the opportunity for monitoring the world's coastlines through seamless data with a high spatiotemporal resolution (Vitousek, Buscombe, et al., 2023).

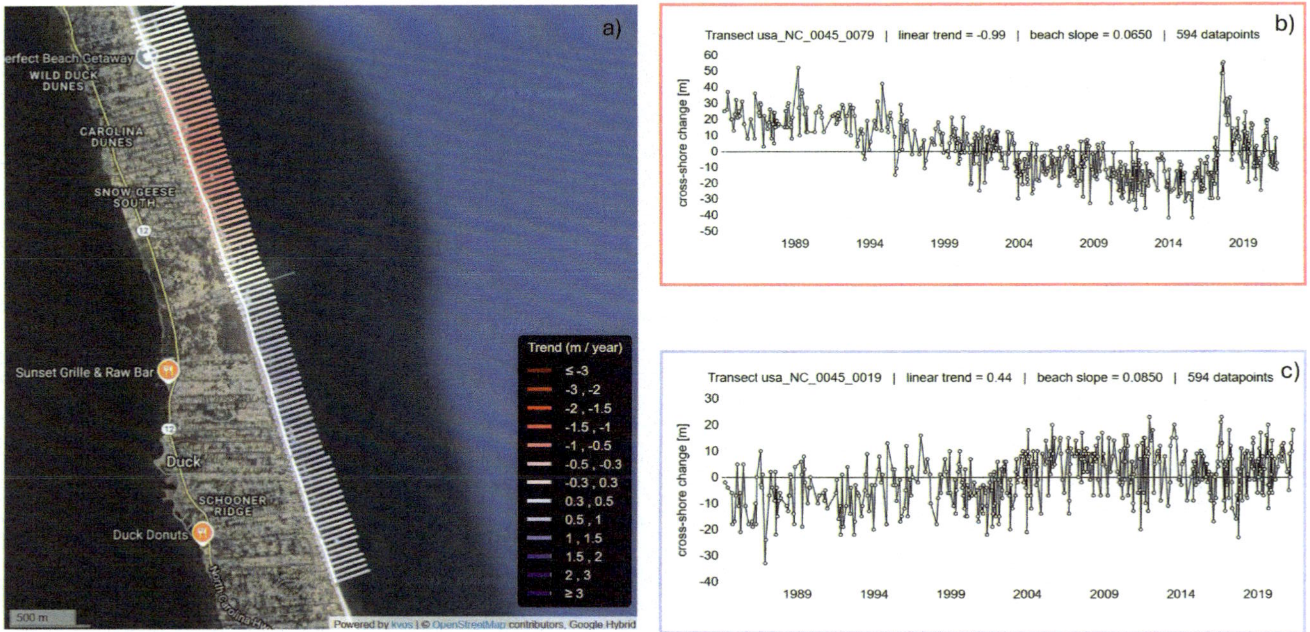

Case Study Figure 2.2 Example shoreline change rates from the mid 1980s to 2020 (a) from shorelines derived from CoastSat for an erosive transect (b) and an accretive transect (c) in Duck, NC.
Image created using the CoastSat WebGIS page: http://coastsat.wrl.unsw.edu.au/

Case study 3: Modeling the influence of compounding coastal and river processes on extreme along-river flooding

Publication: Jane, R. A., Malagón-Santos, V., Rashid, M. M., Doebele, L., Wahl, T., Timmers, S. R., Serafin, K. A., Schmied, C., & Lindemer, C. (2022). A hybrid framework for rapidly locating transition zones: A comparison of event-and response-based return water levels in the Suwannee River FL. *Water Resources Research*, *58*(11), e2022WR032481.

Flooding is one of the costliest hazards in the United States (US), and flood losses will continue to rise as the climate changes (Wing et al., 2022). However, identifying the drivers of flooding in river systems near the coast subject to compound events, where flooding may be generated by two or more processes, such as high river discharge and elevated coastal water level, is challenging. This study evaluates the spatial variation and extent of transition zones, where joint river and ocean processes combine to exacerbate flooding along coastal river systems. By combining multivariate statistical models, hydraulic numerical models, and ML algorithms, we develop an efficient and scalable hybrid modeling framework for quantifying compound flood risk (Case Study Figure 3.1). The framework consists of running a subset of statistically simulated boundary conditions through a numerical model to output along-river water surface elevations. This subset of conditions is then used to generate surrogate models for the efficient extraction of along the river water levels from any combination of upstream discharge and downstream coastal water level. The framework is described in more detail below.

Our study site, the Suwannee River, which flows for approximately 394 km, begins in Georgia and discharges into the Gulf of Mexico in northwest Florida (Case Study Figure 3.2). It is one of the longest rivers in Florida and remains one of the few major US rivers without significant physical alterations such as dams, flow diversions, or navigation projects (Benke, 1990). The river drains an area of 25,770 km^2, primarily in South Georgia and North Florida (Light et al., 2002, p. 124). Fed by runoff from swamps and flatwoods, it merges with the Apalaha, Withlacoochee, and Santa Fe Rivers before reaching the Gulf. The lower basin is sparsely populated, dominated by bottomland hardwood swamps, and tides can reach more than 53 km upriver from the river mouth (Bales et al., 2006).

We use the U.S Army Corps of Engineers Hydrologic Engineering Center's River Analysis System (HEC-RAS) numerical model to estimate flood levels along the Suwannee River. The Suwannee River model used in this work was originally developed by the Suwannee River Water Management District (SRWMD, 2014). The model covers approximately 105 km of river length. HEC-RAS outputs along-river water level by solving the 1-D energy equation at each cross-section with an iterative procedure (Brunner, 2016). In this application, the model is forced to use a river discharge as an upstream input and a coastal water elevation as a downstream input. We modified the river model to begin at the Bell River gauge (USGS 02323000), which is downstream from where the Santa Fe River joins the Suwannee River, and ends in the Gulf, where the Cedar Key tide gauge (NOAA 8727520) is used to represent the downstream coastal model input (Case study Figure 3.2). Typically,

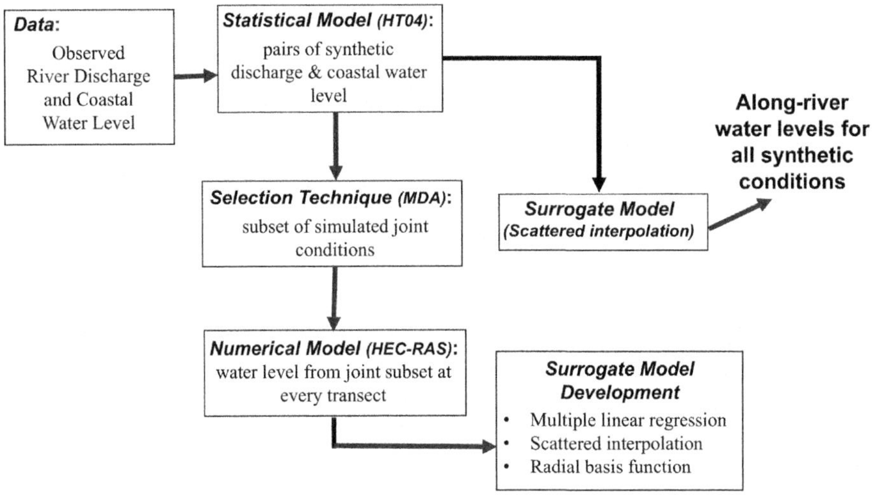

Case Study Figure 3.1 Hybrid compound flood modeling framework that merges statistical, numerical, and machine learning approaches for efficiently producing thousands of estimates of water level along the river from joint river and coastal processes.

you can just enter the upstream and downstream boundary conditions into HEC-RAS that you are interested in for producing an along-river water level. However, we want to understand how many different synthetically simulated joint coastal and inland events (potentially thousands) can combine to cause flooding along the river. These events may not have been captured in the shorter observational records, but are physically capable of occurring. Running this many joint events individually through HEC-RAS might take too long due to computational processing speeds, so we use ML to develop surrogate models to produce results in an efficient manner.

Surrogate models (also known as metamodels, predictive models, or emulators), approximate the response of physics-based models, reducing the need for running full simulations of complex, physics-based models. Essentially, surrogate models are a simplified model of a complex model. The strength of using a surrogate model is that it can make an estimate from new input data in seconds, rather than the hours it would take to run a physics-based model. This study tests three different approaches for developing surrogate models.

To develop a surrogate model at every transect along the river, we first find the optimal joint conditions that represent the bounds of our joint coastal/inland input data. In this study, a selection technique called the maximum dissimilarity algorithm/method (MDA) (e.g., Willett, 1999) is used to select a subset of conditions from the synthetic simulations for running through the HEC-RAS models. The MDA was selected because it chooses joint conditions that are evenly distributed across the space with some points selected in the outline of the data space, and it does a good job at representing boundaries of data (Camus, Mendez, Medina, et al., 2011). The number of conditions chosen for the subset is subjective, representing a tradeoff between running the physics-based model and the performance of the

Case Study Figure 3.2 Suwannee River study area for evaluating compound flooding.

surrogate model. We use 1000 conditions, whereas previous work has ranged from 100 (Camus, Mendez, & Medina, 2011) to over 3000 conditions (Serafin et al., 2019).

The three approaches used for developing surrogate models are multiple linear regression, scattered interpolation, and radial basis functions. The multiple linear regression is the simplest function and is a type of linear regression that uses several independent variables to predict the dependent variable. It has the most restrictive assumptions and assumes residuals are independent and normally distributed. In scattered interpolation, data points are triangulated, and the interpolation is carried out within each triangle. In this application, we use a natural neighbor interpolation and a linear extrapolation method. Radial basis functions, also known as Gaussian kernels, are often used for approximation, interpolation, or pattern recognition. In this work, the radial basis function approximation uses a weighted sum of Gaussian basis functions. The performance of the surrogate models produced from each approach is compared to the HEC-RAS model output and quantified in terms of the Root Mean Square Error (RMSE) and a k-fold cross-validation. After the best performing surrogate model is chosen, a synthetic statistical simulation of joint coastal water level and river discharge representing 10,000 years of storm events under present-day climate conditions is then passed through the surrogate models to extract along-river water height for every joint condition.

We find that the scattered interpolation produces the best surrogate model; however, errors are also small for the radial basis function. Multiple linear regression is the worst-performing surrogate model. Once conditions are passed through the surrogate model, we find that flood transition zones, where both ocean and river processes combine to drive flooding, can be over 40 km in length and can be due to lower discharge or coastal events that combine together. Overall, our findings indicate the importance of local-scale studies for evaluating the spatial variability in compound flooding while providing a tool that produces generalizable findings for characterizing flood transition zones across river systems beyond a specific site.

References

Anderson, D., Bak, A. S., Cohn, N., Brodie, K. L., Johnson, B., & Dickhudt, P. (2023). The impact of inherited morphology on sandbar migration during mild wave seasons. *Geophysical Research Letters, 50*(3), Article e2022GL101219.

Bales, J. D., Tomlinson, S. A., & Tillis, G. (2006). *Flow and salt transport in the Suwannee River Estuary, Florida, 1999–2000: Analysis of data and three-dimensional simulations.* United States Geological Survey. Professional Paper 1656-B, 66.

Bass, B., & Bedient, P. (2018). Surrogate modeling of joint flood risk across coastal watersheds. *Journal of Hydrology, 558*, 159–173.

Bauer, B. O., Davidson-Arnott, R. G., Walker, I. J., Hesp, P. A., & Ollerhead, J. (2012). Wind direction and complex sediment transport response across a beach–dune system. *Earth Surface Processes and Landforms, 37*(15), 1661–1677. https://doi.org/10.1002/esp.3306

Bauer, B. O., Davidson-Arnott, R. G. D., Hesp, P. A., Namikas, S. L., Ollerhead, J., & Walker, I. J. (2009). Aeolian sediment transport on a beach: Surface moisture, wind fetch, and mean transport. *Geomorphology, 105*(1–2), 106–116.

Bauer, B. O., Hesp, P. A., Walker, I. J., & Davidson-Arnott, R. G. (2015). Sediment transport (dis) continuity across a beach–dune profile during an offshore wind event. *Geomorphology, 245*, 135–148. https://doi.org/10.1016/j.geomorph.2015.05.004

Benke, A. C. (1990). A perspective on America's vanishing streams. *Journal of the North American Benthological Society, 9*(1), 77–88.

Birkemeier, W. A., Long, C. E., & Hathaway, K. K. (1996). DELILAH, DUCK94 and SandyDuck: Three nearshore field experiments. In *Coastal engineering 1996* (pp. 4052–4065).

Browder, A. G., & McNinch, J. E. (2006). Linking framework geology and nearshore morphology: Correlation of paleo-channels with shore-oblique sandbars and gravel outcrops. *Marine Geology, 231*, 141–162. https://doi.org/10.1016/j.margeo.2006.06.006

Brunner, G. W. (2016). *HEC-RAS 5.0 reference manual (U.S. Army Corps of Engineers, Institute for Water Resources, Hydrologic Engineering Center) Report No.: CPD-69.*

Buckley, M. L., Buscombe, D., Birchler, J. J., Palmsten, M. L., Swanson, E., Brown, J. A., … Harrison, S. R. (2024). Wave runup and total water level observations from time series imagery at several sites with varying nearshore morphologies. *Coastal Engineering, 193*, Article 104600.

Buscombe, D., Wernette, P., Fitzpatrick, S., Favela, J., Goldstein, E. B., & Enwright, N. M. (2023). A 1.2 billion pixel human-labeled dataset for data-driven classification of coastal environments. *Scientific Data, 10*(1), 46.

Calcraft, K., Splinter, K. D., Simmons, J. A., & Marshall, L. A. (2025). Do LSTM memory states reflect the relationships in reduced-complexity sandy shoreline models. *Environmental Modelling and Software, 183*, Article 106236.

Camus, P., Mendez, F. J., & Medina, R. (2011). A hybrid efficient method to downscale wave climate to coastal areas. *Coastal Engineering, 58*(9), 851–862.

Camus, P., Mendez, F. J., Medina, R., & Cofiño, A. S. (2011). Analysis of clustering and selection algorithms for the study of multivariate wave climate. *Coastal Engineering, 58*(6), 453–462.

Caron, M., Touvron, H., Misra, I., Jégou, H., Mairal, J., Bojanowski, P., et al. (2021). Emerging properties in self-supervised vision transformers. In *Proceedings of the international conference on computer vision.* ICCV).

Castelle, B., Masselink, G., Scott, T., Stokes, C., Konstantinou, A., Marieu, V., & Bujan, S. (2021). Satellite-derived shoreline detection at a high-energy meso-macrotidal beach. *Geomorphology, 383*, Article 107707.

Ciarletta, D. J., Miselis, J. L., Shawler, J. L., & Hein, C. J. (2020). Quantifying thresholds of barrier geomorphic change in a cross-shore sediment partitioning model. *Earth Surface Dynamics Discussions, 2020*, 1–36. https://doi.org/10.5194/esurf-2020-88

Ciarletta, D. J., Shawler, J. L., Tenebruso, C., Hein, C. J., & Lorenzo-Trueba, J. (2019). Reconstructing coastal sediment budgets from beach-and foredune-ridge morphology: A coupled field and modeling approach. *Journal of Geophysical Research, Earth Surface, 124*(6), 1398–1416. https://doi.org/10.1029/2018JF004908

Cipolletti, M. P., Delrieux, C. A., Perillo, G. M., & Piccolo, M. C. (2012). Superresolution border segmentation and measurement in remote sensing images. *Computers and Geosciences, 40*, 87–96.

Cohn, N., Brodie, K. L., Johnson, B., & Palmsten, M. L. (2021). Hotspot dune erosion on an intermediate beach. *Coastal Engineering, 170*, Article 103998.

Cohn, N., Dickhudt, P., & Brodie, K. (2022a). Remote observations of Aeolian saltation. *Geophysical Research Letters, 49*(16), Article e2022GL100066. https://doi.org/10.1016/j.coastaleng.2025.104743

Cohn, N., Dickhudt, P., & Marshall, J. (2022b). In-situ measurement of grain size characteristics within the Aeolian saltation layer on a coastal beach. *Earth Surface Processes and Landforms, 47*(9), 2230–2244.

Conlin, M. P., Wilson, G., Bond, H., Ardağ, D., & Arnowil, C. (2025). Predicting wave runup on composite beaches. *Coastal Engineering, 199*, Article 104743.

da Silva, P. G., Coco, G., Garnier, R., & Klein, A. H. (2020). On the prediction of runup, setup and swash on beaches. *Earth-Science Reviews, 204*, Article 103148.

Davidson-Arnott, R. G., Bauer, B. O., Hesp, P. A., Namikas, S. L., Ollerhead, J. W., & Walker, I. J. (2005). Moisture and fetch effects on aeolian sediment transport rates during a fall storm, Greenwich Dunes, Prince Edward Island. In *Proceedings, Canadian Coastal Conference, Halifax*. Ottawa: National Research Council. *Compact Disk.*

Delgado-Fernandez, I., & Davidson-Arnott, R. (2009). Sediment input to foredunes: Description and frequency of transport events at Greenwich Dunes, PEI, Canada. *Journal of Coastal Research*, 302–306.

Diesing, M., & Stephens, D. (2015). A multi-model ensemble approach to seabed mapping. *Journal of Sea Research, 100*, 62–69.

Dow, H. W., East, A. E., Sankey, J. B., Warrick, J. A., Kostelnik, J., Lindsay, D. N., & Kean, J. W. (2024). Postfire sediment mobilization and its downstream implications across California, 1984–2021. *Journal of Geophysical Research, Earth Surface, 129*(8), Article e2024JF007725. https://doi.org/10.1029/2024JF007725

Duch, W., & Jankowski, N. (1999). Survey of neural transfer functions. *Neural Computing Surveys, 2*(1), 163–212.

Elsayed, S. M., & Oumeraci, H. (2017). Effect of beach slope and grain-stabilization on coastal sediment transport: An attempt to overcome the erosion overestimation by XBeach. *Coastal Engineering, 121*, 179–196. https://doi.org/10.1016/j.coastaleng.2016.12.009

Ezer, T. (2015). Detecting changes in the transport of the Gulf stream and the Atlantic overturning circulation from coastal sea level data: The extreme decline in 2009–2010 and estimated variations for 1935–2012. *Global and Planetary Change, 129*, 23–36. https://doi.org/10.1016/j.gloplacha.2015.03.002

Ezer, T., Atkinson, L. P., Corlett, W. B., & Blanco, J. L. (2013). Gulf Stream's induced sea level rise and variability along the US mid-Atlantic coast. *Journal of Geophysical Research, Oceans, 118*(2), 685–697.

Fitzpatrick, S., Buscombe, D., Warrick, J. A., Lundine, M. A., & Vos, K. (2024). CoastSeg: An accessible and extendable hub for satellite-derived-shoreline (SDS) detection and mapping. *Journal of Open Source Software, 9*(99), 6683.

Gal, N. S., Wallace, D. J., Miner, M. D., Hollis, R. J., Dike, C., & Flocks, J. G. (2021). Influence of antecedent geology on the Holocene formation and evolution of Horn Island, Mississippi, USA. *Marine Geology, 431*, Article 106375. https://doi.org/10.1016/j.margeo.2020.106375

Geisz, J. K., Wernette, P. A., & Esselman, P. C. (2024). Classification of lakebed geologic substrate in autonomously collected benthic imagery using machine learning. *Remote Sensing, 16*(7), 1264. https://doi.org/10.3390/rs16071264

George, E., Lunardi, B., Smith, A., Lehner, J., Wernette, P., & Houser, C. (2021). Storm impact and recovery of a beach-dune system in Prince Edward Island. *Geomorphology, 384*, Article 107721. https://doi.org/10.1016/j.geomorph.2021.107721

George, E., Smith, A., O'Rourke, C., Cherry, N., Sfalcin, A., & Houser, C. (2024). Citizen science monitoring of beach and dune erosion during Hurricane Fiona. *Physical Geography, 45*(4), 382–401.

Green, J., Haigh, I. D., Quinn, N., Neal, J., Wahl, T., Wood, M., Eilander, D., de Ruiter, M., Ward, P., & Camus, P. (2025). A comprehensive review of compound flooding literature with a focus on coastal and estuarine regions. *Natural Hazards and Earth System Sciences, 25*(2), 747–816. https://doi.org/10.5194/nhess-25-747-2025

Gutiérrez, J. M., Cano, R., Cofiño, A. S., & Sordo, C. (2005). Analysis and downscaling multimodel seasonal forecasts in Peru using self-organizing maps. *Tellus A: Dynamic Meteorology and Oceanography, 57*(3), 435–447.

Harley, M. D., & Kinsela, M. A. (2022). CoastSnap: A global citizen science program to monitor changing coastlines. *Continental Shelf Research, 245*, Article 104796.

Holman, R. A., & Guza, R. T. (1984). Measuring run-up on a natural beach. *Coastal Engineering, 8*(2), 129–140.

Holman, R., & Haller, M. C. (2013). Remote sensing of the nearshore. *Annual Review of Marine Science, 5*(1), 95–113.

Huang, W., Murray, C., Kraus, N., & Rosati, J. (2003). Development of a regional neural network for coastal water level predictions. *Ocean Engineering, 30*(17), 2275–2295.

Itzkin, M., Palmsten, M. L., Buckley, M. L., Birchler, J. J., & Torres-Garcia, L. M. (2025). Developing a decision tree model to forecast runup and assess uncertainty in empirical formulations. *Coastal Engineering, 195*, Article 104641.

Jafarzadegan, K., Moradkhani, H., Pappenberger, F., Moftakhari, H., Bates, P., Abbaszadeh, P., Marsooli, R., Ferreira, C., Cloke, H. L., Ogden, F., & Duan, Q. (2023). Recent advances and new frontiers in riverine and coastal flood modeling. *Reviews of Geophysics, 61*(2), Article e2022RG000788.

Janda, C. N., Warrick, J. A., Buscombe, D., & Batiste, S. (2025). Shoreline change of Western Long Island, New York, from satellite-derived shorelines. *Coasts, 5*(1), 2. https://doi.org/10.3390/coasts5010002

Jane, R. A., Malagón-Santos, V., Rashid, M. M., Doebele, L., Wahl, T., Timmers, S. R., Serafin, K. A., Schmied, C., & Lindemer, C. (2022). A hybrid framework for rapidly locating transition zones: A comparison of event-and response-based return water levels in the Suwannee River FL. *Water Resources Research, 58*(11), Article e2022WR032481.

Juan, N. P., & Valdecantos, V. N. (2022). Review of the application of artificial neural networks in ocean engineering. *Ocean Engineering, 259*, Article 111947.

Kang, B., Feagin, R. A., Huff, T., & Durán Vinent, O. (2024). Stochastic properties of coastal flooding events–part 1: Convolutional-neural-network-based semantic segmentation for water detection. *Earth Surface Dynamics, 12*(1), 1–10.

Kim, L. N., Brodie, K. L., Cohn, N. T., Giddings, S. N., & Merrifield, M. (2023). Observations of beach change and runup, and the performance of empirical runup parameterizations during large storm events. *Coastal Engineering, 184*, Article 104357.

Kirillov, A., Mintun, E., Ravi, N., Mao, H., Rolland, C., Gustafson, L., et al. (2023). Segment anything. arXiv:2304.02643.

Konstantinou, A., Scott, T., Masselink, G., Stokes, K., Conley, D., & Castelle, B. (2023). Satellite-based shoreline detection along high-energy macrotidal coasts and influence of beach state. *Marine Geology, 462*, Article 107082.

Lazarus, E., Ashton, A., Murray, A. B., Tebbens, S., & Burroughs, S. (2011). Cumulative versus transient shoreline change: Dependencies on temporal and spatial scale. *Journal of Geophysical Research, Earth Surface, 116*(F2).

Leaman, C., Beuzen, T., & Goldstein, E. B. (2020). *chrisleaman/py-wave-runup: v0.1.10 (v0.1.10)*. Zenodo. https://doi.org/10.5281/zenodo.3629949

LeCun, Y., Boser, B., Denker, J., Henderson, D., Howard, R., Hubbard, W., & Jackel, L. (1989). Handwritten digit recognition with a back-propagation network. *Advances in Neural Information Processing Systems, 2*.

LeCun, Y., Bottou, L., Bengio, Y., & Haffner, P. (1998). Gradient-based learning applied to document recognition. *Proceedings of the IEEE, 86*(11), 2278–2324.

Lee, J. W., Irish, J. L., Bensi, M. T., & Marcy, D. C. (2021). Rapid prediction of peak storm surge from tropical cyclone track time series using machine learning. *Coastal Engineering, 170*, Article 104024.

Lee, G, Nicholls, R. J., & Birkemeier, W. A. (1998). Storm-driven variability of the beach-nearshore profile at Duck, North Carolina, USA, 1981–1991. *Marine Geology, 148*(3-4), 163–177. https://doi.org/10.1016/S0025-3227(98)00010-3

Lee, C. C., Sheridan, S. C., Dusek, G. P., & Pirhalla, D. E. (2023). Atmospheric pattern–based predictions of S2S sea level anomalies for two selected US locations. *Artificial Intelligence for the Earth Systems, 2*(3), Article 220057.

Light, H. M., Darst, M. R., Lewis, L. J., & Howell, D. A. (2002). *Hydrology, vegetation, and soils of riverine and tidal floodplain forests of the lower Suwannee River, Florida, and potential impacts of flow reductions: U.S. Geological Survey Professional Paper 1656-A*.

Lehner, J., Johnson, G., Smith, A., & Houser, C. (2026). Unmasking the topographic complexity of coastal barrier foredunes. *Geomorphology, 505*, Article 110312. https://doi.org/10.1016/j.geomorph.2026.110312

Liu, S., Zeng, Z., Ren, T., Li, F., Zhang, H., Yang, J., et al. (2023). Grounding dino: Marrying dino with grounded pre-training for open-set object detection. arXiv preprint arXiv: 2303.05499.

Liu, Q, Trinder, J, & Turner, I. L. (2017). Automatic super-resolution shoreline change monitoring using Landsat archival data: A case study at Narrabeen–Collaroy Beach, Australia. *Journal of Applied Remote Sensing, 11*(1), 016036-016036.

Longuet-Higgins, M. S., & Stewart, R. W. (1964). Radiation stresses in water waves; a physical discussion, with applications. *Deep-Sea Research and Oceanographic Abstracts, 11*(4), 529–562.

Luijendijk, A., Hagenaars, G., Ranasinghe, R., Baart, F., Donchyts, G., & Aarninkhof, S. (2018). The state of the world's beaches. *Scientific Reports, 8*(1), 6641.

Lunardi, B., Smith, A., George, E., Lehner, J., & Houser, C. (2023). Intra-and Inter-site variability of the beach dune profile and vegetation line: Is there a scale-invariant profile? *Geomorphology, 427*, Article 108618. https://doi.org/10.1016/j.geomorph.2023.108618

Mabrouk, A., Menza, C., & Sautter, W. (2022). *Best practices for ground-truthing and accuracy assessment of lakebed maps in the Great Lakes: A case study offshore the Bayfield Peninsula in Lake Superior*. https://doi.org/10.25923/f1tn-0694

Mathew, S., Davidson-Arnott, R. G., & Ollerhead, J. (2010). Evolution of a beach–dune system following a catastrophic storm overwash event: Greenwich Dunes, Prince Edward Island, 1936–2005. *Canadian Journal of Earth Sciences, 47*(3), 273–290.

McCormack, T., Hopkins, J., Raubenheimer, B., Elgar, S., & Brodie, K. L. (2025). Remote sensing of wave-orbital velocities in the surfzone. *Coastal Engineering, 195*, Article 104631.

McNinch, J. E. (2004). Geologic control in the nearshore: Shore-oblique sandbars and shoreline erosional hotspots, Mid-Atlantic Bight, USA. *Marine Geology, 211*, 121–141. https://doi.org/10.1016/j.margeo.2004.07.006

Merrifield, M. A., Thompson, P. R., & Lander, M. (2012). Multidecadal sea level anomalies and trends in the Western tropical Pacific. *Geophysical Research Letters, 39*(13).

Moore, L. (2000). Shoreline mapping techniques. *Journal of Coastal Research*, 111–124.

Moradi, S., Esselman, P. C., Geisz, J. K., & Roussi, C. (2024). *FishScale*. https://doi.org/10.5066/P13QVR2R. . (Accessed 1 April 2025) Accessed.

Moradi, S., Esselman, P. C., Tilley, A. T., & Geisz, J. K. (2025). *Assessment of submerged aquatic vegetation in Lake Michigan using down-looking AUV-collected imagery*. https://doi.org/10.5066/P1GPDYMM

Moradi, S., Esselman, P. C., Yeager, N. J., Morris, J. M., & Geisz, J. K. (2024). *Down-looking images of Lake Trout (*Salvelinus namaycush*) and round goby (*Neogobius melanostomus*) to support automated biomass estimation*. https://doi.org/10.5066/P9E1DTYB

Moradian, S., AghaKouchak, A., Gharbia, S., Broderick, C., & Olbert, A. I. (2024). Forecasting of compound ocean-fluvial floods using machine learning. *Journal of Environmental Management, 364*, Article 121295.

Morris, J. M., Esselman, P. C., Riseng, C. M., Elgin, A. K., & Rowe, M. D. (2024). Predicting Lake Huron Dreissena spp. spatial distribution patterns from environmental characteristics. *Journal of Great Lakes Research, 50*, Article 102369. https://doi.org/10.1016/j.jglr.2024.102369

Nieves, V., Radin, C., & Camps-Valls, G. (2021). Predicting regional coastal sea level changes with machine learning. *Scientific Reports, 11*(1), 7650.

NOAA. (2022). A NOAA capability for coastal flooding and inundation information and services at climate timescales to reduce risk and improve resilience. https://cpo.noaa.gov/wp-content/uploads/2023/08/NOAA-Coastal-Inundation-at-Climate-Timescales-Whitepaper.pdf.

O'Dea, A., Brodie, K., & Elgar, S. (2021). Field observations of the evolution of plunging-wave shapes. *Geophysical Research Letters, 48*(16), Article e2021GL093664.

Ollerhead, J., Davidson-Arnott, R., Walker, I. J., & Mathew, S. (2013). Annual to decadal morphodynamics of the foredune system at Greenwich Dunes, Prince Edward Island, Canada. *Earth Surface Processes and Landforms, 38*(3), 284–298.

Ollerhead, J., Davidson-Arnott, R., Johnson, P., Walker, I., & Hesp, P. (October 2003). Seasonal variations in a beach and foredune system, Greenwich dunes, Prince Edward Island national park. In *Proceedings, Canadian coastal conference*. Kingston, ON: Queen's University. *Compact Disk*.

Park, J., & Sweet, W. (2015). Accelerated sea level rise and Florida current transport. *Ocean Science, 11*(4), 607–615.

Parker, K., Ruggiero, P., Serafin, K. A., & Hill, D. F. (2019). Emulation as an approach for rapid estuarine modeling. *Coastal Engineering, 150*, 79–93.

Piecuch, C. G., & Ponte, R. M. (2015). Inverted barometer contributions to recent sea level changes along the northeast Coast of North America. *Geophysical Research Letters, 42* (14), 5918–5925.

Poate, T. G., McCall, R. T., & Masselink, G. (2016). A new parameterisation for runup on gravel beaches. *Coastal Engineering, 117*, 176–190.

Power, H. E., Gharabaghi, B., Bonakdari, H., Robertson, B., Atkinson, A. L., & Baldock, T. E. (2019). Prediction of wave runup on beaches using gene-expression programming and empirical relationships. *Coastal Engineering, 144*, 47–61.

Pucino, N., Kennedy, D. M., Carvalho, R. C., Allan, B., & Ierodiaconou, D. (2021). Citizen science for monitoring seasonal-scale beach erosion and behaviour with aerial drones. *Scientific Reports, 11*(1), 3935.

Reif, M. K., Krumwiede, B. S., Brown, S. E., Theuerkauf, E. J., & Harwood, J. H. (2021). Nearshore benthic mapping in the Great Lakes: A multi-agency data integration approach in southwest Lake Michigan. *Remote Sensing, 13*(15), 3026.

Rivas, P. (2020). Introduction to machine learning. In *Deep learning for beginners: A beginner's guide to getting up and running with deep learning from scratch using python* (pp. 3–32). Packt Publishing Ltd.

Roelvink, D., Reniers, A., Van Dongeren, A. P., De Vries, J. V. T., McCall, R., & Lescinski, J. (2009). Modelling storm impacts on beaches, dunes and barrier islands. *Coastal Engineering, 56*(11–12), 1133–1152.

Ruggiero, P., Komar, P. D., McDougal, W. G., Marra, J. J., & Beach, R. A. (2001). Wave runup, extreme water levels and the erosion of properties backing beaches. *Journal of Coastal Research*, 407–419.

Sallenger, A. H. (2000). Storm impact scale for barrier islands. *Journal of Coastal Research*, 890–895.

Seleem, O., Ayzel, G., Bronstert, A., & Heistermann, M. (2022). Transferability of data-driven models to predict urban pluvial flood water depth in Berlin, Germany. *Natural Hazards and Earth System Sciences Discussions, 2022*, 1–2.

Serafin, K. A., & Ruggiero, P. (2014). Simulating extreme total water levels using a time-dependent, extreme value approach. *Journal of Geophysical Research, Oceans, 119*(9), 6305–6329.

Serafin, K. A., Ruggiero, P., Parker, K., & Hill, D. F. (2019). What's streamflow got to do with it? A probabilistic simulation of the competing oceanographic and fluvial processes driving extreme along-river water levels. *Natural Hazards and Earth System Sciences, 19* (7), 1415–1431.

Shawler, J. L., Ciarletta, D. J., Connell, J. E., Boggs, B. Q., Lorenzo-Trueba, J., & Hein, C. J. (2021). Relative influence of antecedent topography and sea-level rise on barrier-island migration. *Sedimentology, 68*(2), 639–669. https://doi.org/10.1111/sed.12798

Sheridan, S. C., & Lee, C. C. (2011). The self-organizing map in synoptic climatological research. *Progress in Physical Geography, 35*(1), 109–119.

Sheridan, S. C., Pirhalla, D. E., Lee, C. C., & Ransibrahmanakul, V. (2017). Atmospheric drivers of sea- level fluctuations and nuisance floods along the mid-Atlantic Coast of the USA. *Regional Environmental Change, 17*, 1853–1861.

Sherman, D. J. (2020). Understanding wind-blown sand: Six vexations. *Geomorphology, 366*, Article 107193.

Short, A. D., & Wright, L. D. (1983). Physical variability of sandy beaches. In *Sandy beaches as ecosystems: Based on the proceedings of the first international symposium on sandy beaches, held in port Elizabeth, South Africa, 17–21 January 1983* (pp. 133–144). Netherlands: Springer.

Simmons, J. A., & Splinter, K. D. (2025). Data-driven shoreline modelling at timescales of days to years. *Coastal Engineering, 197*, Article 104685.

SRWMD. (2014). Retrieved from https://www.mysuwanneeriver.com/DocumentCenter/Index/78.

Stockdon, H. F., Holman, R. A., Howd, P. A., & Sallenger Jr, A. H. (2006). Empirical parameterization of setup, swash, and runup. *Coastal Engineering, 53*(7), 573–588.

Stockdon, H. F., Sallenger Jr, A. H., Holman, R. A., & Howd, P. A. (2007). A simple model for the spatially-variable coastal response to hurricanes. *Marine Geology, 238*(1–4), 1–20.

Straub, J. A., Cialone, M. A., Raubenheimer, B., Brown, J. A., Elko, N., & Brodie, K. L. (2023). The during nearshore event experiment (DUNEX): A collaborative coastal community experiment to address coastal resilience. *Shore and Beach, 91*(3), 23–29.

Sylvester, Z. (2024). *SegmentEveryGrain*. Available at: https://github.com/zsylvester/segmenteverygrain. (Accessed 1 April 2025) Accessed.

Theuerkauf, E. J., Bunting, E. L., Mack, E. A., & Rabins, L. A. (2022). Initial insights into the development and implementation of a citizen-science drone-based coastal change monitoring program in the Great Lakes region. *Journal of Great Lakes Research, 48*(2), 606–613.

United States Federal Geographic Data Committee. (2012). *FGDC-STD-018-2012. Coastal and marine ecological classification standard (CMECS)*. Reston, VA: Federal Geographic Data Committee. Available online: https://repository.library.noaa.gov/view/noaa/27552. (Accessed 6 June 2022) accessed on.

Vitousek, S., Buscombe, D., Vos, K., Barnard, P. L., Ritchie, A. C., & Warrick, J. A. (2023). The future of coastal monitoring through satellite remote sensing. *Cambridge Prisms: Coastal Futures, 1*, Article e10.

Vitousek, S., Vos, K., Splinter, K. D., Erikson, L., & Barnard, P. L. (2023). A model integrating satellite-derived shoreline observations for predicting fine-scale shoreline response to waves and sea-level rise across large coastal regions. *Journal of Geophysical Research, Earth Surface, 128*(7), Article e2022JF006936.

Vitousek, S., Vos, K., Splinter, K. D., Parker, K., O'Neill, A., Foxgrover, A. C., Hayden, M. K., Thomas, J. A., Erikson, L., & Barnard, P. L. (2024). Scalable, data-assimilated models predict large-scale shoreline response to waves and sea-level rise. *Scientific Reports, 14*(1), Article 28029.

Vos, K. (2025). Time-series of shoreline change along the Pacific Rim [Data set]. In *Pacific shoreline erosion and accretion patterns controlled by El Niño/Southern Oscillation, v1.5* (pp. 140–146). Zenodo. https://doi.org/10.5281/zenodo.14889613. Number 16.

Vos, K., Harley, M. D., Splinter, K. D., Simmons, J. A., & Turner, I. L. (2019). Sub-annual to multi- decadal shoreline variability from publicly available satellite imagery. *Coastal Engineering, 150*, 160–174. https://doi.org/10.1016/j.coastaleng.2019.04.004

Vos, K., Harley, M. D., Splinter, K. D., Walker, A., & Turner, I. L. (2020). Beach slopes from satellite-derived shorelines. *Geophysical Research Letters, 47*(14), Article e2020GL088365.

Vos, K., Harley, M. D., Turner, I. L., & Splinter, K. D. (2023). Pacific shoreline erosion and accretion patterns controlled by El Niño/Southern Oscillation. *Nature Geoscience, 16*(2), 140–146. https://doi.org/10.1038/s41561-022-01117-8

Vos, K., Splinter, K. D., Harley, M. D., Simmons, J. A., & Turner, I. L. (2019). CoastSat: A Google earth engine-enabled python toolkit to extract shorelines from publicly available satellite imagery. *Environmental Modelling and Software, 122*, Article 104528. https://doi.org/10.1016/j.envsoft.2019.104528

Wang, G., Wang, X., Wu, X., Liu, K., Qi, Y., Sun, C., & Fu, H. (2022). A hybrid multivariate deep learning network for multistep ahead sea level anomaly forecasting. *Journal of Atmospheric and Oceanic Technology, 39*(3), 285–301.

Warrick, J. A., Buscombe, D., Vos, K., Kenyon, H., Ritchie, A. C., Harley, M. D., … Vitousek, S. (2025). Shoreline seasonality of California's beaches. *Journal of Geophysical Research, Earth Surface, 130*(2), Article e2024JF007836. https://doi.org/10.1029/2024JF007836

Wentworth, C. K. (1922). A scale of grade and class terms for clastic sediments. *The Journal of Geology, 30*, 377–392. https://doi.org/10.1086/622910

Wernette, P., & Houser, C. (2022). Short communication: Evidence for geologic control of rip channels along Prince Edward Island, Canada. *Physical Geography, 43*, 145–162. https://doi.org/10.1080/02723646.2021.1923389

Wernette, P. A., Houser, C., Weymer, B. A., Bishop, M. P., Everett, M. E., & Reece, R. (2018a). Influence of a spatially complex framework geology on island geomorphology. *Marine Geology, 398*, 151–162. https://doi.org/10.1016/j.margeo.2018.01.011

Wernette, P. A., Houser, C., Weymer, B., Bishop, M. P., Everett, M. E., & Reece, R. (2018b). Directional dependency and coastal framework geology: Implications for barrier island resilience. *Earth Surface Dynamics, 6*, 1139–1153. https://doi.org/10.5194/esurf-6-1139-2018

Wernette, P. A., Miller, I. M., Ritchie, A. W., & Warrick, J. A. (2022). Crowd-sourced SfM: Best practices for high resolution monitoring of coastal cliffs and bluffs. *Continental Shelf Research, 245*. https://doi.org/10.1016/j.csr.2022.104799

Wernette, P., Shortridge, A., Lusch, D. P., & Arbogast, A. F. (2017). Accounting for positional uncertainty in historical shoreline change analysis without ground reference information. *International Journal of Remote Sensing, 38*, 3906–3922. https://doi.org/10.1080/01431161.2017.1303218

Weymer, B., Everett, M. E., Wernette, P. A., & Houser, C. (2018). Statistical modeling of long-range-dependent structure of barrier island framework geology and surface geomorphology. *Earth Surface Dynamics, 6*, 431–450. https://doi.org/10.5194/esurf-6-431-2018

Widlansky, M. J., Long, X., & Schloesser, F. (2020). Increase in sea level variability with ocean warming associated with the nonlinear thermal expansion of seawater. *Communications Earth and Environment, 1*(1), 9.

Willett, P. (1999). Dissimilarity-based algorithms for selecting structurally diverse sets of compounds. *Journal of Computational Biology, 6*(3–4), 447–457.

Wing, O. E., Lehman, W., Bates, P. D., Sampson, C. C., Quinn, N., Smith, A. M., Neal, J. C., Porter, J. R., & Kousky, C. (2022). Inequitable patterns of US flood risk in the anthropocene. *Nature Climate Change, 12*(2), 156–162.

Woodworth, P. L., Melet, A., Marcos, M., Ray, R. D., Wöppelmann, G., Sasaki, Y. N., … Merrifield, M. A. (2019). Forcing factors affecting sea level changes at the coast. *Surveys in Geophysics, 40*(6), 1351–1397.

Wright, L. D., & Short, A. D. (1984). Morphodynamic variability of surf zones and beaches: A synthesis. *Marine Geology, 56*(1–4), 93–118.

Xu, Y., Lin, K., Hu, C., Wang, S., Wu, Q., Zhang, L., & Ran, G. (2023). Deep transfer learning based on transformer for flood forecasting in data-sparse basins. *Journal of Hydrology, 625*, Article 129956.

Yuan, L., Chen, D., Chen, Y.-L., Codella, N., Dai, X., Gao, J., et al. (2021). *Florence: A new foundation model for computer vision*. https://doi.org/10.48550/arXiv.2111.11432

Environmental Monitoring and Conservation Using Remote Sensing and Big Data

10

Matthew J. McCarthy [1], Hannah Herrero [2], Stephanie A. Insalaco-Wyner [3], Heidi Tubbs [1] and Assaf Anyamba [1]

[1]Geospatial Science and Human Security Division, Oak Ridge National Laboratory, Oak Ridge, TN, United States; [2]Department of Geography & Sustainability, University of Tennessee, Knoxville, TN, United States; [3]Environmental Studies Program, Southwestern University, Georgetown, TX, United States

1. Introduction

In the past 20 years, the commercialization of space has revolutionized Earth observation (EO), leading to an unparalleled influx of data from orbiting satellites, complementing over 5 decades of observations from government sensors (Berger et al., 2022; Wulder et al., 2022). Today, various sensor platforms image the entire planet daily with high spatiotemporal frequency. These data encompass a wide range of the electromagnetic spectrum, from visible to infrared to radio, providing high-quality images, topography/bathymetry, soil moisture, and gravity-anomaly maps. Routine data collection now includes visible detections at centimeter scales and elevation measurements in millimeters.

The sheer quantity of data presents both opportunities and challenges for remote sensing research. Extracting meaningful insights is critical for managing a world facing severe environmental threats. The United Nations' 17 Sustainable Development Goals (SDGs) seek to address these threats by 2030 and provide a timely and meaningful framework around which the discussion of remote sensing with Big Data for conservation can be organized. SDG decision-makers recognize the utility of remote sensing technologies to contribute toward their success, but the adoption of these technologies has been slow (Estoque, 2020). Further, protected areas like national parks and marine reserves have demonstrated significant local and global conservation benefits, including preventing forest loss, regulating climate, and sequestering carbon (Duncanson et al., 2023), and may represent feasible pathways toward achieving the UN's goals.

This chapter reviews various ways in which remote sensing data have contributed to achieving SDG success. While many SDGs have been or could be addressed with the help of remote sensing data and tools, we focus on SDGs 13, 14, and 15, which address climate, aquatic life, and terrestrial life, respectively, as areas in which remote sensing has been successfully applied and may therefore serve as roadmaps to

Remote Sensing, Big Data, and GeoAI. https://doi.org/10.1016/B978-0-443-26741-3.00010-6

replicate. We review examples of remote sensing research applied for characterizing, monitoring, and predicting the conditions of climate, flora, and fauna in protected areas, beginning with theoretical foundations for each SDG to describe the goal and its global context. The bulk of the chapter describes examples from the literature of remote sensing applied to each SDG by referencing datasets, satellite sensors, methods used, and tangible outcomes. A conclusions section highlights the need, the demonstrated success, and future research areas for remote sensing in an SDG context. Finally, two case studies exemplify in greater detail the utility of satellite-based observations for conservation efforts. The goal of this chapter is to promote broader use of Earth observations and expand protected areas globally to meet the SDGs by the 2030 deadline.

2. Theoretical foundations

2.1 SDG 13: Climate action

Global efforts to limit warming to 1.5°C have proved insufficient as 41 of the 42 indicators assessed by Climate Action Tracker are not on track to meet the 2030 targets (Boehm et al., 2023). Global temperatures are already 1.1°C above pre-industrial levels due to rising greenhouse gas (GHG) emissions. Emissions need to be cut in half by 2030 to meet the GHG targets to stay below 1.5°C and meet the commitments under the Paris Agreement (United Nations, 2023). There are two broad strategies to combat climate change. First, through mitigation by reducing greenhouse gas emissions and carbon sequestration. Second, through adaptation by preparing for the intensifying heatwaves, droughts, flooding, wildfires, sea-level rise and other natural hazards that will result from on-going global warming.

SDG 13 calls for "urgent action to combat climate change and its impacts" (United Nations, 2023), but protected areas are not specifically targeted in many international climate-policy frameworks (Wang et al., 2020). Protected areas are designed for conservation and protection from a variety of stressors, so understanding how protected areas can be used to combat climate change and its impacts should be a priority for decision-makers. For example, Target 13.1 to "strengthen resilience and adaptive capacity to climate-related hazards and natural disasters in all countries" directly aligns with the mandate of protected areas to serve in an adaptive capacity against climate change. Protected areas are one manifestation of Target 13.2, which calls to "integrate climate change measures into national policies, strategies, and planning" (United Nations, 2023). This is exemplified through national policies such as the 1916 National Park Service Organic Act of the United States (US), which established the national parks. The mandate of the US National Park Service (NPS) is a model for integrating climate measures into national policies. While it does not specifically name climate change in the mandate, it does call "to conserve the scenery and the natural and historic objects and the wildlife therein, and to provide for the enjoyment of the same in such a manner and by such means that will leave them unimpaired for the enjoyment of future generations" (National Park Service Organic Act, 1916). The mandate sets the precedent for the NPS and all subsequent protected areas that were modeled after it to conserve natural resources in perpetuity, which now inherently requires integrating climate change predictions into the research and management of protected areas.

Protected areas will experience the impacts of climate change through shrinking glaciers (Marzeion et al., 2014), increased drought (Williams et al., 2020) and wildfire frequency (Abatzoglou & Williams, 2016), and changes to streamflow by snow-cover reduction (Pederson et al., 2011). The average annual temperature increase for US national parks is double the rate compared to the entire US for the period of 1895–2010 (Gonzalez, 2020). This increase in temperature coincides with an earlier timing of leaf-on and blooming (Monahan et al., 2016). Protected areas in northern latitudes, such as Denali National Preserve in Alaska, increased at the astounding rate of 4.3°C per century from the period from 1950–2010. In addition to an increase in average annual temperature there is a decrease in total annual precipitation in national parks (Gonzalez, 2020).

Remotely sensed images provide managers and scientists long-term data that the NPS Inventory and Monitoring networks can use to monitor long-term change associated with climate change. Monitoring will be essential to adaptation for climate change in protected areas. It can be used to develop a baseline upon which change can be measured, to model future scenarios, and to determine if adaptation measures are working (Rannow et al., 2014). One product that is used by the NPS is the Terrestrial Observation and Prediction System (TOPS) (Gross et al., 2006). This system uses information technology, weather and climate forecasting, and ecosystem modeling, to inform management decisions related to natural disasters, human health, and natural resource use (Gross et al., 2006). Through a NASA-NPS collaboration, TOPS was used at Yosemite National Park to forecast snowpack, soil moisture and streamflow, and to explore the impacts of climate change on the parks (Gross et al., 2006). To understand the role of protected areas in climate mitigation via carbon sequestration, research has focused on the role of net primary productivity (NPP), which measures the difference between CO_2 fixed by photosynthesis and CO_2 lost to respiration and is a way to measure carbon sequestration (Cabello et al., 2013). Remote sensing offers vegetation-monitoring tools like the Normalized Difference Vegetation Index (NDVI) to measure the fraction of photosynthetically active radiation intercepted by vegetation, which is the primary determinant of carbon gains (Tucker et al., 1986). In addition to measuring carbon gains, they can also be used to derive primary productivity, seasonality and phenology (Cabello et al., 2013).

2.2 SDG 14: Life below water

SDG 14 encompasses a range of targets and indicators aimed at preserving the rich biodiversity and ecological balance of the oceans while ensuring responsible and sustainable utilization of marine resources. This includes measures to combat overfishing, protect marine habitats, reduce marine pollution, and address ocean acidification. Achieving these objectives is paramount, as marine ecosystems contribute significantly to climate regulation, food security, and the global economy (Maina et al., 2015; da Silveira et al., 2021).

The four most valuable biomes by ecosystem service are coral reefs, coastal wetlands, coastal systems, and inland wetlands (de Groot et al., 2012). While not entirely submerged, these ecosystems all fit within the primary goal of SDG 14: "Conserve and sustainably use the oceans, seas and marine resources for sustainable development" (United Nations, 2023). At the same time as the market value of these ecosystems has come to light, the number of publications mapping reefs, mangroves, seagrasses, wetlands and related habitats has exploded (Avtar et al., 2020). The availability of data and tools to conduct high-resolution mapping with standardized methods over large spatial scales certainly facilitated the advancement, but the ability to cite the dollar amount of services conserved or restored by the groups using such maps has surely been a substantial driver of utilitarian remote sensing for aquatic life. The loss or unsustainable use of disproportionately valuable habitats would have dramatic impacts on the local communities that rely upon them as well as the global economy (de Groot et al., 2012).

2.3 SDG 15: Life on land

The combination of remote sensing, big data, and GeoAI is transforming environmental monitoring and conservation. These technologies offer comprehensive tools for observing and managing terrestrial ecosystems, facilitating precise conservation efforts.

SDG 15, Life on Land, is dedicated to the preservation and sustainable management of terrestrial ecosystems. This goal addresses the urgent need to halt and reverse land degradation, deforestation, and biodiversity loss, focusing on forests, mountains, drylands, wetlands, and other biodiversity-rich landscapes (United Nations, 2023). It aims to restore ecosystems, combat desertification, and ensure sustainable natural resource use, fostering a balance between human activities and the Earth. The international community strives to protect diverse flora and fauna and the livelihoods of communities dependent on healthy terrestrial ecosystems through responsible land practices (United Nations, 2023). A critical strategy for achieving SDG 15 is the use of protected areas. Although there is a global target to reserve 30% of the Earth as protected areas by 2030, progress has stalled since 2015 (IUCN, 2021; United Nations, 2023).

SDG 15 includes 12 targets, each with multiple indicators (United Nations, 2023). For instance, Target 15.1 aims to conserve, restore, and sustainably use terrestrial and freshwater ecosystems by 2020. Its indicators are "Forest area as a proportion of total land area" and "Proportion of important sites for terrestrial and freshwater biodiversity that are covered by protected areas, by ecosystem type." As of 2023, the global forest area has decreased from 31.9% in 2000 to 31.2% in 2020, losing 100 million hectares, with nearly 90% of deforestation due to agricultural expansion. However, sustainable forest management is improving, with increased certified forest areas and forests under management plans, including protected areas (United Nations, 2023).

Satellite remote sensing is crucial for SDG 15, offering essential data for monitoring and evaluating environmental resources like forests, wetlands, and drylands. Satellites with various sensors provide information on land cover, land use, deforestation, and biodiversity changes across large areas (Cochran et al., 2020). This capability enables regular resource monitoring, ecosystem health assessment, and identification of degradation-prone areas. It also tracks human impacts on these resources, which are increasingly threatened by climate change and other human activities (Remelgado et al., 2023).

Remote sensing data are also vital for assessing vegetation productivity, an important sub-indicator of SDG 15.3.1 focused on combating land degradation, especially around protected areas (Ishtiaque et al., 2020). Using vegetation indices derived from remote sensing data, researchers can monitor vegetation productivity trends over time (Herrero, Southworth, et al., 2020; Prince, 2019). By providing timely, comprehensive, and often open-access data, satellite remote sensing empowers decision-makers and stakeholders to develop informed strategies for sustainable land management through protected areas, thus contributing to the achievement of SDG 15 and broader environmental conservation and biodiversity protection goals.

3. Applications

3.1 SDG 13 applications

SDG 13 aims to address climate change and its consequences promptly. It seeks to reduce greenhouse gas emissions through significant measures and mitigate severe environmental risks such as rising sea levels, floods, droughts, wildfires, and heatwaves. Regarding protected areas, particularly those classified under International Union for Conservation of Nature (IUCN, 2023) categories I and II which limit human impact, these areas effectively manage factors like land-use changes within their boundaries. This role parallels that of US National Parks and is crucial for assessing climate change impacts (Dieguez & Paruelo, 2017; Pettorelli et al., 2012). Remote sensing enables comprehensive monitoring of large, remote protected areas, which would otherwise be constrained by resource limitations. Research in climate change remote sensing within these areas covers diverse topics such as the effects of temperature rise, alterations in hydrology, and heightened drought and fire intensity on vegetation phenology (Gillespie et al., 2018; McMenamin et al., 2008; Monahan et al., 2016; O'Leary et al., 2018).

Numerous remote sensing investigations assessing climate change impacts on protected areas utilize the NDVI, which compares reflectance in red and near-infrared bands to estimate photosynthetic activity. NDVI serves as a proxy for phenomena such as phenology, NPP, and growing seasons, measuring the fraction of photosynthetically active radiation absorbed by vegetation (Cabello et al., 2013). Time-series NDVI data are pivotal for monitoring and establishing NPP baselines, crucial for estimating carbon sequestration in parks and protected regions (Cabello et al.,

2013). This index also quantifies the disparity between CO_2 fixed through photosynthesis and that released via respiration (Cabello et al., 2012).

For instance, a study selected three IUCN Category II parks within Spain's National Park Network from a 25-year dataset (1982–2006), employing the Global Inventory Modeling and Mapping Studies (GIMMS) NDVI data to establish an environmental change baseline (Alcaraz-Segura et al., 2009). Results indicated increased radiation absorption and earlier phenological events in these parks (Alcaraz-Segura et al., 2009). Separately, Cabello et al. (2013) utilized Moderate Resolution Imaging Spectroradiometer (MODIS) images from 2001–10 to gauge annual primary productivity via mean Enhanced Vegetation Index (EVI), correlating with carbon gains and phenology (specifically, the date of maximum EVI) across national park networks in Spain, Portugal, and Morocco. Their findings highlighted varied carbon gain patterns, with 86% of Spanish parks exceeding the median, compared to 50% in Morocco and one park in Portugal (Cabello et al., 2013).

Dieguez and Paruelo (2017) applied NDVI to assess climate variable spatial heterogeneity (i.e., variation in parameters across regions) and seasonal dynamics, particularly focusing on carbon gain seasonality across 201 protected areas spanning diverse South American biomes using GIMMS data. They evaluated variables like NDVI annual mean (NDVIm) and its interannual coefficient of variation, alongside mean annual temperature and total annual precipitation from 1982–2012. Significant positive trends ($P < .05$) were observed in tropical and subtropical protected areas, indicating enhanced carbon gains over time, contrasting with negative trends in grassland regions (Dieguez & Paruelo, 2017).

Pettorelli et al. (2012) utilized GIMMS data to investigate the impact of climate change on IUCN categories I and II, employing NDVI as a proxy for NPP. They identified a 27% increase in vegetation greenness from 1982 to 2008, predominantly in Western African units, accompanied by a 9% rise in seasonality. Seasonality was quantified as the annual relative range of NDVI-I (annual integrated NDVI), a normalized measure of seasonal variation. Monahan et al. (2016) conducted a study encompassing 276 natural resource parks under the U.S. National Park Service. They analyzed daily minimum and maximum surface temperatures from "Berkeley Earth" data to investigate phenological changes, focusing on estimates like the start of spring based on climatically modeled dates of first leaf (FLI) and first bloom (FBI) derived from indicator plant species. Their assessment spanned 10–30 years for each park, comparing spring onset timing to historical variability from 1901 to 2012. Among the parks, 76% showed an earlier onset of spring, with 53% experiencing extreme early onsets exceeding 95% of historical norms (Monahan et al., 2016). The advance in spring timing correlated positively with latitude, elevation, and proximity to the coast.

O'Leary et al. (2018) conducted a study at Crater Lake National Park using MODIS imagery spanning from 2001 to 2012. They employed MODIS-derived NDVI to investigate the impacts of snowmelt timing and elevation on five phenological metrics within the evergreen conifer forest. These metrics included green-up (the initiation of seasonal new growth), maximum greenness, senescence, dormancy, and growing-season length. Changes such as rising temperatures, reduced snowpack, and earlier snowmelt have altered hydrological systems, forest dynamics, and fire patterns in montane and high-elevation ecosystems. In these environments, snowpack dynamics, particularly the timing of events like green-up, senescence, and growing-season length, are linked to spring snowmelt timing. Mapping spring snowmelt timing using MODIS data revealed a positive correlation with green-up, although other phenological associations were not statistically significant (O'Leary et al., 2018).

MODIS data was also used in a study by Swanson (2017) to monitor snow cover and vegetation greenness in five national parks in Northern Alaska from 2000 to 2016. To understand the snow season MODIS Terra Snow Cover Daily L3 Global 500 m Grid data was used to map snow cover, snow albedo, and fractional snow cover based on the Normalized Difference Snow Index. Vegetation greenness was studied through NDVI composites, and mean monthly temperature and long-term weather record data were also utilized in combination with the MODIS data. The study found that the end of the continuous snow season and the midpoint of spring green-up had become earlier in certain portions of the study area throughout the time period. There was a strong relationship between spring phenology events and thaw-degree days seen through the MODIS imagery and data collected at weather stations which made it possible to gather the snow-off and green-up dates with historic climate data even back to 1937 (Swanson, 2017). This study finds that with this methodology and combination of data, it should be possible to reconstruct snowmelt and green up dates from long-term weather records calibrated from MODIS data for other areas in the Arctic.

Remote sensing can be a valuable tool in protected areas where weather station data might be sparse, such as Rocky Mountain National Park (RMNP). Collados-Lara et al. (2021) discuss the necessity of considering air temperature to understand the hydrology and management within water resource systems. This task can be difficult due to the scarcity of data and poor accessibility to data in some protected areas. To generate temperature fields in the RMNP, temperature data from 21 SNOTEL stations, Land Surface Temperature (LST) data from MODIS, and elevation data was combined and evaluated through multiple geostatistical approaches. This study proved that both variables, elevation and LST, can explain to some extent temperature variability if used through the geostatistical, multivariate kriging technique with external drift (KED). Merged KED along with these remote sensing and elevation products can be a useful approach when estimating air temperature fields in protected areas with scarce data collection (Collados-Lara et al., 2021).

Additionally, the impacts of droughts on protected areas can be monitored using remote sensing. Gillespie et al. (2018) investigated long-term drought effects exacerbated by climate change in the Santa Monica Mountains National Recreation Area and Channel Islands National Park. They utilized MODIS-derived NDVI to analyze vegetation changes over both long-term and short-term periods from 2000 to 2017. The study noted a significant decline in NDVI in the Santa Monica Mountains National Recreation Area during this period, whereas the Channel Islands, situated offshore, potentially experienced less severe impacts from regional drought conditions.

Melendez-Pastor et al. (2010) also studied environmental changes caused by drought/climate change in the wetlands of "El Hondo" Natural Park in south-eastern Spain with remote sensing methodology. Multitemporal satellite imagery including Landsat 5 Thematic Mapper and Landsat 7 Enhanced Thematic Mapper Plus was used to sample the Natural Park and surrounding area to produce a spectral mixture analysis for 2001 and 2005. The most significant finding was the drying of the park's large water reservoir in 2005, showing that the 2004–05 drought in the area had modified the water cycle and land cover in the reservoir had shifted from standing water to soil coverage. There were also differences witnessed in vegetation inside and outside of the park boundary although land cover is similar, vegetation inside the park saw a decrease but remained unchanged outside the park boundary. This study ultimately proved that there needs to be a balanced solution between agricultural irrigation and wetland conservation during periods of climate induced drought (Melendez-Pastor et al., 2010).

Climate change impacts are also causing declines in amphibian populations as temperatures rise, potentially disrupting their life cycle behaviors and the annual hydrological cycle, including changes to snowmelt patterns. McMenamin et al. (2008) investigated the effects of droughts on wetland hydrology, which have been linked to declines in amphibians within Yellowstone National Park. Utilizing Landsat Thematic Mapper images spanning from 1998 to 2008, they assessed the presence of palustrine wetlands crucial for amphibian habitat in the region. These findings were complemented by long-term weather station data indicating increased spring and summer temperatures alongside reduced annual precipitation. The study highlighted a significant decrease in both the number and diversity of amphibian populations, attributing this decline to regional shifts in hydrological conditions resulting from prolonged climatic changes (McMenamin et al., 2008).

3.2 SDG14 applications

Access to aquatic ecosystems is often challenging (e.g., offshore reefs, dense mangrove forests) and therefore lends itself to remote monitoring (Avtar et al., 2020). This includes monitoring of the water bodies themselves, as has been highlighted recently with record-setting ocean temperatures making global headlines. The publicly available Coral Reef Watch website (www.coralreefwatch.noaa.gov), provided by the US National Oceanic and Atmospheric Administration (NOAA), provides regular satellite-derived products for local, regional, and global monitoring of ocean conditions and predicts risks to coral health for 95% of coral reefs (Liu et al., 2014). While relatively coarse in resolution (i.e., 1–5 km), such products can be paired with high-resolution (i.e., <5 m) commercial-satellite-based maps of benthic habitat to identify areas of suitable substrate and water quality for coral larvae settlement and other reef-conservation efforts (Magris et al., 2015; Wirt et al., 2015). Conditions like sea-surface temperature and turbidity can be gathered from satellite instruments to monitor ongoing conditions and for planning and managing the area (Baird et al., 2016; Jupp et al., 1985; Liu et al., 2003). One study combined WorldView-3 and Landsat-8 satellite imagery with machine learning algorithms to map suitable habitats for a key reef-building coral species within Brazilian marine protected areas (MPAs; da Silveira et al., 2021). By combining remote-sensing-derived seascape characteristics (i.e., depth, slope, rugosity, and aspect) known to influence coral reef zoning and species distribution with turbidity and other environmental variables, they generated a species distribution model and combined it with maps of human use (e.g., recreation, fishing) to map habitat suitability. Such maps could be scaled up, produced operationally, and leveraged for local or regional conservation efforts, as well as for evaluating measures of ongoing conservation efforts, by identifying and predicting areas that are highly suitable for settlement, growth and development of reef-building corals. Similarly, Maina et al. (2015) used QuickBird imagery (2.4 m spatial resolution) covering 686 km^2 of east African coastal protected areas to map benthic habitats. The authors partnered with local MPA managers to rank ongoing or planned activities (e.g., subsistence fishing, tourism, sand mining, oil and gas exploration) in terms of compatibility with conservation so that they could conduct a pairwise comparison of activities relative to their influence on overall conflict. They then created a map of protection status in the study area based on criteria like the age of protection, fishing pressure, and existing gear restrictions that were used to develop management prioritization scenarios to minimize lost fishing opportunities, redistribute fisheries away from overfished reefs, and minimize resource use conflicts while also protecting climate-vulnerable locations. This approach provides a scalable framework for conservation strategies that incorporate holistic considerations of ecosystem service benefits.

Mapping aquatic ecosystems with satellite imagery can be more difficult than on land due to water turbidity, surface sun glint, and other factors that are present in this environment (Howard et al., 2014). Imagery is either chosen carefully to avoid these conditions, or algorithms are applied to correct for such features. The Indian River Lagoon (IRL) (study area map Chapter 1, Case Study 1, Figure 1) system in Florida has been a challenge to study due to its turbid water, but some studies have been successful in mapping seagrass species in the system under challenging conditions (Insalaco et al., 2024; Phlips et al., 2015). Mosquito Lagoon (ML), located on the northern end of the IRL, has been experiencing severe ecosystem decline since 2011. Insalaco et al. (2024), used Landsat imagery from 2000–23 and seagrass training samples in a Deep Neural Network classification to map sparse patches of seagrass in the lagoon and record a substantial loss over the study period attributable to Harmful Algal Blooms (HABs) occurring throughout the IRL. Other studies have used satellite-based remote sensing data in the IRL to track HABs (Bostater & Aziz, 2019; Judice et al., 2020), and have found that rising surface temperatures have resulted in an overabundance of *Aureoumbra lagunensis* (i.e., brown tide) in the IRL, which can be monitored through remote sensing to inform conservation practices (Buskey et al., 1999, 2001; Gobler et al., 2013; Phlips et al., 2015).

Coastal wetlands are similarly difficult to monitor for large-scale slow-moving changes without time-series imagery due to their remote and trackless nature making field assessments time-consuming, expensive, and infrequent. The loss of coastal wetlands has been well-documented throughout the world, but typically in case-study format of localized patterns. To scale up such studies, Giri et al. (2011) mapped global mangrove coverage from the year 2000 using 1000 Landsat images, and noted that only 6.9% are located in protected areas. While Landsat arguably offered the best opportunity for global mapping in the 2010s, Giri (2023) reviewed the state of the art in global mangrove forest monitoring and concluded that cloud computing and the possibility of mapping at resolutions better than 5 m at scale represent the next frontier in the field. To that end, Lizcano-Sandoval et al. (2024) used over 10,000 2-meter resolution commercial satellite images to map wetlands across the entire US Gulf of Mexico coastline to 50 km inland and predict future changes with sea-level rise data from the region. Processing the unprecedented volume of very-high-resolution imagery required the development of automated algorithms run on high-performance computing infrastructure. By leveraging commercial imagery, the authors produced maps with 225 times the resolution of Landsat imagery (i.e., 4 vs. 900 m^2 pixels), and combined the new maps with matching-resolution digital elevation models to overlay the land cover classes and predict change out to 2100 based on local rates of sea-level rise. The results are made freely available to the public via the Google Earth Engine portal and could be used to guide targeted conservation priorities for each state. The project demonstrated that (1) such volumes of data could be acquired for free with licensing through government funding, (2) Big Data solutions exist for processing such data and extracting unprecedented insight, and (3) the results may be openly shared through interactive services for greater impact.

Marine pollution from oil spills, floating plastics, fertilizers and other sources, is an increasing threat to the oceans. Remote sensing technology can detect and track pollutants, which has facilitated responses to the Deepwater Horizon oil spill, fertilizer-induced dead zones, and monitoring of harmful algal blooms. Many satellite sensors have been used successfully for identifying oil slicks, plastic marine debris, and fishing nets with optical and SAR imagery data (Ma et al., 2023).

3.3 SDG 15 applications

SDG 15 aims to "protect, restore, and promote the sustainable use of terrestrial ecosystems, sustainably manage forests, combat desertification, and halt and reverse land degradation and biodiversity loss" (United Nations, 2023). Remote sensing technologies have become instrumental in observing land changes. When combined with advanced techniques like machine learning, these technologies enable the comprehensive monitoring of large ecosystems, such as forests, to identify disturbances like deforestation and fires.

A study by Khan et al. (2017) demonstrated the use of incomplete Landsat imagery and a Deep Convolutional Neural Network to detect changes in forest land-cover in Australia. The study focused on two forest plots within Avon Wilderness Park and Mount Buffalo National Park in Melbourne, Victoria. Remote sensing data were collected to study forest changes caused by harvests and fire incidents. This method successfully detected changes in forest land-cover at a fine temporal resolution, distinguishing local features in forest land cover from the given reflectance data. The results can be used to precisely identify change regions and accurately predict the timing of these changes, enabling routine monitoring and mitigation of land cover changes (Khan et al., 2017). Remote sensing imagery can identify patterns and rates of deforestation and be integrated into models to evaluate the biophysical and socioeconomic drivers behind deforestation (Chowdhury, 2006; Frohn et al., 1996). For example, studies in Central and South America (Mexico and Brazil) used Landsat data to analyze spatial and temporal deforestation patterns, linking these changes to factors such as farming. These models evaluated various variables relevant to deforestation outcomes. In Brazil, a Dynamic Ecological-Land Tenure Analysis (DELTA) model was created from remote sensing data for the Amazon Basin to simulate land-cover patterns induced by social change, revealing changes attributable to migration policy (Frohn et al., 1996). Similarly, in Mexico, agricultural expansion was identified as a key driver of deforestation (Chowdhury, 2006).

Remote sensing is also vital for monitoring phenological variability across landscapes. Norman et al. (2017) utilized a MODIS time series of NDVI to assess phenological changes in the Great Smoky Mountain National Park (GSMNP). This study found that elevation had the strongest impact on phenology, with the time series identifying lags in phenological variability and highlighting the impact of biotic and abiotic disturbances on the landscape. Understanding these variations is crucial for managing biodiversity hotspots vulnerable to extreme changes.

Monitoring vegetation health in national parks has been another significant application of remote sensing. In Southern African national parks, studies have used parameters like NDVI and Directional Persistence to track vegetation changes. Directional Persistence identifies statistically significant changes in vegetation cover at the pixel level, providing a nuanced understanding of vegetation trends (Waylen et al., 2014). Research in Gorongosa National Park and the Greater Limpopo Transfrontier Conservation Area (GLTFCA) found that a decline in precipitation led to decreased NDVI and overall vegetation health (Herrero, Southworth, et al., 2020; Herrero, Waylen, et al., 2020, 2023). A broader study of 79 Southern African national parks using MODIS data identified positive persistence as a dominant trend, which supports findings of global greening with significant implications for biodiversity and biogeochemical cycles (De Jong et al., 2011; Herrero, Southworth, et al., 2020; Herrero, Waylen, et al., 2020, 2023; Olsson et al., 2005; Zhang et al., 2015; Zhu et al., 2016). However, a large browning hotspot in southeastern Africa was also noted, highlighting the need for enhanced vegetation health efforts through collaboration between managers and community partners.

Satellite remote sensing is also effective in evaluating the effectiveness of protected areas. In Nahuel Huapi National Park (NHNP) in Argentina, various remote sensing data products from Google Earth Engine, including Landsat data, were used to evaluate vegetation health and the effectiveness of the national park from 2000 to 2020. This analysis identified greening patterns related to habitat degradation, fragmentation, wildfires, and extreme weather events, providing valuable insights for conservation strategies (Rivarola et al., 2022).

Despite the advantages of remote sensing, big data, and GeoAI for SDG 15, challenges remain. Reith et al. (2021) assessed land degradation in Tanzania using Landsat time series and other land cover data. With over half of the country showing signs of land degradation, the study emphasized the need for continuous monitoring and the integration of diverse data sources to effectively address sustainability challenges. Overall, the integration of remote sensing, big data, and GeoAI is transforming environmental monitoring and conservation efforts. These technologies enable the development of more effective strategies to protect and restore terrestrial ecosystems, significantly contributing to the achievement of SDG 15. Future research should address existing challenges and explore innovative applications to enhance the impact of these technologies on global conservation efforts.

4. Future directions

The future of remote sensing and GeoAI for conservation lies in the successful and efficient utilization of better data and computing resources for improved decision-making. Conservation efforts tend to suffer from insufficient data with which to make decisions, but as we've discussed, space-based observations can supplement or even replace in-situ observations with ever-improving spatial, spectral, temporal and radiometric resolution. As space is increasingly commercialized, opportunities for the remote sensing community to leverage rapidly advancing technology for conservation purposes abound. In the last year alone, several companies have launched and begun distributing data from the first commercial high-resolution hyperspectral satellites (e.g., Planet's Tanager, Orbital Sidekick's GHOSt, and Pixxel's Firefly), which offer 8–30-meter resolution spectral information across hundreds of channels with up to daily repeat acquisitions. Planet has freely distributed dozens of Tanager images to the public (planet.com/data/stac/browser/tanager-core-imagery) (Protected Planet, 2023), and programs like NASA's Commercial Satellite Data Acquisition (CSDA) offer free access to commercial data as a "cost-effective means of supplementing Earth observations collected by NASA, other U.S. Government agencies, and international collaborators" (earthdata.nasa.gov/about/csda). Additionally, open-source missions like NASA's PACE offer hyperspectral satellite data—albeit at much coarser resolution—with high temporal resolution that may compliment commercial data through multimodal or sensor-fusion approaches. Successful acceleration of conservation activities will hinge on how well the research and implementation communities can adopt these new technologies and produce much needed actionable insight from the enhanced information they may provide.

5. Conclusion

Conservation of global anthropogenic and natural ecosystems is at a critical point with climate changes accelerating and time running out to achieve the SDG goals by 2030. Remote sensing could play a critical role in monitoring the Earth for SDG indicators, given that 30 out of 231 social-ecological indicators are directly or indirectly supported by remote sensing (Estoque, 2020). Satellite observations also already contribute to early warning systems for natural disasters, such as wildfires, and monitor the effectiveness of conservation efforts (Barmpoutis et al., 2020). The research described here demonstrates that the ability to routinely monitor, measure, and evaluate vegetation health, water quality and condition, and climate variables at large spatial and fine temporal scales can only be feasibly accomplished with remote sensing technologies.

Additionally, meeting SDG goals and otherwise efficiently enhancing our capacity to conserve natural systems should leverage the opportunities already afforded by protected areas, which serve as a buffer against changing conditions and could play a key role in climate mitigation (Wang et al., 2020). Protected areas are pursuing efforts to reduce emissions as they transition from fossil fuel infrastructure to renewable energy sources, reduce energy consumption, and replace internal combustion engine fleets with electric vehicles (Steuer, 2010; Villalba et al., 2013). Maintaining protected areas and ideally expanding them would provide a thoroughfare to carbon sequestration by exploiting the sequestration capacity of soils, vegetation, the oceans themselves, and the microorganisms residing within (Melillo et al., 2016).

Challenges remain, however, with remote sensing tools as used for conservation goals (Prince, 2019; Reith et al., 2021). Prince (2019) discusses land degradation as a metric for evaluating SDG 15 progress specifically, but states that there is currently no reference from which to relate goal progress to human livelihoods as the current relationship between remotely sensed vegetation indices and ecological services that determine sustainable development omit steps that would establish this connection (Prince, 2019). While remote sensing has demonstrated capacity to aid in the evaluation of SDG progress, there is still work to be done in expanding the analysis to how the health of life on land is linked with human livelihoods.

Case studies

Case study 1: Microsoft AI for conservation

Microsoft has partnered with the Peace Parks Foundation (PPF), which addresses rhino and wildlife poaching in South Africa, to develop remote sensing systems capable of detecting and assessing poaching threats. Rhino poaching in Africa has reached a crisis point with the continent losing three rhinos daily (over 1000 rhinos are killed each year for their horns). Traditional methods of preventing poaching are hampered by the extensive areas rangers must patrol and the frequent false alarms caused by wildlife triggering fence sensors. To combat this, a partnership has formed between PPF and Ezemvelo Kwa-Zulu Natal Wildlife, utilizing Azure's Cloud Computing and Artificial Intelligence (AI) to support anti-poaching efforts.

Recently, the foundation partnered with Microsoft to introduce autonomous poacher detection as an initial step toward creating a "smart park" in Hluhluwe-iMfolozi, KwaZulu Natal, South Africa. This initiative marked the beginning of a technological transformation in conservation practices. PPF's mission is to restore ecosystems through transfrontier conservation areas in southern Africa. The AI for Good initiative enhances strategic planning and timely decision-making in rhino anti-poaching activities. In 2018, PPF piloted the use of camera trap images to gain real-time insights into poaching activities. The system designed to support this initiative focuses on scalability, handling variable inputs, expanding camera installations, and ensuring quick alert turnaround for rangers. The system also emphasizes fault tolerance, ensuring resilience against intermittent failures, and maintains high standards for privacy and security. The technical solution employs a serverless approach with Azure Functions, which allows for cost-effective scaling. An auto-retry mechanism is built in to ensure fault tolerance through retries. The machine learning deployment involves an object detection model, hosted on Azure Machine Learning, Azure Kubernetes Service (AKS), which provides high scalability and fault tolerance.

PPF has since secured funding from two Microsoft programs. One grant from Tech for Social Impact (TSI) supports the foundation's transition to cloud computing. Additionally, PPF obtained three grants from the AI for Earth program, which provides access, training, and resources for using Microsoft Azure and AI tools to tackle environmental challenges. These grants include Azure credits, enabling the use of cloud and AI technologies. Moreover, PPF received support to advance its Smart Parks system, utilizing Microsoft's AI to analyze images from connected camera traps for autonomous poacher detection.

The foundation has also received grants for two additional projects. The first focuses on conservation-minded agriculture, offering an app that helps farmers adopt best practices to increase yields while minimizing habitat impact. The second grant supports the Tracker Academy, affiliated with the South African College of Tourism (SACT), which operates under the auspices of PPF, to combine AI with indigenous animal tracking knowledge, training the next generation of trackers.

Ultimately, the collaboration has resulted in the development of a scalable and extendable pipeline for real-time alerts on poaching activities. This approach can be applied to other domains requiring stability, such as manufacturing and conservation efforts. The potential of digital transformation through AI is significant, with the ability to address global issues like climate change, sustainable farming, and biodiversity conservation.

> *Microsoft technologies enhance our ability to share data, information, and knowledge, facilitating collaboration among various stakeholders in large ecosystem transfrontier conservation, says Craig Beech, New Technology Manager at Peace Parks Foundation. This collaboration enables us to manage landscapes more effectively and make informed decisions for better conservation outcomes. (The Guardian, 2024; Microsoft ISE Developer, 2019; Microsoft News Center, 2018; Microsoft News Center, 2021).*

Case study 2: Identifying Sahel vegetation drivers

The Sahel region bounded by 20°W to 40°E and from 10°N to 20°N and stretching across the southern boundary of the Sahara Desert is an ecologically sensitive area characterized by semi-arid conditions and a significant reliance on seasonal rainfall for vegetation growth. This region was the focus of one of the first field studies and applications of vegetation index "big data" derived from the Advanced Very High-Resolution Radiometer (AVHRR) instrument on-board the NOAA Polar Operational Environmental Satellites series (Tucker et al., 1986, 1994). The region experienced extreme droughts between the late 1960s and early 1980s, raising concerns about desertification and long-term ecological degradation (Lamb, 1982; Nicholson, 1979; Tucker & Nicholson, 1999). Monitoring this vast and challenging environment required comprehensive tools to track land and vegetation changes. Traditional ground-based methods were inadequate for capturing regional-scale trends, necessitating the integration of remote sensing technologies. NDVI emerged as a critical tool for understanding the environmental variability in the Sahel (Tucker, 1979; Tucker et al., 1986). Assembling NDVI time-series data has created information-rich data records that can be used as proxy for land surface response to precipitation variability (Nicholson, 2005; Olsson et al., 2005). The AVHRR instrument has been a cornerstone of environmental monitoring in the Sahel for decades. AVHRR instruments, carried by the NOAA family of polar-orbiting satellites, have collected data continuously since 1981. Long-term record continuity is achieved by calibrating data products from different AVHRR sensors. One such long-term dataset is the GIMMS NDVI dataset, a coarse resolution (0.0833°), bimonthly NDVI data product with global coverage from 1982 to 2022. This case study focuses on several research applications that utilized GIMMS NDVI data to analyze precipitation and vegetation trends across the Sahel from 1981 to 2012.

The first study (Anyamba & Tucker, 2005) examined the changes in vegetation across the Sahel region over a 23-year period from 1981 to 2003. The study revealed a strong relationship between NDVI and rainfall in the Sahel, showing that vegetation growth was primarily driven by seasonal rainfall. The data indicated two distinct periods: a period of below-average NDVI corresponding to severe drought conditions from 1982 to 1993, and a period of gradually rising NDVI, indicating some recovery in vegetation, from 1994 to 2003. The study concluded that, while there had been some recovery in vegetation since the extreme droughts of the early 1980s, the overall conditions remained below the levels observed during the mid-20th century (Anyamba & Tucker, 2005).

The second study (Herrmann et al., 2005) expanded on this work, focusing on trends from 1982 to 2003, and introduced the concept of effective rainfall—the portion of rainfall utilized by vegetation for growth. The analysis confirmed that rainfall was the dominant factor driving vegetation greenness in the Sahel. Specifically, NDVI correlated best with cumulative rainfall over a 3-month period, highlighting the importance of soil moisture retention in supporting vegetation growth.

Finally, the third study (Anyamba et al., 2014) extended this analysis further to cover 32 years, from 1981 to 2012. Utilizing the GIMMS third generation NDVI ($NDVI_{3g}$) dataset published in 2014, the study confirmed the extreme nature of the drought that affected the Sahel in the 1980s, particularly in 1984, which marked a significant minimum in vegetation cover. Subsequent positive trends in $NDVI_{3g}$ data indicated greening across large parts of the Sahel, particularly in the Western and Central Sahel regions. $NDVI_{3g}$ data were compared to three precipitation datasets: African Rainfall Climatology (ARC), Variability Analyses of Surface Climate Observations (VASClimO) Rainfall, and precipitation data from ground stations in sub-Saharan Africa. The correlation between NDVI and rainfall was found to be significant, with the NDVI effectively capturing the spatial and temporal variations in vegetation related to rainfall.

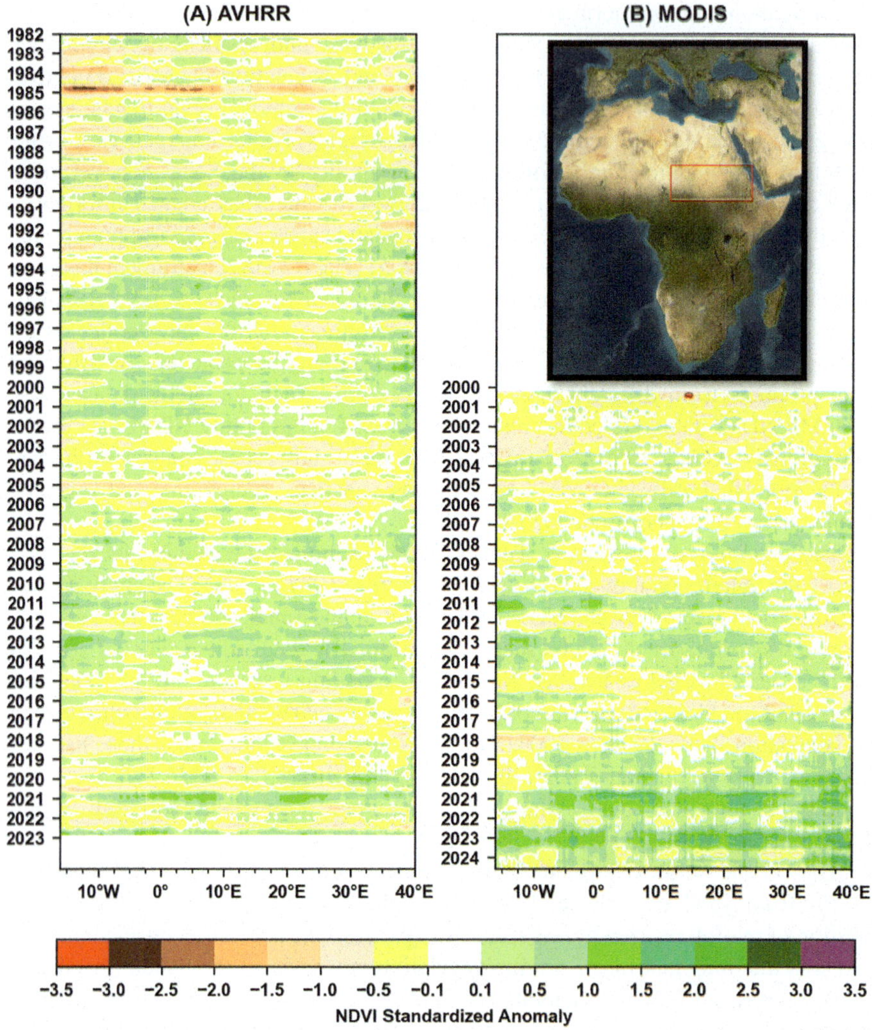

Case Study Figure 2.1 Time-longitude section of monthly NDVI anomaly from January 1982 to July 2024 averaged between 10°N and 20°N and 16°W to 40°E (map inset; ArcMap imagery basemap), expressed as standard deviations from the 2000–20 long-term mean period. (A) AVHRR-derived normalized difference vegetation Index-third generation (NDVI3g) and (B) MODIS/Terra vegetation indices version 6.1 (MOD13C2).

The use of remote sensing big data was key to detecting changes in vegetation across the entire Sahel, and to assessing how these changes correlated with rainfall variability (Case Study Figure 2.1). The analysis not only captured seasonal variations but also long-term shifts that would have been impossible to detect with ground-based observations alone. The ability to capture environmental trends over large areas using remote sensing big data has enabled more effective and timely interventions. For instance, humanitarian organizations can now better predict food security crises based on departures from long-term mean rainfall and vegetation conditions. Additionally, the identification of areas where vegetation recovery has been less pronounced highlights the need for targeted conservation and land management strategies to address local environmental challenges. This large-scale environmental monitoring work continues with utilization of higher resolution big data from the Terra/Aqua MODIS instrument (Fensholt et al., 2009) (Case Study Figure 2.1).

Acronyms

AI	Artificial Intelligence
AKS	Azure Kubernetes Service
ARC	African Rainfall Climatology
AVHRR	Advanced Very High-Resolution Radiometer
DELTA	Dynamic Ecological-Land Tenure Analysis
EO	Earth observation
EVI	Enhanced Vegetation Index
FBI	First bloom
FLI	First leaf
GHG	Greenhouse gas emissions
GIMMS	Global Inventory Modeling and Mapping Studies
GLTFCA	Gorongosa National Park and the Greater Limpopo Transfrontier Conservation Area
GSMNP	Great Smoky Mountain National Park
HABs	Harmful Algal Blooms
IRL	Indian River Lagoon
IUCN	International Union for Conservation of Nature
KED	Kriging technique with external drift
LST	Land Surface Temperature
ML	Mosquito Lagoon
MODIS	Moderate Resolution Imaging Spectroradiometer
MPAs	Marine protected areas
NDVI	Normalized difference vegetation index
$NDVI_{3g}$	GIMMS third generation NDVI
NDVIm	NDVI annual mean
NHNP	Nahuel Huapi National Park
NOAA	National Oceanic and Atmospheric Administration
NPP	Net primary productivity
NPS	National Park Service
PPF	Peace Parks Foundation
RMNP	Rocky Mountain National Park
SACT	South African College of Tourism
SDG	Sustainable Development Goal
TOPS	Terrestrial Observation and Prediction System
TSI	Tech for Social Impact
US	United States
VASClimO	Variability Analyses of Surface Climate Observations

References

Abatzoglou, J. T., & Williams, A. P. (2016). Impact of anthropogenic climate change on wildfire across Western US forests. *Proceedings of the National Academy of Sciences, 113* (42), 11770–11775.

Alcaraz-Segura, D., Cabello, J., Paruelo, J. M., & Delibes, M. (2009). Use of descriptors of ecosystem functioning for monitoring a national park network: A remote sensing approach. *Environmental Management*. https://doi.org/10.1007/s00267-008-9154-y

Anyamba, A., Small, J., Tucker, C. J., & Pak, E. (2014). Thirty-two years of Sahelian zone growing season non-stationary NDVI3g patterns and trends. *Remote Sensing, 6*(4), 3101–3122.

Anyamba, A., & Tucker, C. J. (2005). Analysis of Sahelian vegetation dynamics using NOAA AVHRR NDVI data from 1981-2003. *Journal of Arid Environments, 63*, 596–614.

Avtar, R., Komolafe, A. A., Kouser, A., Singh, D., Yunus, A. P., Dou, J., Kumar, P., Gupta, R. D., Johnson, B. A., Minh, H. V. T., Aggarwai, A. K., & Kurniawan, T. A. (2020). Assessing sustainable development prospects through remote sensing: A review. *Remote Sensing Applications: Society and Environment*. https://doi.org/10.1016/j.rsase.2020.100402

Baird, M. E., Cherukuru, N., Jones, E., Margvelashvili, N., Mongin, M., Oubelkheir, K., Ralph, P. J., Rizwi, F., Robson, B. J., Schroeder, T., Skerratt, J., Steven, A. D. L., & Wild-Allen, K. A. (2016). Remote-sensing reflectance and true colour produced by a coupled hydrodynamic, optical, sediment, biogeochemical model of the Great Barrier Reef, Australia: Comparison with satellite data. *Environmental Modelling and Software, 78*, 79–96. https://doi.org/10.1016/j.envsoft.2015.11.025

Barmpoutis, P., Papaioannou, P., Dimitropoulos, K., & Grammalidis, N. (2020). A review on early forest fire detection systems using optical remote sensing. *Sensors*. https://doi.org/10.3390/s20226442

Berger, K., Machwitz, M., Kycko, M., Kefauver, S. C., van Wittenberghe, S., Gerhards, M., Verrelst, J., Atzberger, C., van der Tol, C., Damm, A., Rascher, U., Herrmann, I., Paz, V. S., Fahrner, S., Pieruschka, R., Prikaziuk, E., Buchaillot, M. L., Halabuk, A., Celesti, M., … Schlerf, M. (2022). Multi-sensor spectral synergies for crop stress detection and monitoring in the optical domain: A review. *Remote Sensing of Environment, 280*, Article 113198. https://doi.org/10.1016/j.rse.2022.113198

Boehm, S., Jeffery, L., Hecke, J., Schumer, C., Jaeger, J., Fyson, C., Levin, K., Nilsson, A., Naimoli, S., Daly, E., Thwaites, J., Lebling, K., Waite, R., Collis, J., Sims, M., Singh, N., Grier, E., Lamb, W., Castellanos, S., … Masterson, M. (2023). *State of climate action 2023. Berlin and Cologne, Germany, San Francisco, CA, and Washington, DC: Bezos Earth fund, climate action tracker, climate analytics, ClimateWorks foundation, New-Climate Institute, the United Nations climate change high-level champions, and world resources institute*. https://doi.org/10.46830/wrirpt.23.00010

Bostater, C. R., & Aziz, S. (2019). Water wave glint corrections, water depth, light attenuation, and WorldView-3 remote sensing algorithms for Indian river lagoon. In *Remote sensing of the ocean, sea ice, coastal waters, and large water regions 2019*. https://doi.org/10.1117/12.2534093

Buskey, E. J., Liu, H., Collumb, C., & Bersano, J. G. F. (2001). The decline and recovery of a persistent Texas brown tide algal bloom in the Laguna Madre (Texas, USA). *Estuaries, 24*, 337–346.

Buskey, E. J., Villareall, T. A., & Lopez-Barreiro, T. (1999). Reconstructing the initiation of the Texas brown tide bloom of Aureoumbra lagunensis from archived samples using an immunofluorescence assay. *Plankton Biology and Ecology, 46*(2), 159–161.

Cabello, J., Fernández, N., Alcaraz, D., et al. (2012). The ecosystem functioning dimension in conservation biology: Insights from remote sensing. *Biodiversity & Conservation, 21*, 3287–3305.

Cabello, J., Lourenço, P., Reyes, A., & Alcaraz-Segura, D. (2013). Ecosystem services assessment of national parks networks for functional diversity and carbon conservation strategies using remote sensing. *Earth Observation of Ecosystem Services, 179*, 180–196.

Chowdhury, R. R. (2006). Driving forces of tropical deforestation: The role of remote sensing and spatial models. *Singapore Journal of Tropical Geography, 27*(1), 82–101. https://doi.org/10.1111/j.1467-9493.2006.00241.x

Cochran, F., Daniel, J., Jackson, L., & Neale, A. (2020). Earth observation-based ecosystem services indicators for national and subnational reporting of the sustainable development goals. *Remote Sensing of Environment, 244*, Article 111796. https://doi.org/10.1016/j.rse.2020.111796

Collados-Lara, A.-J., Fassnacht, S. R., Pardo-Igúzquiza, E., & Pulido-Velazquez, D. (2021). Assessment of high resolution air temperature fields at rocky Mountain National Park by combining scarce point measurements with elevation and remote sensing data. *Remote Sensing, 13*(1). https://doi.org/10.3390/rs13010113

da Silveira, C. B. L., Strenzel, G. M. R., Maida, M., Gaspar, A. L. B., & Ferreira, B. P. (2021). Coral reef mapping with remote sensing and machine learning: A nurture and nature analysis in marine protected areas. *Remote Sensing, 13*, 2907. https://doi.org/10.3390/rs13152907

de Groot, R., Brander, L., van der Ploeg, S., Costanza, R., Bernard, F., Braat, L., Christie, M., Crossman, N., Ghermandi, A., Hein, L., Hussain, S., Kumar, P., McVittie, A., Portela, R., Rodriguez, L. C., ten Brink, P., & van Beukering, P. (2012). Global estimates of the value of ecosystems and their services in monetary units. *Ecosystem Services*. https://doi.org/10.1016/j.ecoser.2012.07.005

De Jong, R., de Bruin, S., de Wit, A., Schaepman, M. E., & Dent, D. L. (2011). Analysis of monotonic greening and browning trends from global NDVI time-series. *Remote Sensing of Environment, 115*, 692–702.

Dieguez, H., & Paruelo, J. M. (2017). Disentangling the signal of climatic fluctuations from land use: Changes in ecosystem functioning in South American protected areas (1982-2012). *Remote Sensing in Ecology and Conservation, 3*(4), 177–189.

Duncanson, L., Liang, M., Leitold, V., Armston, J., Krishna Moorthy, S. M., Dubayah, R., Costedoat, S., Enquist, B. J., Fatoyinbo, L., Goetz, S. J., Gonzalez-Roglich, M., Merow, C., Roehrdanz, P. R., Tabor, K., & Zvoleff, A. (2023). The effectives of global protected areas for climate change mitigation. *Nature Communications*. https://doi.org/10.1038/s41467-023-38073-9

Estoque, R. C. (2020). A review of the sustainability concept and the state of SDG monitoring using remote sensing. *Remote Sensing, 12*(11), 1770.

Fensholt, R., Rasmussen, K., Nielsen, T. T., & Mbow, C. (2009). Evaluation of earth observation based long term vegetation trends—Intercomparing NDVI time series trend analysis consistency of Sahel from AVHRR GIMMS, Terra MODIS and SPOT VGT data. *Remote Sensing of Environment, 113*(9), 1886–1898.

Frohn, R. C., McGwire, K. C., Dale, V. H., & Estes, J. E. (1996). Using satellite remote sensing analysis to evaluate a socio-economic and ecological model of deforestation in Rondônia,

Brazil. *International Journal of Remote Sensing, 17*(16), 3233–3255. https://doi.org/10.1080/01431169608949141

Gillespie, T. W., Ostermann-Kelm, S., Dong, C., Willis, K. S., Okin, G. S., & MacDonald, G. M. (2018). Monitoring changes of NDVI in protected areas of southern California. *Ecological Indicators, 88*, 485–494.

Giri, C. (2023). Frontiers in global mangrove forest monitoring. *Remote Sensing, 15*, 3852. https://doi.org/10.3390/rs15153852

Giri, C., Ochieng, E., Tieszen, L. L., Zhu, Z., Singh, A., Loveland, T., Masek, J., & Duke, N. (2011). Status and distribution of mangrove forests of the world using earth observation satellite data. *Global Ecology and Biogeography, 20*, 154–159.

Gobler, C. J., Koch, F., Kang, Y., Berry, D. L., Tang, Y. Z., Lasi, M., Walters, L., Hall, L., & Miller, J. D. (2013). Expansion of harmful brown tides caused by the pelagophyte, Aureoumbra lagunensis DeYoe et Stockwell, to the US east coast. *Harmful Algae, 27*, 29–41. https://doi.org/10.1016/j.hal.2013.04.004

Gonzalez, P. (2020). Human-caused climate change in United States national parks and solutions for the future. *Parks Stewardship Forum, 36*(2).

Gross, J. E., Nemani, R. R., Turner, W., & Melton, F. (2006). Remote sensing for the national parks. *Park Science, 24*(1), 30–36.

Guardian, The, with Microsoft. “Can AI help endangered species?” Access July 5, 2024. https://www.theguardian.com/advertiser-content/microsoft-ai-for-earth/endangered-species#:~:text=In%20other%20projects%2C%20Microsoft%20has,detect%20and%20evaluate%20poaching%20risks.

Herrero, H. V., Southworth, J., Khatami, R., Insalaco, S., & Muir, C. (2023). Examining the relationship between vegetation decline and precipitation in the national parks of the Greater Limpopo Transfrontier Conservation Area during the 21st century. *Frontiers in Environmental Science, 11*. https://doi.org/10.3389/fenvs.2023.1106849

Herrero, H., Southworth, J., Muir, C., Khatami, R., Bunting, E., & Child, B. (2020). An evaluation of vegetation health in and around Southern African national parks during the 21st century (2000–2016). *Applied Sciences, 10*(7), 2366. https://doi.org/10.3390/app10072366

Herrero, H., Waylen, P., Southworth, J., Khatami, R., Yang, D., & Child, B. (2020). A healthy park needs healthy vegetation: The story of gorongosa National Park in the 21st century. *Remote Sensing, 12*(3), 476. https://doi.org/10.3390/rs12030476

Herrmann, S. M., Anyamba, A., & Tucker, C. J. (2005). Recent trends in vegetation dynamics in the African Sahel and their relationship to climate. *Global Environmental Change, 15*(4), 394–404.

Howard, J., Hoyt, S., Isensee, K., Telszewski, M., & Pidgeon, E. (Eds.). (February 27, 2014). *Coastal blue carbon: Methods for assessing carbon stocks and emissions factors in mangroves, tidal salt marshes, and seagrasses*. CIFOR: International Union for Conservation of Nature (IUCN). https://www.cifor.org/knowledge/publication/5095/.

Insalaco, S., Herrero, H. V., Limber, R., Oliver, C., & Wolfson, W. B. (2024). Monitoring an ecosystem in crisis: Measuring seagrass meadow loss utilizing deep learning in Mosquito Lagoon, Florida. *Photogrammetric Engineering and Remote Sensing*.

Ishtiaque, A., Masrur, A., Rabby, Y. W., Jerin, T., & Dewan, A. (2020). Remote sensing-based research for monitoring progress towards SDG 15 in Bangladesh: A review. *Remote Sensing, 12*(4), 691.

IUCN. (2021). *We need to protect and conserve 30% of the planet: But it has to be the right 30%.* https://www.iucn.org/crossroads-blog/202108/we-need-protect-and-conserve-30-planet-it-has-be-right-30.

IUCN. (2023). https://www.iucn.org/.

Judice, T. J., Widder, E. A., Falls, W. H., Avouris, D. M., Cristiano, D. J., & Ortiz, J. D. (2020). Field-validated detection of Aureoumbra lagunensis brown tide blooms in the Indian river lagoon, Florida, using Sentinel-3A OLCI and ground-based hyperspectral spectroradiometers. *GeoHealth, 4*(6). https://doi.org/10.1029/2019gh000238

Jupp, D. L. B., Mayo, K. K., Kuchler, D. A., Claasen, D. V., Kenchington, R. A., & Guerin, P. R. (1985). Remote sensing for planning and managing the great barrier reef of Australia. *Photogrammetria, 40*(1), 21–42. https://doi.org/10.1016/0031-8663(85)90043-2

Khan, S. H., He, X., Porikli, F., & Bennamoun, M. (2017). Forest change detection in incomplete satellite images with deep neural networks. *IEEE Transactions on Geoscience and Remote Sensing, 55*(9), 5407–5423. https://doi.org/10.1109/tgrs.2017.2707528

Lamb, P. J. (1982). Persistence of Sub-Saharan drought. *Nature, 299*, 46–48.

Liu, G., Heron, S. F., Eakin, C. M., Muller-Karger, F. E., Vega-Rodriguez, M., Guild, L. S., De La Cour, J. L., Geiger, E. F., Skirving, W. J., Burgess, T. F. R., Strong, A. E., Harris, A., Maturi, E., Ignatov, A., Sapper, J., Li, J., & Lynds, S. (2014). Reef-scale thermal stress monitoring of coral ecosystems: New 5-km global products from NOAA coral reef watch. *Remote Sensing*. https://doi.org/10.3390/rs61111579

Liu, G., Strong, A. E., & Skirving, W. (2003). Remote sensing of sea surface temperatures during 2002 barrier reef coral bleaching. *EOS Transactions of the American Geophysical Union, 84*(15), 137–141. https://doi.org/10.1029/2003eo150001

Lizcano-Sandoval, L., Gibeaut, J., McCarthy, M. J., Murray, T., Rueda-Roa, D., & Muller-Karger, F. E. (2024). Vulnerability of wetlands due to projected sea-level rise in the coastal plains of the South and Southeast United States. *Remote Sensing*. https://doi.org/10.3390/rs16122052

Ma, J., Ma, R., Pan, Q., Liang, X., Wang, J., & Ni, X. (2023). A global review of progress in remote sensing and monitoring of marine pollution. *Water, 15*, 3491. https://doi.org/10.3390/w15193491

Magris, R. A., Heron, S. F., & Pressey, R. L. (2015). Conservation planning for coral reefs accounting for climate warming disturbances. *PLoS One, 10*(11), Article e0140828. https://doi.org/10.1371/journal.pone.0140828

Maina, J. M., Jones, K. R., Hicks, C. C., McClanahan, T. R., Watson, J. E. M., Tuda, A. O., & Andrefouet, S. (2015). Designing climate-resilient marine protected area networks by combining remotely sensed coral reef habitat with coastal multi-use maps. *Remote Sensing*. https://doi.org/10.3390/rs71215849

Marzeion, B., Cogley, J. G., Richter, K., & Parkes, D. (2014). Attribution of global glacier mass loss to anthropogenic and natural causes. *Science, 345*(6199), 919–921.

McMenamin, S. K., Hadly, E. A., & Wright, C. K. (2008). Climatic change and wetland desiccation cause amphibian decline in Yellowstone National Park. *Proceedings of the National Academy of Sciences, 105*(44), 16988–16993.

Melendez-Pastor, I., Navarro-Pedreño, J., Gómez, I., & Koch, M. (2010). Detecting drought induced environmental changes in a Mediterranean wetland by remote sensing. *Applied Geography, 30*(2), 254–262. https://doi.org/10.1016/j.apgeog.2009.05.006

Melillo, J. M., Lu, X., Kicklighter, D. W., Reilly, J. M., Cai, Y., & Sokolov, A. P. (2016). Protected areas' role in climate-change mitigation. *Ambio, 45*, 133–145.

Microsoft ISE Developer. (2019). *Preventing rhino poaching with Microsoft azure*. https://devblogs.microsoft.com/ise/preventing-rhino-poaching-though-microsoft-azure/. (Accessed 5 July 2024).

Microsoft News Center. (June 21, 2018). "*Finding solutions to global issues: How Microsoft is empowering nonprofits in South Africa to make a greater impact*". https://news.microsoft.com/en-xm/2018/06/21/finding-solutions-to-global-issues-how-microsoft-is-empowering-non-profits-in-south-africa-to-make-a-greater-impact/. (Accessed 5 July 2024).

Microsoft News Center. (May 14, 2021). *Conservationists turn to data and AI to protect Africa's wildlife*. https://news.microsoft.com/en-xm/features/conservationists-turn-to-data-and-ai-to-protect-africas-wildlife/. (Accessed 5 July 2024).

Monahan, W. B., Rosemartin, A., Gerst, K. L., Fisichelli, N. A., Ault, T., Schwartz, M. D., Gross, J. E., & Weltzin, J. F. (2016). Climate change is advancing spring onset across the US national park system. *Ecosphere, 7*(10), Article e01465.

National Park Service Organic Act. (1916). *The US code. 16 U.S.C. sections 1-4*.

Nicholson, S. E. (1979). Revised rainfall series for the West African subtropics. *Monthly Weather Review, 107*, 473–487.

Nicholson, S. (2005). On the question of the "recovery" of the rains in West African Sahel. *Journal of Arid Environments, 63*, 615–641.

Norman, S., Hargrove, W., & Christie, W. (2017). Spring and autumn phenological variability across environmental gradients of great smoky Mountains national park, USA. *Remote Sensing, 9*(5), 407. https://doi.org/10.3390/rs9050407

O'Leary, D. S., Kellermann, J. L., & Wayne, C. (2018). Snowmelt timing, phenology, and growing season length in conifer forests of Crater Lake national park, USA. *International Journal of Biometeorology, 62*(2), 273–285.

Olsson, L., Eklundh, L., & Ardö, J. (2005). A recent greening of the Sahel—trends, patterns and potential causes. *Journal of Arid Environments, 63*, 556–566.

Pederson, G. T., Gray, S. T., Woodhouse, C. A., Betancourt, J. L., Fagre, D. B., Littell, J. S., & Graumlich, L. J. (2011). The unusual nature of recent snowpack declines in the North American Cordillera. *Science, 333*(6040), 332–335.

Pettorelli, N., Chauvenet, A. L. M., Duffy, J. P., Cornforth, W. A., Meillere, A., & Baillie, J. E. M. (2012). Tracking the effect of climate change on ecosystem functioning using protected areas: Africa as a case study. *Ecological Indicators, 20*, 269–276.

Phlips, E. J., Badylak, S., Lasi, M. A., Chamberlain, R., Green, W. C., Hall, L. M., Hart, J. A., Lockwood, J. C., Miller, J. D., Morris, L. J., & Steward, J. S. (2015). From red tides to green and brown tides:bloom dynamics in a restricted subtropical lagoon under shifting climatic conditions. *Estuaries and Coasts, 38*, 886–904. https://doi.org/10.1007/s12237-014-9874-6

Prince, S. D. (2019). Challenges for remote sensing of the sustainable development goal SDG 15.3.1 productivity indicator. *Remote Sensing of Environment, 234*, Article 111428. https://doi.org/10.1016/j.rse.2019.111428

Protected Planet. (2023). *Discover the world's protected and conserved areas*. https://www.protectedplanet.net/en.

Rannow, S., Macgregor, N. A., Albrecht, J., Crick, H. Q., Förster, M., Heiland, S., & Sienkiewicz, J. (2014). Managing protected areas under climate change: Challenges and priorities. *Environmental Management, 54*, 732–743.

Reith, J., Ghazaryan, G., Muthoni, F., & Dubovyk, O. (2021). Assessment of land degradation in Semiarid Tanzania—using multiscale remote sensing datasets to support sustainable development goal 15.3. *Remote Sensing, 13*(9), 1754. https://doi.org/10.3390/rs13091754

Remelgado, R., Conrad, C., & Meyer, C. (2023). *Limitations in historical satellite archives bias SDG monitoring*.

Rivarola, M. D., Dein, J., Simberloff, D., & Herrero, H. V. (2022). Assessing protected area zoning effectiveness with remote sensing data: The case of Nahuel Huapi national park, Argentina. *Frontiers in Remote Sensing, 3*. https://doi.org/10.3389/frsen.2022.901463

Steuer, C. (2010). Climate friendly parks: Performing greenhouse gas inventories at US national parks and implications for public sector greenhouse gas protocols. *Applied Geography, 30*(4), 475–482.

Swanson, D. K. (2017). Trends in greenness and snow cover in Alaska's Arctic national parks, 2000–2016. *Remote Sensing, 9*(6). https://doi.org/10.3390/rs9060514

Tucker, C. J. (1979). Red and photographic infrared linear combinations for monitoring vegetation. *Remote Sensing of Environment, 8*, 127–150.

Tucker, C. J., Justice, C. O., & Prince, S. D. (1986). Monitoringthe grasslands of the Sahel 1984–1985. *International Journal of Remote Sensing*, 1571–1581.

Tucker, C. J., Newcomb, W. W., & Dregne, H. E. (1994). AVHRR data sets for determination of desert spatial extent. *International Journal of Remote Sensing*, 3547–3565.

Tucker, C. J., & Nicholson, S. E. (1999). Variations in the size of the Sahara desert from 1980 to 1997. *Ambio, 28*, 587–591.

United Nations. (2023). *Sustainable development goal 15*. https://sdgs.un.org/goals/goal15.

Villalba, G., Tarnay, L., Campbell, E., & Gabarrell, X. (2013). A life-cycle carbon footprint of Yosemite national park. *Energy Policy, 62*, 1336–1343.

Wang, Y., Lu, Z., Sheng, Y., & Zhou, Y. (2020). Remote sensing applications in monitoring of protected areas. *Remote Sensing, 12*(9), 1370.

Waylen, P., Southworth, J., Gibbes, C., & Tsai, H. (2014). Time series analysis of land cover change: Developing statistical tools to determine significance of land cover changes in persistence analyses. *Remote Sensing, 6*(5), 4473–4497.

Williams, A. P., Cook, E. R., Smerdon, J. E., Cook, B. I., Abatzoglou, J. T., Bolles, K., … Livneh, B. (2020). Large contribution from anthropogenic warming to an emerging North American megadrought. *Science, 368*(6488), 314–318.

Wirt, K. E., Hallock, P., Palandro, D., & Lunz, K. S. (2015). Potential habitat of *Acropora* spp. on reefs of Florida, Puerto Rico, and the US Virgin Islands. *Global Ecology and Consevation, 3*. https://doi.org/10.1016/j.gecco.2014.12.001

Wulder, M. A., Roy, D. P., Radeloff, V. C., Loveland, T. R., Anderson, M. C., Johnson, D. M., Healey, S., Zhu, Z., Scambos, T. A., Pahlevan, N., Hansen, M., Gorelick, N., Crawfod, C. J., Masek, J. G., Hermosilla, T., White, J. C., Belward, A. S., Schaaf, C., Woodcock, C. E., … Cook, B. D. (2022). Fifty years of landsat science and impacts. *Remote Sensing of Environment, 280*, Article 113195. https://doi.org/10.1016/j.rse.2022.113195

Zhang, K., Kimball, J. S., Nemani, R. R., Running, S. W., Hong, Y., Gourley, J. J., & Yu, Z. (2015). Vegetation greening and climate change promote multidecadal rises of global land evapotranspiration. *Scientific Reports, 5*, Article 15956.

Zhu, Z., Piao, S., Myneni, R. B., Huang, M., Zeng, Z., Canadell, J. G., Ciais, P., Sitch, S., Friedlingstein, P., Arneth, A., et al. (2016). Greening of the Earth and its drivers. *Nature Climate Change, 6*, 791–795.

Further reading

https://www.oceanmind.global/.

https://news.microsoft.com/en-xm/features/conservationists-turn-to-data-and-ai-to-protect-africas-wildlife/.

https://planet.com.

Disaster Management and Remote Sensing Applications

11

Kevin D. Ash
Department of Geography, University of Florida, Gainesville, FL, United States

Key terms and definitions

Disaster: Socially constructed phenomenon stemming from the intersection of a hazard event with exposed and/or vulnerable communities; disaster events overwhelm local capacities to respond and occur with significant negative societal impacts.
Disaster Management: systematic efforts to reduce disaster risks, manage hazard impacts, and support recovery, involving coordinated planning and actions by local and regional governments, non-governmental organizations, international bodies, and local communities.
Hazard: a potentially damaging physical event, phenomenon, or human activity that may cause the loss of life, injury, property damage, social and economic disruption, or environmental degradation.
Resilience: the capacity of individuals, communities, institutions, and systems to anticipate, absorb, adapt to, and recover from the effects of hazardous events in a timely and efficient manner. Resilience emphasizes not only the ability to "bounce back" to pre-disaster conditions, but also the potential to adapt and improve systems and behaviors to reduce future risk.
Risk: the probability of harmful consequences resulting from the interaction between a hazard and the conditions of exposure and vulnerability. It incorporates both the likelihood of the hazard occurring and the potential severity of its impact.
Sendai Framework: framework published by the United Nations that provides a comprehensive global blueprint that emphasizes understanding disaster risk, strengthening disaster governance, investing in risk reduction, and enhancing preparedness for effective response and long-term improvements through recovery, rehabilitation, and reconstruction.
Vulnerability: the characteristics and circumstances of individuals, communities, or systems that contribute to their susceptibility to the damaging effects of a hazard. These may include lower income levels, lack of access to services, poorly designed or maintained infrastructure, or social marginalization.

1. Introduction

In recent decades, numerous disasters have occurred globally in association with natural hazards, including earthquakes, floods, landslides, wildfires, and tropical cyclones, focusing attention on the need for improved disaster management strategies. Rapid

Remote Sensing, Big Data, and GeoAI. https://doi.org/10.1016/B978-0-443-26741-3.00017-9

urbanization, environmental change, and population growth in high-risk regions have all contributed to increasing risk with more people and structures exposed, and with negative impacts often affecting those in the most vulnerable situations (Sterzel et al., 2020). Concurrently, the technological tools that can assist places to prepare for, respond to, and recover from disasters have also continued to develop. There are now more ways to gather large volumes of disaster-related data than ever before (Yu et al., 2018). In this context, remote sensing has emerged as one of the most valuable technologies available to hazard scientists, emergency managers, and policymakers.

Remote sensing (RS) refers to the collection of data about Earth's surface and atmosphere from a distance—typically through satellites, aircraft, or drones—without direct physical contact (Chuvieco, 2020). These technologies provide critical information about hazardous conditions and their aftermath, often in real time or near-real time. RS platforms can monitor rainfall patterns that lead to flooding, detect vegetation stress during drought, track changes in the land surface associated with seismic, volcanic, or other geophysical processes, and provide critical data for many other applications (AghaKouchak et al., 2015; Joyce et al., 2009; Tralli et al., 2005). This capability makes it possible for decision-makers to identify risks, model potential impacts, and mobilize resources quickly and more effectively than ever before (Kaku, 2019; Voigt et al., 2016).

At the same time, the field of disaster management has evolved from a reactive model—focused mostly on post-event emergency response—to a more holistic risk and vulnerability reduction approach that includes prevention, mitigation, preparedness, and long-term recovery (Bosher, 2014; Palliyaguru et al., 2014). This shift has been enabled by technological innovation in the ability to transmit signals rapidly over large distances, for both communication and earth observation purposes. Today, decision-making during disasters increasingly relies on high-resolution satellite imagery, active sensors such as radar and LiDAR (light detection and ranging) for rapid scanning and mapping, increasingly cost-effective unoccupied aerial vehicles (UAVs), and an expanding array of multispectral and hyperspectral sensors and high-performance computer systems (Zhang et al., 2022). These tools allow for detailed mapping and monitoring across a wide range of hazards and geographic contexts, supporting everything from evacuation planning (You et al., 2023) to rescue operations (Lyu et al., 2023), and recovery planning (Marlier et al., 2022).

This chapter summarizes the burgeoning applications of RS used in the study and management of disasters associated with earth system processes and hazards, highlighting how various technologies are used across different types of geophysical, meteorological, and hydrological hazards. By exploring the application of RS tools to real-world disasters, readers will gain a basic understanding of how data-driven insights help reduce risk, save lives, and build more resilient communities.

2. Theoretical foundations

A hazard refers to a physical event, phenomenon, or human activity that could cause death, injury, property or environmental damage, social disruption, and economic loss

(De Angeli et al., 2022). Hazards can be associated with the Earth's systems, or they can be entirely anthropogenic, though often they occur due to complex Earth-human systems interactions (Gill & Malamud, 2017). Common examples of natural hazards include geophysical events like earthquakes, volcanic eruptions, and landslides, as well as meteorological and hydrological phenomena such as floods, droughts, tropical cyclones, and wildfires. Importantly, a hazard is not inherently a disaster—it is a source of potential harm. A disaster, by extension, occurs when a hazard interacts with an exposed population, with some locations and population segments more exposed than others and/or being less able to reduce impacts due to intersecting vulnerability factors (Cutter, 2024). Such disaster events can overwhelm local capacities to respond and result in significant negative societal impacts. However, researchers and practitioners now understand disasters mainly as social phenomena (Perry, 2018). This distinction emphasizes that disasters are not an inevitable result of society-Earth system interactions but occur primarily within, and because of, the social, economic, and political contexts in which people live (Chmutina & Von Meding, 2019).

Two closely related terms, risk and vulnerability, help clarify this relationship further. Risk is the probability of harmful consequences resulting from the interaction between a hazard and the conditions of exposure and vulnerability (Zuzak et al., 2022). It incorporates both the likelihood of the hazard occurring and the potential severity of its impact. Risk increases when hazards are frequent or intense, when populations are highly exposed (e.g., living in floodplains or seismic zones), and when communities are vulnerable due to social, economic, or infrastructural factors. Vulnerability refers to the characteristics and circumstances of individuals, communities, or systems that make them susceptible to the damaging effects of a hazard (Cutter, 2024; Wisner, 2016). These may include lower income levels, lack of access to services, poorly designed or maintained infrastructure, or social marginalization. Understanding these concepts is essential for effective disaster risk reduction, as they enable scientists, planners, and disaster managers to assess where and why impacts are likely to be most severe—and to prioritize mitigation and preparedness efforts accordingly.

Building on the concepts of hazard, risk, and vulnerability, the notion of resilience is central to understanding how communities can withstand and recover from disaster events. Resilience refers to the capacity of individuals, communities, institutions, and systems to anticipate, absorb, adapt to, and recover from the effects of hazardous events in a timely and efficient manner (Cutter et al., 2013). This concept emphasizes not only the ability to "bounce back" to pre-disaster conditions, but also the potential to adapt and improve systems and behaviors to reduce future risk (Cutter, 2016). Enhancing resilience involves reducing exposure and addressing challenges experienced by those in vulnerable situations, improving response mechanisms, and fostering adaptive capacities across multiple societal domains. While many RS studies have contributed to modeling and understanding of risk and vulnerability, there remain many opportunities for RS researchers to contribute to assessments of resilience (Ghaffarian et al., 2018).

This brings us to the field of disaster management, which encompasses the systematic efforts to reduce disaster risks, manage hazard impacts, and support recovery.

Disaster management involves coordinated planning and actions by local and regional governments, non-governmental organizations, international bodies, and local communities (Alexander, 2007). In the United States, the Federal Emergency Management Agency (FEMA) began spearheading a national institutional approach in the 1970s, supporting preparedness and response across states and counties (Phillips et al., 2021). Internationally, the Sendai Framework for Disaster Risk Reduction 2015–30 (Fig. 11.1), led by the United Nations Office for Disaster Risk Reduction (UNDRR) (2020, pp. 1–88), provides a comprehensive global blueprint that emphasizes understanding disaster risk, strengthening disaster governance, investing in risk reduction, and enhancing preparedness for effective response and long-term improvements through recovery, rehabilitation, and reconstruction (UNISDR, 2015).

At the heart of many disaster management frameworks lies the disaster management cycle. This model outlines several interrelated and overlapping phases: mitigation (efforts to reduce hazard impacts before they occur), preparedness (planning and capacity building before an event), response (emergency actions taken during and immediately after a disaster), and recovery (longer-term rebuilding and restoration) (Smith et al., 2023). This model illustrates that disaster management is not confined to post-event response but is an ongoing process of risk reduction and adaptation, with each phase influencing others over time. Institutional initiatives such as the United Nations' Early Warnings for All campaign, which aims to ensure timely and accessible hazard alerts for all people, especially in vulnerable regions, underscore the growing emphasis on proactive disaster management (World Meteorological Organization, 2022). Similarly, rapid response programs that deploy personnel, resources, and technology in the critical hours after a disaster are essential for minimizing loss of life and accelerating recovery. These efforts rely increasingly on timely, accurate, and actionable information—an area where RS technologies have become indispensable.

The application of RS in disaster management has evolved significantly since its emergence in the 1970s. Initial efforts focused on proof-of-concept studies using early earth observation satellite platforms like Landsat to detect and monitor environmental changes tied to hazards such as floods, droughts, and volcanic activity (Howard et al., 1978). Researchers and practitioners in disaster-related fields first envisioned how remote sensing technologies could aid local and regional governments in establishing baseline data and then quickly evaluating the extent and severity of damage after an impactful event (Rush & Holguin, 1975). Other early pilot studies focused on how RS technologies could help lessen the impacts of disasters on food and agricultural systems and people dependent on those resources, for example in the Sahel region of Africa (Barrett, 1979). During this early period, imagery from NOAA satellites quickly became integral to global efforts in detection, monitoring, and forecasting for tropical cyclones (Dvorak, 1975). By the 1990s, this exploratory work expanded into operational uses for rapid assessment, especially as spatial and temporal resolutions improved, and the number and variety of sensors increased (Van Westen, 2000). As disasters became more frequent and complex due to global environmental and urbanization trends, the need for geospatially enabled responses increased, paving the way for further institutional support from organizations like the National Aeronautics

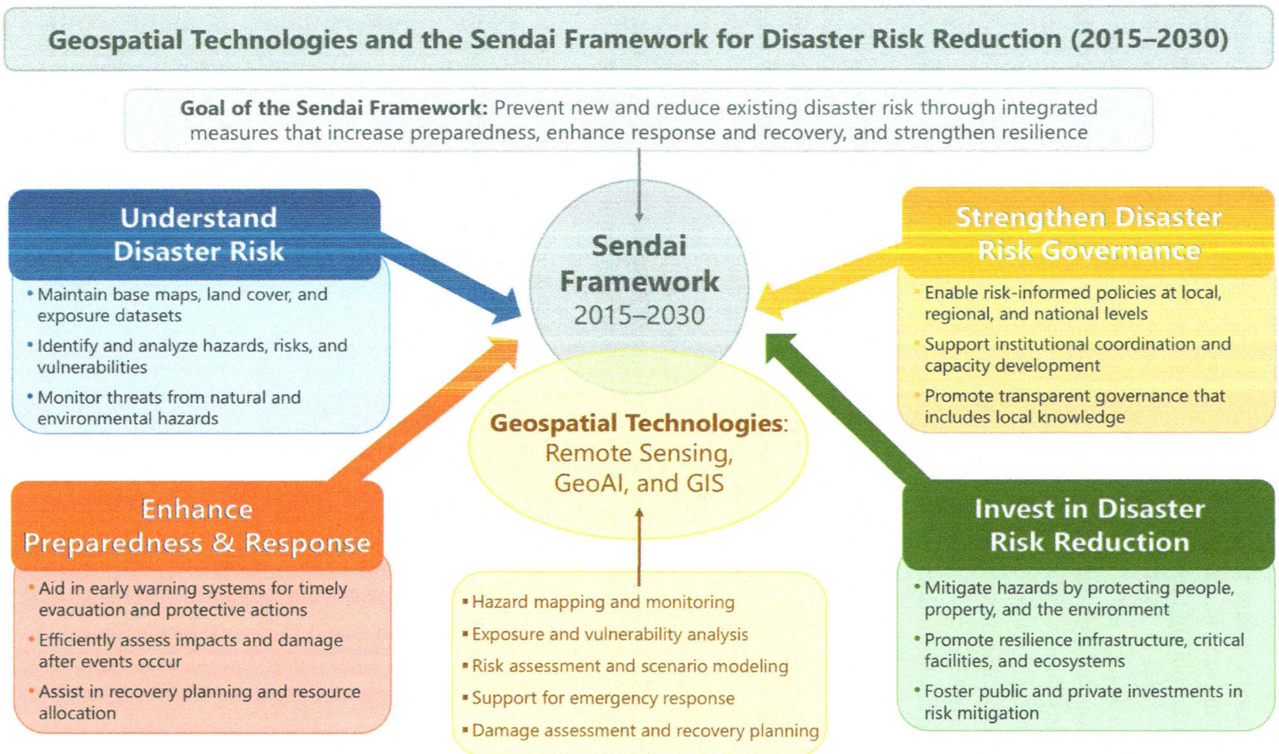

Figure 11.1 Examples of how integration of geospatial technologies supports the goal and four key priorities of the Sendai framework for disaster risk reduction (2015–30).
Adapted by the author from content in the initial Sendai Framework document (United Nations International Strategy for Disaster Reduction (UNISDR). (2015). Sendai framework for disaster risk reduction 2015–2030. UNISDR).

and Space Administration (NASA) in the United States (US), the European Space Agency (ESA), and the United Nations (UN).

Since the 1990s, scholarly focus on RS and disaster management has grown immensely. Based on data from a Web of Science query using the terms "remote sensing" and "disaster management," the number of publications surged from less than five per year during the 1990s to about 40 per year by 2010 to nearly 550 publications on the topic in 2024 alone (Fig. 11.2). This trend reflects both the increasing relevance of hazards and disasters globally, and the expanded availability of computing resources and Earth observation platforms with diverse spatial, spectral, and temporal resolutions. The growth also indicates the mainstreaming of RS within disaster risk science, intersecting with Geographic Information Systems (GIS), real-time alert systems, and more recently with machine learning and Geographic Artificial Intelligence (GeoAI).

There are several prominent institutional programs that support remote sensing for disaster response and humanitarian aid. One example is the United Nations Operational Satellite Applications Program (UNOSAT). UNOSAT offers a centralized system for GIS and RS imagery support to UN agencies and other partners in humanitarian and disaster response (Adriano et al., 2020). UNOSAT cooperates with Space and Major Disasters, an International Charter established in 2000 by major space agencies and industry partners to provide satellite data during major crises, such as recent events including flooding in Nigeria in May/June 2025 and Cyclone Jude in Madagascar and Mozambique in March 2025 (International Charter: Space and Major Disasters, 2025). Initiatives such as the ESA-funded Respond Atlas project aim to extend geospatial tools to the entire humanitarian aid community, especially in regions with limited technological capacity (European Space Agency, 2011). In the US, NASA's Disaster Response Coordination System coordinates geospatial data and information resources across the agency to leverage its RS expertise and catalog in support of disaster management professionals working with communities coping

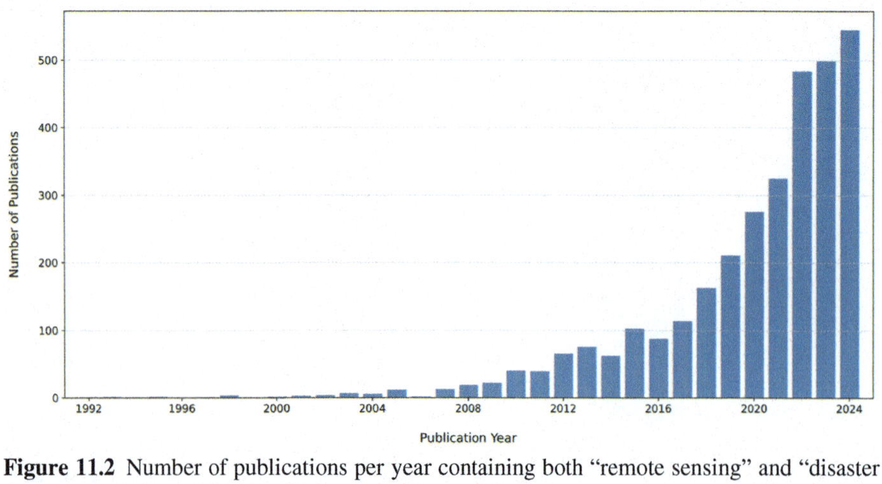

Figure 11.2 Number of publications per year containing both "remote sensing" and "disaster management" from the web of science database, 1992–2024.

with impacts from a range of hazard types, including recent wildfires in Hawaii, windstorms in Texas, and flooding in Brazil (Barnes et al., 2024).

The continued advancement of spaceborne and aerial technologies—combined with web- and cloud-based tools, open data policies, and international frameworks—has contributed to a more widely applicable and efficient model of global disaster risk reduction and response. As we will explore in the remainder of this chapter, RS provides the observational foundation for hazard detection, impact assessment, and response coordination, enhancing all phases of the disaster management cycle and helping societies become more risk-aware, responsive, and resilient.

3. Remote sensing for disaster management of natural hazards

This section focuses on RS applications for natural hazards arising from Earth system processes, including geophysical, meteorological, hydrological, and climate-related hazards. While RS technologies have also been applied to biological, technological, and other anthropogenic or environmental hazards, such applications are beyond the scope of this chapter due to space limitations. Instead, the emphasis here is on hazards such as earthquakes, landslides, droughts, wildfires, tropical cyclones, floods, and others where RS offers critical insights for hazard detection, monitoring, response, and long-term risk reduction. Fig. 11.3 provides a synthesis of the remote sensing platforms and data commonly applied across the disaster types discussed in this section, highlighting areas of methodological overlap as well as hazard-specific applications.

Hazard Types	Remote Sensing Tools						
	Optical (Polar-Orbiting)	SAR	Thermal (Polar-Orbiting)	LiDAR	UAV-Based Imagery	Weather Radar	Geostationary Weather Satellites
Landslides	•	••	○	•	•	○	○
Earthquakes	•	••	○	•	•	○	○
Volcanoes	•	•	••	•	•	○	•
Droughts	••	•	•	○	•	○	••
Wildfires	•	○	••	•	•	•	••
Floods	•	••	○	•	•	••	•
Tropical Cyclones	•	•	•	○	○	•	••
Severe Thunderstorms	•	○	○	○	•	••	•
Tsunamis	•	•	○	•	○	○	○
Snow/Ice/Avalanche	•	••	•	•	•	•	•
Air Quality/Dust	•	○	••	○	○	○	••

•• Primary application • Supporting application ○ Limited or situational application

Figure 11.3 Conceptual overview of the overlap in remote sensing tools used across major natural hazard types discussed in this chapter. The matrix highlights commonalities and differences in the application of optical and thermal imagery from polar orbiting platforms, synthetic aperture radar, LiDAR, UAV-based imagery, and meteorological satellite and ground-based radar remote sensing platforms for disaster monitoring, response, and recovery.

These categories reflect the breadth of RS contributions to disaster management across a wide range of environmental conditions and spatial scales.

3.1 Geohazards

3.1.1 Landslide hazards

RS technologies are indispensable in monitoring and managing landslide hazards, offering tools for monitoring, detecting changes, and preparedness and mitigation for these events. Satellites, UAVs, and ground-based systems form a triad of platforms that address various spatial and temporal scales of landslide activity. Synthetic Aperture Radar (SAR) platforms such as Sentinel-1, Radarsat, and the Advanced Land Observing Satellite (ALOS) Phased Array L-band (PALSAR) have been central to landslide studies (Mondini et al., 2021), with newer SAR platforms such as Capella seeing increased use since 2020 (Prakash et al., 2025). The ability of SAR to measure ground deformation with precision at millimeter-scale makes them critical for identifying slow-moving landslides and areas of slope instability (Tan et al., 2023). Optical sensors, including those aboard Sentinel-2 (Lacroix et al., 2018), Landsat, SPOT (Casagli et al., 2023), and CubeSat constellations (e.g., PlanetScope) (Mazzanti et al., 2020), provide data for detecting vegetation disturbances and surface changes that often portend landslides, or to show the extent of affected areas after landslides occur (as in Fig. 11.4). When combined, radar and optical datasets enhance the

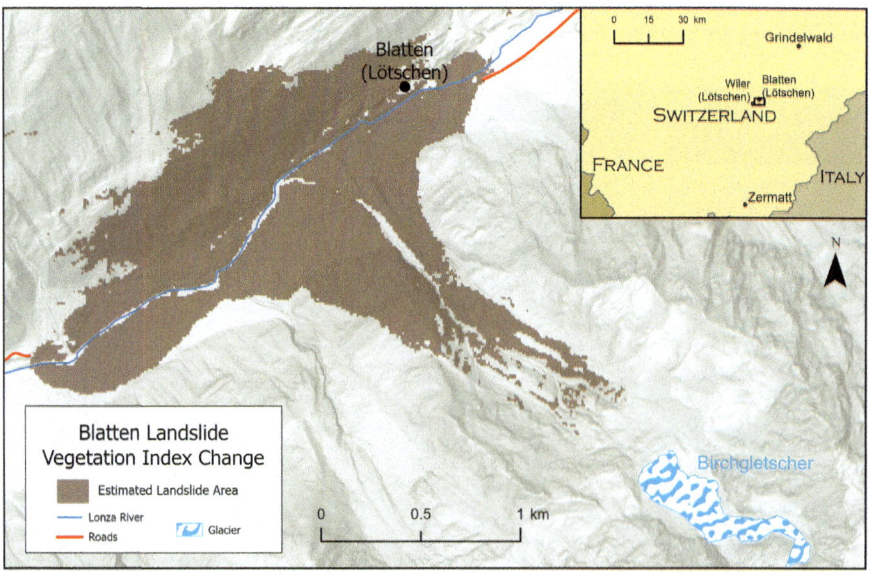

Figure 11.4 Estimated extent of the May 28, 2025, glacier collapse and landslide at Blatten, Switzerland, derived from Sentinel-2 vegetation change detection. Areas in brown indicate where a substantial drop in normalized difference vegetation index (NDVI) occurred immediately after the event, suggesting widespread vegetation loss consistent with slope failure and debris movement.

reliability of landslide detection, the accuracy of risk modeling, and thus the efficacy of short-term warnings and preparedness, as well as longer term landslide mitigation activities.

UAVs are increasingly popular for localized, high-resolution landslide studies. Equipped with photogrammetric and LiDAR sensors, UAVs assist in the generation of detailed 3D models and precise digital elevation models (DEMs) of unstable slopes (Sun et al., 2024). These tools are invaluable for assessing morphological changes and mapping landslide-prone areas. Their rapid deployment capacity makes UAVs particularly effective for post-event analysis and monitoring in inaccessible regions (Schilirò et al., 2025). Geographically, landslide hazards that cause fatalities are concentrated in high-risk areas with favorable factors—steep terrain and frequent seismic and rainfall-induced landslides—including countries in South Asia, Southeast Asia, East Asia, southern Europe, and Central America (Haque et al., 2019).

3.1.2 Earthquake hazards

Earthquake disaster management has benefited significantly from RS technologies, which provide vital tools for monitoring seismic activity, assessing risks, and guiding disaster response. LiDAR collected using aircraft can be particularly useful for producing high-resolution terrain models that reveal structural deformations and fault movements (Scott et al., 2018). SAR platforms, including Sentinel-1 and TerraSAR-X, measure ground deformation with precision through InSAR and Differential InSAR (DInSAR) techniques (Ali et al., 2021; Moro et al., 2017; Suresh & Yarrakula, 2020). Optical sensors aboard satellites like Landsat, MODIS (Moderate Resolution Imaging Spectroradiometer), and ASTER support vegetation monitoring and structural damage assessments (Matin & Pradhan, 2022; Yang & Qi, 2016).

UAVs are invaluable for rapid post-disaster damage evaluations, offering high-resolution imagery to classify building damage and map debris distribution in urban areas (Kerle et al., 2020). Because earthquakes occur with little warning and may cause widespread damage to buildings and infrastructures, limiting the ability to travel by vehicle, UAVs can be essential for timely damage assessment, informing search and rescue efforts and restoration of critical lifelines such as road, utility, and communication networks (Mandirola et al., 2022). Multi-sensor approaches that integrate LiDAR, optical, and SAR data offer comprehensive insights into earthquake impacts with models of landscape and building change (Chiabrando et al., 2017; Li et al., 2024). Geographically, earthquake studies are concentrated in seismic hotspots with greater technical research capacities, such as China, Japan, Iran, Türkiye, and Italy. For example, analysis of pre- and post-earthquake SAR imagery can identify areas where buildings and debris were cleared, and no buildings have yet been rebuilt (Fig. 11.5).

3.1.3 Volcanic hazards

RS has revolutionized volcanic hazard management by providing systematic observations of volcanic activity. Recent volcanic activity in Indonesia, Italy, Iceland, Japan, and Hawaii (US) have been major foci for a growing number of case studies using

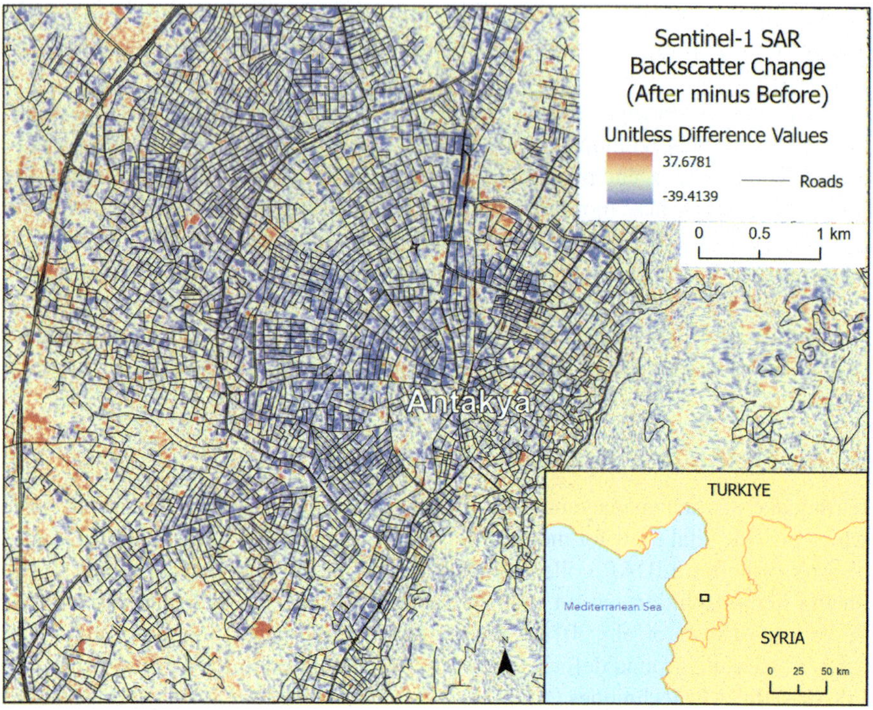

Figure 11.5 Change in radar backscatter intensity in Antakya, Türkiye, derived from Sentinel-1 SAR imagery, showing surface changes following the February 2023 earthquake. Blue areas indicate decreased backscatter, suggesting surface smoothing where buildings likely collapsed or were demolished, and debris was then cleared. Red areas reflect increased backscatter, suggestive of substantial structural disturbance or new buildings placed in previously undeveloped spaces, such as temporary container settlements. The analysis compares SAR acquisitions from just before and approximately 1 year after the earthquake, using the VV polarization band.

RS, and especially UAVs, for disaster management. Optical sensors like Landsat and MODIS monitor thermal anomalies (Coppola et al., 2020; Layana et al., 2020) and ash plumes (Mota et al., 2024). Like other geohazard applications, SAR platforms like Sentinel-1 track ground deformation associated with surface lava flows and subsurface changes in magma volume and location (Valade et al., 2019). Thermal sensors such as MODIS and Visible Infrared Imaging Radiometer Suite (VIIRS) detect radiative fluxes providing global coverage and enhancing the ability to issue volcanic activity alerts (Campus et al., 2022). UAVs provide high-resolution optical imagery, as well as landform change maps using LiDAR data (Gomez et al., 2022; Jordan, 2019; Romàn et al., 2022). Recovery efforts focused on ecosystems are supported by longer term monitoring of geomorphological changes and vegetation regrowth using a mixture of sensors like Sentinel-2 and UAVs (James et al., 2020; Teltscher & Fassnacht, 2018). RS technology has also assisted water quality monitoring and evolution over time in the aftermath of volcanic activity affecting marine environments (Caballero et al., 2022).

3.2 Meteorological and hydrological hazards

3.2.1 Droughts

RS technology is essential for management of meteorological and hydrological hazards, as drought monitoring was one of the earliest disaster management applications of RS technology (Pulwarty & Sivakumar, 2014). Extreme heat episodes and water budget deficits often occur together, especially in summer months. Drought monitoring and management with RS technologies offers critical data on vegetation, soil moisture, and precipitation through a range of advanced sensors and platforms. MODIS and TRMM (Tropical Rainfall Measuring Mission) are widely used, with MODIS tracking vegetation indices like NDVI and Vegetation Condition Index (VCI) (Ha et al., 2023; Wassie et al., 2022) and TRMM provided reliable precipitation estimates for drought monitoring (Brasil Neto et al., 2021), with GPM (Global Precipitation Measurement) providing continued drought coverage after TRMM's mission ended in 2015 (Yu et al., 2022).

Sentinel-1 and Sentinel-2 enhance monitoring capabilities with their frequently used combination of high-resolution radar and optical imagery, supporting soil moisture assessments and detailed vegetation analysis (Bousbih et al., 2018; Urban et al., 2018). Additional widely used spaceborne platforms, such as Landsat (Ghaleb et al., 2015) and geostationary systems like Himawari (Hu et al., 2020) and the Geostationary Operational Environmental Satellites (GOES) series (AghaKouchak et al., 2015), contribute to early warning systems by providing near-real-time passively sensed data. An array of additional technologies, including LiDAR, SAR, UAVs, and CubeSats, enable more localized, higher resolution mapping for discerning hydrological and agricultural drought impacts (Jiao et al., 2021; Su et al., 2019; Van Den Hoek et al., 2019).

3.2.2 Wildfires

Following closely the definition of Tedim and Leone (2020), wildfires refer to unplanned and uncontrolled (initially) combustion of vegetation in open environments, which may be variously called by other names such as bushfires, wildland fires, grassfires, etc. Wildfire causes damage to a range of environments, including sometimes damaging and destroying built environments when they ignite structures in populated regions.

Wildfires are increasingly monitored using advanced RS technologies. Geostationary weather satellites (e.g., the GOES constellation) can detect wildfires in near-real-time using thermal infrared bands (Koltunov et al., 2016), while ground-based weather radar networks aid in fire detection and rapid diagnosis of changes in fire behavior via radar returns from smoke/ash plumes (McCarthy et al., 2019). Platforms like MODIS and Landsat provide critical data for detecting burned areas, monitoring long-term trends in wildfires over large regions, and assessing vegetation recovery (Bright et al., 2019; Giglio et al., 2018; Wang & Zhang, 2020). Sentinel-2's MSI and Sentinel-1's SAR are utilized for high-resolution fire severity mapping and vegetation monitoring (Ban et al., 2020; Gibson et al., 2020) (see example in Fig. 11.6),

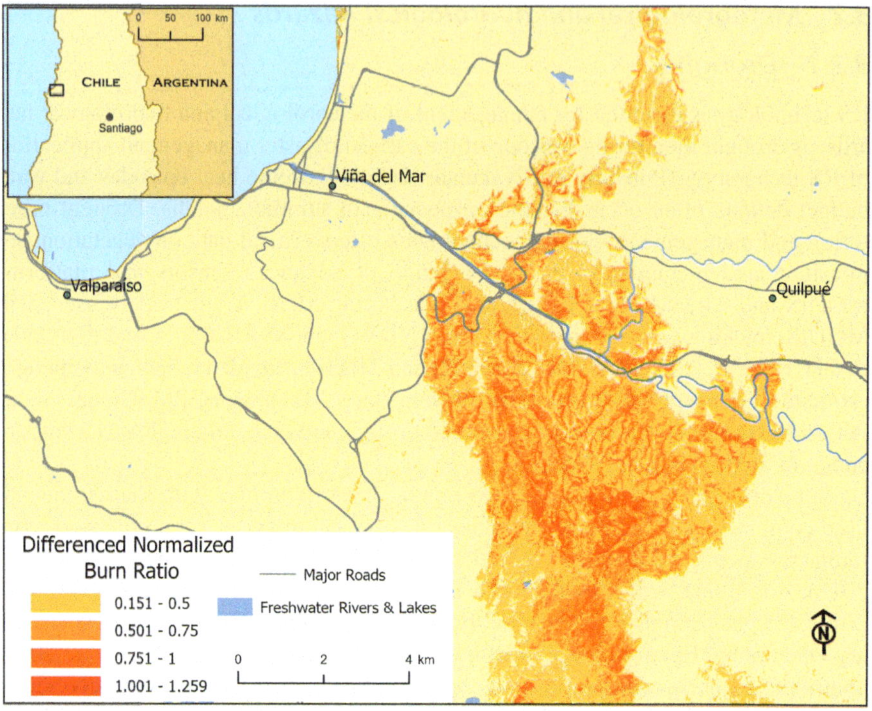

Figure 11.6 Burn severity map of the February 2024 wildfire near Viña del Mar and Quilpué, Chile, derived from Sentinel-2 imagery using the differenced normalized burn ratio (dNBR). The dNBR highlights vegetation loss by comparing near-infrared and shortwave-infrared reflectance before and after the fire. Higher values (orange to red) indicate greater volume of biomass burned.

while higher spatial resolution optical imagery from the WorldView series satellites has allowed identification of individual buildings damaged or destroyed by wildfires (McNamara & Mell, 2022). Recently, high spatial resolution burned area analyses are possible using imagery from a CubeSat constellation such as PlanetScope (Hamilton et al., 2023; Roy et al., 2024). Aerial platforms equipped with thermal sensors, including UAVs, are very useful for near-real-time detection and fire boundary mapping, complementing satellite imagery under cloud cover (Allison et al., 2016). Smaller UAVs are limited in application for rapidly evolving wildfire events accompanied by strong winds, wind shear, and severe air turbulence, because their low mass and relatively slow airspeeds reduce aerodynamic stability and control authority under adverse wind conditions (Wang, Wang, et al., 2019). Limited battery capacity and flight endurance further restrict the ability of UAVs to sustain continuous or repeated observations for complex, rapidly evolving fires (Diaz-Vilor et al., 2025). Also, UAVs flying near active wildfires may be dangerous to aircraft engaging in fire suppression or other critical support activities (Allison et al., 2016).

Emerging technologies, including AI-enabled platforms like Google Earth Engine, enhance wildfire prediction and early warning systems (Tavakkoli Piralilou et al., 2022). The ECOsystem and Spaceborne Thermal Radiometer Experiment on Space Station (ECOSTRESS) from NASA contributes data on ecosystem stress and pre-fire conditions for risk modeling (Pascolini-Campbell et al., 2022). A few countries and regions where wildfire management is especially emphasized include Chile, Australia, and Southeast Asia, with the US and China producing a diverse array of RS applications. Recovery efforts often leverage LiDAR and optical sensors to evaluate structures and biomass loss and monitor ecosystem regeneration (Bolton et al., 2015; Viana-Soto et al., 2022).

3.2.3 Snow, Ice, and Avalanches

RS technologies are essential for snow-related disaster management, providing insights into snow cover, depth, and water equivalent (SWE). The MODIS and VIIRS datasets, when combined, enable global snow cover monitoring over time (Riggs et al., 2017), while SAR platforms like Sentinel-1 and TerraSAR-X excel in wet snow mapping and snow depth estimation (Snapir et al., 2019; Tsai et al., 2019). Advanced Microwave Scanning Radiometer (AMSR-2) and Global Navigation Satellite System (GNSS)-Reflectometry contribute to SWE assessments, enhancing snowpack modeling (e.g., Hu et al., 2024; Ma et al., 2022).

Avalanche management employs SAR for snowpack and terrain monitoring, while optical sensors like Landsat and SPOT assist in hazard mapping with digital terrain models (Denissova et al., 2024; Hafner et al., 2022). Machine learning-generated susceptibility maps enhance risk reduction via preparedness and mitigation activities (Choubin et al., 2020), and SAR and optical imagery guide damage detection and assessments (Vickers et al., 2016).

Key focus regions for avalanches and snow/ice monitoring include the Tibetan Plateau, South and Central Asia, and polar regions. RS supports avalanche modeling, snowmelt flood forecasting, and seasonal resource management. Preparedness benefits from accurate snow property mapping in avalanche-prone regions, while near real-time data can aid in emergency response. Recovery efforts leverage time series data for ecological restoration and infrastructure repair.

3.2.4 Tropical cyclone hazards

RS of both the atmosphere and earth surface is essential to scientific monitoring and disaster management for tropical cyclone (TC) hazards–including the components of wind, storm surge, and flooding rainfall–by providing essential data on their formation, physical characteristics, and impacts. As tropical cyclones form over warm ocean waters, RS platforms provide monitoring of systems as they develop and move. The most critical asset for monitoring is geostationary weather satellites with optical sensors in the visible and infrared spectra (Qian et al., 2024). For monitoring in the western hemisphere and into the Pacific Ocean near the international dateline, NASA and NOAA operate the GOES series of satellites, while the Meteosat series provides coverage for Europe, Africa, and the Indian Ocean region (Ricciardulli et al., 2023). China, Japan, and India also operate series of geostationary weather satellites covering their areas of interest (Higuchi, 2021; Nadimpalli et al., 2020; Zheng et al., 2019). These instruments provide imagery multiple times per hour with the ability to increase temporal resolution if needed to monitor significant tropical systems (Qian et al., 2024), especially when they move near populated coastal regions.

Strong tropical cyclones exhibit well-known cloud patterns in optical satellite imagery, such as an identifiable eye feature and symmetrical organization of the spiral bands, that allow estimation of pressure and maximum wind speed (Olander & Velden, 2019). The areal extent of winds and precipitation associated with a tropical cyclone provides additional information relevant for storm surge and rainfall predictions (Hu & Zou, 2022; Zhuo & Tan, 2021), and infrared imagery facilitates monitoring at night (Ricciardulli et al., 2023). Microwave-band satellite imagery from polar orbiting platforms allows complementary analyses of storm intensity (e.g., using the Defense Meteorological Satellite Program satellites) (Wimmers et al., 2019) and rainfall rates (e.g., using the GPM instrument), the latter of which is key for warnings of freshwater flooding (Kim & Matyas, 2024; Qi et al., 2021). Finally, ground-based weather radar systems provide higher spatiotemporal resolution monitoring of TC winds and heavy rainfall when they move within range of coastlines (Matyas et al., 2018).

Key platforms for land surface monitoring related to tropical cyclone hazards include MODIS, Landsat, and Sentinel satellites, each offering unique capabilities. MODIS is used frequently for post-cyclone vegetation assessments (Cortés-Ramos et al., 2020), while Landsat provides higher spatial resolution multispectral imagery for mapping coastal changes and storm surge impacts (Charrua et al., 2021). During the response and early recovery period, wind damage to buildings can often be identified using high spatial resolution spaceborne and/or aerial imagery through identification of tarpaulins (often blue colored) stretched across rooftops to lessen water damage through compromised sections (Gonsoroski et al., 2023; Xu et al., 2022).

Sentinel-1's SAR capabilities enable flood mapping even in the presence of cloud cover (Hassan et al., 2020), and Sentinel-2 offers multispectral data for vegetation health and land cover analysis (Phiri et al., 2021). Additional microwave platforms such as ALOS-PALSAR contribute to flood mapping and land deformation monitoring (Asbridge et al., 2018; Chen et al., 2017). The Advanced Very High Resolution Radiometer (AVHRR) and VIIRS sensors are key for sea surface temperature monitoring, which is critical for prediction of tropical cyclone genesis and potential for intensification (Govekar et al., 2022).

Geographically, the use of RS data is common in TC-prone nations like India, China, and the United States, where high exposure along extensive coastlines intersects with strong national investments in satellite infrastructure, scientific expertise, and operational disaster management capacity. These countries not only experience frequent TC impacts but also operate or have direct access to advanced Earth observation systems, enabling sustained research, monitoring, and applied use of RS data for TC hazards. For example, researchers and analysts in India used Sentinel and MODIS data for flood mapping and vegetation damage assessments along its coastline

(Dutta et al., 2015; Halder & Bandyopadhyay, 2022). Examples from China include employment of SAR and optical imagery for monitoring of urban flooding and for mapping of coastal erosion (Liu et al., 2022; Wu et al., 2024). In the United States, past studies leveraged Landsat and VIIRS for hurricane damage assessments (Karaer et al., 2022; Zhao et al., 2020). Bangladesh, the Philippines, Japan, Australia, and Vietnam also feature prominently in research and operational uses of RS data for TCs, highlighting their exposure and susceptibility to TCs, and their reliance on a variety of platforms for disaster management to support coping and adaptive responses.

3.2.5 Severe thunderstorm hazards

As with tropical cyclones, RS technologies are essential for understanding and managing severe thunderstorm hazards–tornadoes, thunderstorm wind hazards, and hail—providing critical insights for monitoring, warning systems, estimation of impacts, and rapid response and recovery. Spaceborne and ground-based platforms support monitoring of the atmosphere for weather forecasting and early warning purposes. Geostationary optical sensors, such as the GOES satellites, allow for constant monitoring of synoptic-scale weather systems (e.g., low pressure systems) and atmospheric conditions that could support development of severe thunderstorms. These platforms are part of the suite of tools used by weather forecasters to issue forecasts for severe thunderstorms a few hours to a few days in advance, providing early warning and lengthening the window for preparedness actions (Brotzge & Donner, 2013). Weather satellite instruments are valuable for short-term disaster management when storms begin to form as identifiable features such as very tall, cold cloud tops and intense concentrations of lightning (both observed via infrared sensors) are often indicative of storms capable of producing damaging winds and hail (Cintineo et al., 2022; Hong et al., 2023).

However, tornadoes, damaging wind events, and hail often occur locally with relatively limited spatial extents. Thus, networks of ground-based weather radar complement the RS toolbox for disaster management in developed countries prone to severe thunderstorms such as the US, Canada, Germany, China, and Australia (Allen & Allen, 2016; Durage et al., 2013; Hengstebeck et al., 2018; Xie et al., 2024). These radar systems can scan and provide new data every few minutes when needed (Bodine & Kurdzo, 2018), with adequate coverage extending to about 150 km from each radar site, though higher terrain or other obstacles such as wind power farms may interfere with the signal (Brotzge et al., 2011; Saunders et al., 2018). Recent advancements in radar technology, such as dual polarization capabilities, facilitate improved identification and tracking for thunderstorms capable of producing damaging hailstones (e.g., Skripniková & Řezáčová, 2019). Modern weather radar systems also allow analysts to identify areas where damaging winds or tornadoes occur, using the Doppler effect to estimate wind directions and velocities in specific areas of individual thunderstorms. The most prolific tornado-producing thunderstorms–supercells–are characterized by persistent areas of rotation through their life cycles and often display recognizable radar signatures, allowing analysts and disaster managers to issue timely and specific warnings (Wapler et al., 2016). Dual polarization radar technology even allows detection of lofted debris (from vegetation and buildings), enhancing warning messages with evidence of a likely tornado in progress (Bodine et al., 2014).

In the immediate aftermath of severe thunderstorms, UAVs are increasingly deployed for rapid damage assessment due to the relatively localized damage swaths (Wagner et al., 2025), often in rural locations not easily accessible by roads. In some cases, when damage swaths from windstorms and tornadoes are more pronounced and cover a larger footprint, they can be identified and mapped in satellite imagery and then analyzed over time to monitor progress in recovery (Kingfield & de Beurs, 2017; Wagner et al., 2012). For example, in forested areas, tornado damage swaths are readily identified in optical medium- and high-resolution satellite imagery (e.g., Landsat, Sentinel-2) due to vegetation disruption, and many analysts have used common vegetation indices (e.g., NDVI) to locate and map tornado or windstorm paths (Burow et al., 2020; Shikhov & Chernokulsky, 2018; Yuan et al., 2002). For example, Chen et al. (2021) used Landsat and Sentinel-2 imagery to estimate nearly 650 ha of rubber forest was damaged due to a tornado on Hainan Island in southern China in 2019.

Similar optical RS image analyses are used to identify and estimate crop damage from hailstorms, as recent studies have done for maize/corn, wheat, soybeans, and sunflowers using multispectral imagery from the MODIS, Landsat, and Sentinel-2 satellites (Angearu et al., 2022; Bell et al., 2023; Ha et al., 2022; Schillaci et al., 2022). At finer scales, high resolution aerial and spaceborne optical imagery, in tandem with LiDAR data collected using UAVs, even allows damage assessments at the level of individual parcels and structures, using imagery captured before and after a damaging severe thunderstorm event (Kashani et al., 2015; Womble et al., 2018). Changes in urban tree canopies due to windstorms or tornado impacts can also be detected, even decades after an event, using LiDAR data and change detection analyses of panchromatic aerial images (Blackman & Yuan, 2020).

3.2.6 Tsunamis

Tsunamis are a series of water waves that are generated by large-scale displacement of a water column by seismic or volcanic events or other types of mass movements (Reid & Mooney, 2023). Large tsunamis are rare, but they cause devastating impacts to people and ecosystems in coastal areas when they occur and move inland. Key RS applications for tsunamis are similar to those for storm surges. Analysts utilized SAR platforms like Sentinel-1 and ALOS PALSAR to assess the extent of inundation and surface changes due to erosion and deposition in large tsunami events that occurred in 2004 and 2011 (Iyyappan et al., 2018; Wieland et al., 2016). High spatial resolution optical sensors, including IKONOS and QuickBird (which were both operational from 2001 to 14), complemented SAR for detailed land-use damage assessments (Römer et al., 2012; Wikantika et al., 2007). Airborne LiDAR data can provide high-resolution DEMs critical for risk and vulnerability mapping in tsunami-prone regions.

The Indian Ocean basin–especially Sumatra in Indonesia, Thailand, Sri Lanka, and India–as well as Japan have been key regions of recent focus due to their vulnerabilities as seen in the tragic impacts of devastating earthquake-generated tsunamis in 2004 and 2011 respectively. Disaster risk reduction strategies use DEMs and SAR data to identify high-risk areas and guide coastal zone management. Mitigation involves GIS-based hazard mapping and infrastructure risk assessments. Preparedness efforts incorporate early warning systems enabled by spaceborne and airborne radar altimetry (Hamlington et al., 2012; Mulia et al., 2020) and ground-based high frequency radar systems (Dzvonkovskaya et al., 2018; Lipa et al., 2023; Wang et al., 2023) for tsunami wave detection in near-coastal waters. As large tsunamis hitting coastal areas often raze buildings and saltwater inundation kills vegetation, post-tsunami responses benefit from SAR and optical imagery for rapid damage assessments, while recovery uses satellite and airborne data to monitor vegetation restoration and shoreline morphology.

3.2.7 Floods

Flood events, whether rapid-onset flash floods or slower riverine inundations, pose significant threats globally. RS technologies are indispensable in disaster risk reduction, offering tools to monitor, predict, and manage flood impacts effectively. SAR platforms, like Sentinel-1 and ALOS PALSAR, are vital for delineating flooded areas under all weather conditions (see Fig. 11.7), although their ability to detect water beneath vegetation is limited (Amitrano et al., 2018; Liu & Yamazaki, 2018). Optical sensors, such as Landsat and Pléiades, complement SAR by providing higher resolution imagery for land cover and damage assessments (Cerbelaud et al., 2023; Ounrit et al., 2022). MODIS imagery is especially useful for time series analysis of slower, larger spatial extent river flooding events (Huang et al., 2014). DEMs from sources like SRTM, ASTER, and LiDAR improve terrain analysis, identifying flash flood-prone zones through slope and drainage mapping (Abdelkareem, 2017; Mashaly & Ghoneim, 2018; Muhadi et al., 2020).

RS technologies and products enable flash flood risk reduction by identifying high-risk areas (prevention), guiding urban conservation projects (mitigation), and creating GIS-based vulnerability maps for preparedness. SAR and optical imagery support damage assessments and resource prioritization during response efforts. In recovery, satellite data monitor land cover and hydrological changes to enhance resilience. Recent case studies from India (Singha et al., 2022), Yemen (Al-Aizari et al., 2022), and South Sudan (El-Mahdy et al., 2024) demonstrate the integration of RS with GIS and machine learning to inform flash flood early warning systems and guide infrastructure improvements.

RS also offers critical insights into more slowly emerging, yet highly impactful, floods. For example, seasonal floods such as those in high-latitude climates associated with ice jams, can be managed with the aid of passive and active RS platforms (Plant Lévesque et al., 2024). Optical sensors such as the Sentinel-2 (Goldberg et al., 2020; Möldner et al., 2024), the Landsat series (Sakai et al., 2015), Planet (Cooley et al., 2017), and MODIS (Cooley & Pavelsky, 2016) offer a range of temporal and spatial resolutions for detection and tracking of ice in river systems across the globe (Yang et al., 2020), and weather satellites also contribute coarser spatial resolution imagery

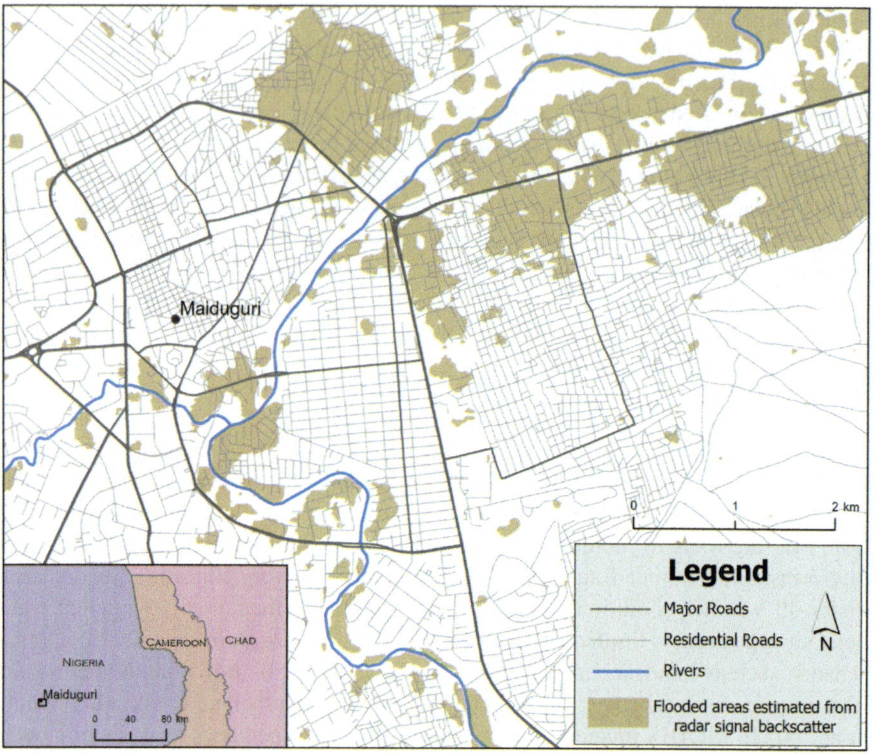

Figure 11.7 Detected flood extent in Maiduguri, Nigeria, based on changes in Sentinel-1 SAR VV backscatter between pre-flood (in August 2024) and flood (mid-September 2024) conditions. Areas shown in brown represent significant radar signal disturbances interpreted as likely flooded surfaces, where water smooths the land surface and reduces backscatter intensity.

for monitoring such seasonal floods (Goldberg et al., 2020). Space-borne radar platforms help to further discriminate ice of different types and stages of development and movement (Lindenschmidt & Li, 2019). UAVs enhance data collection, monitoring, and modeling abilities for more localized applications (Plant Lévesque et al., 2024).

3.2.8 Air quality and pollution

Air pollution studies leverage several RS datasets (MODIS, CALIPSO, OMI, and TROPOMI) for analyzing particulate matter, aerosol concentrations, and trace gases (e.g., Mereuţă et al., 2022; Taloor et al., 2022). Ground-based systems like AERONET validate satellite data, while machine learning models enhance spatial and temporal pollution mapping (El-Nadry et al., 2019). China, India, and industrialized regions in Europe and North America are geographic areas that have been frequently studied in the academic literature, however satellite data facilitates analyses in fast-growing urban regions in Africa and the Middle East as well. Prevention strategies identify pollution hotspots using satellite-derived data (Sharma et al., 2022), which can guide emission control measures. Preparedness uses predictive models to issue air quality warnings, while response efforts deploy near-real-time satellite data to address acute pollution events (Singh et al., 2021).

Increasingly, air quality and pollution studies rely on data fusion approaches that integrate observations from multiple satellite platforms to extend temporal coverage, improve spatial resolution, and enhance the robustness of long-term analyses. For example, MODIS and AVHRR have been widely used to construct multi-decadal aerosol and vegetation-related time series, often in combination with newer sensors to ensure continuity and cross-calibration (Wang, Yuan, et al., 2019). As legacy sensors such as MODIS and AVHRR approach the end of their operational lifetimes, many studies are shifting toward newer instruments, including VIIRS, TROPOMI, and geostationary sensors, which offer improved spatial, spectral, or temporal resolution (Weisz & Menzel, 2020). These newer datasets are frequently harmonized with historical records through statistical adjustment and machine learning-based fusion techniques, allowing researchers and practitioners to maintain long-term monitoring capabilities while taking advantage of advances in sensor technology (Elbestar et al., 2025).

Dust storms and sandstorms are air quality and visibility hazards for which RS technologies are often applied. For example, dust storm monitoring may use MODIS for aerosol optical depth analysis (Gkikas et al., 2021) and Cloud-Aerosol LiDAR and Infrared Pathfinder Satellite Observation (CALIPSO) for vertical aerosol profiles (Liu et al., 2019). Historical climate and weather datasets (e.g., ERA5) integrate satellite and ground data for severity assessments (Bai et al., 2023). Regions such as Inner Mongolia, the Middle East, and Patagonia are key foci for dust storm monitoring and prediction. Machine learning models also enhance dust storm predictions by identifying key relationships and patterns in meteorological and soil data (Alshammari et al., 2022). Prevention and mitigation involve using satellite data to monitor vegetation loss and soil degradation. Preparedness benefits from composite indices (e.g., O'Loingsigh et al., 2014), while response efforts rely on near-real-time weather satellites or MODIS data for public health advisories.

4. Machine learning and geographic artificial intelligence

As RS datasets grow in volume and complexity, traditional analytical methods are increasingly supplemented by more sophisticated computational techniques. Machine learning (ML) and artificial intelligence (AI) have emerged as powerful tools to extract patterns, classify features, and generate predictions from multi-source geospatial data. Their integration into disaster-related RS workflows reflects a broader shift toward automated, scalable, and adaptive analysis (Reichstein et al., 2019; Zhu et al., 2017). Based on a frequency analysis of words and phrases from the nearly 3200 publication abstracts retrieved from Web of Science that included the terms remote sensing and disaster management, machine learning and AI methods are playing an increasingly vital role in disaster management applications. Among the most frequently mentioned terms and techniques are deep learning, various types of neural networks, decision trees and random forest models, and support vector machines (Fig. 11.8). These methods are used to automate complex image classification and prediction tasks, such as detecting hailstorms (Scarino et al., 2023), extracting building damage from satellite imagery (Kerle et al., 2020; Virtriana et al., 2023), and modeling coral reef change in coastal areas (Gapper et al., 2019).

Other machine learning techniques such as predictive modeling, principal component analysis (PCA), and Monte Carlo simulations are also cited across abstracts for analyzing disaster risk and uncertainty. These tools help researchers simulate hazard impacts, prioritize vulnerable zones, and modernize early warning systems. Additionally, cloud-based platforms such as Google Earth Engine are frequently referenced as computational environments where these algorithms are deployed at scale, enabling near–real-time analysis of large, multi-sensor RS datasets (DeVries et al., 2020; Gorelick et al., 2017). Collectively, the publication abstracts confirm a clear shift toward the integration of ML- and AI-driven approaches in RS workflows, offering enhanced analytical capacity, scalability, and decision support for disaster preparedness and response.

Beyond academic research, the adoption of GeoAI within the emergency management sector represents a transition from manual, time-consuming interpretation and integration of RS and other geospatial data toward automated, high-throughput analytical pipelines. Operational agencies are increasingly embracing these tools to bridge the gap between data collection and actionable insight, particularly as disaster-relevant RS datasets continue to grow in volume and complexity (Adriano et al., 2020). Cloud-based environments such as Google Earth Engine enable the deployment of machine learning algorithms that can process massive, multi-sensor datasets at scale, supporting near–real-time analyses that were previously infeasible for many local and regional agencies with limited computing resources (Yang et al., 2022). Practical applications include automated extraction of building damage following earthquakes, floods, and tropical cyclones, which can substantially accelerate post-event needs assessments and resource prioritization (Zheng et al., 2021), as well as AI-driven wildfire detection and fire spread modeling that support evacuation planning and early warning efforts (Tavakkoli Piralilou et al., 2022). Through these AI-enabled workflows, disaster management is increasingly shifting toward more proactive and anticipatory approaches, in which automated systems help identify high-risk zones, simulate hazard impacts, and modernize early warning and decision-support systems.

Looking forward, the continued growth of Earth observation archives, cloud computing, and open-access RS data is expected to further accelerate the integration of GeoAI into disaster management. Emerging directions include expanded use of multi-sensor data fusion to combine optical, radar, thermal, and in situ datasets across longer time horizons, enabling more robust hazard characterization, trend analysis, and impact prediction and assessment (Adriano et al., 2020; Kemper &

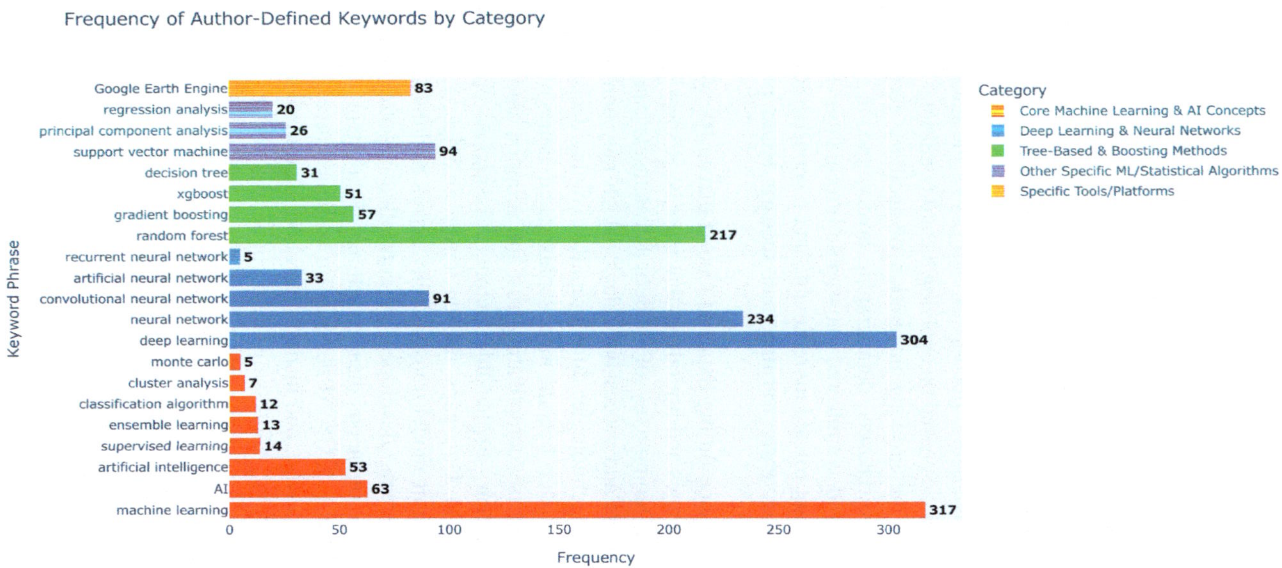

Figure 11.8 Frequencies of machine learning and artificial intelligence terms found in publication abstracts from a web of science search on remote sensing and disaster management. Terms are grouped by conceptual or methodological category.

Kemper, 2020). Advances in deep learning architectures and the emergence of foundation models for Earth observation offer new opportunities to generalize damage detection and hazard classification methods across regions and hazard types, reducing reliance on event-specific model training (Xiao et al., 2025). In parallel, cloud-based geospatial platforms are lowering technical barriers to entry, allowing local governments, humanitarian organizations, and researchers in low-resource settings to apply advanced analytical tools without maintaining extensive local computing infrastructure (DeVries et al., 2020). Building on these developments, agentic AI systems capable of executing analytical workflows may further reduce time delay between observation and analysis by automating data ingestion, model execution, and update cycles as new RS observations become available, keeping humans in the decision loop (Li et al., 2025). At the same time, advances in physical AI and adaptive sensing are enabling more responsive data collection through UAVs and sensor networks that can adjust acquisition strategies based on evolving hazard conditions (Chandran & Vipin, 2024). Together with the growing integration of ancillary geospatial information, including infrastructure datasets, demographic layers, and, in some cases, crowd-sourced observations and volunteered geographic information (Huang et al., 2024), these developments point toward more flexible and responsive RS systems that support anticipatory action, early warning, and scenario-based planning in complex disaster environments.

The increasing reliance on big data and GeoAI-driven RS workflows also introduces important challenges and limitations that must be acknowledged. The growing volume, velocity, and complexity of modern RS datasets can impose substantial computational demands and processing delays, which may constrain the near real-time promise of these technologies during rapidly evolving disaster response phases. As machine learning models become more sophisticated, issues of interpretability and transparency can also arise, complicating trust and adoption among emergency managers who require clear, explainable outputs to support operational decisions (McGovern et al., 2024). Data gaps, uneven sensor coverage, and biases in training datasets may further limit model transferability across regions and hazard contexts, particularly when systems developed in controlled research environments are deployed in noisy, heterogeneous disaster settings (Reichstein et al., 2019). Finally, effective operational use of GeoAI depends on sustained investments in data infrastructure, technical expertise, and institutional coordination, which remain unevenly distributed across regions. Addressing these challenges will be essential to ensure that advances in big data analytics and GeoAI consistently enhance timely, reliable decision-making in disaster management.

5. Conclusion

This chapter has explored the expanding role of RS in disaster management, with a focus on natural hazards. RS technologies—including satellite platforms, UAVs, radar and LiDAR systems—have become indispensable for understanding, anticipating, and responding to a wide range of geophysical, meteorological, hydrological, and climate-related hazards. Across hazards such as earthquakes, landslides, wildfires, tropical cyclones, floods, droughts, snow and avalanche events, and air quality crises, RS supports all phases of the disaster management cycle—from prevention and preparedness to response and long-term recovery. These technologies have also enabled the production of high-resolution hazard maps, real-time impact assessments, and large-scale vulnerability and risk models. In recent years, the integration of machine learning and artificial intelligence has further enhanced the analytical power of RS by enabling automated processing of large volumes of geospatial data for disaster applications.

Looking ahead, disaster management research and practice is increasingly moving beyond hazard monitoring toward predictive and scenario-based GeoAI applications that support anticipatory action and decision support. Continued integration of cloud-based geospatial platforms with machine learning and large language models is expected to simplify data access, automate analytical workflows, and provide coding assistance that lowers technical barriers for local governments and practitioners, particularly in low-resource settings. Ongoing advances in multi-source data fusion—combining RS observations with environmental, infrastructural, and socio-economic datasets—offer the potential to enhance situational awareness and support more holistic assessments of risk and resilience. As GeoAI and big data analytics become more deeply embedded in RS workflows, sustained attention to scalability, interpretability, and operational relevance will be essential to ensure these tools effectively support emergency management practice. Continued collaboration among geospatial scientists, emergency managers, and policymakers will play a critical role in translating technological advances into safer, more resilient, and more sustainable communities.

Case Studies

Case study 1: Detecting Hurricane-related damage using Sentinel-2 imagery in Valdosta, Georgia

In late September 2024, Hurricane Helene tracked northeastward from the Gulf of Mexico (now Gulf of America) and moved inland across the southeastern United States, producing hurricane-force winds and widespread vegetation and structural damage across portions of southern Georgia, including the city of Valdosta. To illustrate how moderate-resolution, open-access satellite imagery can support rapid post-event damage assessment, Sentinel-2 multispectral data were analyzed to identify potential hurricane-related impacts within an urban environment (Case Study Figure 1.1). Despite its 10 m spatial resolution, Sentinel-2 imagery was sufficient to reveal spatial patterns consistent with storm damage, including vegetation loss and possible roof damage indicated by the presence of temporary blue tarpaulins.

This example was generated using Google Earth Engine and relied on simple, yet effective change-detection techniques applied to imagery acquired in the two weeks before and after Hurricane Helene. The Google Earth Engine workflow was

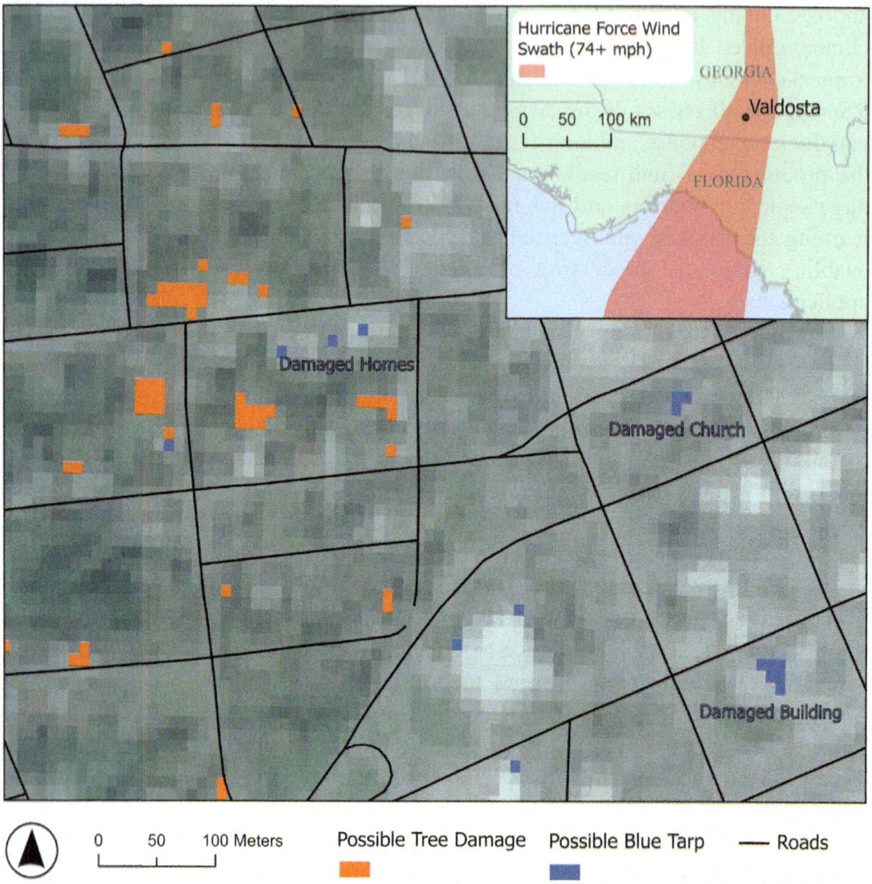

Case Study Figure 1.1 Possible hurricane-related damage in Valdosta, Georgia, following Hurricane Helene (2024). Orange areas show likely tree loss based on NDVI change; blue areas indicate possible blue tarps detected. The base image is post-storm Sentinel-2 true color imagery. Labels mark known damaged structures, and the inset map shows the broader hurricane force wind swath across the region.

developed by the author, with JavaScript coding assistance from a large language model (Google Gemini) used to support script structure and syntax, while the analytical choices and interpretation were performed by the researcher. Vegetation impacts were identified using changes in vegetation indices derived from Sentinel-2 red and near-infrared bands, highlighting areas of likely tree loss or canopy disturbance associated with strong winds. In addition, analysis of spectral differences in visible bands was used to flag localized, blue-colored features consistent with temporary roof coverings, which are commonly deployed following wind damage to prevent further water intrusion.

To provide qualitative validation, selected locations flagged by the satellite-based analysis were cross-checked using Google Street View imagery acquired in February 2025. Visual inspection confirmed the presence of blue tarpaulins at several highlighted locations near downtown Valdosta, including residential structures and a church facility. While this example does not represent a comprehensive damage inventory, it demonstrates how freely available satellite data and cloud-based analysis platforms can support spatially explicit identification of areas warranting closer inspection following a tropical cyclone. Such approaches are particularly valuable for supporting situational awareness, prioritizing field assessments, and informing early recovery planning when on-the-ground information is incomplete or delayed due to widespread power and communication outages.

Case study 2: Mapping tornado damage with Sentinel-2 imagery in the Southeastern U.S

On March 3, 2019, several tornadoes struck southern Alabama in the southeastern United States, resulting in widespread severe damage and causing 23 fatalities near the town of Beauregard. After the storm, a research team from the University of Tennessee assessed the potential of high-resolution Sentinel-2 satellite imagery to identify and quantify damage across three distinct tornado paths (Burow et al., 2020). Their goal was to evaluate how remotely sensed damaged vegetation compared with ground survey estimates and tornado wind fields generated by computer models. The study utilized Normalized Difference Vegetation Index (NDVI) comparisons and Principal Component Analysis (PCA) to detect vegetation changes before and after the tornadoes (Case Study Figure 2.1). NDVI was used to represent vegetation health, while PCA visually enhanced subtle changes in reflectance across multiple bands in the visible and infrared bands of the spectrum. The team correlated these RS metrics with post-storm wind speed estimates from National Weather Service on-the-ground damage survey maps.

They found that NDVI-difference values were strongly correlated with estimated wind speeds, particularly when smoothed to reduce noise. NDVI patterns along the strongest tornado's path resembled the wind and damage distribution scientists expect; wind speeds changed from the center of the tornado outward with the strongest winds near the middle and weaker winds farther out. This suggests that vegetation damage patterns documented by satellite imagery can detect detailed evidence of how the tornado moved and how strong it was at different points in its life cycle. In contrast, the weaker tornadoes in this event showed flatter NDVI profiles, reflecting their lower wind intensities and less visible damage. Importantly, the study revealed areas of likely tornado damage outside of official survey areas, especially in forested regions where ground access was limited. These results underscore the value of using freely available, moderate-resolution satellite imagery for rapid, consistent wind damage assessments—especially in rural or inaccessible areas.

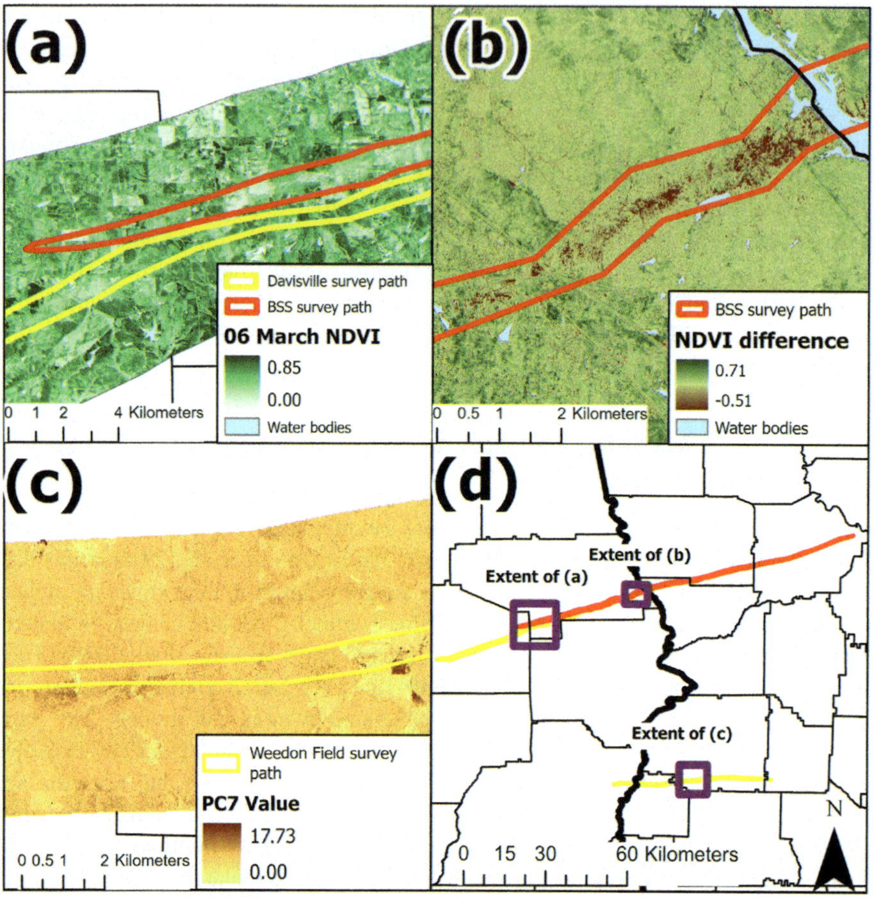

Case Study Figure 2.1 Close-up of remotely sensed damage signatures using three different metrics: (a) Post-event NDVI measured on 6 March, (b) NDVI difference, and (c) PC7. (d) Shows the extent of (a–c) within the tornado paths.

Credit: Recreated by D. Burow and used with permission from: Burow, D., Herrero, H. V., and Ellis, K. N. (2020). Damage analysis of three long-track tornadoes using high-resolution satellite imagery. Atmosphere, 11(6), 613.

Acronyms

3D	Three-Dimensional
AERONET	Aerosol Robotic Network
ALOS	Advanced Land Observing Satellite
AMSR	Advanced Microwave Scanning Radiometer
ASTER	Advanced Spaceborne Thermal Emission and Reflection Radiometer
AVHRR	Advanced Very High Resolution Radiometer
CALIPSO	Cloud-Aerosol LiDAR and Infrared Pathfinder Satellite Observation
DEM	Digital Elevation Model
DInSAR	Differential Interferometric Synthetic Aperture Radar
dNBR	Differenced normalized burn ratio

ERA	European Center for Medium-Range Weather Forecasts Reanalysis
ESA	European Space Agency
FEMA	Federal Emergency Management Agency
GeoAI	Geographic Artificial Intelligence
GIS	Geographic Information System(s)
GNSS	Global Navigation Satellite System
GOES	Geostationary Operational Environmental Satellite
GPM	Global Precipitation Measurement
InSAR	Interferometric Synthetic Aperture Radar
LiDAR	Light Detection and Ranging
ML	Machine Learning
MODIS	Moderate Resolution Imaging Spectrometer
MSI	Multispectral Instrument
NASA	National Aeronautics and Space Administration
NDVI	Normalized Difference Vegetation Index
NOAA	National Oceanic and Atmospheric Administration
OMI	Ozone Monitoring Instrument
PALSAR	Phased Array type L-band Synthetic Aperture Radar
PCA	Principal Component Analysis
RS	Remote Sensing
SAR	Synthetic Aperture Radar
SPOT	Satellite Pour l'Observation de la Terre
SRTM	Shuttle Radar Topography Mission
TRMM	Tropical Rainfall Measurement Mission
TROPOMI	Tropospheric Monitoring Instrument
UAV	Unoccupied Aerial Vehicle
UN	United Nations
UNDRR	United Nations Office for Disaster Risk Reduction
UNOSAT	United Nations Operational Satellite Applications Program
US	United States
VCI	Vegetation Condition Index
VIIRS	Visible Infrared Imaging Radiometer Suite

References

Abdelkareem, M. (2017). Targeting flash flood potential areas using remotely sensed data and GIS techniques. *Natural Hazards, 85*, 19–37.

Adriano, B., Yokoya, N., Xia, J., & Baier, G. (2020). Big earth observation data processing for disaster damage mapping. In *Handbook of big geospatial data* (pp. 99–118). Cham: Springer International Publishing.

AghaKouchak, A., Farahmand, A., Melton, F. S., Teixeira, J., Anderson, M. C., Wardlow, B. D., & Hain, C. R. (2015). Remote sensing of drought: Progress, challenges and opportunities. *Reviews of Geophysics, 53*(2), 452–480.

Al-Aizari, A. R., Al-Masnay, Y. A., Aydda, A., Zhang, J., Ullah, K., Islam, A. R. M. T., … Liu, X. (2022). Assessment analysis of flood susceptibility in tropical desert area: A case study of Yemen. *Remote Sensing, 14*(16), 4050.

Alexander, D. (2007). Disaster management: From theory to implementation. *Journal of Seismology and Earthquake Engineering, 9*(1 and 2), 49–59.

Ali, M., Shahzad, M. I., Nazeer, M., Mahmood, I., & Zia, I. (2021). Estimation of surface deformation due to Pasni earthquake using RADAR interferometry. *Geocarto International, 36*(14), 1630–1645.

Allen, J. T., & Allen, E. R. (2016). A review of severe thunderstorms in Australia. *Atmospheric Research, 178*, 347–366.

Allison, R. S., Johnston, J. M., Craig, G., & Jennings, S. (2016). Airborne optical and thermal remote sensing for wildfire detection and monitoring. *Sensors, 16*(8), 1310.

Alshammari, R. K., Alrwais, O., & Aksoy, M. S. (2022). Machine learning applications to dust storms: A meta-analysis. *Aerosol and Air Quality Research, 22*(12), Article 220183.

Amitrano, D., Di Martino, G., Iodice, A., Riccio, D., & Ruello, G. (2018). Unsupervised rapid flood mapping using Sentinel-1 GRD SAR images. *IEEE Transactions on Geoscience and Remote Sensing, 56*(6), 3290–3299.

Angearu, C. V., Ontel, I., Irimescu, A., Sorin, B., & Dodd, E. (2022). Remote sensing methods for detecting and mapping hailstorm damage: A case study from the 20 July 2020 hailstorm, Baragan Plain, Romania. *Natural Hazards, 114*(2), 2013–2040.

Asbridge, E., Lucas, R., Rogers, K., & Accad, A. (2018). The extent of mangrove change and potential for recovery following severe Tropical Cyclone Yasi, Hinchinbrook Island, Queensland, Australia. *Ecology and Evolution, 8*(21), 10416–10434.

Bai, C., Cai, Z., Yin, X., & Zhang, J. (2023). LSDSSIMR: Large-scale dust storm database based on satellite images and meteorological reanalysis data. *IEEE Journal of Selected Topics in Applied Earth Observations and Remote Sensing, 16*, 10121–10131.

Ban, Y., Zhang, P., Nascetti, A., Bevington, A. R., & Wulder, M. A. (2020). Near real-time wildfire progression monitoring with Sentinel-1 SAR time series and deep learning. *Scientific Reports, 10*(1), 1322.

Barnes, J., Roller, C., Soobitsky, R., Childs-Gleason, L., Rea, P., Kerwin, P., & Golla, N. (July 2024). Informing disaster response: An introduction to the NASA disaster response coordination system. In *49th annual natural hazards research and applications workshop.*

Barrett, E. C. (1979). Satellite remote sensing in hazard monitoring and disaster evaluation. Satellite remote sensing applications in agroclimatology and agrometeorology. In E. C. Barrett (Ed.), *Proceedings of the Joint ESA/FAO/WMO International Training Course held 2-13 October, 1978 in Rome, Italy* (p. 157). European Space Agency. ESA SP-1020.

Bell, J. R., Wisinski, E. F., Molthan, A. L., Schultz, C. J., Gilligan, E., & Sharp, K. G. (2023). Developing a hail and wind damage swath event database from daily MODIS true color imagery and storm reports for impact analysis and applications. *Weather and Forecasting, 38*(9), 1575–1588.

Blackman, R., & Yuan, F. (2020). Detecting long-term urban forest cover change and impacts of natural disasters using high-resolution aerial images and LiDAR data. *Remote Sensing, 12*(11), 1820.

Bodine, D. J., & Kurdzo, J. M. (2018). Ground-based radar technologies for tornado observations. In C. Andronache (Ed.), *Remote sensing of clouds and precipitation* (pp. 65–112). Springer.

Bodine, D. J., Palmer, R. D., & Zhang, G. (2014). Dual-wavelength polarimetric radar analyses of tornadic debris signatures. *Journal of Applied Meteorology and Climatology, 53*(2), 242–261.

Bolton, D. K., Coops, N. C., & Wulder, M. A. (2015). Characterizing residual structure and forest recovery following high-severity fire in the Western boreal of Canada using Landsat time-series and airborne lidar data. *Remote Sensing of Environment, 163*, 48–60.

Bosher, L. (2014). Built-in resilience through disaster risk reduction: Operational issues. *Building Research and Information, 42*(2), 240–254.

Bousbih, S., Zribi, M., El Hajj, M., Baghdadi, N., Lili-Chabaane, Z., Gao, Q., & Fanise, P. (2018). Soil moisture and irrigation mapping in A semi-arid region, based on the synergetic use of Sentinel-1 and Sentinel-2 data. *Remote Sensing, 10*(12), 1953.

Brasil Neto, R. M., Santos, C. A. G., da Costa Silva, J. F. C. B., da Silva, R. M., Dos Santos, C. A. C., & Mishra, M. (2021). Evaluation of the TRMM product for monitoring drought over Paraíba State, northeastern Brazil: A trend analysis. *Scientific Reports, 11*(1), 1097.

Bright, B. C., Hudak, A. T., Kennedy, R. E., Braaten, J. D., & Henareh Khalyani, A. (2019). Examining post-fire vegetation recovery with Landsat time series analysis in three Western North American forest types. *Fire Ecology, 15*(1), 1–14.

Brotzge, J., & Donner, W. (2013). The tornado warning process: A review of current research, challenges, and opportunities. *Bulletin of the American Meteorological Society, 94*(11), 1715–1733.

Brotzge, J., Erickson, S., & Brooks, H. (2011). A 5-yr climatology of tornado false alarms. *Weather and Forecasting, 26*(4), 534–544.

Burow, D., Herrero, H. V., & Ellis, K. N. (2020). Damage analysis of three long-track tornadoes using high-resolution satellite imagery. *Atmosphere, 11*(6), 613.

Caballero, I., Román, A., Tovar-Sánchez, A., & Navarro, G. (2022). Water quality monitoring with Sentinel-2 and Landsat-8 satellites during the 2021 volcanic eruption in La Palma (Canary Islands). *Science of the Total Environment, 822*, Article 153433.

Campus, A., Laiolo, M., Massimetti, F., & Coppola, D. (2022). The transition from MODIS to VIIRS for global volcano thermal monitoring. *Sensors, 22*(5), 1713.

Casagli, N., Intrieri, E., Tofani, V., Gigli, G., & Raspini, F. (2023). Landslide detection, monitoring and prediction with remote-sensing techniques. *Nature Reviews Earth and Environment, 4*(1), 51–64.

Cerbelaud, A., Blanchet, G., Roupioz, L., Breil, P., & Briottet, X. (2023). Mapping pluvial flood-induced damages with multi-sensor optical remote sensing: A transferable approach. *Remote Sensing, 15*(9), 2361.

Chandran, I., & Vipin, K. (2024). Multi-UAV networks for disaster monitoring: Challenges and opportunities from a network perspective. *Drone Systems and Applications, 12*, 1–28.

Charrua, A. B., Padmanaban, R., Cabral, P., Bandeira, S., & Romeiras, M. M. (2021). Impacts of the tropical cyclone Idai in Mozambique: A multi-temporal Landsat satellite imagery analysis. *Remote Sensing, 13*(2), 201.

Chen, R. F., Lee, C. Y., Yin, H. Y., Huang, H. Y., Cheng, K. P., & Lin, C. W. (2017). Monitoring the deep-seated landslides by using ALOS/PALSAR satellite imagery in the disaster area of 2009 Typhoon Morakot, Taiwan. In *Advancing culture of living with landslides: Volume 3 advances in landslide technology* (pp. 239–247). Springer International Publishing.

Chen, B., Yun, T., An, F., Kou, W., Li, H., Luo, H., Yang, C., Wang, Q., Sun, R., & Wu, Z. (2021). Assessment of tornado disaster in rubber plantation in Western Hainan using Landsat and Sentinel-2 time series images. *National Remote Sensing Bulletin, 25*(3), 816–829.

Chiabrando, F., Sammartano, G., & Spanò, A. (2017). A comparison among different optimization levels in 3D multi-sensor models. A test case in emergency context: 2016 Italian earthquake. *The International Archives of the Photogrammetry, Remote Sensing and Spatial Information Sciences, 42*, 155–162.

Chmutina, K., & Von Meding, J. (2019). A dilemma of language: "Natural disasters" in academic literature. *International Journal of Disaster Risk Science, 10*, 283–292.

Choubin, B., Borji, M., Hosseini, F. S., Mosavi, A., & Dineva, A. A. (2020). Mass wasting susceptibility assessment of snow avalanches using machine learning models. *Scientific Reports, 10*(1), Article 18363.

Chuvieco, E. (2020). *Fundamentals of satellite remote sensing: An environmental approach.* CRC press.

Cintineo, J. L., Pavolonis, M. J., & Sieglaff, J. M. (2022). ProbSevere LightningCast: A deep-learning model for satellite-based lightning nowcasting. *Weather and Forecasting, 37*(7), 1239–1257.

Cooley, S. W., & Pavelsky, T. M. (2016). Spatial and temporal patterns in Arctic river ice breakup revealed by automated ice detection from MODIS imagery. *Remote Sensing of Environment, 175*, 310–322.

Cooley, S. W., Smith, L. C., Stepan, L., & Mascaro, J. (2017). Tracking dynamic northern surface water changes with high-frequency planet CubeSat imagery. *Remote Sensing, 9* (12), 1306.

Coppola, D., Laiolo, M., Cigolini, C., Massimetti, F., Delle Donne, D., Ripepe, M., … William, R. (2020). Thermal remote sensing for global volcano monitoring: Experiences from the MIROVA system. *Frontiers in Earth Science, 7*, 362.

Cortés-Ramos, J., Farfán, L. M., & Herrera-Cervantes, H. (2020). Assessment of tropical cyclone damage on dry forests using multispectral remote sensing: The case of Baja California Sur, Mexico. *Journal of Arid Environments, 178*, Article 104171.

Cutter, S. L. (2016). Resilience to what? Resilience for whom? *The Geographical Journal, 182* (2), 110–113.

Cutter, S. L. (2024). The origin and diffusion of the social vulnerability index (SoVI). *International Journal of Disaster Risk Reduction, 109*, Article 104576.

Cutter, S. L., Ahearn, J. A., Amadei, B., Crawford, P., Eide, E. A., Galloway, G. E., … Zoback, M. L. (2013). Disaster resilience: A national imperative. *Environment: Science and Policy for Sustainable Development, 55*(2), 25–29.

De Angeli, S., Malamud, B. D., Rossi, L., Taylor, F. E., Trasforini, E., & Rudari, R. (2022). A multi-hazard framework for spatial-temporal impact analysis. *International Journal of Disaster Risk Reduction, 73*, Article 102829.

Denissova, N., Nurakynov, S., Petrova, O., Chepashev, D., Daumova, G., & Yelisseyeva, A. (2024). Remote sensing techniques for assessing snow avalanche formation factors and building hazard monitoring systems. *Atmosphere, 15*(11), 1343.

DeVries, B., Huang, C., Armston, J., Huang, W., Jones, J. W., & Lang, M. W. (2020). Rapid and robust monitoring of flood events using Sentinel-1 and Landsat data on the Google Earth engine. *Remote Sensing of Environment, 240*, Article 111664.

Diaz-Vilor, C., Barzegaran, M., & Jafarkhani, H. (2025). Multi-UAV energy-efficient wildfire coverage optimization. *IEEE Transactions on Wireless Communications, 24*(10), 8633–8648.

Durage, S. W., Wirasinghe, S. C., & Ruwanpura, J. (2013). Comparison of the Canadian and US tornado detection and warning systems. *Natural Hazards, 66*, 117–137.

Dutta, D., Das, P. K., Paul, S., Sharma, J. R., & Dadhwal, V. K. (2015). Assessment of ecological disturbance in the mangrove forest of Sundarbans caused by cyclones using MODIS time-series data (2001–2011). *Natural Hazards, 79*, 775–790.

Dvorak, V. F. (1975). Tropical cyclone intensity analysis and forecasting from satellite imagery. *Monthly Weather Review, 103*(5), 420–430.

Dzvonkovskaya, A., Petersen, L., Helzel, T., & Kniephoff, M. (2018). High-frequency ocean radar support for Tsunami early warning systems. *Geoscience Letters, 5*(1), 29.

El-Mahdy, M. E. S., Mousa, F. A., Morsy, F. I., Kamel, A. F., & El-Tantawi, A. (2024). Flood classification and prediction in South Sudan using artificial intelligence models under a changing climate. *Alexandria Engineering Journal, 97*, 127–141.

El-Nadry, M., Li, W., El-Askary, H., Awad, M. A., & Mostafa, A. R. (2019). Urban health related air quality indicators over the Middle East and North Africa countries using multiple satellites and AERONET data. *Remote Sensing, 11*(18), 2096.

Elbestar, M., Aly, S. G., Ghannam, R., & Eraqi, H. (2025). AirFusion: Sensor data fusion for air quality monitoring. *International Journal of Remote Sensing*. https://doi.org/10.1080/01431161.2025.2516692 (in press).

European Space Agency. (June 24, 2011). Satellites respond to humanitarian needs. https://esa.int/Applications/Observing_the_Earth/Envisat/Satellites_respond_to_humanitarian_needs.

Gapper, J. J., El-Askary, H., Linstead, E., & Piechota, T. (2019). Coral reef change detection in remote Pacific Islands using support vector machine classifiers. *Remote Sensing, 11*(13), 1525.

Ghaffarian, S., Kerle, N., & Filatova, T. (2018). Remote sensing-based proxies for urban disaster risk management and resilience: A review. *Remote Sensing, 10*(11), 1760.

Ghaleb, F., Mario, M., & Sandra, A. N. (2015). Regional Landsat-based drought monitoring from 1982 to 2014. *Climate, 3*(3), 563–577.

Gibson, R., Danaher, T., Hehir, W., & Collins, L. (2020). A remote sensing approach to mapping fire severity in south-eastern Australia using Sentinel-2 and random forest. *Remote Sensing of Environment, 240*, Article 111702.

Giglio, L., Boschetti, L., Roy, D. P., Humber, M. L., & Justice, C. O. (2018). The collection 6 MODIS burned area mapping algorithm and product. *Remote Sensing of Environment, 217*, 72–85.

Gill, J. C., & Malamud, B. D. (2017). Anthropogenic processes, natural hazards, and interactions in a multi-hazard framework. *Earth-Science Reviews, 166*, 246–269.

Gkikas, A., Proestakis, E., Amiridis, V., Kazadzis, S., Di Tomaso, E., Tsekeri, A., ... Pérez García-Pando, C. (2021). ModIs Dust AeroSol (MIDAS): A global fine-resolution dust optical depth data set. *Atmospheric Measurement Techniques, 14*(1), 309–334.

Goldberg, M. D., Li, S., Lindsey, D. T., Sjoberg, W., Zhou, L., & Sun, D. (2020). Mapping, monitoring, and prediction of floods due to ice jam and snowmelt with operational weather satellites. *Remote Sensing, 12*(11), 1865.

Gomez, C., Setiawan, M. A., Listyaningrum, N., Wibowo, S. B., Hadmoko, D. S., Suryanto, W., ... Malawani, M. N. (2022). LiDAR and UAV SfM-MVS of Merapi volcanic dome and crater rim change from 2012 to 2014. *Remote Sensing, 14*(20), 5193.

Gonsoroski, E., Ahn, Y., Harville, E. W., Countess, N., Lichtveld, M. Y., Pan, K., ... Uejio, C. K. (2023). Classifying building roof damage using high resolution imagery for disaster recovery. *Photogrammetric Engineering and Remote Sensing, 89*(7), 437–443.

Gorelick, N., Hancher, M., Dixon, M., Ilyushchenko, S., Thau, D., & Moore, R. (2017). Google Earth engine: Planetary-scale geospatial analysis for everyone. *Remote Sensing of Environment, 202*, 18–27.

Govekar, P. D., Griffin, C., & Beggs, H. (2022). Multi-sensor sea surface temperature products from the Australian bureau of meteorology. *Remote Sensing, 14*(15), 3785.

Ha, T., Shen, Y., Duddu, H., Johnson, E., & Shirtliffe, S. J. (2022). Quantifying hail damage in crops using sentinel-2 imagery. *Remote Sensing, 14*(4), 951.

Ha, T. V., Uereyen, S., & Kuenzer, C. (2023). Agricultural drought conditions over mainland Southeast Asia: Spatiotemporal characteristics revealed from MODIS-based vegetation time-series. *International Journal of Applied Earth Observation and Geoinformation, 121*, Article 103378.

Hafner, E. D., Barton, P., Daudt, R. C., Wegner, J. D., Schindler, K., & Bühler, Y. (2022). Automated avalanche mapping from SPOT 6/7 satellite imagery with deep learning: Results, evaluation, potential and limitations. *The Cryosphere, 16*(9), 3517–3530.

Halder, B., & Bandyopadhyay, J. (2022). Monitoring the tropical cyclone "Yass" and "Amphan" affected flood inundation using Sentinel-1/2 data and Google Earth engine. *Modeling Earth Systems and Environment, 8*(3), 4317–4332.

Hamilton, D., Gibson, W., Harris, D., & McGath, C. (2023). Evaluation of multi-spectral band efficacy for mapping wildland fire burn severity from PlanetScope imagery. *Remote Sensing, 15*(21), 5196.

Hamlington, B. D., Leben, R. R., Godin, O. A., Gica, E., Titov, V. V., Haines, B. J., & Desai, S. D. (2012). Could satellite altimetry have improved early detection and warning of the 2011 Tohoku tsunami? *Geophysical Research Letters, 39*(15).

Haque, U., Da Silva, P. F., Devoli, G., Pilz, J., Zhao, B., Khaloua, A., … Glass, G. E. (2019). The human cost of global warming: Deadly landslides and their triggers (1995–2014). *Science of the Total Environment, 682*, 673–684.

Hassan, M. M., Ash, K., Abedin, J., Paul, B. K., & Southworth, J. (2020). A quantitative framework for analyzing spatial dynamics of flood events: A case study of super cyclone Amphan. *Remote Sensing, 12*(20), 3454.

Hengstebeck, T., Wapler, K., Heizenreder, D., & Joe, P. (2018). Radar network–based detection of mesocyclones at the German weather service. *Journal of Atmospheric and Oceanic Technology, 35*(2), 299–321.

Higuchi, A. (2021). Toward more integrated utilizations of geostationary satellite data for disaster management and risk mitigation. *Remote Sensing, 13*(8), 1553.

Hong, Y., Trapp, R. J., Nesbitt, S. W., & Di Girolamo, L. (2023). Near-global distributions of overshooting tops derived from Terra and Aqua MODIS observations. *Atmospheric Measurement Techniques*, 1391–1406.

Howard, J. A., Barrett, E. C., & Heilkema, J. U. (1978). The application of satellite remote sensing to monitoring of agricultural disasters. *Disasters, 2*(4), 231–240.

Hu, Y., Che, T., Dai, L., Zhu, Y., Xiao, L., Deng, J., & Li, X. (2024). A long-term daily gridded snow depth dataset for the Northern Hemisphere from 1980 to 2019 based on machine learning. *Big Earth Data, 8*(2), 274–301.

Hu, T., van Dijk, A. I., Renzullo, L. J., Xu, Z., He, J., Tian, S., … Li, H. (2020). On agricultural drought monitoring in Australia using Himawari-8 geostationary thermal infrared observations. *International Journal of Applied Earth Observation and Geoinformation, 91*, Article 102153.

Hu, Y., & Zou, X. (2022). Determining tropical cyclone center and rainband size in geostationary satellite imagery. *Remote Sensing, 14*(14), 3499.

Huang, C., Chen, Y., & Wu, J. (2014). Mapping spatio-temporal flood inundation dynamics at large river basin scale using time-series flow data and MODIS imagery. *International Journal of Applied Earth Observation and Geoinformation, 26*, 350–362.

Huang, X., Wang, S., Yang, D., Hu, T., Chen, M., Zhang, M., Zhang, G., Biljecki, F., Lu, T., Zou, L., Wu, C. Y. H., Park, Y. M., Li, X., Liu, Y., Fan, H., Mitchell, J., Li, Z., & Hohl, A. (2024). Crowdsourcing geospatial data for Earth and human observations: A review. *Journal of Remote Sensing, 4*, 0105.

International Charter: Space and Major Disasters. (2025). *International charter: Space and major disasters*. https://disasterscharter.org/.

Iyyappan, M., Usha, T., Ramakrishnan, S. S., Srinivasa Raju, K., Gopinath, G., Chenthamil Selvan, S., ... Mishra, P. (2018). Evaluation of tsunami inundation using synthetic aperture radar (SAR) data and numerical modeling. *Natural Hazards, 92*, 1419–1432.

James, M. R., Carr, B., D'Arcy, F., Diefenbach, A., Dietterich, H., Fornaciai, A., ... Zorn, E. (2020). Volcanological applications of unoccupied aircraft systems (UAS): Developments, strategies, and future challenges. *Volcanica, 3*(1), 67–114.

Jiao, W., Wang, L., & McCabe, M. F. (2021). Multi-sensor remote sensing for drought characterization: Current status, opportunities and a roadmap for the future. *Remote Sensing of Environment, 256*, Article 112313.

Jordan, B. R. (2019). Collecting field data in volcanic landscapes using small UAS (sUAS)/drones. *Journal of Volcanology and Geothermal Research, 385*, 231–241.

Joyce, K. E., Belliss, S. E., Samsonov, S. V., McNeill, S. J., & Glassey, P. J. (2009). A review of the status of satellite remote sensing and mage processing techniques for mapping natural hazards and disasters. *Progress in Physical Geography, 33*(2), 183–207.

Kaku, K. (2019). Satellite remote sensing for disaster management support: A holistic and staged approach based on case studies in Sentinel Asia. *International Journal of Disaster Risk Reduction, 33*, 417–432.

Karaer, A., Chen, M., Gazzea, M., Ghorbanzadeh, M., Abichou, T., Arghandeh, R., & Ozguven, E. E. (2022). Remote sensing-based comparative damage assessment of historical storms and hurricanes in Northwestern Florida. *International Journal of Disaster Risk Reduction, 72*, Article 102857.

Kashani, A. G., Crawford, P. S., Biswas, S. K., Graettinger, A. J., & Grau, D. (2015). Automated tornado damage assessment and wind speed estimation based on terrestrial laser scanning. *Journal of Computing in Civil Engineering, 29*(3), Article 04014051.

Kemper, H., & Kemper, G. (2020). Sensor fusion, GIS and AI technologies for disaster management. *The International Archives of the Photogrammetry, Remote Sensing and Spatial Information Sciences, XLIII*(B3), 1677–1683.

Kerle, N., Nex, F., Gerke, M., Duarte, D., & Vetrivel, A. (2020). UAV-based structural damage mapping: A review. *ISPRS International Journal of Geo-Information, 9*(1), 14.

Kim, D., & Matyas, C. J. (2024). Classification of tropical cyclone rain patterns using convolutional autoencoder. *Scientific Reports, 14*(1), 791.

Kingfield, D. M., & de Beurs, K. M. (2017). Landsat identification of tornado damage by land cover and an evaluation of damage recovery in forests. *Journal of Applied Meteorology and Climatology, 56*(4), 965–987.

Koltunov, A., Ustin, S. L., Quayle, B., Schwind, B., Ambrosia, V. G., & Li, W. (2016). The development and first validation of the GOES early fire detection (GOES-EFD) algorithm. *Remote Sensing of Environment, 184*, 436–453.

Lacroix, P., Bièvre, G., Pathier, E., Kniess, U., & Jongmans, D. (2018). Use of Sentinel-2 images for the detection of precursory motions before landslide failures. *Remote Sensing of Environment, 215*, 507–516.

Layana, S., Aguilera, F., Rojo, G., Vergara, Á., Salazar, P., Quispe, J., ... Urrutia, D. (2020). Volcanic anomalies monitoring system (VOLCANOMS), a low-cost volcanic monitoring system based on Landsat images. *Remote Sensing, 12*(10), 1589.

Li, Z., Ning, H., Gao, S., Janowicz, K., Li, W., Arundel, S. T., Yang, C., Bhaduri, B., Wang, S., Zhu, A. X., Gahegan, M., Shekhar, S., Ye, X., McKenzie, G., Cervone, G., & Hodgson, M. E. (2025). GIScience in the era of artificial intelligence: A research agenda towards autonomous GIS. *Annals of GIS, 31*(4), 501–536.

Li, Q., Zhang, J., & Jiang, H. (2024). Incorporating multi-source remote sensing in the detection of earthquake-damaged buildings based on logistic regression modelling. *Heliyon, 10*(12).

Lindenschmidt, K. E., & Li, Z. (2019). Radar scatter decomposition to differentiate between running ice accumulations and intact ice covers along rivers. *Remote Sensing, 11*(3), 307.

Lipa, B., Barrick, D., Whelan, C., Losekoot, M., & Parikh, H. (2023). First open-coast HF radar observations of a 2-phase volcanic tsunami, Tonga 2022. *Remote Sensing, 15*(9), 2325.

Liu, W., & Yamazaki, F. (2018). Detection of inundation areas due to the 2015 Kanto and Tohoku torrential rain in Japan based on multi-temporal ALOS-2 imagery. *Natural Hazards and Earth System Sciences, 18*(7), 1905–1918.

Liu, X., Zhang, Y., Liang, C., Yang, Y., Huang, W., Jia, N., & Cheng, B. (2022). Storm surge damage interpretation by satellite imagery: Case review. *Natural Hazards, 112*(1), 349–365.

Liu, D., Zhao, T., Boiyo, R., Chen, S., Lu, Z., Wu, Y., & Zhao, Y. (2019). Vertical structures of dust aerosols over East Asia based on CALIPSO retrievals. *Remote Sensing, 11*(6), 701.

Lyu, M., Zhao, Y., Huang, C., & Huang, H. (2023). Unmanned aerial vehicles for search and rescue: A survey. *Remote Sensing, 15*(13), 3266.

Ma, W., Huang, L., Wu, X., Jin, S., Bai, W., & Li, X. (2022). Evaluation of CYGNSS observations for snow properties, a case study in Tibetan Plateau, China. *Remote Sensing, 14* (15), 3772.

Mandirola, M., Casarotti, C., Peloso, S., Lanese, I., Brunesi, E., & Senaldi, I. (2022). Use of UAS for damage inspection and assessment of bridge infrastructures. *International Journal of Disaster Risk Reduction, 72*, Article 102824.

Marlier, M. E., Resetar, S. A., Lachman, B. E., Anania, K., & Adams, K. (2022). Remote sensing for natural disaster recovery: Lesson learned from Hurricanes Irma and Maria in Puerto Rico. *Environmental Science and Policy, 132*, 153–159.

Mashaly, J., & Ghoneim, E. (2018). Flash flood hazard using optical, radar, and stereo-pair derived dem: Eastern desert, Egypt. *Remote Sensing, 10*(8), 1204.

Matin, S. S., & Pradhan, B. (2022). Challenges and limitations of earthquake-induced building damage mapping techniques using remote sensing images-A systematic review. *Geocarto International, 37*(21), 6186–6212.

Matyas, C. J., Zick, S. E., & Tang, J. (2018). Using an object-based approach to quantify the spatial structure of reflectivity regions in Hurricane Isabel (2003). Part I: Comparisons between radar observations and model simulations. *Monthly Weather Review, 146*(5), 1319–1340.

Mazzanti, P., Caporossi, P., & Muzi, R. (2020). Sliding time master digital image correlation analyses of cubesat images for landslide monitoring: The Rattlesnake Hills landslide (USA). *Remote Sensing, 12*(4), 592.

McCarthy, N., Guyot, A., Dowdy, A., & McGowan, H. (2019). Wildfire and weather radar: A review. *Journal of Geophysical Research: Atmospheres, 124*(1), 266–286.

McGovern, A., Demuth, J., Bostrom, A., Wirz, C. D., Tissot, P. E., Cains, M. G., & Musgrave, K. D. (2024). The value of convergence research for developing trustworthy AI for weather, climate, and ocean hazards. *npj Natural Hazards, 1*(1), 13.

McNamara, D., & Mell, W. (2022). Towards the use of remote sensing for identification of building damage, destruction, and defensive actions at wildland-urban interface fires. *Fire Technology, 58*(1), 641–672.

Mereuţă, A., Ajtai, N., Radovici, A. T., Papagiannopoulos, N., Deaconu, L. T., Botezan, C. S., … Ozunu, A. (2022). A novel method of identifying and analysing oil smoke plumes based on MODIS and CALIPSO satellite data. *Atmospheric Chemistry and Physics, 22*(7), 5071–5098.

Möldner, F., Hentschel, B., & Carstensen, D. (2024). Ice-Jam investigations along the Oder river based on satellite and UAV data. *Water, 16*(10), 1323.

Mondini, A. C., Guzzetti, F., Chang, K. T., Monserrat, O., Martha, T. R., & Manconi, A. (2021). Landslide failures detection and mapping using synthetic aperture radar: Past, present and future. *Earth-Science Reviews, 216*, Article 103574.

Moro, M., Saroli, M., Stramondo, S., Bignami, C., Albano, M., Falcucci, E., ... Wegmüller, U. (2017). New insights into earthquake precursors from InSAR. *Scientific Reports, 7*(1), Article 12035.

Mota, R., Pacheco, J. M., Pimentel, A., & Gil, A. (2024). Monitoring volcanic plumes and clouds using remote sensing: A systematic review. *Remote Sensing, 16*(10), 1789.

Muhadi, N. A., Abdullah, A. F., Bejo, S. K., Mahadi, M. R., & Mijic, A. (2020). The use of LiDAR-derived DEM in flood applications: A review. *Remote Sensing, 12*(14), 2308.

Mulia, I. E., Hirobe, T., Inazu, D., Endoh, T., Niwa, Y., Gusman, A. R., ... Hibiya, T. (2020). Advanced tsunami detection and forecasting by radar on unconventional airborne observing platforms. *Scientific Reports, 10*(1), 2412.

Nadimpalli, R., Srivastava, A., Prasad, V. S., Osuri, K. K., Das, A. K., Mohanty, U. C., & Niyogi, D. (2020). Impact of INSAT-3D/3DR radiance data assimilation in predicting tropical cyclone Titli over the Bay of Bengal. *IEEE Transactions on Geoscience and Remote Sensing, 58*(10), 6945–6957.

Olander, T. L., & Velden, C. S. (2019). The advanced Dvorak technique (ADT) for estimating tropical cyclone intensity: Update and new capabilities. *Weather and Forecasting, 34*(4), 905–922.

O'Loingsigh, T., McTainsh, G. H., Tews, E. K., Strong, C. L., Leys, J. F., Shinkfield, P., & Tapper, N. J. (2014). The dust storm index (DSI): A method for monitoring broadscale wind erosion using meteorological records. *Aeolian Research, 12*, 29–40.

Ounrit, I., Sinnung, S., Meena, P., & Laosuwan, T. (2022). Flash flood mapping based on data from Landsat-8 satellite and water indices. *International Journal on Technical and Physical Problems of Engineering (IJTPE)*, (51), 130–135.

Palliyaguru, R., Amaratunga, D., & Baldry, D. (2014). Constructing a holistic approach to disaster risk reduction: The significance of focusing on vulnerability reduction. *Disasters, 38*(1), 45–61.

Pascolini-Campbell, M., Lee, C., Stavros, N., & Fisher, J. B. (2022). ECOSTRESS reveals pre-fire vegetation controls on burn severity for Southern California wildfires of 2020. *Global Ecology and Biogeography, 31*(10), 1976–1989.

Perry, R. W. (2018). Defining disaster: An evolving concept. In *Handbook of disaster research* (pp. 3–22).

Phillips, B. D., Neal, D. M., & Webb, G. R. (2021). *Introduction to emergency management and disaster science*. Routledge.

Phiri, D., Simwanda, M., & Nyirenda, V. (2021). Mapping the impacts of cyclone Idai in Mozambique using Sentinel-2 and OBIA approach. *South African Geographical Journal, 103*(2), 237–258.

Plante Lévesque, V., Persent, M. A., Lhissou, R., Chokmani, K., Gauthier, Y., & Bernier, M. (2024). How ice mapping can help manage and prevent ice jams: Remote sensing monitoring of the Saint-François River, Québec. *Canadian Journal of Remote Sensing, 50* (1), Article 2391972.

Prakash, N., Bickel, V. T., & Manconi, A. (2025). Mapping landslides on Earth, Moon, and Mars using satellite imagery and deep learning techniques. In *Earth observation applications to landslide mapping, monitoring and modeling* (pp. 147–169). Elsevier.

Pulwarty, R. S., & Sivakumar, M. V. (2014). Information systems in a changing climate: Early warnings and drought risk management. *Weather and Climate Extremes, 3*, 14–21.

Qi, W., Yong, B., & Gourley, J. J. (2021). Monitoring the super typhoon Lekima by GPM-based near-real-time satellite precipitation estimates. *Journal of Hydrology, 603*, Article 126968.

Qian, C., Li, Y., Xu, Y., Wang, X., Zhang, Z., Nie, G., … Zhang, S. (2024). Tropical cyclone monitoring and analysis techniques: A review. *Journal of Meteorological Research, 38*(2), 351–367.

Reichstein, M., Camps-Valls, G., Stevens, B., Jung, M., Denzler, J., Carvalhais, N., & Prabhat. (2019). Deep learning and process understanding for data-driven Earth system science. *Nature, 566*, 195–204.

Reid, J. A., & Mooney, W. D. (2023). Tsunami occurrence 1900–2020: A global review, with examples from Indonesia. *Pure and Applied Geophysics, 180*(5), 1549–1571.

Ricciardulli, L., Howell, B., Jackson, C. R., Hawkins, J., Courtney, J., Stoffelen, A., … Escullar, M. A. G. (2023). Remote sensing and analysis of tropical cyclones: Current and emerging satellite sensors. *Tropical Cyclone Research and Review, 12*(4), 267–293.

Riggs, G. A., Hall, D. K., & Román, M. O. (2017). Overview of NASA's MODIS and visible infrared imaging radiometer suite (VIIRS) snow-cover earth system data records. *Earth System Science Data, 9*(2), 765–777.

Román, A., Tovar-Sánchez, A., Roque-Atienza, D., Huertas, I. E., Caballero, I., Fraile-Nuez, E., & Navarro, G. (2022). Unmanned aerial vehicles (UAVs) as a tool for hazard assessment: The 2021 eruption of Cumbre Vieja volcano, La Palma Island (Spain). *Science of the Total Environment, 843*, Article 157092.

Römer, H., Willroth, P., Kaiser, G., Vafeidis, A. T., Ludwig, R., Sterr, H., & Revilla Diez, J. (2012). Potential of remote sensing techniques for tsunami hazard and vulnerability analysis–a case study from Phang-Nga province, Thailand. *Natural Hazards and Earth System Sciences, 12*(6), 2103–2126.

Roy, D. P., De Lemos, H., Huang, H., Giglio, L., Houborg, R., & Miura, T. (2024). Multi-resolution monitoring of the 2023 maui wildfires, implications and needs for satellite-based wildfire disaster monitoring. *Science of Remote Sensing, 10*, Article 100142.

Rush, M., & Holguin, A. (1975). *Remote sensing utility in a disaster struck urban environment (No. NASA-CR-145696).*

Sakai, T., Hatta, S., Okumura, M., Hiyama, T., Yamaguchi, Y., & Inoue, G. (2015). Use of landsat TM/ETM+ to monitor the spatial and temporal extent of spring breakup floods in the Lena river, Siberia. *International Journal of Remote Sensing, 36*(3), 719–733.

Saunders, M. E., Ash, K. D., & Collins, J. M. (2018). Usefulness of the United States national weather service radar display as rated by website users. *Weather, Climate, and Society, 10* (4), 673–691.

Scarino, B., Itterly, K., Bedka, K., Homeyer, C. R., Allen, J., Bang, S., & Cecil, D. (2023). Deriving severe hail likelihood from satellite observations and model reanalysis parameters using a deep neural network. *Artificial Intelligence for the Earth Systems, 2*(4), Article 220042.

Schilirò, L., Bosman, A., Caielli, G. M., Corazza, A., Crema, S., Di Salvo, C., … Tommasi, P. (2025). The May 2023 Rainstorm-induced landslides in the Emilia-Romagna region (Northern Italy): Considerations from UAV investigations under emergency conditions. *Geosciences, 15*(3), 101.

Schillaci, C., Inverardi, F., Battaglia, M. L., Perego, A., Thomason, W., & Acutis, M. (2022). Assessment of hail damages in maize using remote sensing and comparison with an insurance assessment: A case study in Lombardy. *Italian Journal of Agronomy, 17*(4), 2126.

Scott, C. P., Arrowsmith, J. R., Nissen, E., Lajoie, L., Maruyama, T., & Chiba, T. (2018). The M7 2016 Kumamoto, Japan, earthquake: 3-D deformation along the fault and within the damage zone constrained from differential lidar topography. *Journal of Geophysical Research: Solid Earth, 123*(7), 6138–6155.

Sharma, E., Deo, R. C., Soar, J., Prasad, R., Parisi, A. V., & Raj, N. (2022). Novel hybrid deep learning model for satellite based PM10 forecasting in the most polluted Australian hotspots. *Atmospheric Environment, 279*, Article 119111.

Shikhov, A., & Chernokulsky, A. (2018). A satellite-derived climatology of unreported tornadoes in forested regions of northeast Europe. *Remote Sensing of Environment, 204*, 553–567.

Singh, D., Dahiya, M., Kumar, R., & Nanda, C. (2021). Sensors and systems for air quality assessment monitoring and management: A review. *Journal of Environmental Management, 289*, Article 112510.

Singha, C., Swain, K. C., Meliho, M., Abdo, H. G., Almohamad, H., & Al-Mutiry, M. (2022). Spatial analysis of flood hazard zoning map using novel hybrid machine learning technique in Assam, India. *Remote Sensing, 14*(24), 6229.

Skripniková, K., & Řezáčová, D. (2019). Comparison of radar-based hail detection using single-and dual-polarization. *Remote Sensing, 11*(12), 1436.

Smith, K., Fearnley, C. J., Dixon, D., Bird, D. K., & Kelman, I. (2023). *Environmental hazards: Assessing risk and reducing disaster*. Routledge.

Snapir, B., Momblanch, A., Jain, S. K., Waine, T. W., & Holman, I. P. (2019). A method for monthly mapping of wet and dry snow using Sentinel-1 and MODIS: Application to a Himalayan river basin. *International Journal of Applied Earth Observation and Geoinformation, 74*, 222–230.

Sterzel, T., Lüdeke, M. K., Walther, C., Kok, M. T., Sietz, D., & Lucas, P. L. (2020). Typology of coastal urban vulnerability under rapid urbanization. *PLoS One, 15*(1), Article e0220936.

Su, Y., Wu, F., Ao, Z., Jin, S., Qin, F., Liu, B., … Guo, Q. (2019). Evaluating maize phenotype dynamics under drought stress using terrestrial lidar. *Plant Methods, 15*, 1–16.

Sun, J., Yuan, G., Song, L., & Zhang, H. (2024). Unmanned aerial vehicles (UAVs) in landslide investigation and monitoring: A review. *Drones, 8*(1), 30.

Suresh, D., & Yarrakula, K. (2020). InSAR based deformation mapping of earthquake using Sentinel 1A imagery. *Geocarto International, 35*(5), 559–568.

Taloor, A. K., Singh, A. K., Kumar, P., Kumar, A., Tripathi, J. N., Kumari, M., … Johnson, B. A. (2022). Geospatial technology-based analysis of air quality in India during the COVID-19 pandemic. *Remote Sensing, 14*(18), 4650.

Tan, W., Wang, Y., Huang, P., Qi, Y., Xu, W., Li, C., & Chen, Y. (2023). A method for predicting landslides based on micro-deformation monitoring radar data. *Remote Sensing, 15*(3), 826.

Tavakkoli Piralilou, S., Einali, G., Ghorbanzadeh, O., Nachappa, T. G., Gholamnia, K., Blaschke, T., & Ghamisi, P. (2022). A Google Earth engine approach for wildfire susceptibility prediction fusion with remote sensing data of different spatial resolutions. *Remote Sensing, 14*(3), 672.

Tedim, F., & Leone, V. (2020). The dilemma of wildfire definition: What it reveals and what it implies. *Frontiers in Forests and Global Change, 3*, Article 553116.

Teltscher, K., & Fassnacht, F. E. (2018). Using multispectral Landsat and Sentinel-2 satellite data to investigate vegetation change at Mount St. Helens since the great volcanic eruption in 1980. *Journal of Mountain Science, 15*(9), 1851–1867.

Tralli, D. M., Blom, R. G., Zlotnicki, V., Donnellan, A., & Evans, D. L. (2005). Satellite remote sensing of earthquake, volcano, flood, landslide and coastal inundation hazards. *ISPRS Journal of Photogrammetry and Remote Sensing, 59*(4), 185–198.

Tsai, Y. L. S., Dietz, A., Oppelt, N., & Kuenzer, C. (2019). Remote sensing of snow cover using spaceborne SAR: A review. *Remote Sensing, 11*(12), 1456.

United Nations International Strategy for Disaster Reduction (UNISDR). (2015). *Sendai framework for disaster risk reduction 2015–2030*. UNISDR.

United Nations Office for Disaster Risk Reduction (UNDRR). (2020). *Hazard definition and classification review: Technical report.*

Urban, M., Berger, C., Mudau, T. E., Heckel, K., Truckenbrodt, J., Onyango Odipo, V., … Schmullius, C. (2018). Surface moisture and vegetation cover analysis for drought monitoring in the southern Kruger National Park using Sentinel-1, Sentinel-2, and Landsat-8. *Remote Sensing, 10*(9), 1482.

Valade, S., Ley, A., Massimetti, F., D'Hondt, O., Laiolo, M., Coppola, D., … Walter, T. R. (2019). Towards global volcano monitoring using multisensor sentinel missions and artificial intelligence: The MOUNTS monitoring system. *Remote Sensing, 11*(13), 1528.

Van Den Hoek, J., Getirana, A., Jung, H. C., Okeowo, M. A., & Lee, H. (2019). Monitoring reservoir drought dynamics with Landsat and radar/lidar altimetry time series in persistently cloudy eastern Brazil. *Remote Sensing, 11*(7), 827.

Van Westen, C. J. (2000). Remote sensing for natural disaster management. *International Archives of Photogrammetry and Remote Sensing, 33*(B7/4; PART 7), 1609–1617.

Viana-Soto, A., García, M., Aguado, I., & Salas, J. (2022). Assessing post-fire forest structure recovery by combining LiDAR data and Landsat time series in Mediterranean pine forests. *International Journal of Applied Earth Observation and Geoinformation, 108*, Article 102754.

Vickers, H., Eckerstorfer, M., Malnes, E., Larsen, Y., & Hindberg, H. (2016). A method for automated snow avalanche debris detection through use of synthetic aperture radar (SAR) imaging. *Earth and Space Science, 3*(11), 446–462.

Virtriana, R., Harto, A. B., Atmaja, F. W., Meilano, I., Fauzan, K. N., Anggraini, T. S., … Suminar, W. (2023). Machine learning remote sensing using the random forest classifier to detect the building damage caused by the Anak Krakatau Volcano tsunami. *Geomatics, Natural Hazards and Risk, 14*(1), 28–51.

Voigt, S., Giulio-Tonolo, F., Lyons, J., Kucera, J., Jones, B., Schneiderhan, T., Platzeck, G., Kaku, K., Kumar Hazarika, M., Czaran, L., Li, S., Pedersen, W., Kadiri James, G., Proy, C., Macharia Muthike, D., Bequignon, J., & Guha-Sapir, D. (2016). Global trends in satellite-based emergency mapping. *Science, 353*(6296), 247–252.

Wagner, M., Coniglio, M., Rasmussen, E., Satrio, M., Bodine, D., Candela, D., … Tirone, E. (2025). Harnessing UAS and high-resolution satellite imagery to better characterize wind damage and understand tornado behavior. *Bulletin of the American Meteorological Society, 106*(3), E492–E508.

Wagner, M. A., Myint, S. W., & Cerveny, R. S. (2012). Geospatial assessment of recovery rates following a tornado disaster. *IEEE Transactions on Geoscience and Remote Sensing, 50* (11), 4313–4322.

Wang, Y., Imai, K., Mulia, I. E., Ariyoshi, K., Takahashi, N., Sasaki, K., … Sato, Y. (2023). Data assimilation using high-frequency radar for tsunami early warning: A case study of the 2022 Tonga volcanic tsunami. *Journal of Geophysical Research: Solid Earth, 128*(2), Article e2022JB025153.

Wang, B. H., Wang, D. B., Ali, Z. A., Ting Ting, B., & Wang, H. (2019). An overview of various kinds of wind effects on unmanned aerial vehicle. *Measurement and Control, 52* (7–8), 731–739.

Wang, Y., Yuan, Q., Li, T., Shen, H., Zheng, L., & Zhang, L. (2019). Large-scale MODIS AOD products recovery: Spatial-temporal hybrid fusion considering aerosol variation mitigation. *ISPRS Journal of Photogrammetry and Remote Sensing, 157*, 1–12.

Wang, J., & Zhang, X. (2020). Investigation of wildfire impacts on land surface phenology from MODIS time series in the Western US forests. *ISPRS Journal of Photogrammetry and Remote Sensing, 159*, 281–295.

Wapler, K., Hengstebeck, T., & Groenemeijer, P. (2016). Mesocyclones in Central Europe as seen by radar. *Atmospheric Research, 168*, 112–120.

Wassie, S. B., Mengistu, D. A., & Birlie, A. B. (2022). Agricultural drought assessment and monitoring using MODIS-based multiple indices: The case of North Wollo, Ethiopia. *Environmental Monitoring and Assessment, 194*(11), 787.

Weisz, E., & Menzel, W. P. (2020). Approach to enhance trace gas determinations through multi-satellite data fusion. *Journal of Applied Remote Sensing, 14*(4), Article 044519.

Wieland, M., Liu, W., & Yamazaki, F. (2016). Learning change from synthetic aperture radar images: Performance evaluation of a support vector machine to detect earthquake and tsunami-induced changes. *Remote Sensing, 8*(10), 792.

Wikantika, K., Sinaga, A., Hadi, F., & Darmawan, S. (2007). Quick assessment on identification of damaged building and land-use changes in the post-tsunami disaster with a quick-look image of IKONOS and Quickbird (A case study in Meulaboh City, Aceh). *International Journal of Remote Sensing, 28*(13–14), 3037–3044.

Wimmers, A., Velden, C., & Cossuth, J. H. (2019). Using deep learning to estimate tropical cyclone intensity from satellite passive microwave imagery. *Monthly Weather Review, 147* (6), 2261–2282.

Wisner, B. (2016). Vulnerability as concept, model, metric, and tool. In *Oxford research encyclopedia of natural hazard science*.

Womble, J. A., Wood, R. L., & Mohammadi, M. E. (2018). Multi-scale remote sensing of tornado effects. *Frontiers in Built Environment, 4*, 66.

World Meteorological Organization. (2022). *Early warnings for all: The UN global early warning initiative for the implementation of climate adaptation* (p. 56). https://library.wmo.int/idurl/4/58209.

Wu, G., Liu, Q., Xu, H., & Wang, J. (2024). Modelling the combined impact of sea level rise, land subsidence, and tropical cyclones in compound flooding of coastal cities. *Ocean and Coastal Management, 252*, Article 107107.

Xiao, A., Xuan, W., Wang, J., Huang, J., Tao, D., Lu, S., & Yokoya, N. (2025). Foundation models for remote sensing and earth observation: A survey. *IEEE Geoscience and Remote Sensing Magazine, 13*(4), 297–324.

Xie, J., Zhou, K., Chen, H., Han, L., Guan, L., Wang, M., … Mao, J. (2024). Multi-task learning for tornado identification using Doppler radar data. *Geophysical Research Letters, 51*(11), Article e2024GL108809.

Xu, J., Zeng, F., Liu, W., & Takahashi, T. (2022). Damage detection and level classification of roof damage after Typhoon Faxai based on aerial photos and deep learning. *Applied Sciences, 12*(10), 4912.

Yang, L., Driscol, J., Sarigai, S., Wu, Q., Chen, H., & Lippitt, C. D. (2022). Google Earth engine and artificial intelligence (AI): A comprehensive review. *Remote Sensing, 14*(14), 3253.

Yang, X., Pavelsky, T. M., & Allen, G. H. (2020). The past and future of global river ice. *Nature, 577*(7788), 69–73.

Yang, W., & Qi, W. (2016). Spatial-temporal dynamic monitoring of vegetation recovery after the Wenchuan earthquake. *IEEE Journal of Selected Topics in Applied Earth Observations and Remote Sensing, 10*(3), 868–876.

You, D., Wang, S., Wang, F., Zhou, Y., Wang, Z., Wang, Y., Wang, J., Xiong, Y., & Ji, J. (2023). Monitoring and analyzing the effectiveness of the effective refuge area of emergency shelters by using remote sensing: A case study of Beijing's Fifth Ring Road. *Remote Sensing, 15*(14), 3646.

Yu, L., Leng, G., & Python, A. (2022). A comprehensive validation for GPM IMERG precipitation products to detect extremes and drought over mainland China. *Weather and Climate Extremes, 36*, Article 100458.

Yu, M., Yang, C., & Li, Y. (2018). Big data in natural disaster management: A review. *Geosciences, 8*(5), 165.

Yuan, M., Dickens-Micozzi, M., & Magsig, M. A. (2002). Analysis of tornado damage tracks from the 3 May tornado outbreak using multispectral satellite imagery. *Weather and Forecasting, 17*(3), 382–398.

Zhang, S., Xue, Y., Zhou, X., Zhang, X., Liu, W., Li, K., & Liu, R. (2022). State of the art: High-performance and high-throughput computing for remote sensing big data. *IEEE Geoscience and Remote Sensing Magazine, 10*(4), 125–149.

Zhao, N., Liu, Y., Hsu, F. C., Samson, E. L., Letu, H., Liang, D., & Cao, G. (2020). Time series analysis of VIIRS-DNB nighttime lights imagery for change detection in urban areas: A case study of devastation in Puerto Rico from hurricanes Irma and Maria. *Applied Geography, 120*, Article 102222.

Zheng, G., Liu, J., Yang, J., & Li, X. (2019). Automatically locate tropical cyclone centers using top cloud motion data derived from geostationary satellite images. *IEEE Transactions on Geoscience and Remote Sensing, 57*(12), 10175–10190.

Zheng, Z., Zhong, Y., Wang, J., Ma, A., & Zhang, L. (2021). Building damage assessment for rapid disaster response with a deep object-based semantic change detection framework: From natural disasters to man-made disasters. *Remote Sensing of Environment, 265*, Article 112636.

Zhu, X. X., Tuia, D., Mou, L., Xia, G. S., Zhang, L., Xu, F., & Fraundorfer, F. (2017). Deep learning in remote sensing: A comprehensive review and list of resources. *IEEE Geoscience and Remote Sensing Magazine, 5*(4), 8–36.

Zhuo, J. Y., & Tan, Z. M. (2021). Physics-augmented deep learning to improve tropical cyclone intensity and size estimation from satellite imagery. *Monthly Weather Review, 149*(7), 2097–2113.

Zuzak, C., Mowrer, M., Goodenough, E., Burns, J., Ranalli, N., & Rozelle, J. (2022). The national risk index: Establishing a nationwide baseline for natural hazard risk in the US. *Natural Hazards, 114*(2), 2331–2355.

Machine Learning and GeoAI for Medical Geography

12

Jason K. Blackburn[1, 2], *Hannah Herrero*[3], *Morgan C. Metrailer*[1, 2], *Treenate Jiranantasak*[1, 2], *Bryn Nurczyk*[1, 2], *Ted L. Hadfield*[1, 2], *Michael H. Norris*[2, 4] *and Kathleen A. Alexander*[5, 6]
[1]Spatial Epidemiology & Ecology Research Laboratory, Department of Geography, University of Florida, Gainesville, FL, United States; [2]Emerging Pathogens Institute, University of Florida, Gainesville, FL, United States; [3]Department of Geography & Sustainability, University of Tennessee, Knoxville, TN, United States; [4]Pathogen Analysis and Translational Health Group, School of Life Sciences, University of Hawai'i at Mānoa, Honolulu, HI, United States; [5]Department of Fish and Wildlife Conservation, Virginia Tech, Blacksburg, VA, United States; [6]Chobe Research Institute, CARACAL, Kasane, Botswana

1. Introduction

This chapter is focused on applications of remote sensing, GeoAI in the form of machine learning and genetic algorithms, in both contemporary (MaxEnt in ArcGIS Pro) and legacy modeling systems (Genetic Algorithm for Rule-set Prediction; GARP) for medical geography. Additionally, bioinformatics is used to improve pathogen characterization to optimize models in a medical geography context. Here we focus on how remote sensing data can be integrated into models to predict the distribution of a pathogen, *Bacillus anthracis*–the bacterium that causes anthrax, as a proxy for the disease. In this chapter, we describe ecological niche models (ENMs), also referred to as species distribution models (SDMs), with an emphasis on remotely derived variables (covariates), which have long been used in infectious disease ecology, evolution, and conservation research. While today spatial analysis is seeing rapid growth in machine learning and deep learning approaches to ecological niche modeling, genetic and machine learning algorithms have been common since the late 1990s, with a rapid growth in published studies and modeling approaches and platforms since the mid-2000s. Specific to disease mapping, GARP was used to predict several diseases in the 2000s, including Chagas (Peterson et al., 2002), Filoviruses in Africa (Peterson et al., 2004), Ranavirus (Ron, 2005), and anthrax (Blackburn et al., 2007). More recently, ENM has seen growth in deep learning approaches, employing neural networks (Estopinan et al., 2022; Watts & Worner, 2012). We briefly introduce and define medical geography, then review genetic algorithms for spatial disease prediction grounded in ecological niche theory. A primer on the ecological niche is presented with additional readings for those interested. Readings are also provided on a wide range of remote sensing and ENM uses in medical geography.

Remote Sensing, Big Data, and GeoAI. https://doi.org/10.1016/B978-0-443-26741-3.00001-5

2. Objective

The goal of this chapter is to define ecological niche modeling and describe the modeling process with an emphasis on remote sensing methods deriving environmental datasets. The priority of these ENMs is to predict disease risk on the landscape. As part of this, we illustrate a classical genetic algorithm for prediction from the late 1990s and show updated approaches to harvesting biological information from the model outputs. We introduce an approach to extract climatic/vegetation indices from Google Earth Engine (GEE) and build models within ArcGIS Pro version 3.3 for implementing an ecological niche model experiment directly within a desktop GIS. Typically, ENMs are performed in stand-alone software packages or scripted in R code or Python, often in integrated data environments (IDE), like R Studio and Jupyter Notebooks, respectively. This is to first provide a workflow that can be implemented locally with a modern, moderately powered desktop computer (GARP), and to illustrate the integration of machine learning directly into the desktop Geographic Information System (GIS) environment (MaxEnt).

3. Medical geography

Medical geography is a sub-discipline within geography focused on the relationships between "locale and disease" (Numbers, 2000) or more specifically environmental factors and disease. In the US, many consider the work of May (1950) as instrumental in defining medical geography and its tools. In that work, he defined the "pathological factors" as causative agents—the pathogens (e.g., viruses, bacteria, parasites), the vectors (such as insects and ticks), intermediate hosts (such as an insect or mammal required for part of a pathogen's life cycle), reservoirs (the population maintain the pathogen—such as a specific mammal species for a virus or the soil for pathogens that persist directly, the anthrax bacterium or some parasitic worms), and "man" (more appropriately humans). May (1950) also defined the "geographical factors", which were sub-divided into "physical" (e.g., climatic factors, elevation, wind, or soil, etc..), "human/social" (e.g., population density, behavior, or culture, etc..), and "biological" (e.g., vegetation, animal life, disease prevalence, or blood group [in humans], etc..). Specifically, May argued that disease is related to the geographical features. It is important to note; we present a singular definition of medical geography that aligns with the studies presented in this chapter. For an in-depth definition of the broader realms of modern health and medical geography, readers are pointed to two key textbooks (Emch et al., 2017; Kearns & Collins, 2010). For the purposes of this chapter, we can easily assign remotely sensed variables to each of these three geographical factors. For example, many satellites measure land surface temperature and radiation ("physical"). Likewise, many remote sensing efforts focus of delineating the built environment (such as bare surfaces like parking lots, or physical structures, like buildings). Finally, vegetation indices, such as the normalized difference vegetation index (NDVI) has become a key satellite-derived variable for ecological studies (Petorelli et al., 2011), including disease studies. For example, NDVI has been linked to the timing and intensity of anthrax outbreaks in wildlife in West Texas (Blackburn & Goodin, 2013).

4. Remote sensing in medical geography: A brief overview

Remotely sensed data has long been an important component in the field of medical geography. There is a rich literature on the use of remote sensing to derive spatial and temporal data relating environmental conditions to disease occurrence, beyond ENM approaches. Here we provide several key examples to illustrate the role or remote sensing in several modeling approaches, including machine learning approaches to predicting disease risk.

As a starting point, Hay et al. (1998) applied temporal Fourier processing to 8×8 km covariates of land surface temperature, middle infra-red reflectance, and normalized difference vegetation index (NDVI) from Advanced Very High Resolution Radiometer (AHVRR) from the US and High Resolution Radiometer (HRR) from Europe. These processed variables were used to predict seasonal malaria and were compared to monthly malaria admissions in Kenya. Using a regression framework, NDVI in preceding months, with an NDVI threshold of 0.35–0.40 was required for more than 5percent of the annual malaria cases. These results were then mapped as the number of month malaria admission would be expected in Kenya. Rogers et al. (2002) went on to use similar remote sensing products in a spatial modeling algorithm that applied maximum likelihood discriminant function to predict the global distribution of malaria, a seminal analysis for many ENM approaches.

Around this same time, remote sensing was being applied in local studies to relate disease intensity to the landscape. In one such study, elevation data and Landsat Thematic Mapper satellite imagery was significant in predicting the next year's hantavirus pulmonary syndrome (HPS) when in situ measures of precipitation showed no differences (Glass et al., 2000). In subsequent research related to HPS, HPS was linked to local rodent populations that are reservoirs of a closely related virus, Sin Nombre virus (SNR), and the study revealed satellite imagery could identify environmental conditions associated with SNR prevalence over time (Glass et al., 2002). That same study noted those environmental conditions mapped via remote sensing may identify refugia on the landscape where SNV can persist in local mouse populations. Similar early work linked remotely sensed sea surface temperature and sea surface height, both from AVHRR, to cycles of cholera outbreaks in Bangladesh, caused by the aquatic bacterium *Vibrio cholera* (Lobitz et al., 2000). Today, such modeling continues with improvements in computational power, higher resolution, and a greater number of parameters from available sensors.

5. Geospatial challenges in disease mapping

Estimating disease distributions across space and time is a key priority in medical geography, helping public and veterinary health professionals implement surveillance and control strategies. Mapping diseases has been a key facet of medical geography and epidemiology for nearly 200 years (Howe, 2014). Additionally, the concept of relating pathogens to ecological conditions has been a key part of medical geography (Mayer, 2014). Readers are directed to the Howe and Mayer readings, and several other important chapters originally published in 1986 and revived by the publisher in 2014. For those interested in a longer history of mapping disease distributions, readers are pointed to other seminal works (Cliff & Haggett, 2014; Cliff et al., 2004). Disease reporting is often incomplete, even under ideal conditions, especially for neglected diseases. Predictive modeling helps estimate disease extents and guide interventions. For neglected diseases such as anthrax, underreporting is common, and predictive tools are needed to estimate disease extent (Blackburn et al., 2007) and guide interventions (Blackburn et al., 2015).

Ecological niche modeling provides a framework for estimating disease distributions from incomplete point datasets. The goal is to identify non-random relationships between case locations or environmental confirmations of pathogens and ecological conditions (McNyset, 2005). Point data sources include public health and hospital reports, published literature, and field studies. Environmental covariates are derived from interpolated ground station data (such as soil, precipitation, and temperature) and remotely sensed satellite-borne sensors (such as vegetation, land cover, and surface temperature). These relationships are identified through pattern matching and statistical testing. Advances in GeoAI have significantly improved the accuracy and scalability of these modeling approaches. Today, there are numerous ways to pattern match or statistically predict species distributions including:

- Genetic Algorithm for Rule-set Prediction (GARP), see detailed methods below.
- Maximum Entropy (MaxEnt), see detailed methods below.
- Logistic regression for presence/absence (a part of GARP and often used as a standalone approach). In this framework, locations of cases (disease or vector locations) serve as presence points, and locations without events are absence points. A logistic regression is applied to differentiate presence (1) and absence (0) using a set of environmental covariates. See (Kessler et al., 2018) for an application to model tick distributions in Florida) and discussion on presence versus absence.
- Machine learning with Random Forest–a non-parametric approach using classification and regression trees built using randomly selected bootstrap samples of the training data, which in this chapter represent specific locations of disease reports, here anthrax cases in both case studies, where the number of bootstrap samples is equal to the number of trees selected. Each tree is split by randomly sampling several predictor variables to use at each node and then choosing the best split. Model error estimates are obtained by internal splits of the training data and then predicting the data not used to build a tree, referred to as out-of-bag and aggregating these predictions for each ensemble of trees. See Breiman (2001) for details on Random Forest and Kracalik et al. (2017) for a specific application to modeling anthrax in Ghana.

- Boosted Regression Trees–often described as a best practice for ENMs (Carlson et al., 2019), applies a combination of regression trees and boosting to fit an ensemble of regression trees to differentiate presence and absence records using environmental covariates. The boosting is a means of reducing error or improving model fit from tree to tree and has been shown to improve overall model performance. For a review of the BRT method as an ENM, see Elith et al. (2008). Carlson et al. (2019) applied the approach to model the global distribution of anthrax and Otieno et al. (2021) applied BRTs to model anthrax in Southern Kenya.

6. Theoretical foundations

6.1 Remote sensing for ecological niche models

Remote sensing plays a key role in ENM approaches. Ecological niche models rely on gridded covariates as sources of independent variables in models and many datasets are derived from satellites. Earlier satellite datasets include Advanced Very High Resolution Radiometer (AVHRR), and they were applied in some of the first models of arthropod ENMs, including a discriminant function approach (Rogers et al., 2002). An extensive remote sensing effort by Hay et al. (2006) led to the development of several temporal Fourier processed variables at 8 and 1 km, that were available via DVD provided in a sleave of the book (eventually downloadable), including measures of temperature, precipitation, and NDVI. The Fourier processing allowed for derivation of annual means, minima, maxima, as well as the phase and amplitude of change over the year. These data have been used extensively across the ENM literature, including several anthrax studies (Blackburn et al., 2017; Kracalik et al., 2017). Today these types of processed data are available at much higher spatial resolution data and can be more quickly derived via cloud-based computing, including Google Earth Engine [GEE] (Zhou et al., 2023), which is well described in other chapters throughout this book.

Cloud computing has increased access to these types of datasets and allows for rapid development of bespoke covariate datasets for ENMs. Here, we applied a GEE workflow to derive a high resolution 30m Landsat-8 dataset of multiple covariates to model anthrax in West Texas. For our workflow, we used GEE to develop covariates of interest and export them to GeoTIFF format for use in the desktop ArcGIS Pro 3.3 for model building. Notably, some workflows are available for covariate development and direct ENM implementation within GEE (Zhou et al., 2025).

Ecological niche modeling has been applied across various disease systems and spatial scales. However, few studies have directly examined how model performance changes with varying spatial resolution of covariates or with changes in the spatial extent of predictions. Researchers must balance the geographic extent of a study with the definition of the ecological niche.

6.2 The ecological niche

The niche is not simply a geographic boundary but a set of ecological and biological conditions that determine species survival. Early work by Grinnell (1917) defined the ecological niche as the geographic and ecological conditions necessary to maintain a species. Hutchinson later refined this concept by describing the niche as an n-dimensional hypervolume that captures the ecological and biological requirements of a species (Hutchinson, 1957). It is important to note that in this chapter we focus on the ecological niche as defined by abiotic factors (such as temperature, soil pH, and precipitation) and indices that describe greenness, from tasseled cap (Crist & Cicone, 1984) and biomass (such Normalized Difference Vegetation Index [NDVI] (Petorelli et al., 2011), numerous variables which are derived from remote sensing data. While these can describe the ecological conditions for species presence, they do not consider biological considerations, such as competition between two species for resources that might reduce the available space for a species (Macarthur, 1972). As a starting point for appreciating these concepts, readers are directed to Soberón and Nakamura (2009) for an in-depth review of ecological niche concepts for ENM approaches and Blackburn (2010) for a primer on ENMs for disease ecology. It is critical to note for readers that newer and more sophisticated models do not replace a detailed knowledge of the foundational ecological niche theory and how these ENM approaches are hypothesis tests of those concepts.

6.3 Ecological niche modeling

The evolution of ENMs/SDMs has been driven by the rapid growth of computer processing power (from local desktops to high performance supercomputers) and availability of species occurrence data from efforts to digitize historical data (see [Blackburn et al., 2007]). Additionally, online data repositories for host species and vectors exist, such as the Global Biodiversity Information Facility (GBIF; https://www.gbif.org/) and NatureServe (https://explorer.natureserve.org/) provide access to point locations of species of interest for modeling components of disease transmission.

Although statistical methods vary, the primary function is identifying non-random associations between known disease occurrences and environmental variables to estimate potential geographic distributions (Blackburn et al., 2007; McNyset, 2005). Ecological niche models are rooted in ecological niche theory, which states that no two species occupy the same ecological niche (Blackburn, 2010). When modeling disease distributions with ENMs, the species predicted is often the target pathogen (e.g., bacteria, virus, parasite, or fungus), with laboratory- or case-defined reports serving as a proxy, and the disease results from host infection. These ENMs are especially useful for indirectly transmitted pathogens, such as vector-borne diseases transmitted by insects or ticks or environmentally mediated diseases, those acquired through contact with contaminated water, soil, or animal meat. Vector-borne diseases have been modeled extensively using ENMs. Readers are pointed to several readings to review

ENMs and spatial analysis vector-borne diseases (Cuervo et al., 2023; Molina-Guzmán et al., 2022).

Importantly, ENMs for public health require the model to specify which part of the disease system is being predicted. It is important to understand the mode of transmission to best determine which population to model. We can broadly define diseases as *directly transmitted*, such as influenza or COVID-19, where one host can directly infect another through close contact and virus-infected air droplets (Killingley and Nguyen-Van-Tam 2013; Mehraeen et al., 2021). Alternatively, we define *indirect transmission* as one host being infected by a vector (such as an insect or tick) or direct contact with the pathogen from an environmental source, such as soil or contaminated bones or meat. For example, modeling efforts may focus on rodent hosts or insect vectors for diseases like plague or tularemia (Hightower, 2012; Holt et al., 2009), mosquito vectors of malaria and dengue (Lippi et al., 2019), or the pathogens themselves for environmentally persistent pathogens, such as anthrax (Blackburn et al., 2007) or melioidosis (Limmathurotsakul et al., 2016). Briefly, ENMs detect non-random associations using bioclimatic rules, range rules, or regression trees. Machine learning (ML) algorithms refine these models through iterative statistical evaluations. One early ML tool, GARP, simulates evolutionary processes such as point mutations (changing single values within covariates), deletions (removing variables or rule components), insertions (adding new covariates), and crossover (combining parts of different rules).

7. Data for ENMs in disease mapping

ENMs provide a framework for estimating disease distributions from incomplete point data, identifying relationships between case locations and ecological conditions derived from remote sensing or ground interpolation. Case data sources include:

- Public health and hospital reports
- Published literature
- Field studies

Environmental covariates are derived from:

- Interpolated ground station data (such as precipitation and temperature)
- Satellite sensors (such as vegetation, land cover, and surface temperature)

While the details from Table 12.1 below are the datasets used in this chapter's case studies, they are exemplar of remote sensing datasets that are often used in medical geography studies.

Table 12.1 Environmental covariate descriptions, spatial resolution, and sources of for variables used in ENMs for this case study.

Variable name	Native resolution	Description	Source
		Case study 1	
Soils	1 km	Soil variables, see below at higher resolution	SoilsGrids 1 km
MODIS imagery	250 m	16-day composite NDVI products	NASA
Case study 2			
Soils			
Bulk density	250 m	Indication of soil compaction	Soil grids 2.0
Organic carbon	250 m	Organic content in soil	Soil grids 2.0
Soil pH	250 m	Measure of pH in water at location	Soil grids 2.0
Environmental			
Elevation	30 m	Digital elevation model	U.S. Geological society
Satellite covariates			
NDVI min, mean, & max	30 m	Normalized difference vegetation index: Proxy for vegetation biomass and health	Landsat 8 via google earth engine
TCAP brightness min, mean, & max	30 m	Tasseled cap brightness: Proxy for bare earth and built environments	Landsat 8 via google earth engine
TCAP wetness min, mean, & max	30 m	Tasseled cap wetness: Proxy for moisture	Landsat 8 via google earth engine

8. Genetic algorithms: Machine learning for predicting species distributions

Ecological niche models are typically genetic algorithms or multi-tree algorithms designed to find *non-random patterns between point distributions of species' presence points and environmental covariates* describing ecological conditions that are biologically relevant to the species (Blackburn, 2010). In ENMs, variable space represents ecological conditions captured by raster data, which may be derived from remote sensing or interpolated ground measurements. The goal is to predict statistically meaningful geographical distributions of a target species on the landscape. As stated earlier, ENMs are rooted ecological niche theory and should be applied to individual species, such as a single host, vector, or pathogen. The user needs to understand the biology of the target species well enough to determine which component (host, vector, or pathogen) to use as input point data. Historically, the ecological niche was divided into the *fundamental niche*, which includes all possible conditions under which a species can survive, and the *realized niche*, which reflects the conditions under which a species exists due to competition, habitat fragmentation, and other limiting factors (Hutchinson, 1957). For medical geography, these approaches allow us to apply state-of-the-art covariate development with disease occurrence (or vector/host proxy) data to develop best estimates of disease distributions to inform risk. These models lead to maps that can inform surveillance priorities and guide control, such as educational campaigns, vaccine campaigns, or to guide exclusion areas.

8.1 Genetic algorithm for rule-set prediction (GARP)

GARP (a legacy GeoAI tool) is one of the first wide-spread presence-only modeling algorithm for predicting species distributions and was extensively tested (Anderson et al., 2003); prior to GARP, BIOCLIM was introduced in the mid-1980s and set off a generation of modeling relating species occurrence data to gridded climatic covariates (Booth et al., 2014; Nix, 1986). The GARP modeling system has been defined in detail elsewhere (Martinez-Meyer et al., 2006; Peterson & Cohoon, 1999; Stockwell & Peters, 1999). For this chapter, GARP is an iterative algorithm that aims to identify non-random relationships between point locations of species' occurrence records and environmental covariates (termed coverages in the GARP literature; raster files of climatic or environmental variables). GARP develops rules, which are a series of if/then logic statements written as one of four types (range, negated range, atomic, or logit) to describe either presence or absence of the target species in ecological space. Here are the four types with examples:

1) Range rules, which are often written in many GARP rule-sets, specify envelope(s) with upper and lower bounds for the presence of the species (e.g., IF temperature = [12.1–15.5°C] AND NDVI = [0.10–0.27] THEN species = PRESENCE).
2) In contrast, negated range rules test conditions outside of variable ranges (e.g., IF NOT temperature = [8.5–12.7°C] AND NDVI = [0.15–0.26] THEN species = ABSENCE).
3) More like traditional logistic regression approaches, Logit rules employ a logistic regression function to estimate relationships between occurrences and associated covariates (e.g., IF temperature*0.0065 + NDVI*0.0048 THEN species = PRESENCE). Four
4) The most simplistic rules are atomic and use specific covariate values to relate to occurrence (e.g., IF temperature = 12.5°C AND NDVI = 0.19 THEN species = PRESENCE).

Within a single model run, rules are written and tested internally using random draws of known (actual) presence points and the background space (i.e., pseudo-absences; in GARP, any pixel without a point is a potential pseudo-absence point). A chi-square test built on the predicted and observed values evaluates rule success at predicting presence or absence with the user's pre-defined proportion of input data.

GARP can accept, modify, or delete rules using deletions, insertions, and crossovers, to improve predictive accuracy following the rules of evolution. Each model results in a rule-set of 50 rules. Rules are then mapped to the landscape, and rule outcomes are projected onto the geography of the study area (at the resolution of the raster dataset) to develop a presence/absence map predicting the species' potential geographic distribution. For examples of the relationship between rulesets and geographic predictions see Blackburn (2006), Joyner (2010), and Stockwell (1999). Due to the genetic nature of the rule-set development and random resampling of points, GARP is a random walk through variable space. Because of this, GARP is run as an experiment, commonly fitting up to 200 models and selecting the best 10 through a subset process. As described by Blackburn et al. (2017), a standard GARP experiment should be repeated *n* times to ensure that user splits in training and testing data do not bias models results. As best practice, 10 random divisions of available presence data should be split to the user allowance (e.g., 80% training and 20% external testing) and 10 experiments created and evaluated. To perform these divisions and expedite the process, Haase et al. (2021) developed an R package (GARPTools) to automate this process for pre-experiment data preparation and accuracy metric evaluation (https://github.com/cghaase/GARPTools). For GARP, experiment success is based on overall prediction accuracy (low omission of true occurrence data) and high area under the curve (AUC) scores (McNyset, 2005). The GARPTools package automates this after models are constructed in the DesktopGARP standalone GUI (the version that write rule-sets is available here: https://github.com/jkblackburn/DesktopGARP1.1.3). Throughout its history, GARP has been shown to perform well across the spectrum of species' prevalence on the landscape from rare to common making it useful for management oriented studies focused on relating geographic potential to management or conservation needs (Peterson et al., 2007).

8.2 MaxEnt: An ecological niche modeling application rooted in maximum entropy

Another type of commonly used ENM is MaxEnt. Originally, MaxEnt was released as a standalone software package, like DesktopGARP, and is still available (https://biodiversityinformatics.amnh.org/open_source/maxent/). Additionally, it is now available in ArcGIS Pro, meaning desktop GIS analysts can apply it directly within the GIS environment rapidly and evaluate model accuracy and spatial patterns; this limits the need for an IDE.

The goal of a MaxEnt experiment is to predict the potential geographic distribution of a species of interest by estimating the relationship between point locations of species occurrence and features describing the environment (covariates), many derived from remote sensing platforms. These covariates are discretized, or converted to pixels, to represent the possible space for the distribution we are modeling. The mathematics are based on the theory of maximum entropy, which can be summarized with the quote "The best approach is to ensure that the approximation satisfies any constraints on the unknown distribution that we are aware of, and that subject to those constraints, the distribution should have maximum entropy (Jaynes, 1957)" as quoted in (Phillips et al., 2006). The most straightforward review of the statistics within the MaxEnt modeling algorithm are presented in a few key papers (Elith et al., 2011; Merow et al., 2013). In short, we are using incomplete point data of species occurrence to estimate the range of conditions that support the species so we can maximize the relationship between covariates and those point locations to predict across the landscape. Statistical relationships are found between features and points to maximize the estimate of the known conditions and apply those across the landscape by fitting the final model, after internal parameter optimization, to the collection of discretized (raster) covariates. The outcomes are maps of habitat suitability, or probability of presence, variable contributions to models, and statistical evaluations of model fit. Probability of presence is used with cutoffs to define presence or absence of the species at a given location.

MaxEnt is a presence-only approach, meaning we allow the application to select some proportion of the background pixels outside of the presence pixels to make comparisons between features and locations. These are often called *pseudo-absence* points. The idea is to capitalize on what is known, actual presence, and exhaustively search background points. MaxEnt is an iterative model using penalty functions and optimization routines to make the best estimates of the approximate distribution relating features to points. Variable contributions in the form of jackknife histograms and response curves are useful for extracting biological information from models. Final maps are probability estimates of presence.

Case studies

Case study 1: Time-specific ecological niche modeling reveals environmental reservoirs for pathogen persistence: A case study with bacillus anthracis *in chobe national park, Botswana*

Background

In this study, we apply the GARP modeling system to model the distribution of anthrax in wildlife in Chobe National Park (CNP), Botswana. Anthrax, caused by *Bacillus anthracis*, is an important zoonosis, with several foci across Africa. In Botswana, few anthrax outbreaks were reported historically, with outbreaks in the Linyanti Delta reported in the 1960s (Pienaar, 1961) and 1990s (though disjunct from the park data presented here). Until 2004, there were no reports in CNP, though reports are common in mixed wildlife and livestock in the Caprivi Strip of Namibia to the north across the Chobe River (Beyer et al., 2012; Pienaar, 1961). Anthrax is widespread and frequent in livestock and humans in neighboring Zimbabwe (Chikerema et al., 2012). Several papers have applied ENMs to model anthrax in Zimbabwe, including a current day study using the MaxEnt modeling approach (Chikerema et al., 2013) along with a current distribution and future projection using climate change scenarios and an ENM ensemble approach, where several modeling approaches were employed (John et al., 2024). After a large wildlife outbreak in CNP in 2004, anthrax occurred annually, with varying intensity, though 2009. Intermittent, sporadic cases have been reported in CNP since. Intervention in such systems relies on carcass detection during surveys and decontamination efforts, primarily carcass burning (which kills the anthrax spores). Such efforts are expensive and labor intensive. In other systems, like Etosha National Park, Namibia, west of CNP, surveillance is routine, but control was abandoned and mathematical models argue the importance of anthrax in contributing meat to key vertebrate scavengers (Getz, 2009). The CNP anthrax situation requires insights into the ecology and geography of the disease to inform surveillance and control. CNP encompasses more than 11,000 km^2, much of which is difficult to traverse, prompting the need for predictive models to map disease across the park. This system also provides an opportunity to examine the spatial distribution of a novel pathogen during an establishment period. Recently, fundamental and realized niches were re-defined as the *establishment* and *persistence* niches, respectively, with sustained population success in the persistence niche (Holt, 2009). Using this as our theoretical framework, we hypothesize that areas predicted from cases when the pathogen was established (in 2004) would show high overlap to case distributions and ENMs built for subsequent years.

Methods

An illustrated workflow for this study is presented in Case Study Figure 1.1.

Field and laboratory data: Anthrax case mapping

We constructed a wildlife anthrax geospatial database for CNP using historical data compiled by KAA (Case Study Figure 1.1). Precise GPS locations from wildlife carcasses discovered during park surveillance from 2004–08 were cataloged and mapped. Cases were documented as part of efforts to identify and decontaminate carcasses to reduce future anthrax risk and delimit the extent of the disease, as it had been undocumented in CNP prior to 2004. Information on the estimated date of death, date found, wildlife species and sex (when carcass condition was adequate for determination) were recorded. When carcasses were fresh, blood microscopy was used to determine the presence of chains of *Bacillus anthracis*-like rods, a classical diagnostic for anthrax outbreaks in the absence of laboratory confirmation using modern techniques, like polymerase chain reaction (PCR; Case Study Figures 1.1 and 1.2). To build ENMs, we split the outbreak data by year to develop a separated data set for the 5 years from 2004 to 2008 (Case Study Figures 1.1–1.3).

Remotely sensed environmental variables

For this study, our goal was to develop an ENM experiment for 2004 using case data and environmental data from that year on each of the next four consecutive years. Toward this, we downloaded 250m resolution MODIS 16-day NDVI data from NASA (https://ladsweb.modaps.eosdis.nasa.gov/missions-and-measurements/products/MOD13Q1#overview). For each year, 2004–08, we downloaded all available images and

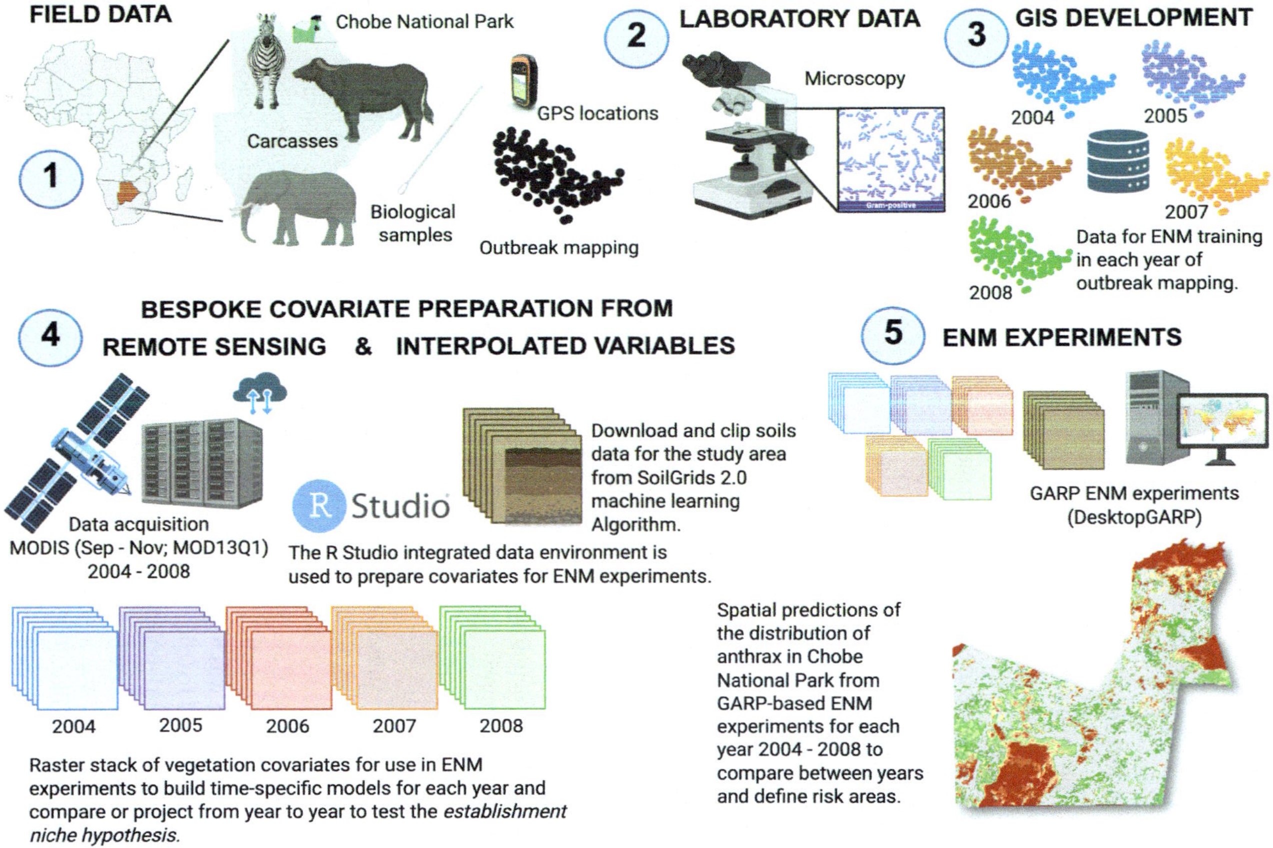

Case Study Figure 1.1 A conceptual diagram applying time-specific ecological niche modeling to anthrax data confirmed through field and laboratory work. In this work flow, field data (1) provide GPS locations of anthrax cases; (2) laboratory work confirms the pathogens; (3) annual case data are derived in a GIS from field data; (4) a remote sensing workflow derives a dataset of soils variables and annual NDVI metrics for 2004–08, with the start and end dates of reported cases used to define anthrax season; (5) ecological niche modeling experiments are performed for 2004, projected onto subsequent annual climate data with accuracy metrics used to compare between years.

derived the mean, min, max, range, and standard deviation of each for the images most closely captured to the first and last reported case in each year (Case Study Figures 1.1–1.3, 2.1). Additionally, we used soils variables at 1 km from the SoilGrid (pH, organic content, % sand, and % vertisols) (Hengl et al., 2014) and resampled them to match the 250 m MODIS. These soils variables have been used in several anthrax ENM studies (e.g., Blackburn et al., 2017; Kracalik et al., 2017).

Statistical analysis

Ecological niche modeling experiments

Ecological niche modeling experiments were performed using DesktopGARP (DG; Case Study Figures 1.1–1.3, 2.1, 2.2) version 1.1.3 (http://www.lifemapper.org/desktopgarp/). Here, we used case data from field efforts to develop annual ecological niche model (ENM)-based predictions for each of those years of *B. anthracis* across CNP (2004–08) (Case Study Figure 1.2A). We built ENMs using time-matched normalized difference vegetation index data (NDVI) and soils variables (pH, organic content, % sand, and % vertisols). For 2004, we developed 10 experiments, selected the best performing best subset from those based on accuracy metrics (Blackburn et al., 2017) and then projected that experiment on the time-specific covariate set for each subsequent year. We then validated those projections with that year's cases to derive accuracy metrics (AUC, Omission, and Commission following [Blackburn et al., 2007]). We used the Warren's I metric to evaluate niche overlap between projections and native years following Warren et al. (2010), with values close to one indicating niche overlap. Warren's I is a diversity index applied to ENM-defined predictions to measure similarity. Warren's I values are plotted as histograms to compare across years. To illustrate overlap, we overlaid all years and examined areas of greatest overlap to define where *establishment* (2004 estimates) and *persistence* (multiple years) overlapped.

Results

The best performing subset for 2004 was experiment 8 of 10 (Case Study Figure 1.2B). All accuracy metrics for the 2004 models and subsequent accuracy metrics from 2004 projected onto other years, calculated with cases from the projected year, are summarized in Case Study Table 1.1. The models from 2004 predicted between ~79% and ~98% of all cases from each year 2005–08, based on average omission; average omission was highest for predictions from 2004-05. The AUC for each year ranged from 0.917 to 0.96, suggesting 2004 models predicted the distributions of subsequent years well. All potential distributions from 2004 onto subsequent years are mapped in Case Study Figure 1.2C–F.

Across years, there was a high amount of overlap between projected and native years. 2004 was most like 2006 and 2008. The least overlap was between 2004 and 2005, with the lowest Warren's I statistic (Case Study Figure 1.2G). Broadly, all years predicted northern Chobe, southwestern Chobe, and east central Chobe. The greatest variability was in the interior of the park, and this variability is most likely linked to

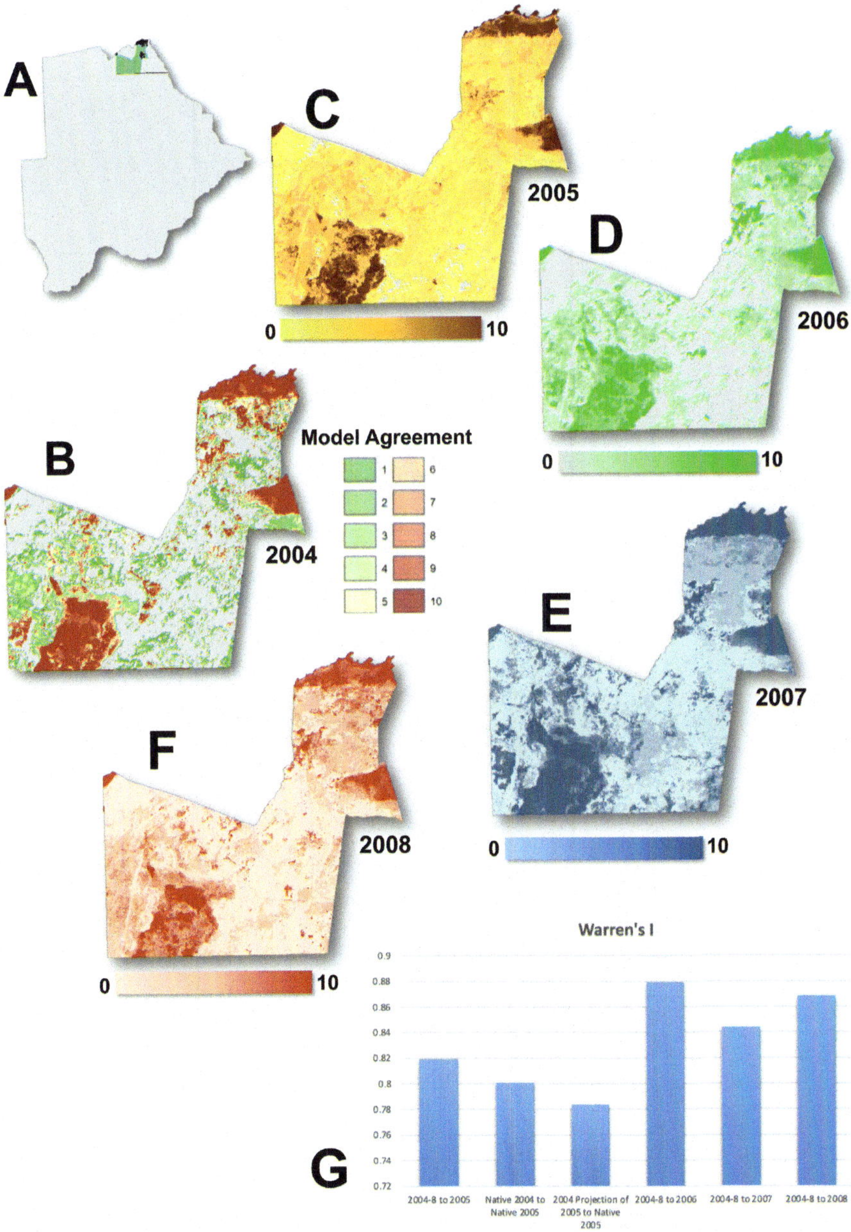

Case Study Figure 1.2 The location and distribution of anthrax points in chobe national park, botswana (A). The predicted distribution of anthrax in chobe in 2004 using that year's case data and covariates (B). Projections of the 2004 model onto the covariates of each 2005 (C), 2006 (D), 2007 (E), and 2008 (F). The warren's I statistic measures overlap of ENM predictions, with values closer to one indicating high overlap (G). Note all year were greater than 0.78, with most years above 0.8, indicating relative high overlap between cases in a year and the projection of the 2004 model onto that year's covariates.

Case Study Table 1.1 GARP experiment accuracy metrics for all 10 experiments run for 2004 using the 2004 cases and environmental covariates. The 2004-8 experiment was chosen to project onto each year (2005–08) and accuracy metrics were derived from that year's case locations. (*AUC*, area under the curve; *SE*, standard error).

Scenario	GARP experiment	AUC	z-score	SE	Total commission	Average commission	Total commission	Avg commission
Native 2004								
	2004–08	**0.903**	**16.56**	**0.022**	**13.66**	**28.33**	**0.03**	**4.48**
Projections								
2004–08 to 2005		0.917	13.19	0.024	2.19	29.395	0	20.16
2004–08 to 2006		0.933	21.8	0.013	4.23	22.39	0	15.12
2004–08 to 2007		0.911	18.11	0.019	15.21	35.84	0	2.09
2004–08 to 2008		0.96	15.74	0.015	5.17	22.46	0	3.75

changes in vegetation moisture associated with inter-annual variability of rains. Previously, we illustrated long-term averaged annual variability in the north of Botswana (Alexander et al., 2012). Here our models are detecting these changes at higher spatial and temporal resolution.

Across all years, there was general model agreement that the Chobe River waterfront in the northeast of the park, an area south of there on eastern border between the Tchinga Pan and Mokororo Pan, the Savuti Marsh and areas south to the park boundary, and a small area in the northwest corner near Linyanti are areas of suitable habitat for *Bacillus anthracis* (Case Study Figure 1.3). There was some variation from year to year across these predicted areas suggesting that changes in the remotely sensed NDVI from year to year influence the boundaries of predicted areas with an overall core area identified by reclassifying the GARP outputs from each year as absent (0–5 models predicted presence in a pixel) or present (6 or more models predicted pixels present). This cutoff follows Morris et al. (2016) as the moderate risk definition for an anthrax model of western Montana. The areas of overlap represent the extent of

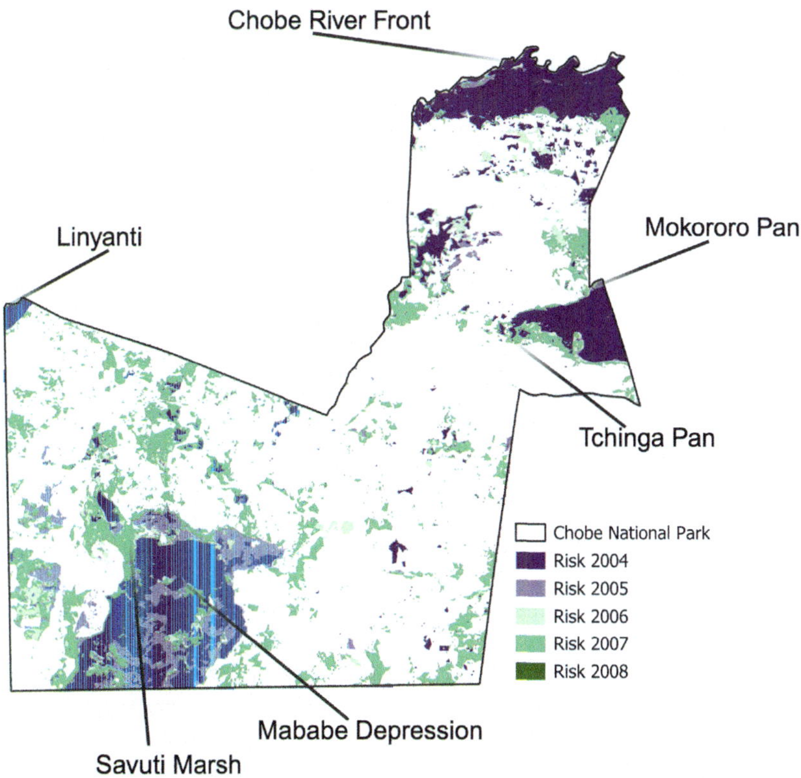

Case Study Figure 1.3 Overlapping estimates of anthrax risk from the native model of 2004 and projected years 2005–08 using the 2004 model. In this figure, areas of high overlap (purple colors) reflect the distribution of the persistence niche in areas overlapping the 2004 establishment distribution.

the persistence niche can be used to prioritize surveillance and education of park personnel and local community members of areas most at risk for future anthrax outbreaks.

Conclusions

Here we show that the initial introduction of anthrax into the park 2004 could be used to successfully model potential distributions for each of the next 4 years as the pathogen established in the park. Inter-annual variability was detected, particularly 2005, but by the later years, there was high agreement spatially and supported by high Warren's I values. While preliminary, these results suggest that modeling anthrax in areas of recent introduction (*establishment*) successfully predict areas of *persistence*; in other words our results are a direct test of the *establishment* and *persistence* niche defined by Holt (2009). Uses of this approach should be considered with other environmental pathogens. Here, such maps can be used to prioritize anthrax surveillance and or control strategies with the goal of reducing personnel and resources for future carcass searching or decontamination efforts.

Case study 2: Building ecological niche models of specific **bacillus anthracis** *lineages in west texas to improve risk predictions*

Background

In this study, we apply ecological niche modeling to predict the potential distribution of anthrax in Texas, an active anthrax zone in the Continental US with sustained cases in wildlife and livestock from year to year, with major outbreaks in some years (Blackburn & Goodin, 2013), and occasional spillover into the human population (Sidwa et al., 2020).

For this study, we apply a type of genetic fingerprinting (genotyping) to strains of *Bacillus anthracis* bacteria isolated from animal cases as a means of differentiating populations of *B. anthracis* in West Texas. Here our genotyping is based on a 25-Genetic marker system. The framework is known as the multi-locus variable number tandem repeat analysis (MLVA) and is well described elsewhere (Keim et al., 2004). Here, MLVA-25 is a tool for discriminating a group of *B. anthracis* strains to specific lineages. Phylogenetically, *B. anthracis* is divided into canonical groups, originally 12, more recently more, using canonical single nucleotide polymorphisms (SNPs) (Bruce et al., 2020). Within these canSNP groups, strains can be further characterized using MLVA profiles (Keim et al., 2004). We apply the MLVA-25 genotyping scheme of Lista et al. (2006) frequently used to characterize *B. anthracis* locally (within outbreaks). Previous studies have confirmed that the Ames-like (A3.b) and Vollum-like (A4) canSNP lineages of *B. anthracis* are established and circulate in wildlife and livestock populations in West Texas (Simonson et al., 2009; Mullins, 2013; Blackburn, Hadfield, et al., 2014; Mullins et al., 2015). Additionally, they appear to have distinct geographies, with no available evidence of co-circulation within a single ranch (of those sampled to date; Blackburn, unpublished data).

We used the MaxEnt algorithm, as implemented in ArcGIS Pro and nine covariates frequently used in other studies to predict two specific lineages of *Bacillus anthracis* recovered and genotyped using the MLVA-25 framework to examine if different MLVA-25 groups exhibit spatial and ecological differences.

Methods

An illustrated workflow for this study is presented in Case Study Figure 2.1.

This study focused on the Texas Anthrax Triangle (TAT; Case Study Figure 2.2A), an area of west Texas—primarily on the Edwards Plateau and Southern Plain, with sustained and repeat anthrax in livestock and wildlife. The TAT has poorly defined boundaries (Blackburn, Hadfield, et al., 2014). Colloquially, the TAT encompasses the space between Eagle Pass, Uvalde, and Ozona. However, of the 15 counties reporting anthrax since 2000, 10 are beyond this triangle (Sidwa et al., 2020). Here we apply an ENM approach to better predict the distribution of anthrax within the Enzootic Zone defined by Blackburn, Hadfield et al. (2014). While this area is smaller than the extent of reported counties in more recent work, outbreak data with known *B. anthracis* MLVA-25 profiles are most concentrated within the Enzootic Zone.

Sample and spatial data collection

Carcasses from mixed livestock and wildlife outbreaks in West Texas were located via ground searches using foot patrols, ATVs, or trucks. Carcass detection was aided by visual sightings, odors, and turkey vulture activity (*Cathartes aura*). GPS coordinates were recorded for spatial analysis. Samples from 2005 included various body parts and soil beneath carcasses, after 2005, we switched to collecting mostly nasal turbinates and ecological samples around carcasses including flies, rocks, vegetation, and carcass hide swabs (Blackburn, McNyset et al., 2014).

Genetic analyses

MLVA-25

Isolates collected between 2001 and 2010 were previously genotyped using classical MLVA-25 sequencing, following protocols from Lista et al. (2006), adapted by Mullins et al. (2015). Fragment sizing was performed on an ABI 3730 platform with minor modifications to PCR chemistry and primer labeling. VNTR sizing was analyzed using GeneMapper software. For any isolates from 2016 or later, genotyping was performed in silico within a next generation sequencing approach using assembled genomes and the same primer sets described in the original studies. Those methods were described in detail for an anthrax study in Vietnam (Metrailer et al., 2023). MLVA-25 types were assigned to each isolate and used here to define either A4/Vollum or A3.b/Ames lineages for input into ENM experiments.

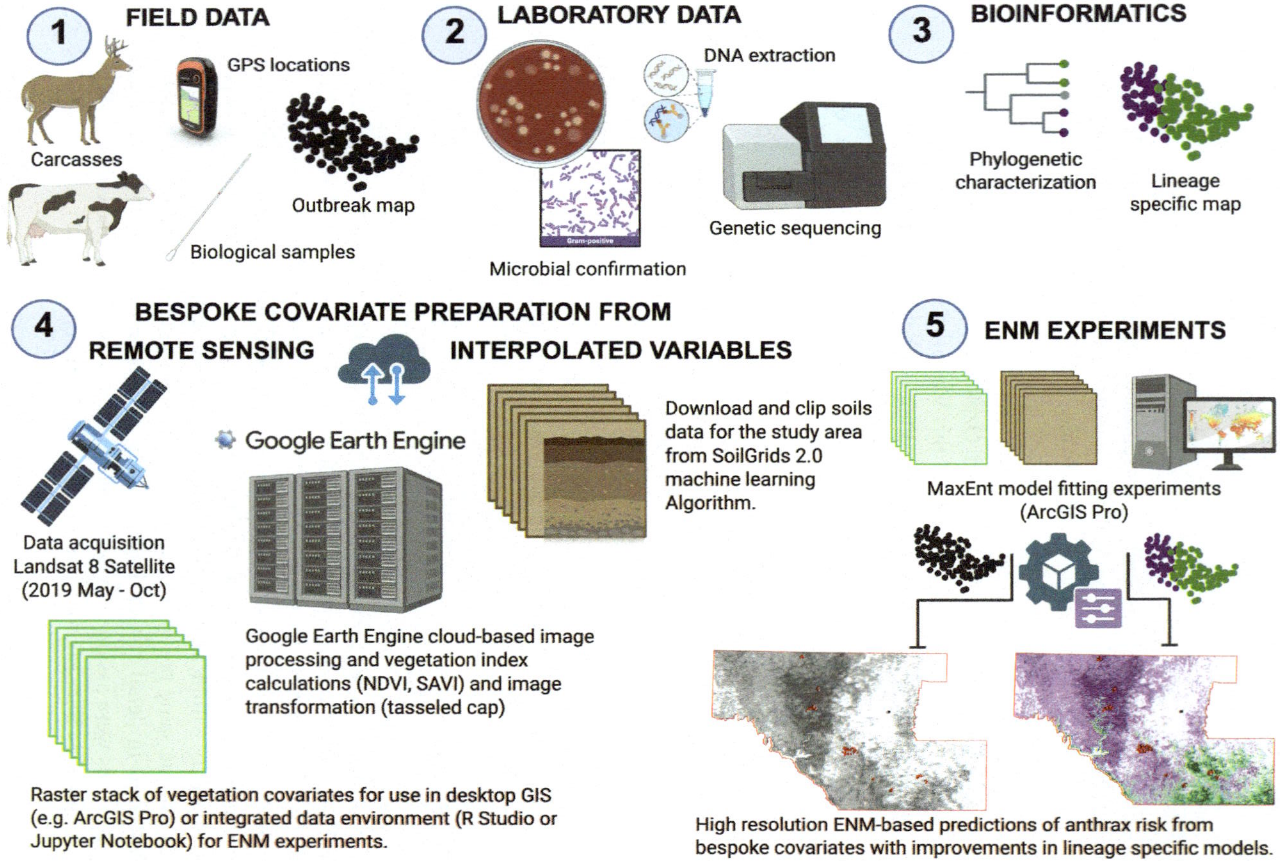

Case Study Figure 2.1 A workflow for processing field data (1) and laboratory data (2) to map anthrax. Here strains were genotyped with MLVA-25 and divided into two groups for ENM models (*bacillus anthracis* vollum and ames) (3). A google earth engine (GEE) pipeline was used to derive landsat 8-Based NDVI and tasseled cap metrics along with available soils data (4). Data from 1–4 are combined in desktop GIS environment to develop ENMs for each vollum and ames to compare to an all outbreak model (5).

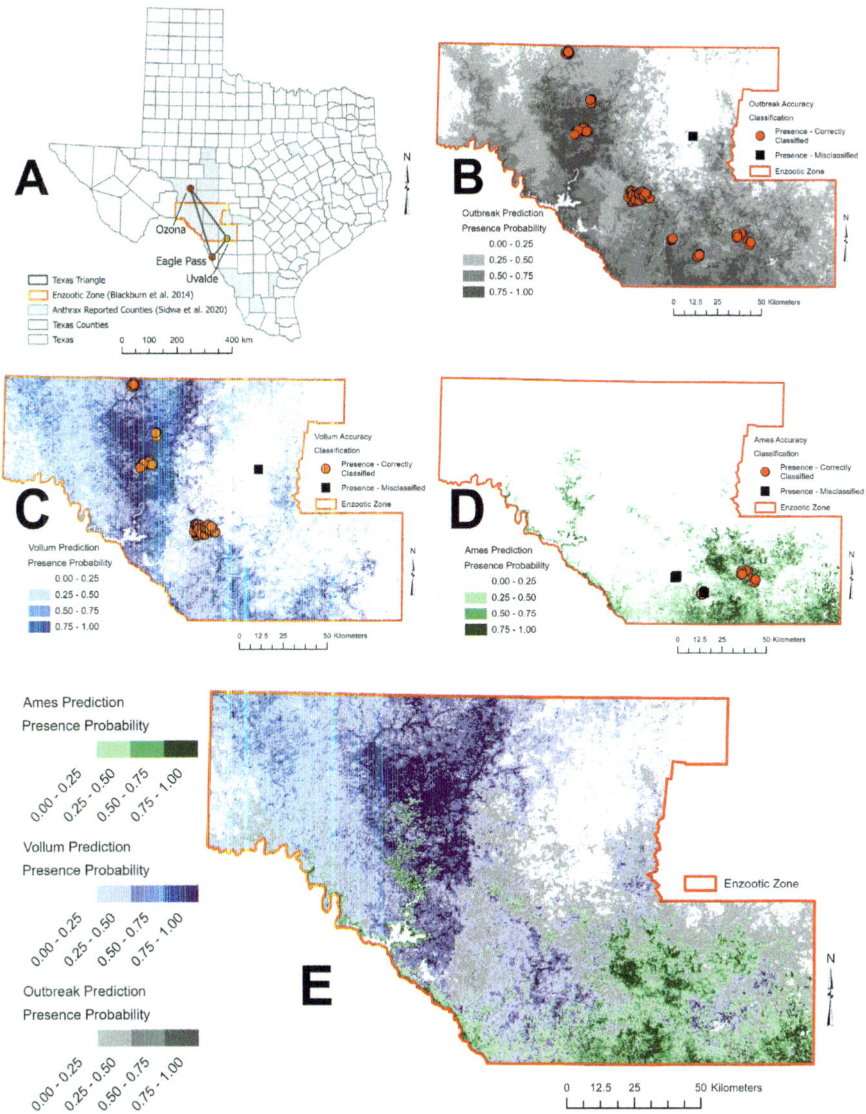

Case Study Figure 2.2 The texas anthrax triangle and enzootic zone of west texas (A). The predicted distribution of anthrax in the enzootic zone using all available case data excluding any MLVA-25 data (B). The predicted distribution of the *bacillus anthracis* vollum lineage using the same covariates (C). The predicted distribution of the *B. anthracis* ames lineage in west texas (D). Anthrax distribution in texas is best defined by a mosaic of vollum and ames models with some areas predicted by the outbreak model, with each lineage showing distinct geographies (E).

Remotely sensed and modeled environmental covariates

Environmental data were prepared using each a GEE approach for deriving season- and year-specific indices from Landsat eight and pulling soils data from the SoilsGrids 2.0 database, which is derived from another machine learning algorithm (Poggio et al., 2021). We used the Enzootic Zone as an Area of Interest in GEE and pulled cloud-filtered Landsat eight data from May to October 2019 and derived mean, minimum, maximum NDVI and Tasseled Cap-based brightness, greenness, and wetness; 2019 was the last major anthrax year reported. All code was written in Javascript (and later Python) and all variables derived in GEE and output as GeoTIFFs for use in a GIS. These data were then combined in ArcGIS Pro 3.3 and input into the MaxEnt modeling algorithm now available within ArcGIS Pro. Covariates are defined in Table 12.1 (above).

MaxEnt experiments

For this study, we used the same covariates and ran three model experiments. First, we modeled all carcass locations (outbreak), without an MLVA-25 based definition. Next, we modeled each Vollum/A4 and Ames-like/A3b, using carcass locations with known strains. We held covariates constant for all three models and used model metrics, specifically AUC and omission (false negative) rates to compare model success for each experiment. For our experiments, we used the linear model, no spatial thinning, with the relative weight of presence to background at 100. We used the complementary log-log link function (CLOGLOG), as recommended (Phillips et al., 2017), and a presence probability cutoff of 0.25 (determined through experimentation). We projected our models to the native 30 m resolution of Landsat eight matching our vegetation indices resolution. We allowed predictions outside of the data ranges and performed random resampling to allow for three groups to ensure each model was split into random testing and training for model evaluation. We then mapped each output (outbreak, Vollum, Ames) separately and in an overlay to compare predictions and produce a mosaic of *B. anthracis* distribution for the Texas Enzootic Zone at 30 m resolution for covariates associated with an epizootic year (2019).

Map production

Colors for model outputs and point features were selected by choosing color ramps that allow for differentiation when uploaded to the Coblis–Color Blindness Simulator (Coblis—Color Blindness Simulator–Colblindor) as individual model outputs and the mosaic of all three outputs. All maps were produced in ArcGIS Pro.

Results

Since this study was meant to illustrate the integration of a machine learning approach into ArcGIS Pro desktop GIS, we captured the model calculation time and report the computing power used in Case Study Table 2.1.

Accuracy metrics for each experiment are summarized in Case Study Table 2.2.

Case Study Table 2.1 Model run times for MaxEnt experiments to predict the potential geographic distribution of *bacillus anthracis,* the anthrax pathogen in west texas using 30m spatial resolution covariates. All experiments were performed using ArcGIS Pro 3.3.0 on a dell precision 7865 tower with (1) AMD ryzen threadripper PRO 5955WX processor with 16-Cores and 64.0 GB RAM. All experiments were to use 90% of the available cores in the environment of the MaxEnt toolbox.

Experiment	Model run time
Outbreak	1 Minute 33 Seconds
Vollum	1 Minute 45 Seconds
Ames	2 Minutes 23 Seconds

Case Study Table 2.2 Samples sizes available for modeling and accuracy metrics from each of three MaxEnt experiments to predict the potential geographic distribution of *bacillus anthracis,* the anthrax pathogen, across the enzootic zone of west texas.

		Accuracy metrics	
Experiment	**Carcasses (n)**	**Omission rate**	**AUC***
Outbreak	184	0.24	0.732
Vollum	119	0.45	0.77
Ames	65	0.35	0.9271

The distribution of anthrax based on the Outbreak model shows two large and distinct areas of high suitability, one in the north central area and one in the south central (Case Study Figure 2.2B). Generally, models predict a relatively large area with an AUC of 0.73 (an adequate model) and omission of ~24%. The highest rate of misclassification in the model was in cases from the central and northern outbreak locations. *AUC = area under the curve.

The distribution of the Vollum model showed a higher AUC than the outbreak model (0.77) with higher omission (~44.5%) with many of the misclassified points in the central part of the Enzootic Zone (Case Study Figure 2.2C). The highly suitable areas for Vollum are primarily in the north central region of the Enzootic Zone with a strong gradient of suitability in the west compared to east. In contrast to the Vollum distribution, the Ames distribution showed high suitability in the south of the Enzootic Zone and some overlap of the central part, where Vollum was also predicted suitable (Case Study Figure 2.2D). The AUC for this experiment was 0.92, a very high value, indicative of a very good model. The omission rate for Ames was ~35%, less than Vollum, but greater than the outbreak experiment.

When considered together (Case Study Figure 2.2E), the distribution of anthrax is a mosaic of suitable Vollum habitat in the north and Ames habitat in the south of the study area. There are some areas predicted by each of the three experiments not predicted in all experiments, suggesting that a mosaiced approach is preferred to a single experiment relying solely on AUC and omission to select "the best" output. It is important to note several features from these models. First, none of the three experiments were "ideal", and none predicted all the point data with low omission. While, the outbreak model had the lowest omission rate overall, it had the lowest of the three AUC values and did not directly overlap with the prediction of either the Vollum or Ames lineage-specific models. The relatively low AUC rates in the first two experiments are partially due to predicting a large portion of the landscape (Lobo et al., 2008; McNyset, 2005; Peterson et al., 2008), but that does not explain the high AUC for Ames. At the same time, all three experiments were limited to a single set of covariates for an "apples-to-apples" comparison. Efforts are underway to examine a larger set of covariates for each population separately, a process that can take advantage of GeoAI.

Discussion

We applied a remote sensing framework in Google Earth Engine to derive bespoke seasonal Landsat-8 covariates and added those to measures of soils conditions (also derived from a separate machine learning approach) to develop a set of independent variables for modeling. This workflow is fast and flexible and reduces data downloads before processing. It is also code-driven, meaning it is quickly repeatable and editable to adapt to other landscapes for similar covariate development. We chose MaxEnt in the ArcGIS Pro platform because it provides immediate access to a GeoAI for ENMs for analysts trained in desktop GIS environments. One advantage of this platform is that it allows users to prepare and edit the necessary point data and compile the required raster images for model preparation (after GEE was used to derive them). This makes it an ideal environment for training students or analysts new to GIS and/or machine learning algorithms. The same workflow presented here for developing these models can be implemented in an open-source environment with minor modifications. First, QGIS can be used for data preparation and editing. Second, the standalone implementation of MaxEnt can be used for running and evaluating the models, and QGIS can be used again for model production. Additionally, the same workflow can be performed using the R modeling platform, programming language, and packages such as dismo for executing the ecological niche model. In either of these two alternatives, the maximum entropy model can be replaced with a random forest or boosted regression tree approach (or other approaches, like neural networks [Estrada-Peña & de la Fuente, 2024]). There are also GEE-based MaxEnt workflows available (Zhou et al., 2025).

Conclusion

Here we have framed our literature review and case studies on map disease risk using GeoAI and modern genetic tools for characterizing pathogens. These are tools for the modern medical geographer, including the remote sensing approaches for characterizing the landscape with raster data to feed into ecological niche models. This chapter is meant to demonstrate several analysis workflows that combine cloud computing-based remote sensing for environmental data preparation with GIS and IDE processes for preparing data and running machine learning algorithms to explore complex relationships between species presence and sets of environmental covariates. In this chapter, we have presented two examples predicting the geographic distribution of *Bacillus anthracis*, the causative agent of anthrax, and provided the reader with extensive examples from the literature on other pathogens or disease systems. The examples presented here serve as starting points for medical geographers and remote sensors alike to develop new experiments and covariate development to aid in mapping disease risk.

9. Future directions

Ecological niche models are evolving as new machine learning and deep learning models become available for pattern matching between covariates and point distributions. The work presented here demonstrates that ENMs for environmentally maintained pathogens can be improved with bespoke remotely sensed covariates, increasing spatial resolution, and matching covariates to outbreak timing. With GEE and other IDEs for managing big data, and new remote platforms, such as Planet 1 m resolution, ENMs can be refined to very small parts of the landscape. This will improve model accuracy for land managers and public health personnel. Integrating ENMs with phylogenetic data will also allow for additional ecological analyses such as pathogen habitat fragmentation measures or geneflow mapping; these can improve our understanding of pathogen persistence and diversity.

References

Alexander, K. A., Blackburn, J. K., Vandewalle, M. E., Pesapane, R., Baipoledi, E. K., & Elzer, P. H. (2012). Buffalo, bush meat, and the zoonotic threat of brucellosis in botswana. *PLoS One, 7*(3).

Anderson, R. P., Lew, D., & Peterson, A. T. (2003). Evaluating predictive models of species' distributions: Criteria for selecting optimal models. *Ecological Modelling, 162*(3), 211–232.

Beyer, W., Bellan, S., Eberle, G., Ganz, H. H., Getz, W. M., Haumacher, R., Hilss, K. A., Kilian, W., Lazak, J., & Turner, W. C. (2012). Distribution and molecular evolution of bacillus anthracis genotypes in namibia. *PLoS Neglected Tropical Diseases, 6*(3).

Blackburn, J. K. (2006). *Evaluating the spatial ecology of anthrax in North America: Examining epidemiological components across multiple geographic scales using a GIS-based approach.*

Blackburn, J. (2010). Integrating geographic information systems and ecological niche modeling into disease ecology: A case study of bacillus anthracis in the united states and mexico. In E. W. S. K. P. O'Connell, A. Sulakvelidze, & L. Bakanidze (Eds.), *Emerging and endemic pathogens: Advances in surveillance, detection, and identification* (pp. 59–88). Springer.

Blackburn, J. K., & Goodin, D. G. (2013). Differentiation of springtime vegetation indices associated with summer anthrax epizootics in west texas, USA deer. *Journal of Wildlife Diseases, 49*(3), 699–703.

Blackburn, J. K., Hadfield, T. L., Curtis, A. J., & Hugh-Jones, M. E. (2014). Spatial and temporal patterns of anthrax in white-tailed deer, odocoileus virginianus, and hematophagous flies in west texas during the summertime anthrax risk period. *Annals of the Association of American Geographers, 104*(5), 939–958.

Blackburn, J. K., Kracalik, I. T., & Fair, J. M. (2015). Applying science: Opportunities to inform disease management policy with cooperative research within a one health framework. *Frontiers in Public Health, 3*.

Blackburn, J. K., Matakarimov, S., Kozhokeeva, S., Tagaeva, Z., Bell, L. K., Kracalik, I. T., & Zhunushov, A. (2017). Modeling the ecological niche of bacillus anthracis to map anthrax risk in kyrgyzstan. *The American Journal of Tropical Medicine and Hygiene, 96*(3), 550–556.

Blackburn, J. K., McNyset, K. M., Curtis, A., & Hugh-Jones, M. E. (2007). Modeling the geographic distribution of bacillus anthracis, the causative agent of anthrax disease, for the contiguous united states using predictive ecologic niche modeling. *The American Journal of Tropical Medicine and Hygiene, 77*(6), 1103–1110.

Blackburn, J. K., Mullins, J. C., Van Ert, M., Hadfield, T., O'Shea, B., & Hugh-Jones, M. E. (2014). The necrophagous fly anthrax transmission pathway: Empirical and genetic evidence from a wildlife epizootic in west texas 2010. *Vector Borne and Zoonotic Diseases, 14*(8), 576–583.

Booth, T. H., Nix, H. A., Busby, J. R., & Hutchinson, M. F. (2014). BIOCLIM: The first species distribution modelling package, its early applications and relevance to most current MAXENT studies. *Diversity and Distributions, 20*(1), 1–9.

Breiman, L. (2001). Random forests. *Machine Learning, 45*(1), 5–32.

Bruce, S. A., Schiraldi, N. J., Kamath, P. L., Easterday, W. R., & Turner, W. C. (2020). A classification framework for bacillus anthracis defined by global genomic structure. *Evolutionary Applications, 13*(5), 935–944.

Carlson, C. J., Kracalik, I. T., Ross, N., Alexander, K. A., Hugh-Jones, M. E., Fegan, M., Elkin, B. T., Epp, T., Shury, T. K., Zhang, W., Bagirova, M., Getz, W. M., & Blackburn, J. K. (2019). The global distribution of bacillus anthracis and associated anthrax risk to humans, livestock and wildlife. *Nature Microbiology, 4*(8), 1337–1343.

Chikerema, S., Murwira, A., Matope, G., & Pfukenyi, D. (2013). Spatial modelling of bacillus anthracis ecological niche in zimbabwe. *Preventive Veterinary Medicine, 111*(1), 25–30.

Chikerema, S. M., Pfukenyi, D., Matope, G., & Bhebhe, E. (2012). Temporal and spatial distribution of cattle anthrax outbreaks in zimbabwe between 1967 and 2006. *Tropical Animal Health and Production, 44*(1), 63–70.

Cliff, A., & Haggett, P. (2014). Disease diffusion: The spread of epidemics as a spatial process. *Medical Geography (Routledge Revivals)*, 84–125.

Cliff, A. D., Haggett, P., & Smallman-Raynor, M. (2004). *World atlas of epidemic diseases*. London: Arnold.

Crist, E. P., & Cicone, R. C. (1984). Application of the tasseled cap concept to simulated thematic mapper data. *Photogrammetric Engineering & Remote Sensing, 50*. ISSN: 0099-1112.

Cuervo, P., Artigas Bascur, P., Lorenzo Morales, J., Bargues Castelló, M. D., & Mas-Coma, S. (2023). *Ecological niche modelling approaches: Challenges and applications in vector-borne diseases.*

Elith, J., Leathwick, J., & Hastie, T. (2008). A working guide to boosted regression trees. *Journal of Animal Ecology, 77*(4), 802–813.

Elith, J., Phillips, S. J., Hastie, T., Dudík, M., Chee, Y. E., & Yates, C. J. (2011). A statistical explanation of MaxEnt for ecologists. *Diversity and Distributions, 17*(1), 43–57.

Emch, M., Root, E. D., & Carrel, M. (2017). *Health and medical geography*. Guilford Publications.

Estopinan, J., Servajean, M., Bonnet, P., Munoz, F., & Joly, A. (2022). Deep species distribution modeling from sentinel-2 image time-series: A global scale analysis on the orchid family. *Frontiers in Plant Science, 13*.

Estrada-Peña, A., & de la Fuente, J. (2024). Machine learning algorithms for the evaluation of risk by tick-borne pathogens in Europe. *Annals of Medicine, 56*(1).

Getz, W. M. (2009). Disease and the dynamics of food webs. *PLoS Biology, 7*(9).

Glass, G. E., Cheek, J. E., Patz, J. A., Shields, T. M., Doyle, T. J., Thoroughman, D. A., Hunt, D. K., Enscore, R. E., Gage, K. L., & Irland, C. (2000). Using remotely sensed data to identify areas at risk for hantavirus pulmonary syndrome. *Emerging Infectious Diseases, 6*(3), 238.

Glass, G. E., Yates, T. L., Fine, J. B., Shields, T. M., Kendall, J. B., Hope, A. G., Parmenter, C. A., Peters, C., Ksiazek, T. G., & Li, C.-S. (2002). Satellite imagery characterizes local animal reservoir populations of sin nombre virus in the southwestern united states. *Proceedings of the National Academy of Sciences, 99*(26), 16817–16822.

Grinnell, J. (1917). The niche-relationships of the california thrasher. *The Auk, 34*(4), 427–433.

Haase, C. G., Yang, A., McNyset, K. M., & Blackburn, J. K. (2021). GARPTools: R software for data preparation and model evaluation of GARP models. *Ecography, 44*(12), 1790–1796.

Hay, S. I., Snow, R. W., & Rogers, D. J. (1998). Predicting malaria seasons in kenya using multitemporal meteorological satellite sensor data. *Transactions of the Royal Society of Tropical Medicine and Hygiene, 92*(1), 12–20.

Hay, S. I., Tatem, A., Graham, A., Goetz, S., & Rogers, D. (2006). Global environmental data for mapping infectious disease distribution. *Advances in Parasitology, 62*, 37–77.

Hengl, T., De Jesus, J. M., MacMillan, R. A., Batjes, N. H., Heuvelink, G. B., Ribeiro, E., Samuel-Rosa, A., Kempen, B., Leenaars, J. G., & Walsh, M. G. (2014). SoilGrids1km—global soil information based on automated mapping. *PLoS One, 9*(8).

Hightower, J. (2012). Examining the distribution of francisella tularensis, the causative agent of tularemia. In *Ukraine using ecological niche modeling*.

Holt, R. D. (2009). Bringing the hutchinsonian niche into the 21st century: Ecological and evolutionary perspectives. *Proceedings of the National Academy of Sciences, 106*(Suppl. ment_2), 19659–19665.

Holt, A., Salkeld, D., Fritz, C., Tucker, J., & Gong, P. (2009). Spatial analysis of plague in california: Niche modeling predictions of the current distribution and potential response to climate change. *International Journal of Health Geographics, 8*(1), 38.

Howe, G. M. (2014). Disease mapping. In *Medical Geography (Routledge Revivals)* (pp. 35–63). Routledge.

Hutchinson, G. (1957). Concluding remarks. Cold springs harbor symp. quant. *Biol, 22*, 415–427.

Jaynes, E. T. (1957). Information theory and statistical mechanics. *Physical Review, 106*(4), 620.

John, L., Shekede, M., Gwitira, I., Mazhindu, A., Pfukenyi, D., & Chikerema, S. (2024). Modelling climate change impacts on the spatial distribution of anthrax in zimbabwe. *BMC Public Health, 24*(1), 632.

Joyner, T. A. (2010). *Ecological niche modeling of a zoonosis: A case study using anthrax and climate change in Kazakhstan*. MSc thesis (p. 150). Gainesville, FL, USA: University of Florida.

Kearns, R., & Collins, D. (2010). Health geography. In *A companion to health and medical geography* (pp. 15–32).

Keim, P., Van Ert, M., Pearson, T., Vogler, A., Huynh, L., & Wagner, D. (2004). Anthrax molecular epidemiology and forensics: Using the appropriate marker for different evolutionary scales. *Infection, Genetics and Evolution, 4*(3), 205–213.

Kessler, W. H., Blackburn, J. K., Sayler, K. A., & Glass, G. E. (2018). Estimating the geographic distribution of host-seeking adult amblyomma americanum (Acari: Ixodidae) in florida. *Journal of Medical Entomology, 56*(1), 55–64.

Killingley, B., & Nguyen-Van-Tam, J. (2013). Routes of influenza transmission. *Influenza and other respiratory viruses, 7*, 42–51.

Kracalik, I. T., Kenu, E., Ayamdooh, E. N., Allegye-Cudjoe, E., Polkuu, P. N., Frimpong, J. A., Nyarko, K. M., Bower, W. A., Traxler, R., & Blackburn, J. K. (2017). Modeling the environmental suitability of anthrax in ghana and estimating populations at risk: Implications for vaccination and control. *PLoS Neglected Tropical Diseases, 11*(10).

Limmathurotsakul, D., Golding, N., Dance, D. A., Messina, J. P., Pigott, D. M., Moyes, C. L., Rolim, D. B., Bertherat, E., Day, N. P., & Peacock, S. J. (2016). Predicted global distribution of burkholderia pseudomallei and burden of melioidosis. *Nature Microbiology, 1*(1).

Lippi, C. A., Stewart-Ibarra, A. M., Loor, M. F. B., Zambrano, J. E. D., Lopez, N. A. E., Blackburn, J. K., & Ryan, S. J. (2019). Geographic shifts in aedes aegypti habitat suitability in ecuador using larval surveillance data and ecological niche modeling: Implications of climate change for public health vector control. *PLoS Neglected Tropical Diseases, 13*(4).

Lista, F., Faggioni, G., Valjevac, S., Ciammaruconi, A., Vaissaire, J., Le Doujet, C., Gorgé, O., De Santis, R., Carattoli, A., & Ciervo, A. (2006). Genotyping of bacillus anthracis strains based on automated capillary 25-loci multiple locus variable-number tandem repeats analysis. *BMC Microbiology, 6*(1), 33.

Lobitz, B., Beck, L., Huq, A., Wood, B., Fuchs, G., Faruque, A., & Colwell, R. (2000). Climate and infectious disease: Use of remote sensing for detection of vibrio cholerae by indirect measurement. *Proceedings of the National Academy of Sciences, 97*(4), 1438–1443.

Lobo, J. M., Jiménez-Valverde, A., & Real, R. (2008). AUC: A misleading measure of the performance of predictive distribution models. *Global Ecology and Biogeography, 17*(2), 145–151.

Macarthur, R. (1972). Geographical ecology: Patterns in the distribution of species. *Harper & Row, 269*.

Martinez-Meyer, E., Peterson, A. T., Servín, J. I., & Kiff, L. F. (2006). Ecological niche modelling and prioritizing areas for species reintroductions. *Oryx, 40*(4), 411–418.

May, J. M. (1950). Medical geography: Its methods and objectives. *Geographical Review, 40* (1), 9–41.

Mayer, J. (2014). Ecological associative analysis. In *Medical Geography (Routledge Revivals)* (pp. 64–83).

McNyset, K. (2005). Use of ecological niche modelling to predict distributions of freshwater fish species in kansas. *Ecology of Freshwater Fish, 14*(3), 243–255.

Mehraeen, E., Salehi, M. A., Behnezhad, F., Moghaddam, H. R., & SeyedAlinaghi, S. (2021). Transmission modes of COVID-19: A systematic review. *Infectious Disorders-Drug TargetsDisorders), 21*(6), 27–34.

Merow, C., Smith, M. J., & Silander, J. A., Jr. (2013). A practical guide to MaxEnt for modeling species' distributions: What it does, and why inputs and settings matter. *Ecography, 36* (10), 1058–1069.

Metrailer, M. C., Hoang, T. T. H., Jiranantasak, T., Luong, T., Hoa, L. M., Pham, Q. T., Hung, T. T. M., Huong, V. T. L., Pham, T. L., & Ponciano, J. M. (2023). Spatial and phylogenetic patterns reveal hidden infection sources of bacillus anthracis in an anthrax outbreak in son la province, vietnam. *Infection, Genetics and Evolution, 114*.

Molina-Guzmán, L. P., Gutiérrez-Builes, L. A., & Ríos-Osorio, L. A. (2022). Models of spatial analysis for vector-borne diseases studies: A systematic review. *Veterinary World, 15*(8), 1975.

Morris, L. R., Proffitt, K. M., Asher, V., & Blackburn, J. K. (2016). Elk resource selection and implications for anthrax management in Montana. *Journal of Wildlife Management, 80*(2).

Mullins, J. (2013). *Combining genetic diversity and spatio-temporal data to characterize the spatial ecology of anthrax across multiple scales*. University of Florida.

Mullins, J. C., Ert, M., Hadfield, T., Nikolich, M. P., Hugh-Jones, M. E., & Blackburn, J. K. (2015). Spatio-temporal patterns of an anthrax outbreak in white-tailed deer, odocoileus virginanus, and associated genetic diversity of bacillus anthracis. *BMC Ecology, 15*(1), 1.

Nix, H. A. (1986). A biogeographic analysis of australian elapid snakes. *Atlas of elapid snakes of Australia, 7*, 4–15.

Numbers, R. L. (2000). Medical science before scientific medicine: Reflections on the history of medical geography. *Medical History, 44*(S20), 217–220.

Otieno, F. T., Gachohi, J., Gikuma-Njuru, P., Kariuki, P., Oyas, H., Canfield, S. A., Blackburn, J. K., Njenga, M. K., & Bett, B. (2021). Modeling the spatial distribution of anthrax in southern kenya. *PLoS Neglected Tropical Diseases, 15*(3).

Peterson, A. T., Carroll, D. S., Mills, J. N., & Johnson, K. M. (2004). Potential mammalian filovirus reservoirs. *Emerging Infectious Diseases, 10*, 2073–2081.

Peterson, A., & Cohoon, K. (1999). Sensitivity of distribution prediction algorithms to geographic completeness. *Ecological Modeling, 117*, 159–164.

Peterson, A. T., Papeş, M., & Eaton, M. (2007). Transferability and model evaluation in ecological niche modeling: A comparison of GARP and maxent. *Ecography, 30*(4), 550–560.

Peterson, A. T., Papes, M., & Soberón, J. (2008). Rethinking receiver operating characteristic analysis applications in ecological niche modeling. *Ecological Modelling, 213*(1), 63–72.

Peterson, A. T., Sánchez-Cordero, V., Beard, C. B., & Ramsey, J. M. (2002). Ecologic niche modeling and potential reservoirs for chagas disease, mexico. *Emerging Infectious Diseases, 8*(7), 662.

Petorelli, N., Ryan, S., Mueller, T., Bunnefeld, N., Jedrzejewska, B., Lima, M., & Kausrud, K. (2011). The normalized difference vegetation index (NDVI) in ecology: A decade of unforseen successes. *Climate Research, 46*, 15–27.

Phillips, S. J., Anderson, R. P., Dudík, M., Schapire, R. E., & Blair, M. E. (2017). Opening the black box: An open-source release of maxent. *Ecography, 40*(7), 887–893.

Phillips, S., Anderson, R., & Schapire, R. (2006). Maximum entropy modeling of species geographic distributions. *Ecological Modelling, 190*(3–4), 231–259.

Pienaar, U. de V. (1961). A second outbreak of anthrax amongst game animals in the kruger national park. 5th June to 11th October, 1960. *Koedoe, 4*(1), 4–17.

Poggio, L., De Sousa, L. M., Batjes, N. H., Heuvelink, G. B., Kempen, B., Ribeiro, E., & Rossiter, D. (2021). SoilGrids 2.0: Producing soil information for the globe with quantified spatial uncertainty. *Soil, 7*(1), 217–240.

Rogers, D. J., Randolph, S. E., Snow, R. W., & Hay, S. I. (2002). Satellite imagery in the study and forecast of malaria. *Nature, 415*(6872), 710–715.

Ron, S. (2005). Predicting the distribution of the amphibian pathogen batrachochytrium dendrobatidis in the new world1. *Biotropica, 37*(2), 209–221.

Sidwa, T., Salzer, J. S., Traxler, R., Swaney, E., Sims, M. L., Bradshaw, P., O'Sullivan, B. J., Parker, K., Waldrup, K. A., & Bower, W. A. (2020). Control and prevention of anthrax, texas, USA, 2019. *Emerging Infectious Diseases, 26*(12), 2815.

Simonson, T. S., Okinaka, R. T., Wang, B., Easterday, W. R., Huynh, L., U'Ren, J. M., Dukerich, M., Zanecki, S. R., Kenefic, L. J., & Beaudry, J. (2009). Bacillus anthracis in china and its relationship to worldwide lineages. *BMC Microbiology, 9*, 1–11.

Soberón, J., & Nakamura, M. (2009). Niches and distributional areas: Concepts, methods, and assumptions. *Proceedings of the National Academy of Sciences, 106*(Suppl. ment_2), 19644–19650.

Stockwell, D. (1999). The GARP modelling system: Problems and solutions to automated spatial prediction. *International Journal of Geographical Information Science, 13*(2), 143–158.

Stockwell, D., & Peters, D. (1999). The GARP modelling system: Problems and solutions to automated spatial prediction. *International Journal of Geographical Information Science, 13*(2), 143–158.

Warren, D. L., Glor, R. E., & Turelli, M. (2010). ENMTools: A toolbox for comparative studies of environmental niche models. *Ecography, 33*(3), 607–611.

Watts, M. J., & Worner, S. P. (2012). Using artificial neural networks to predict the distribution of bacterial crop diseases from biotic and abiotic factors. *Computational Ecology and software, 2*(1), 70.

Zhou, J., Menenti, M., Jia, L., Gao, B., Zhao, F., Cui, Y., Xiong, X., Liu, X., & Li, D. (2023). A scalable software package for time series reconstruction of remote sensing datasets on the google earth engine platform. *International Journal of Digital Earth, 16*(1), 988–1007.

Zhou, L., Zhang, Y., Chen, Z., Bai, G., Dong, X., Zhang, Y., & Hou, G. (2025). An entire-process MaxEnt framework for habitat suitability modeling on google earth engine: A case study of the oriental white stork in eastern mainland china. *Journal of Environmental Management, 386*.

Ethical and Legal Considerations in Remote Sensing and Big Data: Privacy, Justice, and Governance in the GeoAI EraS

13

Jane Southworth[1], *Erin Bunting*[2], *Cerian Gibbes*[3] *and Mia M. Bennett*[4]
[1]Department of Geography, University of Florida, Gainesville, FL, United States; [2]Department of Geography, Environment, and Spatial Sciences, Michigan State University, East Lansing, MI, United States; [3]Department of Geography & Environmental Studies, University of Colorado Colorado Springs, Colorado Springs, CO, United States; [4]Department of Geography, University of Washington, Seattle, WA, United States

Key terms and definitions

Systems of power: Systems of power are the persistent ways that authority and resources are exercised and organized across societies. Systems of power shape who makes decisions, who gets access to resoruces, and who sets norms.
Cloud-first data infrastructures: Cloud-first data infrastructures store and serve geospatial data in ways designed for cloud computing, allowing users to access, process, and analyze large datasets in the cloud rather than downloading and processing locally.
Human bycatch: Human bycatch is the incidental capture of people in remote sensing data.
Surveillance: Surveillance is systematic monitoring of personal details for purposes such as influence, management, protection, or control.
Critical remote sensing (CRS): Critical remote sensing is an approach that treats remote sensing as a political and social endeavor as well as technical science. Critical remote sensing considers how choices about data, methods, and deployment reproduce or challenge power, and how harms and benefits are distributed.
Ethical imperatives: Ethical imperatives are actions considered to be moral necessities, i.e., what we ought to do.
Ethical risks: Ethical risks are foreseeable ways that actions can produce harm, even if unintended.
Geospatial data pipelines: Geospatial data pipelines are end-to-end workflows that ingest, clean, transform, integrate, analyze, and publish geospatial data and products.
Institutional decision processes: Institutional decision processes are the formal and informal procedures organizations use to turn information into action. This includes how who authorizes action and how decisions are justified.
Feature extraction: Feature extraction is the process of identifying and delineating meaningful objects or patterns from geospatial data.
Governance models: Governance models describe how oversight and accountability are organized within and across institutions.

Remote Sensing, Big Data, and GeoAI. https://doi.org/10.1016/B978-0-443-26741-3.00004-0

Sociotechnical practice: Sociotechnical practice is work produced through the inseparable interaction of social systems (people, norms, institutions) and technical systems (data, algorithms, infrastructures).

Publication pathways: Publication pathways are the routes through which geospatial outputs become available to audiences. These include a range of formats such as papers, repositories, dashboards, and policy briefs. Each format may have a different intended audience, different levels of access, and correspondingly have ethical consequences.

Consent at scale: Consent at scale refers to the difficulty of obtaining meaningful informed consent from all people potentially affected when data collection occurs.

Equity: Equity is the pursuit of fairness in outcomes and processes with intentional consideration given to existing inequalities and power differences.

Sovereignty: Sovereignty is the authority of a people or polity to govern themselves. Data sovereignty refers to when data is owned and governed by the people or organization that generated and/or stores it.

Traditional Knowledge (TK): Traditional Knowledge refers to knowledge systems held and developed by Indigenous peoples and local communities.

Global commons: The global commons are domains viewed as belonging to all humanity.

Uncertainty quantification: Uncertainty quantification is the practice of measuring and characterizing how certain, or not, we are about a model's outputs or estimates.

Uncertainty maps: Uncertainty maps are spatial products that visualize where predictions or classifications are more versus less reliable.

Global North/Global South: Global North/Global South are terms often used to describe uneven development and power in the world system. Although generally separating the Northern and Southern Hemisphere, the terms are not strictly geographic and instead primarily account for differences in wealth, institutional capacity, and historical influence over global governance.

Knowledge production hierarchies: Knowledge production hierarchies are the uneven structures that shape which theories, methods, institutions, and data become authoritative.

Counter-mapping: Counter-mapping is mapping undertaken to challenge dominant representations of space and produce alternative spatial narratives.

1. Why this chapter matters now

1.1 Linking pixels to people, places, and power in the era of GeoAI

In an era of big data and GeoAI, this chapter explores how large geospatial datasets represent the physical world and how this representation actively shapes decision-making, social outcomes, and systems of power. Over the past decade, higher spatial resolutions, sub-daily revisit capabilities, and cloud-first data infrastructures have fundamentally changed what remote sensing (RS) can observe, how often it can make observations, and who can access those observations. This makes RS, in many contexts, intimate. What used to be "land cover" is now a driveway, a clinic car park, a protest, or a roof: things people care about, which regulators notice, and which courts will increasingly have to parse. This shift cuts across domains. In conservation, the same drone or satellite feed used to count nests or patrol park boundaries also captures "human bycatch", or people incidentally recorded while monitoring wildlife, and can be folded into deterrence campaigns that make fear a conservation tool (York et al., 2023). In urban services and water management, sub-meter imagery plus building segmentation can help generate lists of "unserved households." Yet these same lists can identify "illegal connections" if the frame changes; the line between inclusion and enforcement is a single query away (Kochupillai et al., 2022). In agriculture, parcel-level monitoring and yield modeling improve resilience, while also exposing sensitive commercial strategies at field scale. In security applications, routine tasking has eroded traditional temporal barriers, while synthetic aperture radar (SAR) expands observation beyond optical constraints, reframing surveillance around not just what is visible, but how frequently and reliably it can be observed.

That intimacy is not only technical: it is political. As Bennett et al. (2024) argue, RS is not a view from nowhere. The six-step "critical RS" framework makes visible the human decisions behind every product: who is exposed by a map (Exposure), who is consulted (Engagement), who gains capacity (Empowerment), who is taught to interpret (Teaching), who can access the data and tools (Accessibility), and who is acknowledged and held responsible for impacts (Acknowledgment). York et al. (2023) document how the same systems are framed alternately as ethical imperatives or ethical risks depending on whose harms and benefits are centered: surveillance can be stewardship, or militarization; "ecosystem monitoring" can be social sorting when it lands on human bodies (York et al., 2023).

Meanwhile, adoption by public institutions is accelerating. National mapping agencies are piloting and operationalizing GeoAI, integrating machine learning and AI techniques with geospatial data pipelines and institutional decision processes for feature extraction, change detection, and quality control, and facing the organizational consequences: governance models, skills, risk registers, and alignment with evolving AI policy (Kausika & van Altena, 2025). Cloud platforms make "planet-scale" computing accessible, but terms of service, licensing, and cost controls quietly define who can compute at scale and who cannot, posing a distributional issue as much as a technical one (Kausika & van Altena, 2025).

Public expectations are also shifting. In a US survey, most respondents were unaware that commercial satellite imagery is available but expressed discomfort when briefed, especially with frequent imaging. The acceptability of satellite imagery varied sharply by use case: search-and-rescue scored very differently than employer monitoring or celebrity tracking (McAmis et al., 2024). Comfort with spatial resolution did not translate into comfort with temporal frequency, and demography (e.g., gender and political affiliation) affected which uses were seen as overreach. At the same time, the National Oceanic and Atmospheric Administration (NOAA) has eased rapid-revisit restrictions and broadened imaging permissions, signaling a policy trend toward capability growth alongside calls for responsible use (McAmis et al., 2024).

RS is also deeply social. While often understood as a technical field focused on acquiring and processing data from satellites or aircraft, its very purpose has always been embedded in human agendas ranging from national security, agricultural planning, disaster response, to environmental management. Every satellite launched, every image acquired, and every classification scheme applied reflects societal priorities and values (Bennett, 2023). For instance, the origins of RS technologies, which emerged out of the United States and Soviet Union during the Cold War, were rooted in geopolitical surveillance and later adapted for scientific monitoring in response to growing concern for the environment, particularly at a global scale. RS has always been a sociotechnical practice that mediates relationships between people and the environment. However, the emergence of GeoAI has amplified and transformed the social dimensions of RS into explicitly sociolegal ones. The automation of analysis, interpretation, and possibly even decision-making has shifted the field from one of observation to one tightly intertwined with judgment and governance. Teams can no longer assume that pixels are "just pixels" (Pickles, 1995). Choices about resolution, cadence, training labels, validation protocols, and publication pathways are choices about exposure, consent at scale, equity of access, and trust.

1.2 The core tension: Maximizing social benefit while minimizing Harm—Privacy, equity, sovereignty, and trust

Every chapter in this book demonstrates the social value of GeoAI: disaster response, biodiversity monitoring, climate-health analytics, and more resilient water and food systems. Yet the very properties that make GeoAI powerful—scale, speed, and specificity—also magnify harm when systems fail, are misused, or are repurposed. The tension between benefit and harm is not abstract. It surfaces in everyday research decisions: what to collect, how to label, when to publish, who to consult, and who ultimately gains from the work.

Privacy is often the most visible ethical boundary. Following Bennett et al.'s (2024) recommendations, a useful test is this: if we were conducting this inquiry as fieldwork on the ground, would we be granted permission? If not, what changes when the same observation is conducted from above? In some contexts, open visibility may be ethically defensible, for example, documenting atrocities or environmental violations. In others, restraint, obfuscation, aggregation, or delayed release may be more appropriate (Bennett et al., 2024). Kochupillai et al. (2022) illustrate how quickly seemingly technical classification decisions can become stigmatizing: morphological labels such as "slum," employer monitoring through vehicle counts, or disease inference from parking lot imagery. They argue for reframing analytic questions toward "ethical opportunities" that meet policy needs without creating collateral harm (Kochupillai et al., 2022). The same imagery feed can support stewardship or enable surveillance. Ethical acceptability depends on purpose, actor, temporal cadence, and the downstream consequences of use. Searching for a missing hiker is widely regarded as legitimate in many contexts; verifying whether an employee who called in sick has a car at home is often considered intrusive. These distinctions are contextual and culturally contingent, but they are not trivial.

This dual-use character of GeoAI deserves careful attention and reflects broader concerns in AI ethics about beneficial technologies being repurposed in ways that produce unintended harm (Floridi et al., 2018). In GeoAI contexts, tools designed for conservation, disaster mitigation, or public health can be repurposed for coercive monitoring, discriminatory enforcement, or political control. The problem is not hypothetical: scale and automation lower the threshold for such repurposing. Once high-resolution monitoring systems are operational, governance determines whether they function as public goods or instruments of exclusion. As the U.S. National Space Council (2023) observes, "space data ethics" differ from ordinary data ethics precisely because dual use, cross-border implications, and national security concerns complicate open-by-default assumptions. Ethical design must therefore consider not only intended use, but plausible misuse, and build in safeguards accordingly.

Equity and sovereignty extend these concerns. The CARE Principles of Collective Benefit, Authority to Control, Responsibility, and Ethics translate into concrete practices: recording permits, attaching Traditional Knowledge (TK) and Biocultural (BC) labels, enabling embargo and obfuscation where appropriate, linking datasets to approvals, and acknowledging positionality. They also require design choices: consultation with Indigenous Peoples and local communities (IPLCs), selection of product languages, and governance structures that persist beyond a grant cycle (O'Brien et al., 2024). Sovereign interests do not map neatly onto global data commons. Open access may advance science, but it can also undermine community control or exacerbate geopolitical tensions (National Space Council Users' Advisory Group, 2023). Balancing openness with sovereignty is therefore not a procedural detail, but an ethical commitment.

Trust is built, or eroded, in technical details. It is not sufficient to assert that "the model works." Spatial autocorrelation can inflate accuracy metrics; random cross-validation can exaggerate reliability; imbalanced classes can obscure minority harms; and unreported uncertainty can mislead decision-makers. For end users, the map becomes the narrative. Publishing confident boundaries without transparent uncertainty communicates a level of certainty that may not be warranted (Koldasbayeva et al., 2024). For communities, process is equally important. Engagement that is extractive or symbolic erodes legitimacy; engagement that shapes research questions, methods, and outputs strengthens it (Bennett et al., 2024; O'Brien et al., 2024).

1.3 A running example: Mapping urban change for delivery services

Take an example to which we will return throughout this chapter. Imagine a city partnering with a university lab to map building-level change and informal settlement growth to plan water connections and waste pickup. The technical plan? Fuse sub-meter imagery with archival basemaps; fine-tune a segmentation model; run it quarterly; and publish a web map to target service expansions.

Every step presents legal and ethical choices. Resolution and cadence determine exposure: will the cadence enable "hotspotting" of evictions or raids? Label choices set narratives: "informal" versus "unserved" matters for dignity and policy response (Kochupillai et al., 2022). Validation design matters: if we evaluate with random cross-validation, we will likely overstate performance; spatial variation and uncertainty maps are needed to avoid over-promising (Koldasbayeva et al., 2024). Engagement matters, too: have residents' associations been consulted, and is there a pathway for corrections? CARE matters: is the work on Indigenous land, and if so, are permits logged, labels applied, and approvals recorded (O'Brien et al., 2024)? Governance matters: has the city completed a data protection impact assessment; is there documentation for model and data lineage; who can access the raw outputs (Kausika & van Altena, 2025)? Publication matters: should we publish a de-identified, aggregated index rather than a clickable, address-level map; do we delay release during sensitive periods (Bennett et al., 2024)? The second example we will use in this chapter focuses on the rapid convergence of RS and GeoAI and the ethical, legal, and social challenges within the context of a coastal citizen science project across the Great Lakes region of the U.S.Specifically, through a National Science Foundation funded project, we looked to develop a coastal, drone-based, citizen science project that merged the fields of RS and geomorphology with a community-centric ethics driven collaboration.

We use these examples throughout the chapter to show how "small" technical decisions are also sociolegal choices. Over time, these choices accumulate into an exposure profile, or the pattern of legal, ethical, and social risks a system creates. That profile may be defensible to regulators, courts, and affected individuals/communities, or it may fail to meet prevailing legal standards, ethical normals, or societal expectations.

2. Core ethical frames for RS and big data

We draw on three core frameworks to guide how ethics in RS and GeoAI can move from theory to practice: critical RS (CRS), justice in RS, and Indigenous data sovereignty. Together, these perspectives reflect a broader shift in geospatial science toward more reflexive, accountable, and community-centered forms of observation and governance. Rather than treating ethics as an external constraint, they ask fundamental questions about power, consent, and knowledge production. Who controls data? How is consent secured? Whose knowledge is recognized as legitimate?

Taken together, these frameworks challenge top-down models of Earth observation and emphasize participatory approaches in which Indigenous peoples and local communities, which have chosen to engage in the RS process, shape how tools are designed, applied, and governed. By centering community priorities and respecting privacy, knowledge systems, and data sovereignty, RS can move toward practices grounded in accountability, collaboration, and fairness.

2.1 Critical remote sensing as a practical scaffold

A useful way to make ethics operational in RS is to treat it as a team practice that runs across the project lifecycle. RS has traditionally been rooted in positivist epistemologies, which emphasize the value of empirical observation for all scientific inquiry. Yet the field is increasingly re-evaluated through critical theoretical lenses. CRS interrogates not just the technologies and methodologies of Earth observation, but also the power relations, ethical dilemmas, and socio-political implications of RS approaches (Kroth, 2025; Lappe et al., 2025). In contrast to extractive models, CRS emphasizes the co-production of data and embodied knowledge, while challenging dominant technoscientific narratives. Through this lens, CRS is not only a critique of technological systems but also a reimagination of how we observe and relate to the Earth. The idea of RS offering a "view from nowhere" erases the history and sociopolitical dynamics that shape the field. For example, Bennett, et al. (2022) highlight that RS history has been significantly shaped by geopolitics and that geopolitical agendas can serve to reinforce hierarchies between the Global North and Global South. Hierarchies in knowledge production can also be reinforced. Pritchard et al. (2022) examine how conservation initiatives often privilege data generated by satellites and global institutions over local ecological knowledge, which can undermine local autonomy and environmental justice. CRS acknowledges the reductive nature of commonly used methodologies and seeks to center diverse ways of knowing. In practice, CRS can yield new forms of representing the Earth.

Bennett and colleagues offer a six-step framework that is simple to remember and apply in practice: Exposure, Engagement, Empowerment, Teaching, Accessibility, Acknowledgment (Fig. 13.1; Bennett et al., 2024). The spirit of the framework is pragmatic. It asks what your work does to people and places, how you will attempt to engage those affected, who gains lasting capacity, who learns to interpret, who can access the tools and data, and how you will own the consequences.

Exposure is the first gate. It asks a permission question most projects skip. If you were doing this as fieldwork, would those affected grant consent? If not, what changes when you do research from above? The answer might be to proceed, modify the plan, or simply abstain. Practical moves include scoping what will be made visible, reducing spatial resolution or temporal frequency when visibility can cause harm, obfuscating sensitive features, aggregating outputs, delaying release during volatile periods, and designing consent pathways for local or Indigenous partners where feasible (Bennett et al., 2024).

Engagement is substantive, not ceremonial. The foundation of engagement is to consult early and co-produce if possible. Increasingly, knowledge co-production is seen as vital for advancing science that responds to societal issues. Co-production offers an empowering approach that strengthens the connections between theoretical frameworks, practical understanding, and real-world action. Cinnamon and Noth (2023) use an open data approach to develop understandings of informal settlements in Cape Town, underscoring the need for co-produced spatial knowledge. For urban change mapping, this means meeting with resident associations, utilities, and civil society before choosing labels and generating outputs. It means inviting critique of the

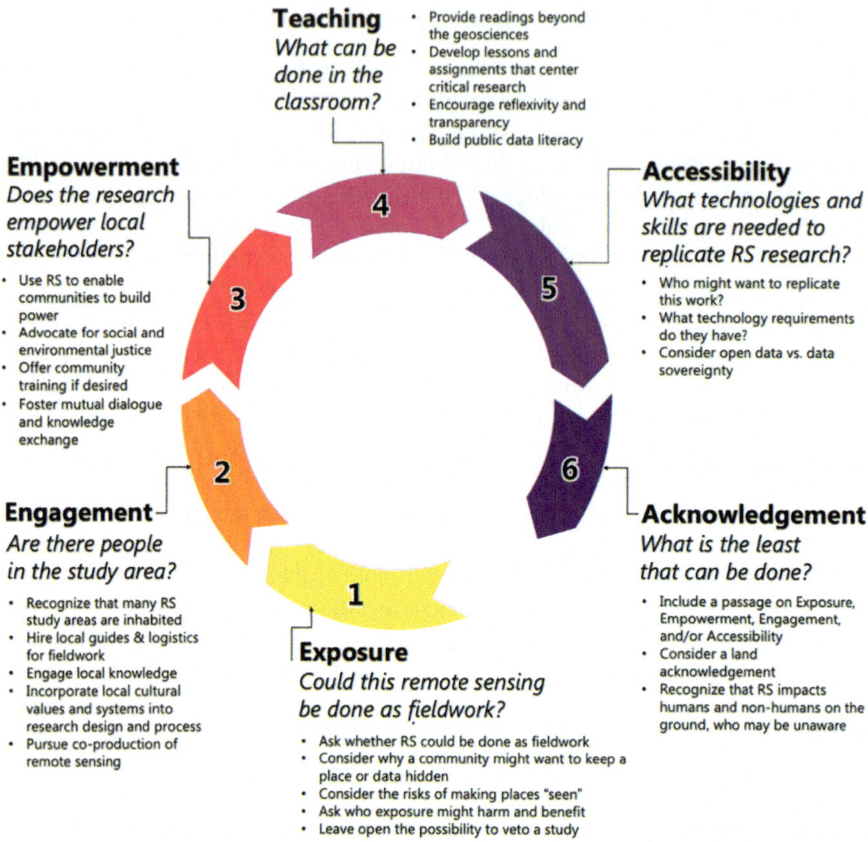

Figure 13.1 The six-step framework centered on exposure, engagement, empowerment, teaching, accessibility, and acknowlesgement, with key questions from RS scientists in their methods and practice.
From: Bennett, M. M., Gleason, C. J., Tellman, B., Alvarez León, L. F., Friedrich, H. K., Ovienmhada, U., & Matthews, A. J. (2024). Bringing satellites down to Earth: Six steps to more ethical remote sensing. *Global Environmental Change Advances*, *2*, 100003. https://doi.org/10.1016/j.gecadv.2023.100003

question, the cadence, and the way the map will be used. For conservation, it means listening to park communities and rangers about what monitoring does on the ground, for people as well as wildlife. Good engagement changes design, while bad engagement erodes legitimacy (Bennett et al., 2024).

Empowerment is about who gains agency. Teams should plan for local capacity building so that communities and institutions can use data for their own aims. This can be as simple as training sessions and shared toolkits and as ambitious as building local pipelines and including community representatives in steering groups. Segarra et al. (2024) frame RS as a tool for environmental justice. In their work, they emphasize how communities are using RS to counter state narratives, track pollution, and resist dispossession. In doing so, RS becomes a method for control and resistance. The point is to avoid extractive relationships where knowledge and capability flow in only one direction (Bennett et al., 2024).

Teaching extends empowerment. It is the work of making outputs interpretable to non-specialists, demystifying uncertainty and limits, and supporting critical use. In practice, this means explaining results clearly, using plain language, showing where predictions are uncertain spatially, and investing the time to read, discuss, and share findings not only among researchers but also with the communities being observed (Bennett et al., 2024).

Accessibility governs who can obtain data and tools. Open data and open models are strong defaults, but openness is not an absolute good. Responsible teams think through the exposure consequences of fully open release, consider tiered access where sensitive content is involved, especially in conflict zones (Bennett et al. 2022, 2024), and avoid designs that require expensive, gated platforms that exclude less resourced partners. Accessibility also includes inclusive design that is sensitive to language choice, formatting, and disabilities (Bennett et al., 2024).

Acknowledgment involves stating the researchers' positionalities and responsibilities. Teams should explicitly declare where they stand, whose knowledge they relied on, and how they will respond if harms or errors emerge. This can come in the form of including ethics statements in outputs, logging decision rationales, and intentionally striving to foster dialog post-publication. The statement is not a shield: rather, it is an invitation to scrutiny and a commitment to act (Bennett et al., 2024).

2.2 Justice in RS

Justice in RS builds upon CRS by asking not only how data is produced, but how algorithmic systems distribute risk, visibility, and benefit. In this sense, we can speak of "geospatial algorithmic justice": the obligation to ensure that automated classification, prediction, and monitoring systems do not reproduce inequities, obscure minority harms, or concentrate power in the hands of those who control sensing infrastructure (Flynn et al., 2022). Questions in conservation RS often come down to who is counted, whose harms and benefits matter, and whose interpretive frames dominate. Justice requires attention not only to outcomes, but to design choices embedded in models, release strategies, and narratives. Three tensions stand out. Many of these tensions emerge when rapidly expanding technical capabilities outpace the governance systems designed to guide their use. High-resolution sensing, automated classification, and large-scale monitoring expand what can be seen and inferred from geospatial data, but without corresponding safeguards they can also amplify risk. Fig. 13.2 conceptualizes this relationship as a "risk surface." As technical capability increases, the potential for harm grows when governance maturity remains low. In these conditions, systems designed for conservation, planning, or monitoring can drift toward coercive surveillance, human bycatch, or other forms of dual-use, or military and civilian, applications. When governance mechanisms develop alongside technical capability, however, the same systems can support more responsible and socially grounded applications.

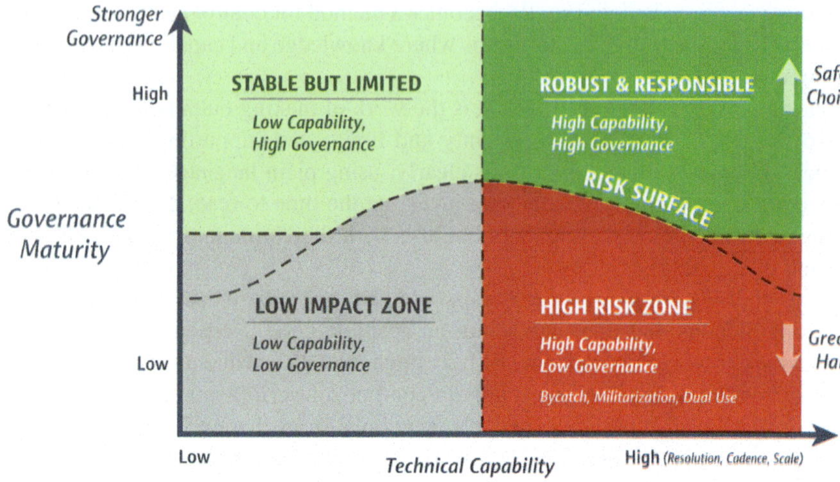

Figure 13.2 The risk surface of GeoAI and remote sensing: Ethical risk increases when technical capability (resolution, cadence, and analytic scale) grows faster than governance maturity, creating a high-risk zone where powerful sensing systems operate without sufficient safeguards.

The first is surveillance versus stewardship. York and colleagues review how surveillance can be justified as stewardship or critiqued as coercion depending on context and perspective. They document the distinction between intentional surveillance and human bycatch, or the unintentional capture of people when systems are aimed at animals. They show how the same technologies used to deter poaching through fear can alternately inflame conflict by normalizing militarized responses or empower communities through counter mapping and resistance (York et al., 2023). Within conservation, some authors frame remote monitoring as an ethical imperative to protect endangered species. Others show how surveillance campaigns produce fear and hostility among local people, and how rangers and police sometimes weaponize this visibility against marginalized groups. In an urban planning context, Wang et al. (2023) raise questions about automated pattern recognition and its influence on urban planning. While efficient, these techniques flatten diverse urban experiences into abstract typologies that can further stigmatize communities. Hancock et al. (2024) extends this to maritime surveillance, detailing how RS supports the documentation of human rights abuses at sea but also contributes to a growing digital enclosure of the oceans, as governments and private firms increasingly aggregate and control satellite imagery, vessel tracking data, and analytic outputs, shaping how ocean space is monitored and governed. This reveals a potentially ambivalent "politics of pixels" where RS can contribute to both advocacy and exploitation. The question of justice focuses on whose safety counts and who bears the cost. Teams should treat narratives that frame surveillance and monitoring as tools to deter illegal or harmful behavior with caution, avoid relying on fear as a motivator, and weigh community impacts alongside conservation gains (York et al., 2023).

Informed consent at scale is difficult in these contexts, but that difficulty is not a license to ignore it. Inform communities, adjust design to reduce human capture, commit to policies for handling incidental human imagery, and plan for requests from authorities. When illegal activity is captured, it is important to recall that many such activities are tied to livelihoods and that turning a monitoring program into an enforcement program can cause disproportionate harm to already marginalized groups (York et al., 2023).

The second tension in conservation RS is militarization. RS is often linked to military discourse more than many practitioners acknowledge. Advances in RS technologies have long been connected to military and geopolitical agendas, and this history shapes how these tools are perceived. The relationship carries associations with surveillance, control, and warfare that, if not explicitly addressed, can influence how RS projects are interpreted and deployed. For example, drones still carry strong associations with warfare, while satellite monitoring of deforestation or water extraction can be used to justify aggressive enforcement strategies. York et al. (2023) cite cases where militarized conservation escalates tensions and puts people at risk, particularly in regions with recent histories of conflict. These dynamics can worsen relationships between governments and local communities and, in some cases, place individuals in direct danger.

The data itself may be valuable, but its benefits are rarely distributed evenly. How the data is used, and who ultimately controls its application, matters greatly. Even when scientists produce data and methods for benign purposes, their eventual use often falls outside the researchers' control. For this reason, teams should make an effort to sketch out plausible use scenarios and openly discuss their potential benefits and harms, particularly with marginalized groups who are often most vulnerable to misuse of monitoring technologies. The practical takeaway is to prototype non-coercive uses first, pair monitoring with restorative and rights-based approaches, and question whether a project unintentionally normalizes coercion (York et al., 2023). Decisions about release cadence, spatial precision, and narrative framing all matter. Choosing stewardship over surveillance may mean publishing ecosystem health indicators rather than maps that facilitate patrol targeting, or pairing monitoring programs with benefit initiatives designed alongside local communities. The Critical RS steps of Exposure and Engagement provide practical checkpoints for identifying these risks early. If an exposure assessment suggests that new visibility will likely justify coercive enforcement, teams should modify the design or reconsider the project altogether. If engagement reveals fear or distrust among affected communities, that feedback should guide redesign (Bennett et al., 2024; York et al., 2023).

Justice also requires mechanisms for learning when harms occur. Even carefully designed systems can produce unintended consequences: misclassification that disproportionately affects a community, visibility that invites enforcement rather than support, or narratives that stigmatize rather than inform. Ethical practice therefore extends beyond initial design into ongoing monitoring, feedback, and revision. Teams should establish clear channels for reporting harm, document downstream uses of data and outputs when possible, and periodically audit model performance across spatial, demographic, and institutional contexts. When misuse or equity failures are detected, responses should include transparent acknowledgment, corrective action, and, when necessary, redesign or withdrawal of the product. Justice in GeoAI is therefore not a one-time checklist but an ongoing commitment to governance and accountability.

The third tension is bycatch and consent. RS bycatch occurs when data is unintentionally captured outside the intended scope of observation. For example, when monitoring wildlife corridors using high-resolution satellite or drone imagery, analysts may inadvertently capture individuals gathering firewood or crossing protected areas for subsistence activities. If such imagery is later shared with enforcement agencies without context, what began as ecological monitoring can escalate into legal or economic harm for already vulnerable groups. Similar dynamics arise in other domains. In urban planning, satellite imagery used to monitor infrastructure or city growth may inadvertently reveal informal settlements or unpermitted housing developments (Cinnamon & Noth, 2023). Although these communities were not the intended focus of observation, they may become targets of eviction or enforcement simply because their presence becomes visible from space. In agricultural monitoring, systems designed to assess crop health may also reveal details about undocumented labor or land-use practices that conflict with state regulations. In each case, incidental visibility can expose individuals and communities to legal, economic, or social penalties.

These risks mean that teams must plan in advance for how incidental human imagery will be handled. Clear policies should address the treatment of bycatch, including avoiding publication of identifiable scenes and resisting pressure to share raw imagery feeds without due process. Applying Exposure and Engagement early in project design can reduce the likelihood of harm by limiting resolution, clarifying intended use boundaries, and establishing response protocols before incidental capture occurs. Consent at scale is difficult in practice, but teams can still take meaningful steps by informing communities where possible, adjusting system design to reduce human capture, and committing in advance to how requests from authorities will be handled (York et al., 2023).

2.3 Indigenous data sovereignty and the CARE principles in practice

Indigenous data sovereignty articulates a simple but demanding norm. Indigenous peoples have rights to control data about their lands, communities, and knowledge; those rights do not vanish when data is derived from space rather than on the ground. This view aligns with the United Nations Declaration on the Rights of Indigenous Peoples (UNDRIP), which affirms rights to land, culture, and self-representation (UNDRIP, 2007). The CARE Principles anchor Indigenous data sovereignty in the following four notions: Collective Benefit, Authority to Control, Responsibility, Ethics. While CARE provides a widely recognized international framework, Indigenous data governance is articulated through multiple regionally grounded principles. In Canada, for example, the Ownership, Control, Access, and Possession principles (OCAP) assert that First Nations communities own and control data about their peoples and territories (First Nations Information Governance Centre, 2020). Métis data governance principles similarly emphasize self-determination, relational accountability, and community-defined benefit. Inuit Qaujimajatuqangit (IQ) and the National Inuit Strategy on Research stress that research involving Inuit Nunangat must align with Inuit values, decision-making structures, and priorities, and that Inuit must lead or meaningfully govern research affecting their lands and knowledge systems (Inuit Tapiriit Kanatami, 2018). Recognizing these frameworks underscores that Indigenous data sovereignty is not a single doctrine but a set of evolving, place-based governance systems that must be engaged contextually (Carroll et al., 2019). O'Brien and colleagues show how to turn these values into actions (O'Brien et al., 2024).

Collective benefit means designing so that data ecosystems function in ways that enable Indigenous peoples and local communities to derive benefit. For teams and repositories, this translates into product types that are useful to communities, not only to external agencies; into consultations that define what counts as useful; and into language and visualization choices that meet local needs. For example, this could be achieved by including Indigenous languages in metadata and interfaces and creating views that are meaningful to community governance rather than only to technical audiences (O'Brien et al., 2024).

Bunting et al., 2026 highlight, in the development of a coastal citizen science project across the Great Lakes region of the U.S., that consultation with communities during data collection helps produce actionable data, broader community buy-in to the project, and tangible outcomes beyond traditional academic channels (Bunting et al., 2026). For example, in Chikaming Township, Michigan, collaboration with the community and research team, in consultation with lawyers at the University of Michigan, resulted in township ordinance aimed at minimizing coastal erosion and therefore banning coastal armoring (Fig. 13.3). This case illustrates how community-engaged data practices can shape not only scientific understanding but also local legal frameworks governing land and resource use.

Authority to control asks who gets to say yes or no. Practical acts include recording permits and approvals, requesting and storing documentation that shows consent, and attaching labels to datasets and products so that rights and protocols travel with the data. Technology choices include enabling granular data embargoes, protecting sensitive features through generalization or masking, and building access control mechanisms that allow communities to approve or reject specific requests. A simple repository improvement is to link permits and approvals to datasets via persistent identifiers and to render those links on landing pages for transparency (O'Brien et al., 2024).

Responsibility is both process and posture. Teams need to keep logs of where data went and for what purpose, publish ethics statements, and commit to responding to harms and errors. Project routines should include deciding who will respond to data inquires (e.g., if law enforcement requests incidental human imagery), and how to treat scenes that show illegal activity tied to livelihoods. Responsibility also means budgeting for education and capacity building equitably, not only for core technical work (O'Brien et al., 2024).

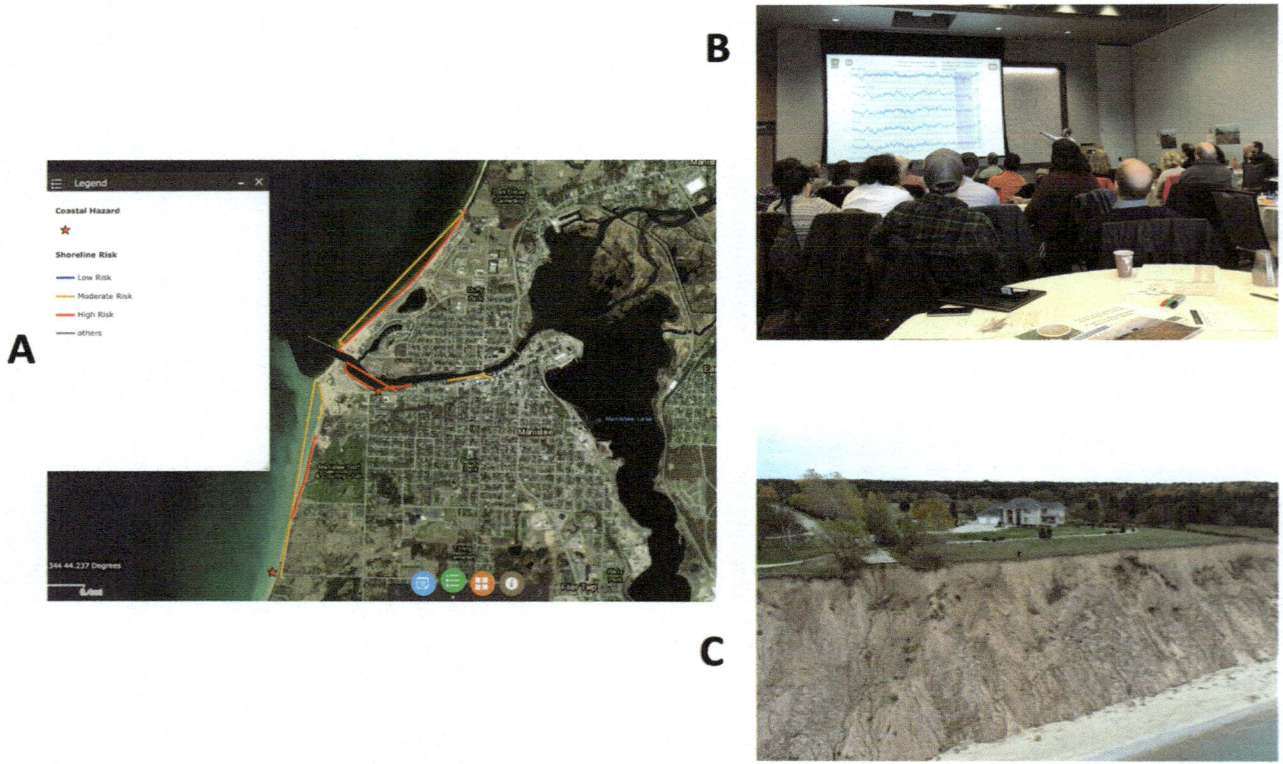

Figure 13.3 Community engaged events across the Great Lakes shoreline. (A) Community-driven data collection to determine where to monitor and collect drone imagery due to coastal erosion, (B) community workshop in Marquette, MI to include the local voice in collection and output efforts, (C) image of extreme bluff erosion on Lake Michigan shoreline highlighting the need for coastal monitoring (Bunting et al., 2026).

Achieving consent at scale is a challenge that brings about many privacy and justice risks. RS projects that span Indigenous lands or affect Indigenous communities are not exempt from sovereignty claims simply because data is collected from above. Indigenous data sovereignty articulates a clear norm. The CARE Principles make this practical, spelling out Collective Benefit, Authority to Control, Responsibility, and Ethics. O'Brien and colleagues translate these values into repository and team actions across situational awareness, outreach, protocols, and technology. They list 47 specific activities that are immediately implementable, from attaching TK and BC Labels to datasets so that rights and protocols travel with the data, to enabling embargoes and granular access controls, to recording permits and approvals and linking them visibly to datasets and products (O'Brien et al., 2024).

Authority to control the use of data becomes concrete in access control mechanisms that allow communities to approve or reject specific requests, in provenance systems that link datasets to approvals, and in visible documentation of assumptions and constraints on landing pages. Responsibility appears in continuity plans, so that commitments persist beyond a grant, and in education budgets that are allocated equitably. Ethics appears in the choice to blur, delay, consult, and acknowledge positionality. For teams working on Indigenous lands or with Indigenous communities, this is not a box-ticking exercise. It is a set of design and governance decisions that helps determine whether the work meets Free, Prior, and Informed Consent (FPIC) obligations and delivers benefit without harm (O'Brien et al., 2024). Ultimately, these outcomes can truly only be decided by the Indigenous community itself, and even then, views within the community are likely to differ. FPIC in a GeoAI context requires more than notification. "Free" means that consent is given without coercion or structural pressure, including economic dependency. "Prior" requires consultation before imagery collection, model deployment, or publication decisions, not after outputs are finalized. "Informed" requires clear explanation of what data will be collected, how it will be processed, who may access it, potential downstream uses, and plausible risks. "Consent" requires a meaningful opportunity to say no or to condition participation, including the ability to withdraw or renegotiate terms as projects evolve. In RS contexts where data may be collected from above, operationalizing FPIC often requires proactive outreach, transparent scoping of sensing activities, negotiated agreements on resolution and cadence, and shared governance over release decisions. Acknowledging the diversity of Indigenous viewpoints is a crucial aspect of working with Indigenous communities, which are not monoliths.

2.4 *Infrastructure, environmental burden, and power*

GeoAI and RS can feel distant and intangible: images from space, models running in the cloud. Yet their operation depends on physical infrastructures that consume energy, water, land, and materials. These infrastructures have environmental and social footprints that are unevenly distributed. Large-scale model training and storage require significant electricity and water for cooling; data centers are frequently sited in regions with lower land costs, weaker regulatory regimes, or limited political leverage (Strubell et al.,

2019). As a result, communities that may not directly benefit from advanced GeoAI applications can nonetheless bear environmental burdens associated with their operation.

The energy intensity of large-scale model development has been documented across AI domains (Strubell et al., 2019), and while geospatial models differ in architecture and scale, similar concerns arise when training and deploying high-resolution, data-intensive systems. Expanding access to high-resolution imagery and large models can support conservation, disaster preparedness, and environmental justice. At the same time, the carbon intensity of large models, the water demands of cooling systems, and the extractive supply chains behind GPUs and sensors raise questions about sustainability and distributive justice. Ethical GeoAI practice therefore includes attention to infrastructure choices: selecting computational approaches proportionate to the question, reporting energy and resource implications where feasible, favoring energy-efficient architectures when possible, and considering local environmental impacts in decisions about data center partnerships and cloud use (Jegham et al., 2025; Zewe, 2025). Choices about model size, retraining frequency, and cloud dependency are therefore not only technical optimizations but ethical design decisions with environmental consequences.

Power asymmetries also shape who controls these infrastructures. A small number of corporations and governments operate the majority of high-capacity cloud platforms and satellite constellations (AAG IT Support). This concentration influences who can conduct large-scale analysis, who sets access terms, and whose priorities shape system design. Justice in GeoAI thus extends beyond datasets and models to the political economy of sensing itself. Transparency about infrastructure dependencies, advocacy for equitable access, and collaboration with institutions in lower-resource settings are part of aligning geospatial innovation with broader commitments to equity and environmental responsibility.

2.5 Space data ethics as a complementary lens

Space data has properties that do not fit neatly within ordinary data ethics. It is often dual use, crosses borders, implicates sovereignty and benefit sharing, and can expose people and places without their knowledge. The U.S. National Space Council argues that space data ethics is distinct from business data ethics and research ethics. For example, data minimization is central to many data protection regimes, but in space science, maximizing observation is often critical for discovery and for future uses. Open release is a strong default in many communities, but full openness at high resolution and frequency can enable misuse by adversaries or escalate conflict. As Wang et al. (2023) describe, careful consideration should be given to the trade-off between the principles of open access and the disclosure of details. Benefit sharing is essential across jurisdictions, yet obligations and rights do not map simply onto a global commons (National Space Council Users' Advisory Group, 2023). Again, the citizen science program described in Bunting et al. (2026) higlights the slippery slope of collection and utilization of space data without thought to data ethics. Across the Great Lakes study sites, land ownership boundaries and disputes routinely complicated decisions about where imagery could be collected. While the imagery was only being used to quantify the impact of high and low water level, the images revealed development, management regimes/practices, and personal property.

This lens brings into focus the tradeoffs teams need to consider in the research process. When planning a daily tasking campaign over human landscapes, it is important to think through potential dual-use implications. Who else might be able to use these images, for what purposes, with what consequences? When publishing city-level maps of informal settlements, it is worth asking if fully open release is necessary, or whether sharing summarized information or delaying release could still support service delay while reducing risk to affected communities. Consider logging, access fees, and audience vetting not as barriers, but as risk controls. Many respondents surveyed in the U.S. expressed discomfort with high temporal frequency and found that acceptability depended strongly on use case. This is consistent with the space ethics argument for careful release rather than open by default (McAmis et al., 2024; National Space Council Users' Advisory Group, 2023).

3. The legal landscape: What governs GeoAI in RS

3.1 A short primer on data protection for RS and GeoAI teams

Most RS projects intersect with general data protection law once people become visible, either directly (e.g., in high-resolution imagery) or indirectly (e.g., through inferences about households, movements, and services used). The starting point is to treat location data as personal data whenever it can be linked, or is reasonably likely to be linked, to an identifiable person. This does not require a name or face. This covers imaging of places where people live, work, receive services, or move, even when individuals are not the intended target. The first operational step is to identify and document a lawful basis before collecting or publishing products that can reveal individuals or small groups. Public bodies often rely on a "public task" basis ground in statutory mandates, while private actors often rely on "legitimate interests," and consent is generally impractical at scale. Consent is rarely workable in large-scale RS contexts, particularly where data are collected passively or retrospectively.

Once a lawful basis is established, several core obligations follow: purpose limitation, data minimization, transparency, and security by design. Purpose limitation requires teams to define, document, and adhere to a specific analytic objective, rather than repurposing imagery or derived data for unrelated uses without further assessment. Data minimization means limiting spatial resolution, attributes, retention periods, and publication formats to what is necessary for that objective. Transparency entails explaining, in accessible terms, what data are being collected, what inferences are being generated, and how outputs may be used/shared. Finally, security by design requires embedding safeguards into workflows from the outset, such as access controls, audit logs, and systematic re-identification risk assessments.

For high-risk processing, such as systematic monitoring of publicly accessible areas, a data protection impact assessment (DPAI) is appropriate (e.g., Table 13.1). The purpose is to anticipate who is exposed, assess harms, and specify mitigation

Table 13.1 Quick DPIA starter for RS projects.

DPIA checklist	What to document	Purpose/Outcome
1. Purpose and lawful basis	Describe project purpose, intended benefits, who may be affected, and lawful basis if personal data could be processed (e.g., public task, legitimate interests)	Clarifies justification and legal grounding
2. Description of processing	Note sensors and data sources, spatial resolution, temporal cadence, geographic coverage, transformations, models, and outputs	Provides a clear technical overview of data activities
3. Necessity and proportionality	Explain why RS is required, identify alternatives considered, and justify chosen resolution and cadence using minimization-by-design principles	Demonstrates appropriate and proportionate use
4. Risks and impacts	Identify foreseeable harms such as privacy exposure, stigmatization, militarization risks, enforcement misuse, human bycatch, and impacts on indigenous lands or communities	Supports early identification of ethical and social risks
5. Controls and safeguards	List design controls such as reduced resolution or cadence, aggregation, obfuscation, delayed release, access logging, audience vetting, uncertainty communication, and policies for handling human bycatch or authority requests (Bennett et al., 2024; York et al., 2023)	Documents risk mitigation measures
6. Engagement	Identify stakeholders, communities, indigenous partners, and institutions; describe consultation plans that can influence design; record capacity-building and training plans (Bennett et al., 2024)	Ensures inclusive and responsive project design
7. CARE and data sovereignty	Record approvals required, FPIC steps if applicable, TK and BC labels, permits, protocols, access restrictions, and embargo requirements (O'Brien et al., 2024)	Supports indigenous data governance and ethical compliance
8. Publication decisions	State release approach based on purpose, actors, resolution, cadence, potential misuses, and rationale for open, tiered, or delayed access	Guides responsible dissemination
9. Residual risk and sign-off	Document remaining risks, approval authorities, approval dates, and review schedule	Provides accountability and ongoing oversight

measures. In the GeoAI context, those controls usually include resolution and cadence choices, obfuscation or aggregation of sensitive features, access controls and logging for raw assets, uncertainty reporting, and documented handling of incidental "human bycatch". The critical RS "Exposure" step provides a useful framing device for DPIA scoping: if this were conducted as fieldwork, would permission be required, and what changes when observation occurs from above (Bennett et al., 2024)?

Minimization by design is the practical rule of thumb: use the lowest resolution and revisit frequency consistent with the policy objective; preference indices and aggregates to address-level maps; and blur or delay release of sensitive features, particularly in volatile or conflict settings. These are not only ethical choices but documented risk mitigations within a DPIA. National mapping agencies increasingly formalize such gates through policy alignment, risk registers, and human-in-the-loop oversight embedded in technical pipelines (Kausika & van Altena, 2025). Public attitudes reinforce this approach: acceptance varies sharply by use case and declines with higher temporal frequency, arguing for proportionate, context-specific release rather than blanket openness at full cadence (McAmis et al., 2024).

3.2 Remote sensing specific policy: NOAA licensing and the democratization versus control debate

RS operates inside a moving policy envelope. In the U.S., NOAA has eased or removed several legacy constraints on commercial imaging, including restrictions against rapid revisit times and access limitations over certain areas. The current policy enables imaging and distribution over most of the globe and reduces temporal constraints, reflecting a broader trend toward capability growth and market support. These changes have practical implications for teams because they bring the possibility of more frequent, higher resolution imaging as an operational default. They also sharpen the responsibility to decide what to release, to whom, and when, because the policy gate has moved outward while social expectations remain cautious (McAmis et al., 2024).

Many people were unaware that commercial satellite imagery was available, and once informed, they expressed discomfort with frequent imaging. Preferences varied by purpose and actor, and significant demographic differences appeared. These findings do not dictate legal obligations, but they do illuminate the social license space and suggest that democratization and control are not simple opposites. There is a legitimate public interest in broad access for disaster response, accountability, and science. There is also a legitimate public concern about persistent, high frequency imaging that enables surveillance, profiling, or coercion. A balanced governance approach is to favor openness for low risk, high benefit cases, and to consider tiered access, delayed release, aggregation, or paywalls for higher risk cases, coupled with logging and audit so that access can be reviewed.

Teams can operationalize this by building a release rubric into project plans. List the use case, actor classes, resolution and cadence, and the plausible misuses. Decide whether a public map, a rate-limited API, or restricted access is appropriate, whether aggregation or obfuscation is warranted, and whether to delay release during elections or other sensitive windows. If a dataset contains lands inhabited by Indigenous Peoples (whether this status has been recognized by the state), confirm approvals and attach labels that state local protocols for access and use (O'Brien et al., 2024). If a map could be used for deterrence campaigns in conservation, reconsider the framing and the cadence considering evidence that fear-based messaging can harm vulnerable groups and inflame conflict (York et al., 2023). Document these decisions as part of a DPIA or a project ethics note so that reviewers and future partners can follow the rationale.

3.3 Contracts, terms, and platform governance

Much of what is legally permissible in practice is defined not by statute but by contract. Platform terms of service, licensing and intellectual property clauses, and service level agreements shape what can be done with RS data and models. These also vary immensely across jurisdictions, which RS studies–many of which are cross-border–need to take into account. Additionally, cloud platforms that host imagery and compute provide enormous capability, but they also set rules about scraping, redistribution, derived products, and serving to third parties. Costs and quotas create practical access gates that have ethical valence because they determine who can compute at scale and who cannot.

Repositories and data portals are not neutral either. Their licenses determine if users can combine datasets, if derivative works can be shared, and how attribution travels. For Indigenous lands and knowledge, generic open licenses can conflict with local protocols. The CARE aligned approach is to attach community driven labels and notices, record approvals, support embargoes and granular access controls, and display rights and responsibilities wherever the data appears. O'Brien and colleagues list specific technology actions teams and repositories can take now, from metadata fields that carry permits and proper use statements, to access control and actor registers that track who has rights to approve and who has accessed restricted content (O'Brien et al., 2024). National mapping agencies also exist within this contractual web. They negotiate access to commercial imagery, set downstream licensing for national basemaps, and decide which platform ecosystems to endorse. Kausika and van Altena show that governance, policy alignment, and cross functional teams are as important as models, because without them organizations cannot meet legal and ethical expectations at speed (Kausika & van Altena, 2025).

For teams, a practical takeaway is to read the relevant license for data or repository before you design a product. Questions to consider include: Can you redistribute the data, can you train models on it, can you serve tiles to third parties, can you sell services based on it, and do any clauses conflict with community protocols or repository labels? If you cannot meet your ethical obligations under the default license, then consider renegotiating terms, choosing different sources, or redesigning outputs. Write data use agreements with partners that encode the CARE commitments you have made, and ensure continuity plans so that those commitments survive beyond project end (O'Brien et al., 2024).

RS inherently crosses borders, enabling planetary-scale science, accountability, and disaster response. It also creates predictable tensions that vary by justification. National Space Council Users' Advisory Group (2023) notes that space data are often dual use, benefit sharing is widely endorsed but difficult to operationalize, and national priorities and security concerns do not always fit smoothly with resources and systems that are shared by the world. Traditional data ethics tools such as consent and data minimization do not fully translate to this context because the public stakes and adversarial risks differ. Space-specific ethical frameworks are therefore needed to complement general data protection practices, recognizing that openness can both advance the public good and create harm, that benefit sharing requires concrete mechanisms, and that national security concerns may justify limited or delayed release without defaulting to secrecy (National Space Council Users' Advisory Group, 2023).

As described above, much can go wrong when a surveillance framing dominates. Militarization, coercion, and bycatch harms can follow in the wake of well-intentioned monitoring, particularly when cross-border actors supply tools and narratives without deep engagement on the ground. Bennett and colleagues show how to turn that insight into practice via the critical RS steps of Exposure and Engagement. Together these works suggest rules of thumb for cross-border projects. Consult local institutions and communities before tasking and modify design to reduce human exposure. Clarify approvals and attach labels and notices that travel with data. Consider tiered release and delayed disclosure when public interest can be satisfied without exposing sensitive sites or groups. When partnering with national security actors, set clear limits around how outputs will be used and require oversight mechanisms where possible. Document the rationale for release choices so that reviewers can see how you balanced openness and restraint considering your project's purpose (York et al., 2023; Bennett et al., 2024; O'Brien et al., 2024; National Space Council Users' Advisory Group, 2023).

3.4 Practical implications for teams

Pulling these strands together, the legal landscape is less a maze than a set of recurring questions that can be built into project routines:

- **i.** Document a lawful basis whenever data protection law applies. For high-risk processing, record your purpose, minimization choices, and safeguards in a DPIA. Use the Exposure step to focus DPIA questions, and CARE principles to design controls that respect authority and collective benefit (Bennett et al., 2024; O'Brien et al., 2024).
- **ii.** Align RS practice with evolving policy. Know where your tasking and release sit relative to NOAA's current rules in the United States and to other national regimes where you operate. Use release rubrics to decide what to publish openly, what to aggregate or obfuscate, and what to delay. Treat public acceptability evidence as an input when designing product lines. When in doubt, favor proportionate release tied to purpose rather than maximal release by habit (McAmis et al., 2024; National Space Council Users' Advisory Group, 2023).
- **iii.** Prepare for AI-specific duties. Map your pipeline to likely risk classes and be ready to document data governance, validation, uncertainty, and human oversight. The scientific good practice urged by Koldasbayeva and colleagues becomes regulatory good practice under emerging AI regimes. If your validation inflates performance through spatial leakage, or if you do not quantify and communicate uncertainty, fix that now before a regulator or user forces the issue later (Koldasbayeva et al., 2024).
- **iv.** Treat contracts, licenses, and platform terms as part of ethics. If a license conflicts with community protocols, do not hide behind it— change it or change your plan. Record

approvals and attach labels so that rights travel. Avoid designs that rely on gated platforms where possible or build mechanisms to ensure equitable access if a specific platform is needed. Negotiate data use agreements that encode your commitments to CARE and to transparency (Kausika & van Altena, 2025; O'Brien et al., 2024).

v. Acknowledge that space is global and dual use. Release for benefit, but release thoughtfully. Document benefit sharing pathways, consider partners beyond your usual network, and build transparency around your choices so that others can critique and learn. Where national security claims are valid, prefer delay and aggregation to secrecy when that will achieve the protective purpose with less collateral harm. Where security is invoked to justify coercive uses, interrogate the framing and propose alternate designs that meet conservation or service goals without enabling surveillance harms (National Space Council Users' Advisory Group, 2023; York et al., 2023; Bennett et al., 2024).

This is not an exhaustive legal treatise. It is a workable backbone that teams can utilize now.

4. Where things go wrong in GeoAI and remote sensing

4.1 Algorithmic fairness and validity in GeoAI for remote sensing

Fairness is not only about who is captured, but also about who is correctly captured, and how truth claims are made. The standard mistakes in geospatial modeling are well documented. Sampling bias and imbalanced classes can produce models that perform well on majority classes while failing on minority classes (Braun, 2021). Spatial autocorrelation can inflate scores because nearby pixels are similar, and random cross validation can overstate reliability when training and test sets are not spatially independent. Leakage of spatial information is a quiet error that appears when validation splits do not respect geographic structure. Koldasbayeva and colleagues synthesize these issues and offer clear guidance. Use spatial cross validation, not random splits, when evaluating models on spatial data. Quantify uncertainty, visualize it, and communicate it alongside predictions. Avoid false certainty from maps by publishing error budgets and by resisting esthetic choices that imply spurious precision. Recognize that imbalanced data is typical in environmental contexts and apply resampling and cost sensitive methods with care (Koldasbayeva et al., 2024).

These are not academic formalities. They have direct ethical weight. If a city deploys a building segmentation model to identify unserved households and the validation inflates performance, people will be missed. If uncertainty is not quantified and communicated, a confident boundary line could be used to justify eviction or denial of services. If imbalanced classes hide minority harms, a model could perform well overall while failing the very groups the project is intended to help. Responsible teams should adopt spatial cross validation and uncertainty quantification as routine, publish uncertainty maps, and include plain language explanations of limits and error sources. This is consistent with the critical RS emphasis on Teaching and Accessibility, because it equips nonspecialists to interpret models properly and avoid overuse (Bennett et al., 2024; Koldasbayeva et al., 2024).

Critiques of satellite-based mapping of informal settlements make these issues concrete. Kochupillai and Taubenböck (2023) show how labeling choices, training data representativeness, and fear narratives shape downstream consequences. Morphological classifications such as "slums", often derived from visual density or roof type, carry stigma and can produce material harm. Training data that represent a single region may fail elsewhere. Narrative frames that celebrate fear as a deterrent can justify coercion. The authors argue for reframing questions toward ethical opportunities that meet policy needs without stigmatization. For example, these could involve mapping green space deficits or sanitation gaps rather than labeling neighborhoods as slums, and designing indices that guide service delivery rather than enforcement. The researchers emphasize that even when technical performance is good, maps can create false certainty if uncertainty and limits are not communicated (Kochupillai & Taubenböck, 2023; Kochupillai et al., 2022).

4.2 Political economy and access

RS and GeoAI capabilities are not evenly distributed. Commercial satellites and cloud platforms enable large-scale imaging and computation, but power asymmetries shape who can use them, for what, and with what consequences. York and colleagues show that RS can reinforce or challenge power relations in conservation and livelihoods, depending on who controls data and narratives. Bennett and colleagues emphasize that accessibility goes beyond open release, encompassing language, format, and local computing capacity. Kausika and van Altena highlight adoption challenges in national mapping agencies: governance, risk registers, policy alignment, and cross-functional teams are needed to make GeoAI work in practice. Without these, platform choices and terms of service silently determine who can compute at scale (Bennett et al., 2024; Kausika & van Altena, 2025; York et al., 2023).

The tension between democratization and control is clear. Broad access supports science, disaster response, and accountability, yet high-frequency, high-resolution feeds can enable surveillance, coercion, or commercial exploitation. Spaceborne data's dual-use nature means responsible release often requires tiered access, aggregation, delayed cadence, and logging. Teams can build release rubrics that account for purpose, actor, resolution, cadence, and plausible misuse, balancing open science with Indigenous data sovereignty, privacy, and equitable access to tools and training for local institutions (National Space Council Users' Advisory Group, 2023; Bennett et al., 2024; Kausika & van Altena, 2025).

4.3 Environmental and social benefits, and why the risks matter

Identification and attention given to risks does not hinder progress: rather, it can align technological progress with responsible use. Disasters, conservation, and climate monitoring are domains where RS and GeoAI deliver rapid public value quickly. The question is not whether to deploy them, but how to do so without inflicting predictable harm.

This chapter, and recent work referenced here, show how this alignment can be operationalized. Bennett and colleagues apply Exposure, Engagement, and Accessibility to disaster monitoring, producing outputs that inform responders while reducing exposure through aggregation and careful release. Kochupillai and colleagues demonstrate how reframing questions can produce policy relevant outputs that guide service delivery rather than enforcement, for example by mapping environmental exposures rather than stigmatizing neighobhoods. At the policy level, there is a clear tension: broad access to RS supports science, disaster response, and accountability, yet many datasets are dual use. Responsible release therefore requires balancing openness with proportionate restraint, weighing benefit sharing against foreseeable adversarial or coercive uses (Bennett et al., 2024; Kochupillai et al., 2022; National Space Council Users' Advisory Group, 2023).

In conservation, rights-based approaches that avoid coercion, plan for incidental human exposure, and engage communities can shift monitoring toward stewardship. In climate services, rigorous uncertainty quantification and transparent communication build trust in projections that shape adaptation and finance. In urban services, spatial cross-validation and equity-aware design reduce the risk that models systematically underserve marginalized groups. Traditionally, governance structures (e.g., clear mandates, risk registers, and human-in-the-loop oversight) help sustain both institutional capability (the ability to produce reliable geospatial products) and institutional legitimacy (public trust that they are used in the public interest). These protections, however, are politically contingent; responsible practice must be embedded institutionally rather than assumed.

Several practical rules help manage risk. Treat resolution and cadence as ethical variables. Anticipate incidental human exposure and define handling protocols before collection. Consult meaningfully and adjust design in response. Where Indigenous lands or knowledge are involved, implement CARE and ensure rights travel with data. Validate models spatially, quantify and visualize uncertainty, and publish error budgets. Balance openness with context-sensitive safeguards, and document benefit-sharing pathways.

Applications that exacerbate inequality will persist and failures will occur. Yet the same infrastructures can support search and rescue, equitable service delivery, community stewardship, and accountable climate adaptation. Outcomes are not predetermined; they are shaped by design choices, governance arrangements, and release decisions. What matters is whether teams anticipate foreseeable harms and institutionalize routines to prevent them. When these practices are integrated into pipelines and policy, RS and GeoAI can expand environmental and social benefits while reducing, rather than deepening, existing inequities (York et al., 2023; McAmis et al., 2024; O'Brien et al., 2024; Koldasbayeva et al., 2024; Kochupillai & Taubenböck, 2023; Kochupillai et al., 2022; Bennett et al., 2024; Kausika & van Altena, 2025; National Space Council Users' Advisory Group, 2023).

5. Governance in practice, turning principles into routines

5.1 Project lifecycle gates paired to the six step and CARE frames

Good governance in GeoAI and RS is less a code of conduct and more a set of intentional lifecycle gates. Each gate asks a small number of questions tied to specific pipeline choices, and together they turn values into routines. The six step framework of Exposure, Engagement, Empowerment, Teaching, Accessibility, Acknowledgment provides a practical scaffold for these questions, while the CARE Principles translate sovereignty and justice commitments into actions that can be checked, logged, and revisited over time (Bennett et al., 2024; O'Brien et al., 2024). Fig. 13.4 summarizes this approach by mapping a single set of lifecycle gates onto the six-step framework, with an overlay indicating where CARE commitments are most salient at each stage. Rows represent decision points in a typical GeoAI or remote sensing project; columns represent governance commitments that persist across the lifecycle, and the cells provide short prompts that translate abstract principles into routine practice. The figure is intended as a visual guide rather than a checklist, showing how the same values are enacted differently as a project moves from scoping through operations.

At the scoping gate, teams decide whether and how to proceed, and exposure assessment is primary. Teams identify who and what will become visible, who could be harmed, what misuses are foreseeable, and what resolution and cadence are truly necessary. Obfuscation or aggregation is planned where warranted, and decisions are documented in an ethics note or, where required, a data protection impact assessment. In conservation and livelihoods contexts, this gate includes establishing a policy for handling incidental human imagery in raw scenes and derived products. Where work spans Indigenous lands, FPIC processes are triggered where applicable, and permissions and community protocols are designed to travel with data through labels and notices or are treated as constraints on whether the data can travel at all. Scoping is also where collective benefit is defined in practical terms, and where project goals should be modified if benefits are one sided or likely to impose disproportionate burdens on vulnerable groups (Bennett et al., 2024; O'Brien et al., 2024; York et al., 2023).

At the design gate, engagement becomes specific and consequential. Teams consult early and co-produce where possible, allowing engagement to shape labels, outputs, cadence, release formats, and intended uses. In urban change mapping, this can involve meeting with resident associations and utilities before finalizing classes and outputs and inviting critiques of how maps will circulate. In conservation contexts, it can involve listening to rangers and park adjacent communities about patrol visibility and the risks associated with surveillance narratives. For Indigenous lands, approvals and protocols are recorded, and products are designed to support Indigenous governance needs rather than only external agencies. Design is also the appropriate stage to budget for capacity building and teaching, so local institutions can interpret uncertainty maps and error bounds, and so accessibility is understood as responsible access rather than default openness (Bennett et al., 2024; O'Brien et al., 2024; York et al., 2023).

At the acquisition gate, permissions and protocols attach to datasets in concrete ways. Teams record permits and approvals, attach TK Labels or BC Labels when relevant, set embargo rules and access controls for raw assets and derived products, and enforce minimization choices at ingestion rather than relying on downstream filtering at publication. Actor registers and access logs make authority to approve and to view

Lifecycle Gate	#1. Exposure	#2. Engagement	#3. Empowerment	#4. Teaching	#5. Accessibility	#6. Acknowledgement	CARE Emphasis
Scoping	Who and what become visible; resolution and cadence minimized	Early community or stakeholder consultation planned	Collective benefit defined	Limits anticipated	Need for aggregation or obfuscation identified	Positionality considered	Authority to Control; Collective Benefit
Design	Exposure reduced through design choices	Co-production shapes outputs	Capacity building budgeted	Communication about uncertainty planned	Tiered access considered	Community contributions planned	Collective Benefit; Responsibility
Acquisition	Human bycatch anticipated	Protocols confirmed	Rights and responsibilities travel with data	Use constraints documented	Embargoes and masking enabled	Approvals recorded	Authority to Control; Ethics
Development	Bias and spatial leakage assessed	Domain experts involved	Fairness impacts reviewed	Plain-language limits written	Error and uncertainty made visible	Labels reviewed for stigmatization	Responsibility; Ethics
Publication	Exposure managed through release decisions	Feedback invited	Benefits distributed	Uncertainty maps published	Aggregation or delayed release applied	Credit and positionality stated	Ethics; Collective Benefit
Operations	Drift and new exposure monitored	Ongoing engagement	Sustained capacity building	Updates and limits communicated	Access logs maintained	Complaints and recourse tracked	Responsibility; Ethics

Figure 13.4 Lifecycle gates mapped to the six-step framework and CARE principles.

restricted content visible and support later auditing. Where incidental human imagery is likely, this gate includes publishing a clear policy for how such scenes will be handled, including how requests from authorities will be managed (O'Brien et al., 2024; York et al., 2023).

At the development gate, scientific validity becomes an ethical commitment. Teams use spatial cross-validation rather than random splits for spatial data, quantify and visualize uncertainty, and prepare error budgets. Training data are audited for imbalance and representativeness; resampling or cost sensitive approaches are applied with care, and labels are reviewed for stigmatization risk. Plain language explanations of limits and error sources are built into deliverables, so Teaching is not an afterthought. If a model will shape service allocation or enforcement and performance is inflated through spatial leakage, the resulting harms are predictable, and correction belongs at this stage rather than after deployment (Bennett et al., 2024; Kochupillai & Taubenböck, 2023; Koldasbayeva et al., 2024).

At the publication gate, Accessibility and Acknowledgment become visible. Teams decide what can be released openly, what requires aggregation or obfuscation, and what should be delayed or tiered by audience. Provenance is made legible by linking data, code, and papers; uncertainty maps are published alongside predictions; and ethics statements explain exposure, mitigations, and available recourse if harm occurs. In conservation contexts, deterrence narratives and fear framing are avoided, and in urban services, labels that stigmatize neighborhoods are rejected. Community contributions are credited and critique is invited, reinforcing that acknowledgment is part of legitimacy rather than a matter of etiquette alone (Bennett et al., 2024; O'Brien et al., 2024; York et al., 2023).

At the operations gate, governance becomes a recurring practice rather than a one-time review. Teams monitor model drift, track data quality changes, maintain access logs, respond to complaints, and feed incidents back into scoping and design so that subsequent cycles improve. Empowerment continues through sustained capacity building and joint evaluation with partners, not only through initial training at launch (Kausika & van Altena, 2025; O'Brien et al., 2024). Since these lifecycle decisions must persist beyond individual teams and project timelines, the next section turns to repositories and data portals as the infrastructure where governance commitments are stored, enforced, and made visible at scale.

5.2 Repository level responsibilities and tooling

Repositories and portals are where governance either becomes real or remains rhetorical. CARE aligned metadata, labels that carry protocols, consent and permit tracking, embargo and obfuscation controls, provenance links among papers, data, and code, and clear roles for approval and complaint response are governance infrastructure, not optional add-ons (O'Brien et al., 2024). Practically, this begins with metadata fields that record permits, approvals, and proper use statements, surfaced on landing pages; TK Labels and BC Labels where relevant, implemented so protocols travel with the data; and persistent identifiers that link datasets to approvals and derived products. Consent and permit checks should be part of ingestion, not a separate narrative; actor registers and access logs support authority to control in routine operations. Repositories should also support granular embargoes and simple masking workflows, with clear documentation of what has been blurred, why, and under what conditions access can be expanded. Finally, provenance should be bidirectional, so a user can move from paper to data to code and back, which supports equity and teaching by reducing reliance on insider knowledge (O'Brien et al., 2024).

5.3 Organizational adoption in national mapping agencies, universities, and NGOs

Institutions operationalizing GeoAI need policy alignment, governance models, risk registers, cross functional teams, and machine learning operations (MLOps) for traceability. Kausika and van Altena show how national mapping agencies face immediate scaling constraints when pilots lack governance, and how risk registers and traceability become prerequisites for public trust and policy compliance (Kausika & van Altena, 2025).

A workable model typically has a strategic layer for policy alignment and oversight of high-risk products; an operational layer where product teams run lifecycle gates and record decisions; and a technical layer where MLOps enforces traceability through model registries, data lineage, and standardized validation and uncertainty reporting. Risk registers sit between strategic and operational layers, and should cover exposure risks, legal risks, scientific validity risks, platform and dependency risks, and governance risks such as missing community protocols. MLOps then makes it harder to cut corners by capturing training data versions, spatial cross validation methods, uncertainty reporting, drift monitoring, and change logs in ways product owners and oversight bodies can interpret (Kausika & van Altena, 2025; Koldasbayeva et al., 2024). Cross-functional teams and training are essential to make these routines consistent, including training reviewers to detect human bycatch, train access approvers to interpret labels and notices, and train analysts to read uncertainty maps and error budgets honestly (Bennett et al., 2024; O'Brien et al., 2024; York et al., 2023).

5.4 Practitioner routines, checklists that travel

Checklists do not replace judgment, but they can reduce the friction of doing the right thing under time pressure. Table 13.2 was developed to guide and serve as a starting point for community engagement in this context. A useful consolidation is to anchor routines to three familiar moments: (1) before you collect, (2) before you publish, and (3) before you deploy.

Before you collect, run an exposure assessment; decide resolution and cadence; plan obfuscation or aggregation; initiate FPIC where applicable; prepare labels and notices so protocols travel with data; publish a human bycatch policy appropriate to the context, including how requests from authorities will be handled; document decisions in an ethics note or a DPIA where required (Bennett et al., 2024; O'Brien et al., 2024; York et al., 2023).

Table 13.2 Community engagement checklist.

Checklist step	Key actions	Purpose/Outputs
1. Identify stakeholders	Identify residents' associations, utilities, local government, civil society groups, indigenous communities, and representatives where applicable	Ensure inclusive participation and representation of affected groups
2. Prepare plain-language Briefings	Explain project purpose, what will be visible, resolution and cadence, intended map uses, limits of inference, and how to ask questions or lodge complaints	Build transparency, trust, and informed participation
3. Co-design labels and outputs	Convene sessions to review labeling choices and narrative framing; avoid stigmatizing terms; agree on useful formats, indices, and aggregation levels	Reduce harm and improve relevance and usability of outputs
4. Gather concerns and Suggestions	Document privacy, security, and reputational concerns; collect suggested safeguards and design changes	Identify risks early and incorporate community knowledge
5. Modify project design	Record specific changes made in response to engagement (e.g., lower cadence, blur sensitive locations, aggregate outputs)	Demonstrate responsiveness and risk mitigation
6. Record meetings and approvals	Maintain minutes, decisions, FPIC steps and approvals for indigenous lands, consent and permit documentation where applicable (O'Brien et al., 2024)	Provide accountability and compliance documentation
7. Budget for capacity building	Plan training sessions, toolkits, workshops on interpreting uncertainty and map limits, and ongoing support	Strengthen stakeholder understanding and long-term use
8. Publish engagement summary	Share who was consulted, what changes were made, safeguards implemented, and contact channels	Promote transparency and public accountability
9. Plan ongoing engagement	Establish feedback channels, provide regular updates, and conduct co-evaluation after product launch	Support adaptive management and sustained collaboration

Before you publish, validate with spatial cross-validation; quantify uncertainty and publish uncertainty maps and error budgets; apply a release rubric that considers purpose, actor, resolution, cadence, and plausible misuses; implement tiered access or delayed release where warranted; ensure provenance links among data, code, and papers; include ethics and acknowledgment statements that make exposure, mitigations, and recourse legible, and avoid stigmatizing labels and coercive narratives (Bennett et al., 2024; Koldasbayeva et al., 2024; O'Brien et al., 2024; York et al., 2023).

Before you deploy and operate, register models with training data versions and change logs; capture data lineage and dependencies; monitor drift; train human reviewers and approvers to apply protocols consistently; maintain a pause or rollback pathway; keep risk registers current; resource teaching and capacity building so end users interpret outputs within limits, and sustain engagement so feedback updates products over time (Bennett et al., 2024; Kausika & van Altena, 2025; Koldasbayeva et al., 2024; O'Brien et al., 2024; York et al., 2023).

These routines can appear as one-page visuals, but in practice they belong as templates in repositories, model registries, and product catalogs. Over time, institutions can refine them based on incidents and share them with partners. This practice raises the baseline across projects and makes governance less about policing and more habitual.

6. Building an ethically and legally robust remote sensing and GeoAI workflow

6.1 Standardizing ethical RS reporting, joining norms to validation

A core next step is to standardize ethical reporting in RS so CARE and critical RS commitments sit alongside validation and uncertainty reporting. In practice, this means a shared reporting template that requires exposure assessment and engagement summaries, labels and protocols linked in metadata, and release decisions documented through a rubric, alongside spatial cross validation, uncertainty maps, and error budgets. Bennett et al. and O'Brien et al. provide governance scaffolding, and Koldasbayeva et al. provide validation expectations; the opportunity is to join these into a template that repositories, journals, and institutions can adopt for product launches and publications (Bennett et al., 2024; Koldasbayeva et al., 2024; O'Brien et al., 2024; York et al., 2023).

6.2 A golden path, from principles to institutional defaults

Section 5 outlined a practical gate-based approach; the forward direction is to make that approach a default workflow rather than a project specific choice. A golden path does not add new ethics requirements so much as it standardizes where decisions are made, how they are recorded, and how governance travels through repositories, model registries, and product catalogs. The institutional value is consistency; teams know what must be decided at scoping, design, development, publication, and operations, and oversight bodies can audit decisions without reconstructing them from memory (Bennett et al., 2024; Kausika & van Altena, 2025; Koldasbayeva et al., 2024; McAmis et al., 2024; O'Brien et al., 2024; York et al., 2023).

6.3 Reporting and transparency, documentation that users can read

Transparency is also a usability requirement. Model cards and datasheets adapted for RS can standardize purpose, data sources, resolution and cadence, representativeness, spatial cross validation methods, performance metrics, uncertainty maps, known limits, narrative framing and labeling choices, release rationales, governance commitments such as CARE and the six step framework, and contacts for recourse. Treat uncertainty as a first-class product and documentation as a public service, because these are the mechanisms through which nonspecialists and local institutions can verify claims and use outputs within limits (Koldasbayeva et al., 2024).

6.4 Contracts, licenses, platform terms, and premortems for revocation

Much of what is permissible is quietly shaped by contracts, licenses, and platform terms. A practical safeguard is a premortem that asks what happens if access must be revoked, if licenses conflict with community protocols, if platforms change terms, or if misuse requires pausing a service. Record trigger conditions, design access controls and logging so revocation is possible at granular levels and build continuity plans and alternative sourcing options into partnership agreements. This is not pessimism; it is responsible planning that makes rapid response feasible without surprises (Kausika & van Altena, 2025; O'Brien et al., 2024).

6.5 Platform governance and equity

Capability is mediated by platforms. Computing power, storage, pricing, quota schemes, language support, and licensing determine who can do GeoAI, and who can only consume its outputs. The research and practice question is how to translate governance, risk registers, and traceability into platform design, so equity is built in, including constraints that allow benefit sharing and derivative work without violating community protocols (Kausika & van Altena, 2025).

6.6 Human versus planetary subjects, consent, representation, archives

RS projects have human subjects and planetary subjects, meaning landscapes, ecosystems, and Earth systems that are observed, classified, archived, and reused across time; consent at scale for humans is difficult, and many planetary subjects cannot consent at all. Bennett et al. (2024) emphasize exposure assessment and permission questions at scoping, while the National Space Council highlights that space data ethics must complement data protection and research ethics because sovereignty, dual use, and benefit sharing create tensions that minimization alone cannot resolve. The agenda ahead includes defensible proxies for representation and consent, archives that carry protocols into future uses, rules for reuse, deletion, or embargo when communities request it, and governance mechanisms such as dataset review boards for sensitive collections. Archives are often where governance is most durable, and they require ethics embedded now, so future reuse does not erase today's commitments (Bennett et al., 2024; National Space Council Users' Advisory Group, 2023; O'Brien et al., 2024).

6.7 From principles to institutional practice: Embedding ethics in remote sensing and GeoAI

In practice, these frameworks are meant to be lived, not merely cited. Before collection, teams should conduct an exposure assessment, determine resolution and cadence with minimization in mind, and document mitigation strategies (Bennett et al., 2024). Before finalizing design, they should engage substantively with affected communities and Indigenous partners where relevant, invite critique of labels, outputs, and intended uses, and modify plans accordingly (Bennett et al., 2024; O'Brien et al., 2024). Prior to collection or publication, CARE principles should be operationalized through permits, approvals, TK or BC Labels, and repository controls that enable embargo, obfuscation, and visible provenance (O'Brien et al., 2024).

Throughout the lifecycle, stewardship should guide design choices: avoid fear-based framing, plan for human bycatch, and establish clear policies for handling incidental human imagery (York et al., 2023). At the point of release, space data ethics calls for careful consideration of tiered access, delayed dissemination, audience vetting, and clear documentation of release rationales (National Space Council Users' Advisory Group, 2023). Taken together, these routines turn ethical commitments into repeatable practice across scoping, design, development, publication, and operations. The result is not simply better compliance, but maps and models that are more durable, accountable, and socially grounded.

The trajectory outlined here is not toward layering additional ethical requirements onto RS and GeoAI, but toward institutional coherence. The aim is to align reporting norms, validation standards, repository practices, contractual safeguards, and platform governance into a default workflow in which ethical reasoning is visible, auditable, and durable over time. Fig. 13.5 illustrates this convergence, showing how reporting, process, and governance mechanisms integrate into an ethically and legally robust RS and GeoAI workflow. Standardized reporting situates exposure assessments and CARE commitments alongside spatial cross-validation and uncertainty mapping. A "golden path" workflow ensures that decisions made at scoping, design, development, publication, and operations are recorded and reviewable rather than reconstructed after harm occurs. Transparency mechanisms: including model cards, datasheets, uncertainty maps, and release rationales transform documentation into a public good, enabling non-specialists and affected communities to understand limitations as well

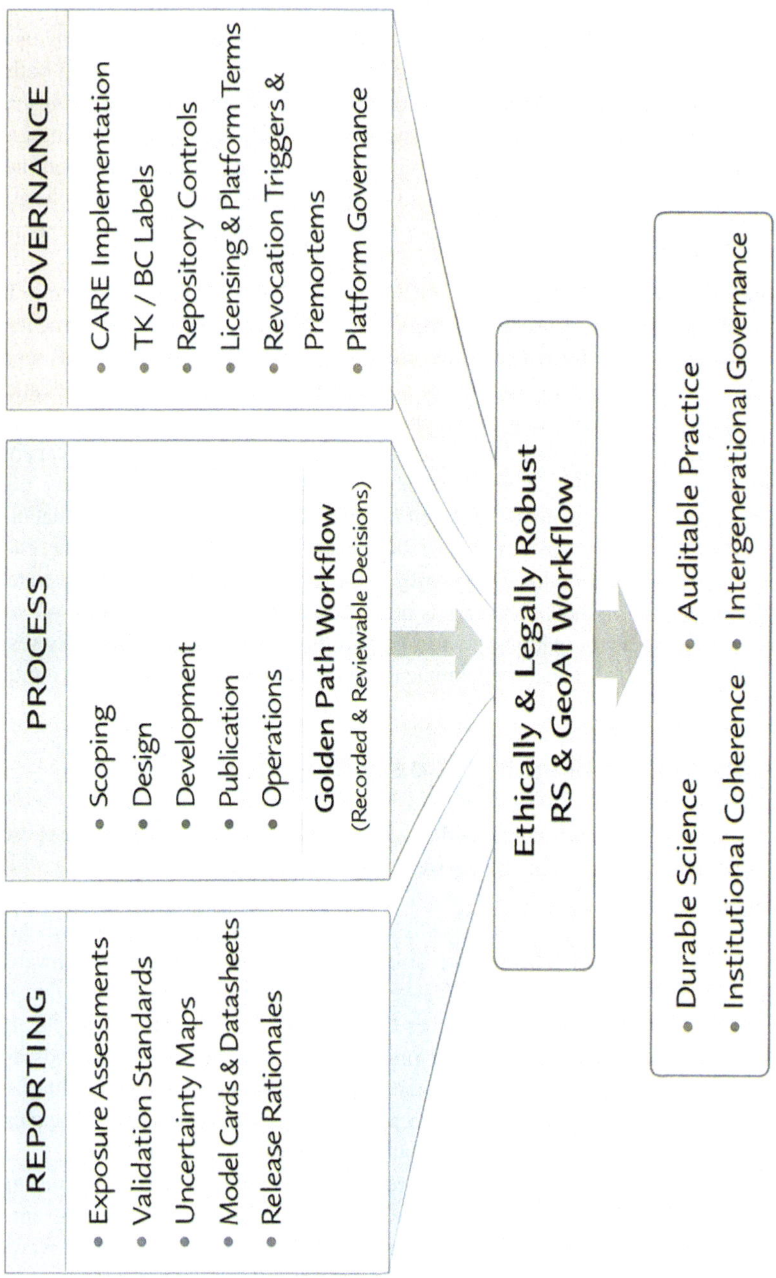

Figure 13.5 Building an ethically and legally robust remote sensing and GeoAI workflow.

as capabilities. Legal and contractual infrastructures, such as licensing terms, platform governance rules, revocation triggers, and premortems, convert abstract risk awareness into operational readiness.

Embedding equity within these infrastructures ensures that governance travels with data and models rather than dissipating at the point of release. Recognizing both human and planetary subjects further expands the scope of responsibility. Landscapes, ecosystems, and archives persist beyond individual project timelines, meaning governance must be embedded at the level of repositories and reuse rather than confined to single studies. Consent, representation, benefit sharing, and protocol inheritance therefore require institutional mechanisms capable of enduring across generations of reuse.

Taken together, these elements move RS and GeoAI from ethical aspiration to institutional practice. Ethical robustness emerges when exposure assessment, documentation, validation, contracts, platform design, and archival governance are not treated as separate concerns but as integrated components of a single lifecycle. This reframes validation to include exposure, documentation to include accountability, and release to include governance. The outcome is not slower science but more durable science: science capable of sustaining trust while navigating increasingly powerful analytic capabilities. Ethical robustness, in this sense, is not a constraint on innovation; it is the condition that allows innovation to persist. In this way, the risk surface introduced earlier in the chapter (Fig. 13.2) finds its institutional response in the governance workflow illustrated in Fig. 13.5. In that sense, the GeoAI turn is not only about more powerful ways of seeing the Earth. It is also about learning how to govern those new forms of visibility responsibly, so that the pixels we analyze continue to serve the people and places they represent.

7. Conclusion: Bringing it back together

In a book focused on what is now possible with GeoAI and remote sensing, we paused to ask a different question: what should we do, and how should we do it? The shift we describe is not simply technical. As GeoAI moves from pixels to people, places, and power, many of the most important decisions are simultaneously scientific, social, ethical, and legal. Throughout this chapter, we offered practical ways to carry that awareness through the full project lifecycle. Critical remote sensing provides six steps that make exposure and engagement deliberate rather than incidental. The CARE Principles translate sovereignty and justice into concrete practices. We examined the legal landscape in plain language and identified where risks tend to surface, from privacy and militarization to algorithmic validity and platform gatekeeping. Finally, we showed how principles can become routine through structured workflows and institutional defaults.

The commercial satellite privacy case reminds us of something simple but important: acceptability depends on purpose and cadence. The same sensor can seem protective or invasive depending on who is using it and why. Many respondents were less concerned with how clearly imagery could see than with how often it was collected. They distinguished sharply between search and rescue and employer monitoring, between documenting wildfire damage and tracking individuals. Even where regulation has become more permissive, social license remains conditional and context dependent. The practical implication is straightforward: design for purpose, calibrate frequency and resolution proportionately, and treat release decisions as ethical choices rather than purely technical ones.

The EU AI Act illustrates a parallel development. Practices that many responsible teams already follow: risk assessment, documentation, transparency, validation, human oversight, are increasingly becoming formal legal obligations when systems directly affect people. For teams already conducting exposure checks, publishing uncertainty maps, validating with spatially aware methods, and writing clear documentation, compliance may feel less like a new burden and more like the formal recognition of good practice.

If there is one idea to carry forward, it is that governance works best when it becomes routine. A structured "golden path" workflow turns values into habits. Exposure gates prompt teams to ask permission questions and minimize harm before data is collected. Engagement gates ensure consultation is substantive and informs design. Legal gates document lawful basis and record protocols that travel with data. Technical gates require appropriate validation and transparent reporting of limits. Publication gates make provenance visible and normalize ethics statements and acknowledgments. These are not bureaucratic hurdles; they are the mechanisms through which trust is built and sustained.

Institutions need not start from scratch. Repositories can support embargo options, link permits to datasets, and attach rights labels so that governance persists beyond publication. National mapping agencies can integrate policy alignment and risk registers with operational workflows and human-in-the-loop oversight. Platforms can treat computing access, storage, and model deployment as matters of public interest as well as pricing strategy. Small structural changes, implemented consistently, can have lasting effects.

Looking ahead, several needs stand out. Ethical commitments and scientific validation should appear together in standard reporting templates, so that exposure assessments and uncertainty maps are as routine as accuracy metrics. Platform governance must recognize how quotas, licensing terms, and access rules shape who can compute and who benefits. Benefit sharing should be operationalized in system design, through tiered access and release models that reflect community and place-based contexts. And archival governance deserves greater attention. Data collected today may be reused decades later. Designing protocols for reuse, revision, or removal when communities request it is part of responsible stewardship.

Every chapter in this book demonstrates the capacity of GeoAI and RS to strengthen disaster response, conservation, disease spread, and landscape management. The work presented here is not intended to slow that progress, but to support it. Conduct exposure assessments early. Document release decisions. Validate carefully and communicate uncertainty clearly. Ensure that rights and protocols travel with data. Consult before collecting and before publishing, and be willing to adjust when concerns are raised. Explain systems in language that affected communities can understand.

RS has always been social; GeoAI has made that reality more visible. As AI models increasingly shape how we interpret the Earth's surface, the ethical dimensions of our work become harder to ignore. The goal is not perfection, but care. The GeoAI turn can and should lean toward stewardship and inclusion rather than surveillance and exclusion. The maps will still be made. The question is whether they are made in ways that build lasting trust.

Case studies

Case study 1: Commercial satellite imagery and privacy: What the public says

Why this matters: Since SPOT was launched in 1986, commercial satellites have captured increasingly detailed, frequent, and affordable imagery of the Earth. These capabilities support disaster response, science, insurance, and accountability. They also raise everyday privacy questions. This case study pairs one clear piece of evidence, a U.S. survey of 99 respondents, with recent U.S. policy changes that have made commercial imaging and distribution more permissive. The combined lesson is straightforward: the same sensor can be judged acceptable or unacceptable depending on purpose, actor, and imaging frequency. We translate those findings into a practical design tool for teams building or releasing satellite-derived products (McAmis et al., 2024).

What respondents think

Awareness gaps. Most respondents did not know that commercial satellite companies sell images to individuals and organizations. When they were told what commercial satellites can do, people were surprised by the capability and availability. The starting point is that many users, residents, and officials you work with will not have an accurate picture of what is possible, or how easy it is to buy images or task collection. That makes early and plain communication crucial to ethical practice (McAmis et al., 2024).

Spatial versus temporal comfort. The survey separated two variables: spatial resolution and temporal frequency. Many respondents were comparatively comfortable with moderate-to-fine spatial resolution. For example, where cars are visible or people appear as small figures. Comfort dropped sharply, however, as revisit frequency increased. Daily or near-daily imaging was far less acceptable than occasional capture. In other words, how often you look matters more than how finely you see, particularly once people recognize that frequent imaging can be assembled into timelines of personal or group activity (McAmis et al., 2024).

Use case acceptability varies. The same sensor can be seen as helpful or harmful depending on purpose and actor. When offered pairs of scenarios and asked to choose whether both should be allowed or neither, respondents drew clear lines. Many were comfortable with using images to locate a missing hiker, few supported using images to check whether an employee who called in sick had a car at home. Using car counts at hospitals to infer disease was more acceptable than using car counts at individual homes for employee monitoring. Celebrity tracking in backyards was generally seen as unacceptable, while confirming items lost to a fire for insurance purposes was viewed by some as reasonable. These are real world, recognizable examples, and they show that blanket policies about "satellites are fine" or "satellites are invasive" will fail. People weigh purpose and actor heavily when judging acceptability (McAmis et al., 2024).

Demographic splits. The survey found differences by gender and political affiliation. In general men were more accepting than women of imaging in specific locations, and Republicans were more accepting than Democrats for several scenarios, especially where the actor was a romantic partner or an employer. There is no single public. This means that when you work with communities, workplaces, or jurisdictions, you should expect and plan for different comfort levels. Treat engagement and communication as work to be done, not a disclaimer to be added at the end (McAmis et al., 2024).

NOAA policy context, capability growth and responsibility

While the survey was clarifying public expectations, US policy was shifting. NOAA has eased legacy constraints on commercial imaging, including some rapid revisit restrictions and limits over certain areas. The current posture enables imaging and distribution over most of the globe and reduces temporal constraints. Commercial companies are offering higher spatial resolution and more frequent collection. Combined, policy and industry practice are moving toward capability growth and market support. This is good for many public interest uses. It also means that the burden is now on teams and institutions to decide what to release, and when, considering public comfort, use case acceptability, and foreseeable misuses (McAmis et al., 2024).

Legal and platform implications

Consent at scale. You cannot obtain individual consent from everyone who might be visible. That is true for city centers, industrial sites, parks, and rural roads. A workable approach is risk-based governance that treats comfort and acceptability as inputs to design. If you are imaging places where people live or receive services, decide and document your lawful basis, most public bodies will rely on public tasks, many private actors on legitimate interests, record minimization choices and controls in a data protection impact assessment if risk is high. If cadence is frequent or if an actor's purpose is controversial, do not assume that consent is implied because collection is done from above. Many uses will be acceptable, some will not. The survey shows that people care (McAmis et al., 2024).

Logging and audit. A great deal of thought must go into determining who has access to data and how and by whom requests for data are reviewed. Build access logs for raw assets and sensitive derived products, use audience vetting or request forms where appropriate, and plan for audit by oversight bodies or the public. This is an equity tool, and it is a safety net when a product is misused.

Paywalls versus democratization. The survey asked a simple question about access models, and many respondents favored paid access that depends on use case rather than free access for everyone. That is not a directive, yet it shows a public appetite for proportionate controls where harm is plausible. Many argue convincingly for open access to support science and disaster response. Others argue, also convincingly, for careful release where products can be used to profile, surveil, or coerce. The truth sits in between. Open does not mean harmless. Use release rubrics that consider purpose, actor, resolution, cadence, and plausible misuses, and set controls accordingly. Consider tiered access, delayed release, or aggregation for higher risk cases. The permissive policy environment and commercial capability growth make careful release your responsibility, not only NOAA's (McAmis et al., 2024).

Platforms matter. Computing and storage are often hosted by platforms, and terms of service quietly define what you can do. Read them. If a platform's terms conflict with community protocols, renegotiate, choose different sources, or redesign outputs. Build architecture that allows revocation if misuse occurs, and record trigger conditions so that teams can act quickly without surprises.

A practical mini tool, the use case acceptability matrix

You can turn the surveys lessons into a picture you can use at scoping and at release. Draw a matrix (see Case Study Figure 1.1). Put resolution and frequency on the y-axis. Put purpose and actor on the x-axis. Use traffic light colors to show zones of acceptability for your project, with green for acceptable with standard safeguards, amber for acceptable with enhanced safeguards or constraints, red for unacceptable or pause.

Fill it with your own expected uses. For example, search and rescue by public safety, green at moderate spatial resolution and moderate cadence, amber at very fine resolution or very high cadence because of incidental exposure and sensitive scenes, with safeguards such as blurring people and delaying release. Employer monitoring of individual homes, red at any resolution and cadence. Insurance confirmation of fire losses for a specific household, amber at fine resolution and low cadence with consent and documentation, red at high cadence. Hospital car counts to infer disease levels, amber to green at moderate resolution and reasonable cadence with aggregation, not house level tiles. Celebrity tracking in backyards, red.

For each amber or green cell, note the safeguards. Minimize data by design using the lowest resolution and frequency that meets the purpose. Blur or mask sensitive sites and publish aggregated indices instead of address-level tiles. Log access, vet audiences, apply delayed release when needed, and provide a rationale. Include uncertainty maps, plain-language limits, and an ethics statement explaining exposures. Consult stakeholders before finalizing design and release.

Run the matrix at scoping with your team, then revisit it at release. Use it as part of a data protection impact assessment where that is appropriate and include it in your ethics note or product documentation so that reviewers and users can see your thinking. This is not a static table. Refine it based on incidents and feedback.

This is what the survey points to in everyday terms. People want cadence to be reasonable. People want safeguards. The policy and industry context is permissive. Your design and release choices are where ethics and law are made real (McAmis et al., 2024).

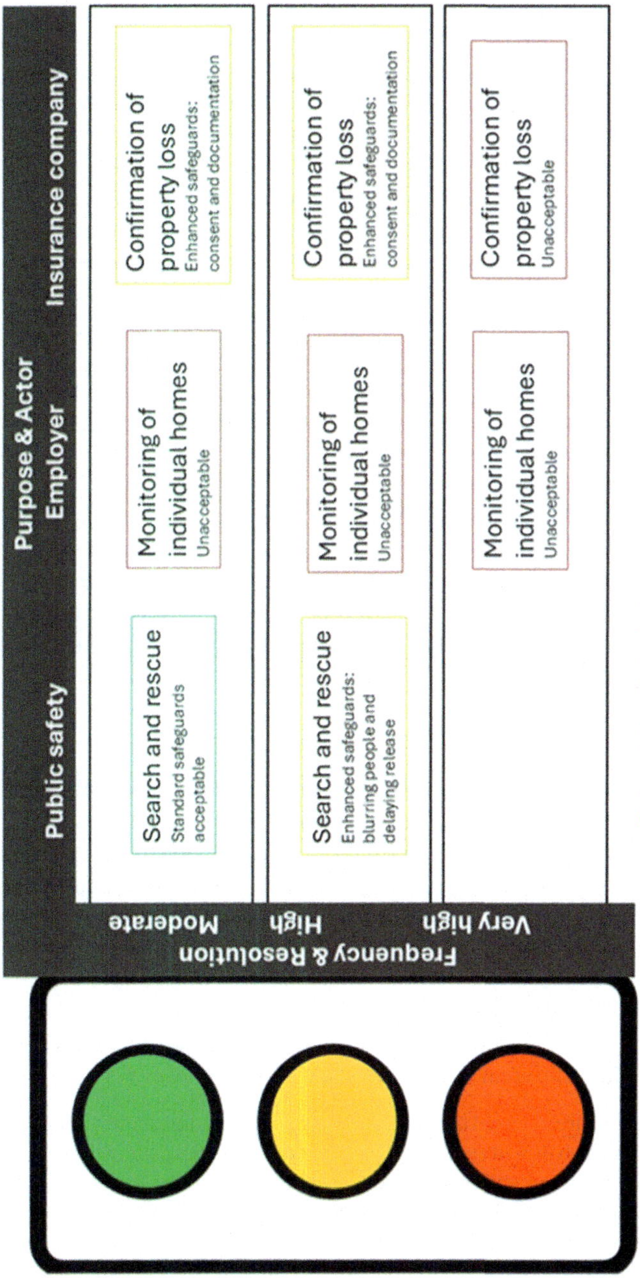

Case Study Figure 1.1 Example acceptability matrix.

Case study 2: The EU AI Act, RS, and GeoAI

What the Act does in plain English

The EU Artificial Intelligence Act is a new law that sets basic rules for how AI systems should be built, documented, and overseen when they could affect people's safety or fundamental human rights. It sorts AI into four risk tiers. Unacceptable risk systems, such as social scoring or real-time remote biometric identification in public spaces by law enforcement except in narrowly defined emergencies, are prohibited. High-risk systems, which include AI used in critical infrastructure, access to public services, law enforcement, migration, and the administration of justice, are allowed but must meet clear obligations. Limited risk systems have lighter transparency duties, for example chatbots that must tell users they are interacting with AI. Minimal risk systems do not face special rules (Matuszczyk et al., 2025; Parliament of the European Union & Council of the European Union, 2024).

For high-risk AI, the Act asks providers to do five main things. First, run a risk management process across the life cycle, identify foreseeable harms, and put controls in place. Second, practice strong data governance, use training and test data that are relevant, representative, and as free of errors as possible, and check for biases that could harm people. Third, write technical documentation that explains what the system does, how it was built, what data it used, how it was validated, and what limits it has. Fourth, make the system transparent and overseen by people, provide clear instructions for use, ensure human oversight is possible and effective, and communicate performance and constraints. Fifth, build for accuracy, robustness, and cybersecurity, monitor the system after deployment, and report serious incidents to authorities.

Where RS and GeoAI meet the Act

Since many RS-based GeoAI systems do not directly identify individuals, land cover maps and cloud masks are often treated as minimal risk. However, risks can still arise from how maps are framed or interpreted. For example, depicting all of Greenland as barren may reinforce the notion that it is uninhabited, or classifying a farmer's occasionally flooded land simply as "river" could erase land rights. The key determinant of risk is not just the data type, but how it is used and the narratives it supports. For example, the same building segmentation pipeline is minimal risk when used for scientific analysis, limited risk when used to inform general urban policy, and high risk when used to decide access to public services or to target enforcement. The following examples further illustrate how different uses map to high-risk categories defined in the Act:

- ➢ Critical infrastructure. AI used to manage or operate critical digital infrastructure or utilities can be high risk. A change detection model directly used to make decisions about water supply or electricity routing will need to meet high-risk obligations.
- ➢ Access to public services. AI used by public authorities to assess eligibility for benefits or services falls into high risk. A settlement growth map used by a municipality to decide which households receive a connection is likely high risk, even if humans review and approve the decisions.

➢ Law enforcement and migration. AI used by police to profile or assess risk, or by migration authorities to examine eligibility, is high risk. A satellite feed used to identify target areas for patrols would raise high risk issues, and other parts of the Act prohibit some uses outright.
➢ Administration of justice. AI used to research and interpret facts or apply law, or in alternative dispute resolution, is high risk. RS evidence fed through a GeoAI pipeline for land use disputes could fall into this category.

In practice, this means that teams building GeoAI for RS should map their pipeline to Annex III use cases as early as possible. If their intended use sits in a high-risk category, they should prepare to meet obligations for risk management, data governance, documentation, transparency, human oversight, performance, and post market monitoring (Matuszczyk et al., 2025; Parliament of the European Union & Council of the European Union, 2024).

A mini scenario: Urban change detection for service planning

Imagine that a city department wants to use building segmentation modeling and change detection (e.g., land use or cover change over time) on sub-meter imagery to plan water and waste service expansions. The university lab that maintains the pipeline acts as the provider if it places the system on the market or into service, while the department is the deployer that uses the system in practice.

Risk classification. The team checks Annex III and finds that AI used by public authorities to assess access to services is a high-risk use. They decide to proceed but adopt the Act's obligations.

Risk management. They set up a lightweight risk register at project start. The scoping entry notes exposure risks, such as stigmatization and misuse for enforcement, and lists controls, such as publishing block level indices rather than house level tiles, blurring culturally sensitive sites, and delaying release during elections. The design entry commits to community consultation on labels and outputs, and to a human in the loop decision process at the department. The operations entry records the plan for monitoring drift, handling complaints, and pausing if issues are detected.

Data governance. They audit training and test data for representativeness across neighborhoods, look for imbalances and sampling gaps, and document any limitations. They check for biases that could lead to systematic under-detection in particular districts. They write down data sources, resolution, cadence, and any preprocessing steps, and they plan to attach notices to dataset records that explain approvals and proper use where applicable (Parliament of the European Union & Council of the European Union, 2024; Future of Life Institute, 2024; Matuszczyk et al., 2025).

Validation and uncertainty. They validate with spatial cross validation, not random splits. They prepare uncertainty maps as first class outputs. They write plain language explanations of limits and error sources, and they include those in documentation and public materials.

Technical documentation. They assemble a model card and a datasheet adapted for RS. It includes purpose, use context, data sources, validation methods, performance metrics under spatial cross validation, uncertainty maps, limits of inference, release choices and rationales, and governance commitments such as community consultation. They store this card in the model registry and share it with the department.

Transparency and human oversight. They draft instructions for use that explain how to interpret indices and uncertainty, how to combine them with field assessments, and how to escalate cases where the map conflicts with local knowledge. They set up a decision workflow in the department that requires human review and approval for any allocation decision.

Accuracy, robustness, cybersecurity. They record their performance targets, plan for monitoring model drift, and keep the system's dependencies updated. They set up access controls and logs for raw assets and sensitive layers, and they document incident response triggers.

Post market monitoring and reporting. After deployment, they monitor unexpected errors, complaints, and model drift. They publish a contact point for residents to raise issues. If a serious incident occurs, for example if the system's outputs are used for enforcement against agreed protocols or if a release causes harm, they record the incident, respond, and report to the appropriate authority as the Act requires.

A few nuances deserve attention

Provider versus deployer duties. The Act places most obligations on providers, yet deployers must also use the system within its constraints, ensure human oversight, and avoid uses that are prohibited or fall outside the documented purpose.

Documentation is not a burden; it is a bridge. The model card and datasheet make the system legible to non-specialists, including residents, councilors, and oversight bodies. Uncertainty maps and plain language prevent false certainty and help humans make better decisions.

Bias checks are practice, not politics. Representativeness, spatial cross validation, and uncertainty mapping are good scientific practice. Under the Act they become good legal practice.

Public communication matters. The department should publish a plain explanation of what the map does, what it does not do, how decisions are made, and how people can appeal. This is part of transparency and trust-building.

Release choices are ethical and legal. Using block level indices rather than house level data is both ethical and the legal choice as it lowers exposure risks while ensuring the data are used proportionately.

You can adapt this mini scenario to other RS applications. The core moves are the same: Decide if your use is high risk, run a risk management process, practice good data governance, validate properly, write documentation, ensure human oversight, and monitor after deployment.

Closing thoughts: The EU AI Act puts familiar responsibilities into law. Many teams already practice good habits, such as consulting communities, validating with spatial cross validation, and publishing uncertainty maps. The Act's obligations for high risk uses risk management, data governance, documentation, transparency, oversight, performance, and monitoring align with those habits. The practical advice to lay readers is simple. Know your use case, map it to risk, build the documentation you would want to read as a citizen, and treat RS GeoAI as a system that affects people, not only pixels (Parliament of the European Union & Council of the European Union, 2024; Future of Life Institute, 2024; Matuszczyk et al., 2025).

Acronyms used and definitions

AI4EO: Artificial Intelligence for Earth Observation, the use of AI to process and interpret satellite and other EO data.
API: Application Programming Interface, a technical interface that allows software to access data or services programmatically.
BC Labels: Biocultural Labels, notices that communicate protocols for access and use of biocultural data tied to Indigenous lands and resources.
CARE: Collective Benefit, Authority to Control, Responsibility, Ethics, a set of Indigenous data governance principles for responsible data use.
CV: Cross-Validation, a model evaluation technique; in geospatial work, spatial CV is used to avoid inflated performance due to spatial autocorrelation
DOI: Digital Object Identifier, a persistent identifier used to reference datasets, papers, or other digital objects.
DPIA: Data Protection Impact Assessment, a formal assessment to identify and mitigate risks to people from high-risk data processing.
EU: European Union, a political and economic union of member states in Europe.
EU AI Act: European Union Artificial Intelligence Act, the EU law that regulates AI systems based on risk and sets obligations for high-risk uses.
FPIC: Free, Prior and Informed Consent, the right of Indigenous peoples to give or withhold consent to projects that affect them and their territories.
GDPR: General Data Protection Regulation, the EU's data protection law governing the processing of personal data.
IP: Intellectual Property, legal rights related to creations such as data, software, models, and documentation.
MLOps: Machine Learning Operations, practices and tooling that manage model versioning, data lineage, validation, monitoring, and traceability
NGO: Non-Governmental Organization, a civil society organization operating independently of government
NMA: National Mapping Agency, a national institution that maintains authoritative geospatial data and mapping services
NOAA: National Oceanic and Atmospheric Administration, the US federal agency that licenses and regulates commercial remote sensing.
SAR: Synthetic Aperture Radar, a radar imaging technology that can capture the Earth's surface day or night and through clouds.
TK Labels: Traditional Knowledge Labels, community-driven notices that communicate protocols for access and use of Indigenous knowledge
ToS: Terms of Service, platform or service rules that govern permitted uses, redistribution, and access.

References

AAG IT support, n.d. https://aag-it.com/the-latest-cloud-computing-statistics/. Last Accessed: March 8, 2026.

Bennett, M. M. (2023). Chinese sociotechnical imaginaries of Earth observation: From sight to foresight. *Big Data & Society, 10*(2).

Bennett, A., Bittner, J., & Mitchell, K. (2022). The politics of pixels: A review and agenda for critical remote sensing. *Annals of the American Association of Geographers, 112*(1), 1–17.

Bennett, M. M., Gleason, C. J., Tellman, B., Alvarez León, L. F., Friedrich, H. K., Ovienmhada, U., & Matthews, A. J. (2024). Bringing satellites down to Earth: Six steps to more ethical remote sensing. *Global Environmental Change Advances, 2.* https://doi.org/10.1016/j.gecadv.2023.100003

Bennett, M. M., Van Den Hoek, J., Zhao, B., & Prishchepov, A. V. (2022). Improving satellite monitoring of armed conflicts. *Earth's Future, 10*(9). e2022EF002904.

Braun, A. C. (2021). More accurate less meaningful? A critical physical geographer's reflection on interpreting remote sensing land-use analyses. *Progress in Physical Geography: Earth and Environment, 45*(5), 706–735.

Bunting, E. L., Theuerkauf, E. J., & Mack, E. A. (2026). Embracing community science to study broadscale landscape change: Successes, failures, challenges, issues of sustainability, and advice. *Journal of Land Use Science, 21*(1), 95–109. https://doi.org/10.1080/1747423X.2026.2627704

Carroll, S. R., Rodriguez-Lonebear, D., & Martinez, A. (2019). Indigenous data governance: Strategies from United States native nations. *Data Science Journal, 18*, 31.

Cinnamon, J., & Noth, T. (2023). Spatiotemporal development of informal settlements in Cape Town, 2000 to 2020: An open data approach. *Habitat International, 133.*

First Nations Principles of OCAP®. (n.d.). The first nations information Governance Centre. Retrieved March 3, 2026, from https://fnigc.ca/ocap-training/.

Floridi, L., Cowls, J., Beltrametti, M., Chatila, R., Chazerand, P., Dignum, V., Luetge, C., Madelin, R., Pagallo, U., Rossi, F., & Schafer, B. (2018). AI4People—An ethical framework for a good AI society: Opportunities, risks, principles, and recommendations. *Minds and Machines, 28*(4), 689–707.

Flynn, C., Guha, A., Majumdar, S., Srivastava, D., & Zhou, Z. (2022). *Towards algorithmic fairness in Space-Time: Filling in Black holes (arXiv:2211.04568).* arXiv. https://doi.org/10.48550/arXiv.2211.04568

Future of Life Institute. (2024). *High-level summary of the AI Act.* https://artificialintelligenceact.eu/high-level-summary.

Hancock, J., Hui, R., Singh, J., & Mazumder, A. (June 2024). Trouble at Sea: Data and digital technology challenges for maritime human rights concerns. In *Proceedings of the 2024 ACM conference on fairness, accountability, and transparency* (pp. 988–1001).

Jegham, N., Abdelatti, M., Koh, C. Y., Elmoubarki, L., & Hendawi, A. (2025). *How hungry is AI? Benchmarking energy, water, and carbon footprint of LLM inference (arXiv: 2505.09598).* arXiv. https://doi.org/10.48550/arXiv.2505.09598

Kausika, B. B., & van Altena, V. (2025). GeoAI in topographic mapping: Navigating the future of opportunities and risks. *ISPRS International Journal of Geo-Information, 14*(8), 313. https://doi.org/10.3390/ijgi14080313

Kochupillai, M., Kahl, M., Schmitt, M., Taubenböck, H., & Zhu, X. X. (2022). Earth observation and artificial intelligence: Understanding emerging ethical issues and opportunities.

IEEE Geoscience and Remote Sensing Magazine, 10(4), 90–119. https://doi.org/10.1109/MGRS.2022.3208357

Kochupillai, M., & Taubenböck, H. (2023). Conducting ethically mindful Earth observation research: The case of slum mapping. In *2023 IEEE International Geoscience and Remote Sensing Symposium (IGARSS)* (pp. 1937–1940). IEEE. https://doi.org/10.1109/IGARSS52108.2023.10281725

Koldasbayeva, D., Tregubova, P., Gasanov, M., Zaytsev, A., Petrovskaia, A., & Burnaev, E. (2024). Challenges in data driven geospatial modeling for environmental research and practice. *Nature Communications, 15*. https://doi.org/10.1038/s41467-024-55240-8

Kroth, L. (2025). Remote sensing and feminist critique: Reappropriations of sensing across distance. *Environment and Planning F, 4*(1), 3–19.

Lappe, R., Calovi, M., & Winther, H. (2025). The view from somewhere: What remote sensing still has to learn about the satellite gaze. *Progress in Physical Geography: Earth and Environment.*

Matuszczyk, N., Barnes, C. R., Roy, S., Gupta, R., Ozel, B., & Mitra, A. (2025). *From bias to accountability, how the EU AI Act confronts challenges in European GeoAI auditing*. arXiv preprint arXiv, 2505.18236.

McAmis, R., Sim, M., Bennett, M., & Kohno, T. (2024). Over fences and into yards: Privacy threats and concerns of commercial satellites. *Proceedings on Privacy Enhancing Technologies, 2024*(1), 379–396. https://doi.org/10.56553/popets-2024-0022

National Space Council Users' Advisory Group. (2023). Space data ethics: The next frontier in responsible leadership, recommendation on space data ethics. https://www.nasa.gov/wp-content/uploads/2024/02/white-paper-space-data-ethics-2023-12-01-final-002.pdf?emrc=65e3abb9b5644. (Accessed 6 May 2026).

O'Brien, M., Duerr, R., Taitingfong, R., Martinez, A., Vera, L., Jennings, L. L., Downs, R. R., Antognoli, E., ten Brink, T., Halmai, N. B., Carroll, S. R., David-Chavez, D., Hudson, M., & Buttigieg, P. L. (2024). Earth science data repositories: Implementing the CARE principles. *Data Science Journal, 23*, 37. https://doi.org/10.5334/dsj-2024-037

Parliament of the European Union, & Council of the European Union. (2024). Regulation (EU) 2024/1689 of the European Parliament and of the Council of 20 August 2024 on Artificial Intelligence and amending certain Union legislative acts. Official Journal of the European Union. https://eur-lex.europa.eu/eli/reg/2024/1689/oj/eng.

Pickles, J. (Ed.). (1995). *Ground truth: The social implications of geographic information systems*. Guilford Press.

Pritchard, R., Sauls, L. A., Oldekop, J. A., Kiwango, W. A., & Brockington, D. (2022). Data justice and biodiversity conservation. *Conservation Biology, 36*(5).

Segarra, J., González-Fernández, A., Osorno-Covarrubias, J., & Couturier, S. (2024). The role of critical remote sensing in environmental justice struggles. *Progress in Environmental Geography, 3*(3), 185–211.

Strubell, E., Ganesh, A., & McCallum, A. (2019). Energy and policy considerations for deep learning in NLP. In *Proceedings of the 57th annual meeting of the Association for computational linguistics* (pp. 3645–3650). https://aclanthology.org/P19-1355/.

UNDRIP. (2007). https://www.un.org/development/desa/indigenouspeoples/wp-content/uploads/sites/19/2018/11/UNDRIP_E_web.pdf.

Wang, J., Fleischmann, M., Venerandi, A., Romice, O., Kuffer, M., & Porta, S. (2023). EO+ morphometrics: Understanding cities through urban morphology at large scale. *Landscape and Urban Planning, 233*.

York, N. D. L., Pritchard, R., Sauls, L. A., Enns, C., & Foster, T. (2023). Justice and ethics in conservation remote sensing: Current discourses and research needs. *Biological Conservation, 287*. https://doi.org/10.1016/j.biocon.2023.110319

Zewe, A. (2025). *Explained: Generative AI's environmental impact*. MIT News.

Further reading

Rogers, S. (September 25, 2018). National Inuit strategy on Research—Implementation plan. *Inuit Tapiriit Kanatami*. https://itk.ca/national-inuit-strategy-on-research-implementation-plan/.

How Citizen Science and New Technologies can Integrate with Big Data

14

Di Yang[1], *Xiao Huang*[2], *Jessica Mitchell*[3], *Yaqian He*[4,5], *Peder Nelson*[6] *and Russanne Low*[7]
[1]Department of Geography, University of Florida, Gainesville, FL, United States; [2]Department of Environmental Sciences, Emory University, Atlanta, GA, United States; [3]Spatial Analysis Lab, University of Montana, Missoula, MT, United States; [4]Department of Geosciences, University of Arkansas, Fayetteville, AR, United States; [5]Department of Geography, Indiana University Bloomington, Bloomington, IN, United States; [6]College of Earth, Ocean, and Atmospheric Sciences, Oregon State University, Corvallis, OR, United States; [7]Institute for Global Environmental Strategies, Arlington, VA, United States

Key terms and definitions

Citizen Science: Public participation in scientific research, often through data collection or analysis
Crowdsourcing: Obtaining data or services from a large group of people, typically via the internet
Volunteered Geographic Information (VGI): Geographic information collected and shared voluntarily by individuals
Mobile Sensing: Use of mobile devices (e.g., smartphones) to collect and transmit environmental data
Data Fusion: Process of integrating multiple data sources to produce more consistent, accurate, and useful information
Internet of Things (IoT): Network of interconnected devices capable of collecting and exchanging data
Cloud-Based Platforms: Online services for storing, processing, and analyzing large volumes of citizen science and remote sensing data
Ethical Data Management: Practices ensuring privacy, security, and appropriate use of citizen-generated data

1. Introduction

1.1 Background

The historical context of citizen science and big data represents a fascinating convergence of public participation and technological advancement. Citizen science, with roots tracing back to the 18th century, has evolved from amateur naturalists

Remote Sensing, Big Data, and GeoAI. https://doi.org/10.1016/B978-0-443-26741-3.00014-3

contributing to scientific knowledge to a global movement empowering million to engage in research (Silvertown, 2009). The advent of the internet in the late 20th century marked a turning point, enabling projects like the Christmas Bird Count to transition from local to global scales (Bonney et al., 2009). A pivotal moment in this evolution was the conceptualization of Volunteered Geographic Information (VGI) by Goodchild et al. (2007), which recognized the potential of citizens as "sensors" in generating geographic data. This framework laid the groundwork for numerous citizen science initiatives in the geospatial domain. Concurrently, the concept of big data emerged in the early 2000s, characterized by the three versus: volume, velocity, and variety (Laney, 2001). As computational power increased exponentially, the ability to process and analyze vast datasets became a reality. This technological revolution set the stage for a new era of citizen science, where public participation could be leveraged to generate and analyze big data at unprecedented scales, particularly in geographic and environmental contexts.

The relevance of integrating citizen science with new technologies in today's data-driven world cannot be overstated. As we grapple with global challenges such as climate change, biodiversity loss, and urban development, the need for comprehensive, real-time data has never been greater (Dickinson et al., 2012). Smartphones, equipped with sensors and GPS capabilities, have transformed every user into a potential data collector (Kitchin, 2014). Machine learning algorithms can now process and validate large volumes of citizen-generated data, enhancing its reliability and usability in scientific research (Kelling et al., 2015). The integration of citizen science with big data technologies offers a powerful approach to addressing complex environmental and societal issues. It not only amplifies the scale and scope of data collection but also fosters public engagement in scientific processes, potentially leading to more informed decision-making and policy formulation (Haklay, 2013). As we progress further into the 21st century, this integration promises to democratize science, enhance our understanding of global systems, and provide timely insights for sustainable development and environmental conservation.

1.2 Objective

The primary goal of this chapter is to explore the synergies between citizen science, new technologies, and big data, unveiling the transformative potential of their integration in scientific research and environmental monitoring. We aim to elucidate how the convergence of these domains is reshaping our approach to data collection, analysis, and interpretation in the geospatial sciences. By examining cutting-edge projects and methodologies, we seek to demonstrate how citizen-generated data, when combined with advanced technologies and big data analytics, can provide unprecedented insights into environmental processes and societal patterns. Furthermore, this chapter will critically assess the challenges and opportunities presented by this integration, offering a balanced perspective on its current state and future prospects.

In addressing specific geospatial challenges, we focus on three key areas: data collection, integration, and long-term sustainability. The challenge of data collection in citizen science projects is multifaceted, encompassing issues of data quality, spatial and temporal coverage, and participant engagement (Wiggins et al., 2011). We will explore innovative approaches to these challenges, such as the use of mobile apps with built-in data validation algorithms and gamification elements to enhance data quality and user participation (Prestopnik & Crowston, 2012). The integration of citizen science data with other big data sources, including satellite imagery and IoT sensor networks, presents both technical and conceptual challenges. We will examine methodologies for data fusion and harmonization, addressing issues of scale, accuracy, and interoperability (Fritz et al., 2017). Finally, the long-term sustainability of citizen science initiatives in the big data era is crucial for their continued relevance and impact. We will discuss strategies for maintaining participant engagement, ensuring data continuity, and developing robust technological infrastructures to support ongoing data collection and analysis (West & Pateman, 2016). By addressing these challenges, we aim to provide a comprehensive understanding of how citizen science can effectively leverage new technologies and big data approaches to advance geospatial research and environmental monitoring.

2. Theoretical foundations

2.1 Citizen science

Citizen science, broadly defined as public participation in scientific research, has undergone a remarkable evolution over the past decades. Traditionally, it referred to amateur naturalists contributing observations to scientific endeavors, but it has since expanded to encompass a wide array of disciplines and methodologies (Bonney et al., 2009). The term "citizen science" itself was coined in the mid-1990s, though the practice predates the terminology by centuries (Silvertown, 2009). Today, citizen science projects range from local biodiversity monitoring to global climate change studies, engaging millions of participants worldwide.

The evolution of citizen science has been profoundly influenced by technological advancements, particularly the internet and mobile technologies. These tools have facilitated the collection, transmission, and analysis of data at unprecedented scales (Newman et al., 2012). The concept of Volunteered Geographic Information (VGI), introduced by Goodchild (2007), further expanded the scope of citizen science in the geospatial domain, recognizing the potential of citizens as "sensors" in generating valuable geographic data.

The significance of citizen science in modern scientific research cannot be overstated. It has emerged as a powerful approach to address complex, large-scale scientific questions that would be infeasible or prohibitively expensive using traditional research methods alone (Theobald et al., 2015). Citizen science projects have contributed to numerous scientific discoveries, from identifying new species to tracking the impacts of climate change (Cooper et al., 2014). Moreover, citizen science plays a crucial role in science education and public engagement, fostering scientific literacy and environmental stewardship among participants (Bonney et al., 2016). By democratizing the scientific process, citizen science also has the potential to bridge the gap between scientific communities and the general public, potentially leading to more informed decision-making and policy formulation (Haklay, 2013).

2.2 Big data in scientific research

Big Data, a term that gained prominence in the early 2000s, refers to datasets that are too large or complex to be dealt with by traditional data-processing application software. It is characterized by the "3 vs.": volume (scale of data), velocity (speed of data generation and processing), and variety (diversity of data types and sources) (Laney, 2001). As technology has advanced, additional characteristics have been proposed, including veracity (uncertainty of data) and value (benefit of data) (Demchenko et al., 2013).

In scientific research, Big Data has emerged from various sources, including high-throughput scientific instruments, sensor networks, and large-scale simulations. The Earth observation sector, for instance, generates petabytes of data annually from satellite imagery and ground-based sensors (Ma et al., 2015). Genomics and bioinformatics have also seen an explosion in data volume, with next-generation sequencing technologies producing vast amounts of genetic data (Marx, 2013). The significance of Big Data in advancing scientific knowledge and decision-making is profound. It enables researchers to identify patterns and correlations that would be impossible to detect with smaller datasets, leading to new insights and discoveries across various scientific domains (Fan et al., 2014). In climate science, Big Data analytics have improved our understanding of global climate patterns and enhanced the accuracy of climate models (Hampton et al., 2013). In healthcare, the analysis of large-scale genomic and clinical data has accelerated personalized medicine and drug discovery (Murdoch & Detsky, 2013). Moreover, Big Data has transformed the scientific method itself, giving rise to data-driven approaches that complement traditional hypothesis-driven research. This paradigm shift has led to new fields such as computational social science and digital humanities, where large-scale data analysis provides novel insights into human behavior and cultural patterns.

2.3 Technology-enabled mass participation

Technology-enabled mass participation in citizen science refers to the use of modern digital technologies to facilitate large-scale public involvement in scientific research. This approach leverages smartphones, web platforms, and other digital tools to enable data collection, processing, and analysis by a vast number of participants across wide geographic areas (Newman et al., 2012). Key technologies include mobile apps with built-in sensors, online crowdsourcing platforms, and machine learning algorithms for data validation and analysis.

The evolution of these technologies has dramatically expanded the scope and scale of citizen science projects. For instance, smartphone apps like eBird and iNaturalist have transformed biodiversity monitoring, allowing users to easily record and share species observations globally (Sullivan et al., 2014). Web-based platforms such as Zooniverse have enabled millions of volunteers to contribute to tasks ranging from galaxy classification to historical document transcription (Simpson et al., 2014). The significance of technology-enabled mass participation in scaling up data collection and analysis is multifaceted. Firstly, it allows for the collection of data at spatial and temporal scales that would be impossible for professional scientists alone, providing unprecedented coverage for environmental and ecological monitoring (Theobald et al., 2015). Secondly, it enables real-time data collection and analysis, crucial for tracking rapidly changing phenomena such as wildlife migrations or disease outbreaks.

Furthermore, these technologies enhance data quality through built-in validation mechanisms and the application of machine learning algorithms to detect and correct errors. They also facilitate the integration of citizen-generated data with professional scientific datasets, enhancing the overall value and impact of citizen science contributions.

3. Methodologies

3.1 Mass data collection through citizen science

Mass data collection through citizen science refers to approaches that enable large-scale public participation in scientific data gathering. This methodology harnesses the collective effort of volunteers, often referred to as "citizen scientists," to collect data at spatial and temporal scales that would be infeasible for professional scientists alone (Bonney et al., 2014). The advent of digital technologies has dramatically expanded the scope and scale of citizen science data collection, transforming how we approach scientific research and environmental monitoring.

The application of mass data collection in citizen science typically involves three main approaches: mobile apps, distributed computing, and crowdsourced mapping. Mobile applications have become a cornerstone of modern citizen science, allowing participants to easily record and submit observations using their smartphones. For instance, eBird, a global bird monitoring project, uses a mobile app that enables birders to log their sightings, complete with location data and timestamps (Sullivan et al., 2014). These apps often incorporate built-in data validation mechanisms and species identification guides to improve data quality. Distributed computing projects, on the other hand, utilize the idle processing power of volunteers' personal computers to analyze large datasets. SETI@home, one of the pioneers in this field, uses volunteers' computers to search for signs of extraterrestrial intelligence in radio telescope data (Anderson et al., 2002). This approach has been particularly effective in fields like astronomy and climate modeling, where vast amounts of data need to be processed. Crowdsourced mapping represents another powerful application of mass data collection. Projects like OpenStreetMap engage volunteers in creating and updating detailed maps of the world (Haklay & Weber, 2008). This approach has been particularly valuable in areas where official mapping data is scarce or outdated and has proven crucial in disaster response scenarios.

The success of these approaches relies heavily on user-friendly interfaces, clear protocols, and effective training materials to ensure data quality. Moreover, gamification elements and real-time feedback mechanisms are often incorporated to maintain participant engagement and motivation (Prestopnik & Crowston, 2012). As technology continues to evolve, new methods of mass data collection are likely to emerge, further expanding the potential of citizen science to contribute to scientific knowledge and environmental stewardship.

3.2 Data integration and harmonization

Data integration and harmonization in the context of citizen science and big data encompass techniques for merging and aligning diverse datasets to create a coherent, unified resource that can support more comprehensive analyses and insights. This process is particularly crucial when combining citizen-generated data with professional scientific datasets or remote sensing data, as these sources often differ in scale, precision, and collection methodologies.

The integration of citizen science data with other big data sets presents unique challenges and opportunities. Citizen-generated data can vary widely in quality, format, and completeness, depending on the expertise of the contributors and the tools used for data collection. Professional scientific data, while often adhering to strict protocols, may lack the spatial or temporal coverage that citizen science can provide. Remote sensing data, such as satellite imagery, offers broad coverage but may lack the fine-grained detail of ground-based observations (See et al., 2016). Techniques for data integration and harmonization have evolved significantly in recent years, driven by advances in big data technologies and machine learning. Statistical methods such as data fusion and ensemble modeling can combine multiple data sources while accounting for their respective uncertainties (Swanson et al., 2016). Machine learning algorithms are increasingly used to identify and correct systematic biases in citizen-generated data, improving its compatibility with professional datasets (Kelling et al., 2015). The application of these techniques is particularly evident in the combination of citizen observations with remote sensing data. For example, in phenology studies, citizen-reported observations of plant flowering times have been successfully integrated with satellite-derived vegetation indices to provide a more comprehensive understanding of climate change impacts on ecosystems (Rosemartin et al., 2014). In urban planning, the integration of crowdsourced data on urban infrastructure with satellite imagery has led to more accurate and up-to-date urban maps.

Geospatial data integration plays a crucial role in this process, with Geographic Information Systems (GIS) serving as a fundamental tool for aligning data from different sources based on their spatial attributes. Techniques such as spatial interpolation and co-kriging can help fill gaps in spatial coverage and leverage the strengths of different data sources to improve overall accuracy (Zhang et al., 2014). As the field continues to evolve, new approaches to data integration and harmonization are likely to emerge, further enhancing our ability to leverage the complementary strengths of citizen science, professional research, and remote sensing in addressing complex scientific and societal challenges (Fig. 14.1).

Figure 14.1 Core methodologies and frameworks for mass data collection, integration, and long-term sustainability in citizen science initiatives.

3.3 *Long-term program sustainability*

Long-term program sustainability in citizen science refers to the strategies and practices employed to maintain citizen science initiatives over extended periods, ensuring their continued relevance, effectiveness, and impact. This aspect is crucial for the success of citizen science projects, as many scientific questions and environmental monitoring efforts require data collection and analysis over years or even decades (West & Pateman, 2016).

The sustainability of citizen science programs encompasses several key elements: data management, infrastructure development, and participant engagement. Effective data management is fundamental to long-term sustainability, ensuring that data collected over time remains accessible, usable, and relevant. This involves implementing robust data storage systems, establishing clear metadata standards, and developing protocols for data quality control and validation (Wiggins et al., 2013). Cloud-based storage solutions and distributed database systems have become increasingly important in managing the large volumes of data generated by citizen science projects. Infrastructure development is another critical component of long-term sustainability. This includes not only the technical infrastructure for data collection and analysis but also the organizational structures needed to support ongoing operations. Many successful long-term citizen science projects have established formal partnerships with academic institutions, government agencies, or non-profit organizations to provide stable institutional support.

Participant engagement is perhaps the most challenging aspect of maintaining long-term citizen science programs. Strategies for sustaining volunteer involvement over time include providing regular feedback on the impact of their contributions, offering opportunities for skill development and social interaction, and adapting project tasks to match participants' interests and abilities (Rotman et al., 2012). Some projects have successfully implemented tiered participation models, allowing volunteers to take on increasing levels of responsibility and leadership within the project (Bonney et al., 2009). The application of these sustainability strategies is evident in long-running citizen science projects like the Christmas Bird Count, which has been conducted annually for over a century. Its longevity can be attributed to a combination of robust data management practices, strong institutional support from the Audubon Society, and a dedicated community of volunteer birders (Dunn et al., 2005). Emerging technologies are also playing a role in enhancing the sustainability of citizen science programs. Machine learning algorithms can help automate data validation processes, reducing the workload on project managers and improving data quality over time. Social media and online community platforms are being leveraged to maintain volunteer engagement and facilitate knowledge sharing among participants. As citizen science continues to evolve, ensuring the long-term sustainability of these programs will be crucial for realizing their full potential in advancing scientific knowledge, informing policy decisions, and fostering public engagement with science.

4. Discussion

4.1 Analysis

Comparative analysis of different citizen science approaches in big data contexts reveals a spectrum of methodologies, each with unique strengths and limitations. Traditional observational projects, such as the Christmas Bird Count, have produced valuable long-term datasets but face challenges in data quality control and spatial coverage (Dunn et al., 2005). In contrast, technology-enabled approaches leveraging mobile apps and sensors have dramatically expanded the scale and scope of data collection. Projects like eBird have successfully combined user-friendly interfaces with sophisticated data validation algorithms, transforming ornithological research (Sullivan et al., 2014). Distributed computing initiatives like SETI@home represent another distinct approach, harnessing volunteers' computational power to process vast datasets (Anderson et al., 2002). Meanwhile, crowdsourced mapping projects such as OpenStreetMap demonstrate the potential for collaborative data creation in the geospatial domain (Haklay & Weber, 2008).

Technology-enabled citizen science presents both challenges and opportunities. Data quality remains a primary concern, as participant expertise and equipment variability can introduce biases (Kosmala et al., 2016). Privacy and ethical considerations also come to the fore, particularly when collecting sensitive location-based data (Resnik et al., 2015). However, these challenges are balanced by significant opportunities. The integration of artificial intelligence and machine learning offers promising avenues for improving data validation and analysis (Kelling et al., 2015). Gamification techniques have shown potential for enhancing participant engagement and data quality (Prestopnik & Crowston, 2012). Furthermore, the development of low-cost, open-source sensors is democratizing access to sophisticated data collection tools, enabling more diverse participation in citizen science (Volten et al., 2018).

As technology continues to evolve, new approaches are emerging that blur the lines between traditional categories of citizen science, such as virtual and augmented reality technologies opening up possibilities for immersive citizen science experiences (Çöltekin et al., 2020).

4.2 Implications

The integration of citizen science, new technologies, and big data has the potential to fundamentally transform scientific research across numerous disciplines. This convergence enables the collection and analysis of data at unprecedented scales, offering insights into complex phenomena that were previously beyond the reach of traditional research methods. In ecology and biodiversity research, the combination of citizen-collected observations with satellite imagery and environmental sensor networks is providing a more comprehensive understanding of ecosystem dynamics and species distributions (Chandler et al., 2017). Public health research has been revolutionized by citizen-driven big data, with participatory epidemiology projects enabling real-time disease surveillance and the identification of environmental health risks (Smolinski et al., 2015). In the geosciences, citizen science is enhancing our ability to monitor and respond to natural hazards, with crowdsourced earthquake detection systems providing crucial early warning capabilities (Kong et al., 2016).

Looking to the future, the prospects for citizen-driven big data in addressing global challenges are promising. Climate change research stands to benefit significantly from the integration of citizen science data with climate models, potentially improving our ability to predict and mitigate climate impacts at local and regional scales (Howe et al., 2017). The democratization of science through citizen participation also has profound implications for science policy and public engagement, potentially leading to more informed policy decisions and greater public support for scientific research (Bonney et al., 2016). However, realizing these potential benefits will require addressing significant challenges, including ensuring data interoperability, developing robust data governance frameworks, and maintaining long-term participant engagement (Wiggins et al., 2013). As we look to the future, the continued evolution of artificial intelligence, Internet of Things technologies, and advanced data analytics promises to further expand the frontiers of citizen-driven science in addressing complex global challenges.

Case studies

Case study 1: OpenStreetMap–land use crowdsourced mapping

OpenStreetMap (OSM), a volunteer-driven initiative to create a free, editable map of the world, has revolutionized geographic data collection and analysis. This case study examines how OSM, through its web and mobile platforms for collaborative mapping, was leveraged to map and analyze land use patterns in the Southeastern United States (SEUS), a region characterized by diverse landscapes including heavily managed forestlands, developed agriculture, and multiple metropolitan areas.

The SEUS presents unique challenges for land use mapping due to its complex and rapidly changing landscape. Traditional mapping methods often struggle with the scale and frequency of updates required to capture these dynamics accurately. OSM, with its global community of contributors and real-time updating capabilities, offers a promising solution to these challenges. In this study, researchers utilized OSM data as a cornerstone for mapping land use across a 34,000 km^2 area of the SEUS. The project leveraged OSM's comprehensive global map dataset, which has been built through collaborative efforts of volunteers worldwide. The implementation of this citizen science approach involved accessing historical OSM data through the OSM API, providing a temporal dimension to the land use analysis that is often missing from traditional mapping efforts.

The growth of OSM contributions in the study area was remarkable. In 2013, there were only 1199 building objects mapped in the region. By October 2018, this number had exploded to 66,488, demonstrating the rapid expansion of the OSM dataset (Yang et al., 2023). This growth not only increased the quantity of data available but also improved its quality and comprehensiveness, as more contributors brought local knowledge to the mapping process. Researchers developed an innovative methodology that combined this crowdsourced OSM data with Earth observation datasets and cloud-based computing. They created a crosswalk dictionary to reclassify OSM tags into five land use categories: cropland/rangeland, commercial/industrial, managed forest, residential areas, and water body. This reclassification allowed for the standardization of the diverse OSM tags into a coherent land use classification scheme. The OSM-derived data served as training samples for a Random Forest classifier, which was then applied to various Earth observation datasets, including Landsat imagery, global forest canopy height data, and nighttime satellite imagery (Case Study Figure 1.1). This integration of citizen science data with remote sensing shows the potential for synergies between different data sources in environmental monitoring and analysis.

The results of this approach were impressive. The Random Forest classifier, trained on OSM-derived samples, achieved high accuracy levels, with an out-of-bag (OOB) overall accuracy close to 97% for both 2013 and 2018 classifications, and Kappa values of 0.97 (Yang et al., 2023). These high accuracy levels demonstrate the potential of citizen-contributed data to produce reliable land use maps, challenging the notion that only professional data collection can yield high-quality results. The OSM-based approach revealed significant land use transitions in the SEUS between 2013 and 2018. Residential areas increased from 4.93% to 8.87% of the study area, while commercial/industrial areas grew from 0.59% to 2.97% (Yang et al., 2023). These findings highlight the rapid urbanization and development occurring in the region, information that is crucial for urban planners, policymakers, and environmental managers. Furthermore, the integration of OSM-derived land use maps with existing forest management classification data allowed for novel insights into the spatial relationships between different forest management types and surrounding land uses. For instance, the study found that production forests showed strong spatial relationships with croplands, particularly within a 1,000-m buffer zone (Yang et al., 2023).

The impact of this OSM-based approach extends beyond academic research. The methodology developed in this study demonstrates the potential for OSM to support

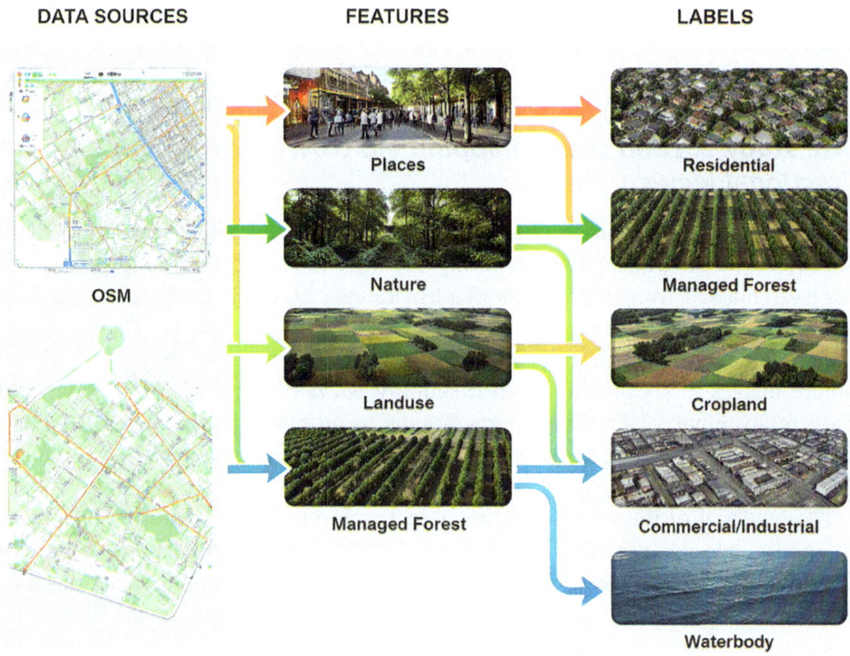

Case Study Figure 1.1 Workflow of OSM training sample setting.

disaster response, urban planning, and navigation services. In disaster response scenarios, up-to-date land use information is crucial for effective resource allocation and risk assessment. For urban planning, the ability to track rapid land use changes can inform more responsive and adaptive policies. Navigation services can benefit from the detailed and frequently updated land use information to provide more accurate and context-aware routing. However, the study also revealed some limitations of using OSM data. Data quality and completeness varied across the study area and between different feature types. There's potential for geographic bias, with more populated areas receiving more detailed mapping. Additionally, some land use types with similar spectral signatures posed challenges for classification. Despite these challenges, this case study illustrates the significant potential of integrating citizen science data from OSM with Earth observation datasets for regional land use mapping and analysis. As OSM continues to grow and evolve, its value for scientific research, environmental monitoring, and land management is likely to increase further. The success of this method in the SEUS context suggests its applicability to other regions facing similar land use mapping challenges. Future research could focus on developing methods to address data quality variabilities in OSM, exploring ways to incentivize contributions in under-mapped areas, and investigating the potential of machine learning to further enhance the integration of citizen science data with remote sensing for land use analysis.

In conclusion, this case study demonstrates how OSM, a volunteer-driven global mapping initiative, can be leveraged to create sophisticated, accurate, and up-to-date land use maps. It shows the power of citizen science in addressing complex environmental monitoring challenges and opens new avenues for integrating crowdsourced data into scientific research and practical applications.

Case study 2: Land cover mapping via crowdsourced multi-directional views

The GLOBE Observer Land Cover program exemplifies the growing potential of citizen science in environmental monitoring. By the end of 2021, this innovative initiative had amassed over 30,000 sets of observations globally, marking a significant milestone in crowdsourced data collection (Huang et al., 2023). The program's distinctive feature is its multi-directional data-collecting protocol, where participants capture images in six directions: east, south, west, north, upwards, and downwards (Case Study Figure 2.1). This approach aims to provide a comprehensive view of land cover types, offering a level of detail that single-direction imagery cannot match (Amos et al., 2020).

The potential of this multi-directional approach prompted researchers to develop an end-to-end classification framework, designed to link user-uploaded multi-directional views with their corresponding land cover labels. This framework served as a testbed for exploring various deep learning architectures and image fusion strategies in the context of land cover classification (Huang et al., 2023). The research team selected four popular deep learning architectures for their investigation: VGG (Simonyan & Zisserman, 2014), ResNet (He et al., 2016), MobileNet (Howard et al., 2017), and EfficientNet (Tan & Le, 2019). Each of these architectures has proven effective in various image classification tasks. To integrate information from multiple directional views, the researchers implemented three distinct image fusion strategies: early fusion, late fusion, and score fusion (Huang et al., 2023).

The investigation also sought to understand the impact of involving different numbers of directional views in the classification process, testing scenarios ranging from a single directional view up to all four cardinal direction views. This approach allowed for quantification of the benefits of additional viewpoints in land cover classification accuracy (Huang et al., 2023). The dataset comprised 9254 complete observations, each including photos from all four cardinal directions and a user-provided land cover label. The researchers reclassified the land cover types into seven major categories: herbaceous vegetation, urban, closed forest, woodland, cultivated land, barren land, and open water (Huang et al., 2023). This reclassification helped focus the study on broad, distinct land cover types while maintaining a diverse range of categories.

The results of this comprehensive study were enlightening. EfficientNet emerged as the top-performing architecture across all tested scenarios, regardless of the number of directional views involved or the fusion strategy employed. This finding suggests that EfficientNet's unique architecture, which balances network depth, width, and resolution, is particularly well-suited to the complexities of land cover classification from multi-directional views (Huang et al., 2023; Tan & Le, 2019).

One of the most significant findings was the consistent improvement in classification performance as more directional views were incorporated into the process

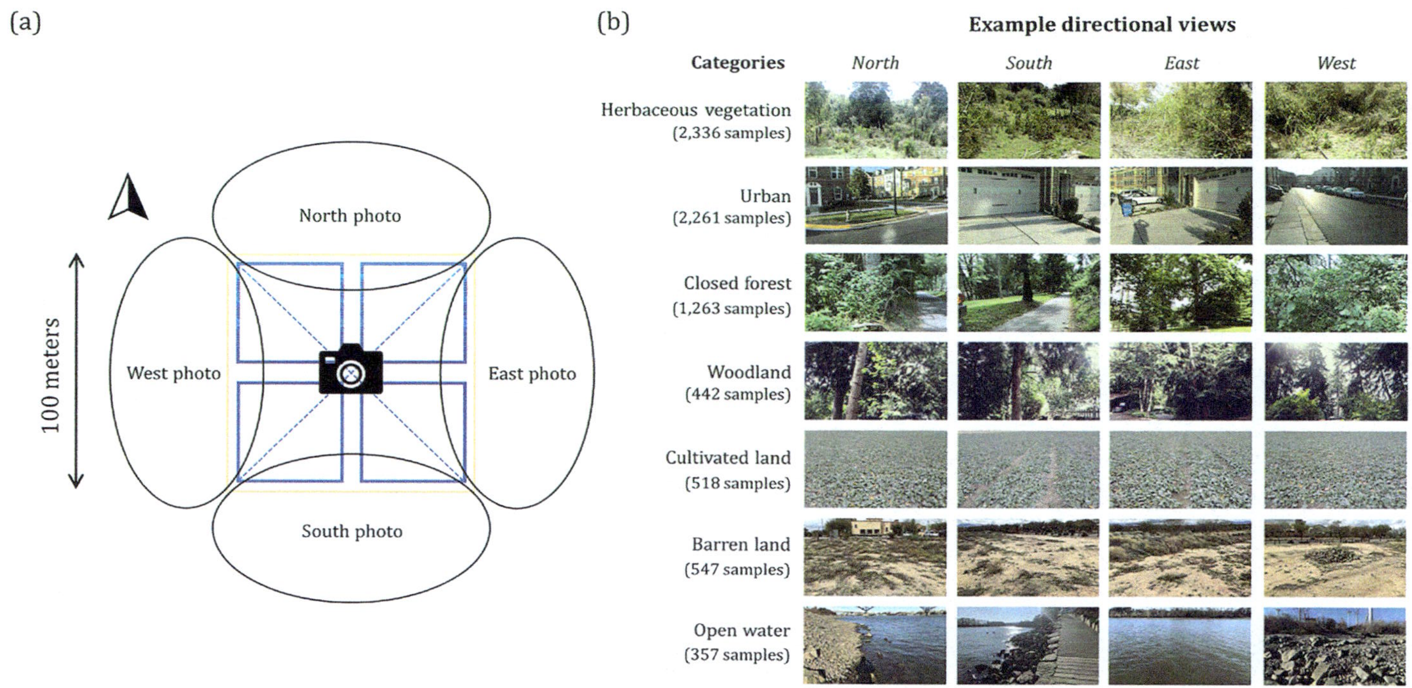

Case Study Figure 2.1 Illustration of the GLOBE observer multi-directional data collection protocol, (a) the sampling protocol for GLOBE observer land cover; (b) example directional views for seven reclassified land cover categories.

(Case Study Figure 2.2). This trend was particularly pronounced for the late fusion and score fusion strategies. This provides empirical evidence for the value of the multi-directional approach employed by the GLOBE Observer program (Huang et al., 2023).

Among the fusion strategies tested, late fusion emerged as the most effective. When using EfficientNet with all four directional views, the late fusion strategy achieved a weighted F1 score of 0.712, compared to 0.628 with a single view. This marked improvement underscores the potential of multi-directional views when combined effectively (Huang et al., 2023). The study also revealed interesting patterns in the classification accuracy of different land cover types. Closed forest, herbaceous vegetation, urban areas, and barren land were identified with higher accuracy compared to woodland and cultivated land. This disparity highlights the varying levels of distinctiveness in their visual characteristics and points to areas where further refinement of the classification approach might be beneficial (Huang et al., 2023).

A highlight of the study was the analysis of similarity between pairs of directional views across different land cover types (Case Study Figure 2.3). Closed forest environments showed the highest cross-view similarity, while open water areas displayed the lowest. This finding reflects the inherent visual characteristics of these environments and has implications for how multi-directional data is processed for different land cover types (Huang et al., 2023).

The implications of this research extend far beyond academic interest. By demonstrating the significant potential of multi-directional views in improving land cover classification accuracy, this study provides strong support for the continued use and expansion of multi-directional view sampling protocols in crowdsourced land cover mapping projects (Fritz et al., 2017; Huang et al., 2023). The improved classification accuracy achieved through this multi-directional approach has real-world applications. More accurate land cover maps can enhance our understanding of land use changes over time, providing valuable insights for urban planners, conservationists, and climate scientists (Wulder et al., 2018). This case study exemplifies the power of combining citizen science with advanced machine learning techniques. It demonstrates how the efforts of volunteer participants can be amplified by sophisticated data analysis methods to produce results that rival or even surpass those of traditional scientific data collection methods (Bonney et al., 2014; Huang et al., 2023).

In conclusion, the GLOBE Observer Land Cover program and the associated research into multi-directional view classification represent a significant advancement in crowdsourced environmental monitoring. By leveraging citizen scientists and advanced machine learning techniques, this approach is paving the way for more accurate, comprehensive, and timely monitoring of Earth's changing landscapes.

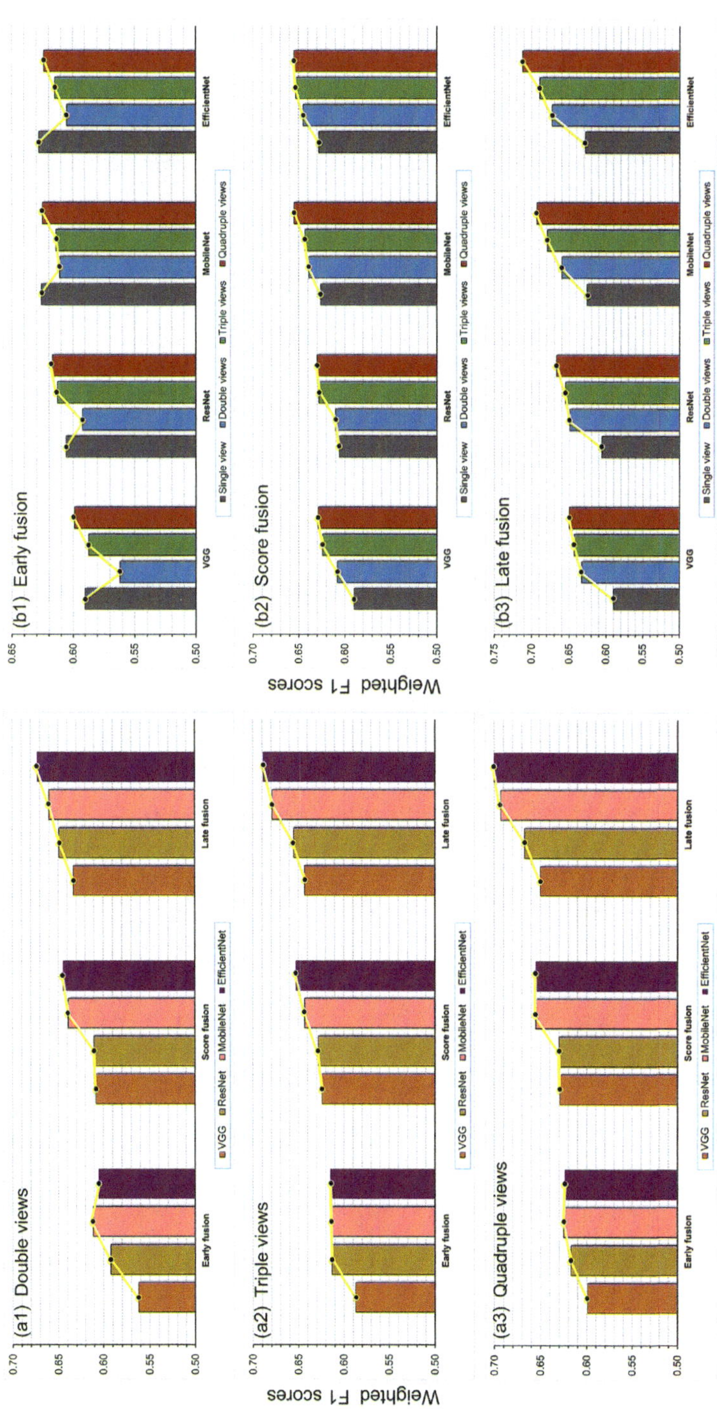

Case Study Figure 2.2 Performance comparison of different deep learning architectures - weighted F1 scores for investigated deep learning architectures using different levels of directional view involvement under different fusion strategies.

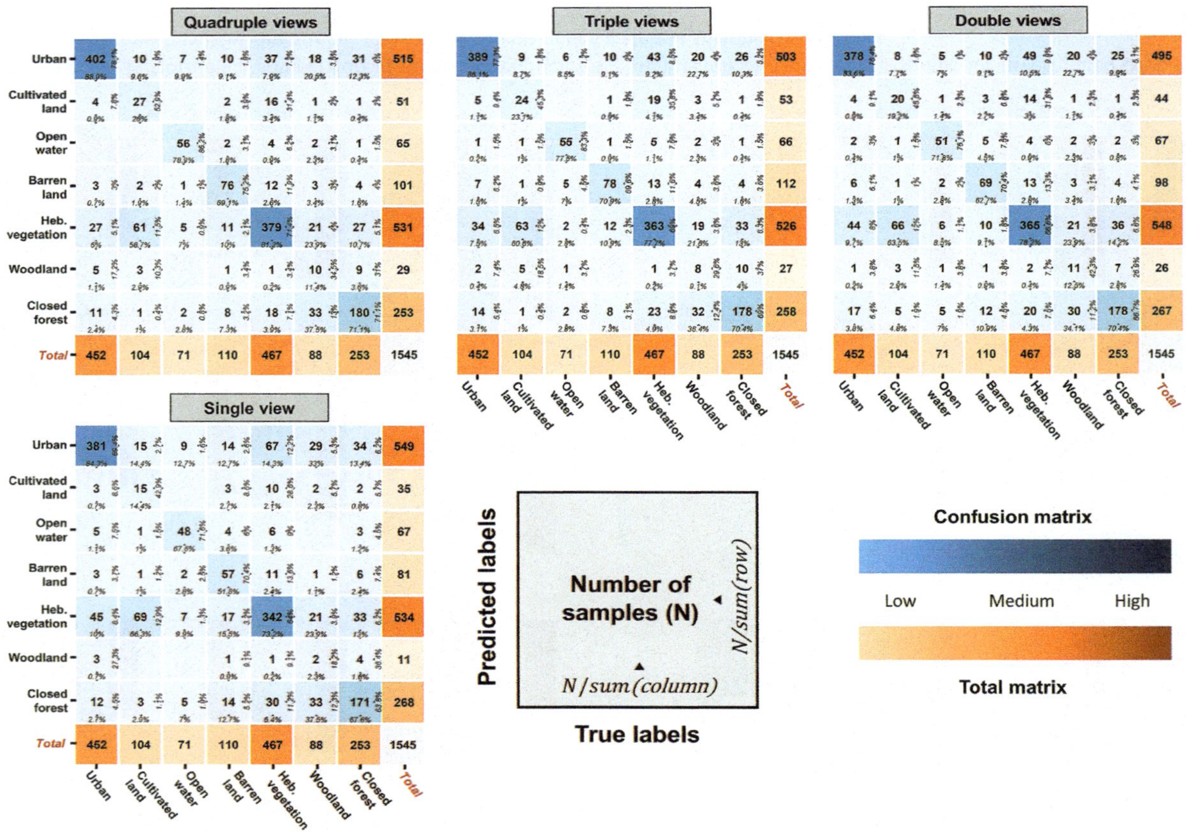

Case Study Figure 2.3 Confusion matrix from efficientNet with different levels of directional view involvement under the late fusion strategy.

5. Conclusion

5.1 Summary

The integration of citizen science with new technologies and big data has ushered in a transformative era for geospatial research and environmental monitoring. This chapter has explored the dynamic intersection of these fields, revealing several key takeaways:

1. Evolution of Citizen Science: From its roots in amateur naturalism to global, technology-driven initiatives, citizen science has become a powerful tool for addressing complex scientific questions at unprecedented scales.
2. Technological Enablement: Mobile devices, AI, and cloud computing have revolutionized data collection and analysis in citizen science projects, enabling real-time, large-scale participation.
3. Data Integration Challenges: Merging citizen-generated data with professional datasets and remote sensing data presents both technical and conceptual challenges, but also offers opportunities for more comprehensive insights.
4. Methodological Innovations: New approaches in mass data collection, data integration, and long-term program sustainability are enhancing the reliability and impact of citizen science initiatives.
5. Diverse Applications: From land use mapping to biodiversity monitoring and disaster response, citizen science is proving valuable across numerous disciplines.
6. Machine Learning Advancements: The application of advanced algorithms, particularly in image classification tasks, is significantly improving the accuracy and efficiency of citizen science data analysis.

5.2 Closing thoughts

Future directions for technology-enabled citizen science in the big data era:

1. Enhanced AI Integration: Further development of AI and machine learning algorithms will likely improve data validation, analysis, and integration processes, potentially automating complex tasks and reducing bias in citizen-generated data.
2. Immersive Technologies: The emergence of virtual and augmented reality technologies may create more engaging and intuitive interfaces for citizen scientists, potentially increasing participation and data quality.
3. IoT and Sensor Networks: The expansion of the Internet of Things and low-cost sensor technologies may enable more sophisticated and continuous data collection by citizen scientists.
4. Blockchain for Data Integrity: Implementing blockchain technologies could enhance data traceability and security in citizen science projects, addressing concerns about data quality and trustworthiness.
5. Ethical Frameworks: As citizen science projects grow in scale and impact, developing robust ethical guidelines for data collection, privacy, and use will become increasingly important.

Potential of citizen-driven science to address global challenges:

Citizen science, empowered by new technologies and big data approaches, holds immense potential for addressing global challenges. Its ability to mobilize large numbers of participants, collect data at unprecedented scales, and integrate diverse data sources makes it a powerful tool for tackling complex issues like climate change, biodiversity loss, and urban development.

The democratization of science through citizen participation not only enhances our collective ability to monitor and understand environmental changes but also fosters a more scientifically literate and engaged public. This increased engagement can lead to more informed decision-making and policy formulation, crucial for addressing global challenges effectively. As we continue to refine methodologies, improve technologies, and expand the scope of citizen science initiatives, we move closer to realizing the full potential of this approach. The future of citizen-driven science is not just about collecting more data, but about creating a global community of informed, engaged individuals working together to understand and protect our planet.

In conclusion, the integration of citizen science with new technologies and big data represents a paradigm shift in how we conduct scientific research and environmental monitoring. As we face increasingly complex global challenges, this approach offers a promising path forward, harnessing the collective power of human observation, technological innovation, and data-driven insights to create a more sustainable and resilient future.

References

Amos, H. M., Starke, M. J., Rogerson, T. M., Colón Robles, M., Andersen, T., Boger, R., … Schwerin, T. G. (2020). GLOBE observer data: 2016–2019. *Earth and Space Science, 7*(8).

Anderson, D. P., Cobb, J., Korpela, E., Lebofsky, M., & Werthimer, D. (2002). SETI@ home: An experiment in public-resource computing. *Communications of the ACM, 45*(11), 56–61.

Bonney, R., Cooper, C. B., Dickinson, J., Kelling, S., Phillips, T., Rosenberg, K. V., & Shirk, J. (2009). Citizen science: A developing tool for expanding science knowledge and scientific literacy. *BioScience, 59*(11), 977–984.

Bonney, R., Phillips, T. B., Ballard, H. L., & Enck, J. W. (2016). Can citizen science enhance public understanding of science? *Public Understanding of Science, 25*(1), 2–16.

Bonney, R., Shirk, J. L., Phillips, T. B., Wiggins, A., Ballard, H. L., Miller-Rushing, A. J., & Parrish, J. K. (2014). Next steps for citizen science. *Science, 343*(6178), 1436–1437.

Chandler, M., See, L., Copas, K., Bonde, A. M., López, B. C., Danielsen, F., … Turak, E. (2017). Contribution of citizen science towards international biodiversity monitoring. *Biological Conservation, 213*, 280–294.

Çöltekin, A., Lochhead, I., Madden, M., Christophe, S., Devaux, A., Pettit, C., … Hedley, N. (2020). Extended reality in spatial sciences: A review of research challenges and future directions. *ISPRS International Journal of Geo-Information, 9*(7), 439.

Cooper, C. B., Shirk, J., & Zuckerberg, B. (2014). The invisible prevalence of citizen science in global research: Migratory birds and climate change. *PLoS One, 9*(9).

Demchenko, Y., Grosso, P., De Laat, C., & Membrey, P. (2013). Addressing big data issues in scientific data infrastructure. In *2013 international conference on Collaboration technologies and systems (CTS)* (pp. 48–55). IEEE.

Dickinson, J. L., Shirk, J., Bonter, D., Bonney, R., Crain, R. L., Martin, J., Phillips, T., & Purcell, K. (2012). The current state of citizen science as a tool for ecological research and public engagement. *Frontiers in Ecology and the Environment, 10*(6), 291–297.

Dunn, E. H., Francis, C. M., Blancher, P. J., Drennan, S. R., Howe, M. A., Lepage, D., … Smith, K. G. (2005). Enhancing the scientific value of the christmas bird count. *The Auk, 122*(1), 338–346.

Fan, J., Han, F., & Liu, H. (2014). Challenges of big data analysis. *National Science Review, 1* (2), 293–314.

Fritz, S., See, L., Perger, C., McCallum, I., Schill, C., Schepaschenko, D., … Lesiv, M. (2017). A global dataset of crowdsourced land cover and land use reference data. *Scientific Data, 4* (1), 1–8.

Goodchild, M. F. (2007). Citizens as sensors: The world of volunteered geography. *Geojournal, 69*(4), 211–221.

Goodchild, M. F., Fu, P., & Rich, P. (2007). Sharing geographic information: An assessment of the geospatial one-stop. *Annals of the Association of American Geographers, 97*(2), 250–266.

Haklay, M. (2013). Citizen science and volunteered geographic information: Overview and typology of participation. In *Crowdsourcing geographic knowledge* (pp. 105–122). Dordrecht: Springer.

Haklay, M., & Weber, P. (2008). OpenStreetMap: User-generated street maps. *IEEE Pervasive Computing, 7*(4), 12–18.

Hampton, S. E., Strasser, C. A., Tewksbury, J. J., Gram, W. K., Budden, A. E., Batcheller, A. L., … Porter, J. H. (2013). Big data and the future of ecology. *Frontiers in Ecology and the Environment, 11*(3), 156–162.

He, K., Zhang, X., Ren, S., & Sun, J. (2016). Deep residual learning for image recognition. In *Proceedings of the IEEE conference on computer vision and pattern recognition* (pp. 770–778).

Howard, A. G., Zhu, M., Chen, B., Kalenichenko, D., Wang, W., Weyand, T., … Adam, H. (2017). *Mobilenets: Efficient convolutional neural networks for mobile vision applications*. arXiv preprint arXiv:1704.04861.

Howe, J., Reichert, T., & Quintel, A. (2017). Citizen science for climate change research: Opportunities and challenges. *Citizen Science: Theory and Practice, 2*(1).

Huang, X., Yang, D., He, Y., Nelson, P., Low, R., McBride, S., … Guarraia, M. (2023). Land cover mapping via crowdsourced multi-directional views: The more directional views, the better. *International Journal of Applied Earth Observation and Geoinformation, 122*.

Kelling, S., Fink, D., La Sorte, F. A., Johnston, A., Bruns, N. E., & Hochachka, W. M. (2015). Taking a 'Big data' approach to data quality in a citizen science project. *Ambio, 44*(4), 601–611.

Kitchin, R. (2014). Big data, new epistemologies and paradigm shifts. *Big Data & Society, 1*(1).

Kong, Q., Allen, R. M., Schreier, L., & Kwon, Y. W. (2016). MyShake: A smartphone seismic network for earthquake early warning and beyond. *Science Advances, 2*(2).

Kosmala, M., Wiggins, A., Swanson, A., & Simmons, B. (2016). Assessing data quality in citizen science. *Frontiers in Ecology and the Environment, 14*(10), 551–560.

Laney, D. (2001). 3D data management: Controlling data volume, velocity, and variety. *META Group Research Note, 6*(70), 1.

Ma, Y., Wu, H., Wang, L., Huang, B., Ranjan, R., Zomaya, A., & Jie, W. (2015). Remote sensing big data computing: Challenges and opportunities. *Future Generation Computer Systems, 51*, 47–60.

Marx, V. (2013). Biology: The big challenges of big data. *Nature, 498*(7453), 255–260.

Murdoch, T. B., & Detsky, A. S. (2013). The inevitable application of big data to health care. *JAMA, 309*(13), 1351–1352.

Newman, G., Wiggins, A., Crall, A., Graham, E., Newman, S., & Crowston, K. (2012). The future of citizen science: Emerging technologies and shifting paradigms. *Frontiers in Ecology and the Environment, 10*(6), 298–304.

Prestopnik, N. R., & Crowston, K. (2012). Citizen science system assemblages: Understanding the technologies that support crowdsourced science. In *Proceedings of the 2012 iConference* (pp. 168–176).

Resnik, D. B., Elliott, K. C., & Miller, A. K. (2015). A framework for addressing ethical issues in citizen science. *Environmental Science & Policy, 54*, 475–481.

Rosemartin, A. H., Crimmins, T. M., Enquist, C. A., Gerst, K. L., Kellermann, J. L., Posthumus, E. E., ... Weltzin, J. F. (2014). Organizing phenological data resources to inform natural resource conservation. *Biological Conservation, 173*, 90–97.

Rotman, D., Preece, J., Hammock, J., Procita, K., Hansen, D., Parr, C., ... Jacobs, D. (2012). Dynamic changes in motivation in collaborative citizen-science projects. In *Proceedings of the ACM 2012 conference on computer supported cooperative work* (pp. 217–226).

See, L., Mooney, P., Foody, G., Bastin, L., Comber, A., Estima, J., ... Rutzinger, M. (2016). Crowdsourcing, citizen science or volunteered geographic information? The current state of crowdsourced geographic information. *ISPRS International Journal of Geo-Information, 5*(5), 55.

Silvertown, J. (2009). A new dawn for citizen science. *Trends in Ecology & Evolution, 24*(9), 467–471.

Simonyan, K., & Zisserman, A. (2014). *Very deep convolutional networks for large-scale image recognition*. arXiv preprint arXiv:1409.1556.

Simpson, R., Page, K. R., & De Roure, D. (2014). Zooniverse: Observing the world's largest citizen science platform. In *Proceedings of the 23rd international conference on world wide web* (pp. 1049–1054).

Smolinski, M. S., Crawley, A. W., Baltrusaitis, K., Chunara, R., Olsen, J. M., Wójcik, O., ... Brownstein, J. S. (2015). Flu near you: Crowdsourced symptom reporting spanning 2 influenza seasons. *American Journal of Public Health, 105*(10), 2124–2130.

Sullivan, B. L., Aycrigg, J. L., Barry, J. H., Bonney, R. E., Bruns, N., Cooper, C. B., ... Kelling, S. (2014). The eBird enterprise: An integrated approach to development and application of citizen science. *Biological Conservation, 169*, 31–40.

Swanson, A., Kosmala, M., Lintott, C., & Packer, C. (2016). A generalized approach for producing, quantifying, and validating citizen science data from wildlife images. *Conservation Biology, 30*(3), 520–531.

Tan, M., & Le, Q. (2019). Efficientnet: Rethinking model scaling for convolutional neural networks. In *International conference on machine learning* (pp. 6105–6114). PMLR.

Theobald, E. J., Ettinger, A. K., Burgess, H. K., DeBey, L. B., Schmidt, N. R., Froehlich, H. E., ... Parrish, J. K. (2015). Global change and local solutions: Tapping the unrealized potential of citizen science for biodiversity research. *Biological Conservation, 181*, 236–244.

Volten, H., Devilee, J., Apituley, A., Carton, L., Grothe, M., Keller, C., ... Wesseling, J. (2018). Enhancing national environmental monitoring through local citizen science. In *Citizen science* (pp. 337–352). UCL Press.

West, S., & Pateman, R. (2016). Recruiting and retaining participants in citizen science: What can be learned from the volunteering literature? *Citizen Science: Theory and Practice, 1* (2).

Wiggins, A., Bonney, R., Graham, E., Henderson, S., Kelling, S., LeBuhn, G., … Newman, G. (2013). Data management guide for public participation in scientific research. *DataONE: Albuquerque, NM, USA*.

Wiggins, A., Newman, G., Stevenson, R. D., & Crowston, K. (January 2011). Mechanisms for data quality and validation in citizen science. In *2011 IEEE seventh international conference on e-Science workshops* (pp. 14–19). IEEE.

Wulder, M. A., Coops, N. C., Roy, D. P., White, J. C., & Hermosilla, T. (2018). Land cover 2.0. *International Journal of Remote Sensing, 39*(12), 4254–4284.

Yang, D., Fu, C. S., Herrero, H. V., Southworth, J., & Binford, M. (2023). Linking forest management to surrounding lands: A citizen-based approach towards the regional understanding of land-use transitions. *Frontiers in Remote Sensing, 4*.

Zhang, J., Atkinson, P. M., & Goodchild, M. F. (2014). *Scale in spatial information and analysis*. CRC Press.

Further reading

"Big Data: A Revolution That Will Transform How We Live, Work, and Think" by Viktor Mayer-Schönberger and Kenneth Cukier

"Citizen Science: Innovation in Open Science, Society and Policy" edited by Susanne Hecker et al.

Cornell Lab of Ornithology's Citizen Science Blog: https://www.birds.cornell.edu/home/citizen-science-blog/

NASA GLOBE Observer website: https://observer.globe.gov/

OpenStreetMap Wiki: https://wiki.openstreetmap.org/

Shen, L., Huang, Z., Wong, S. W., Liao, S., & Lou, Y. (2020). A holistic evaluation of smart city performance in the context of China. *Journal of Cleaner Production, 200*, 667–679.

Shirk, J. L., Ballard, H. L., Wilderman, C. C., Phillips, T., Wiggins, A., Jordan, R., … Bonney, R. (2012). Public participation in scientific research: A framework for deliberate design. *Ecology and Society, 17*(2).

"The Rightful Place of Science: Citizen Science" by Darlene Cavalier and Eric B. Kennedy

Building AI Literacy Across the Curriculum: Integrating GeoAI and Big Data to Prepare Future Geospatial Leaders

15

Jane Southworth, Kati Migliaccio and Gabriela Hamerlinck
Department of Geography, University of Florida, Gainesville, FL, United States

Key terms and definitions

Artificial intelligence (AI): Computational systems capable of performing tasks that typically require human intelligence, including learning, reasoning, pattern recognition, and decision-making.

AI literacy: A set of competencies that enable individuals to understand, use, evaluate, and engage with AI technologies critically, ethically, and effectively across disciplinary contexts.

AI across the curriculum: A university-wide educational model that embeds AI concepts, tools, and ethical considerations across disciplines rather than confining AI education to technical fields.

GeoAI (Geospatial artificial intelligence): An interdisciplinary field that integrates artificial intelligence methods with geospatial data, GIS, and remote sensing to analyze spatial patterns and solve complex geographic problems.

Big data: Large, complex, and often heterogeneous datasets that require advanced computational methods, including AI and machine learning, for storage, processing, and analysis.

Experiential learning: A pedagogical approach emphasizing hands-on, applied learning experiences—such as research projects, labs, and industry collaborations—to deepen conceptual understanding and skill development.

AI literacies framework: A structured model defining core categories of AI competencies, including knowing and understanding AI, using and applying AI, evaluating and creating AI, and AI ethics, supported by enabling skills.

Knowing and understanding AI: Foundational literacy focused on understanding what AI is, how it works, and key concepts such as machine learning, neural networks, and data-driven decision-making.

Using and applying AI: Practical literacy emphasizing the ability to employ AI tools and methods to solve problems, analyze data, and support decision-making across domains.

Evaluating and creating AI: Advanced literacy involving critical assessment of AI systems for effectiveness, bias, and limitations, as well as the design and development of new AI models or applications.

AI ethics: The examination of moral, social, and societal implications of AI, including issues of fairness, accountability, transparency, privacy, and responsible use.

Remote Sensing, Big Data, and GeoAI. https://doi.org/10.1016/B978-0-443-26741-3.00013-1

Enabling AI skills: Supporting competencies—such as data literacy, computational thinking, and domain knowledge—that facilitate effective engagement with AI technologies.
Workforce readiness: The alignment of educational outcomes with current and future labor market demands, ensuring students possess relevant technical, analytical, and ethical skills.
AI certificate program: A structured academic credential designed to provide students from diverse disciplines with foundational and applied AI knowledge without requiring extensive prior technical training.
GeoAI certificate program: A specialized undergraduate certificate focused on applying AI and big data methods to geospatial analysis, remote sensing, and spatial decision-making.
ICURE (Industry coupled undergraduate research experience): A course model that integrates industry-defined problems with undergraduate research, emphasizing interdisciplinary collaboration and real-world AI applications.
Undergraduate research experience (URE): A mentored, research-intensive learning opportunity that immerses students in authentic scholarly inquiry, often using AI and geospatial tools to address real-world challenges.
Interdisciplinary education: An educational approach that integrates perspectives, methods, and knowledge from multiple disciplines to address complex problems that cannot be solved within a single field.

1. Introduction

1.1 Background

1.1.1 AI in society

The 21st century has witnessed transformative changes in both personal and professional spheres, driven largely by the expansion of Artificial Intelligence (AI). Broadly understood as computational systems capable of thinking and acting rationally or even mimicking human behaviors (Russell & Noruig, 1995), AI now permeates various fields, from medicine to literature, making AI literacy an essential competence for future generations (Kandlhofer et al., 2016; Luckin et al., 2022; Ng et al., 2021). Higher education institutions are uniquely positioned to foster this literacy, equipping students with the knowledge and skills required to navigate an increasingly AI-driven world (Cantú-Ortiz et al., 2020; Dai et al., 2020).

Despite AI's growing influence, its integration into higher education curricula, particularly beyond traditional STEM fields, remains inconsistent (Cantú-Ortiz et al., 2020; Dai et al., 2020; Ma & Siau, 2018). The inconsistencies may be attributed to resources for AI integration, training of faculty and staff on AI applications, and/or a culture that sees value in AI integration. Recent research underscores the importance of AI literacy as a fundamental learning outcome for all students, whether at the post-secondary or K-12 levels (Kandlhofer et al., 2016; Luckin et al., 2022; Ng et al., 2021; Su et al., 2022). Incorporating AI into the curriculum not only enhances students' knowledge and skills but also prepares them for the evolving demands of the 21st-century workplace (Kandlhofer et al., 2016; Long & Megerko, 2020;

Ng et al., 2021). AI is increasingly integrated into everyday life, from voice recognition to automated decision-making systems. A recent study found that although 77% of devices people use incorporate AI, only 33% of consumers recognize they are using AI, indicating a significant gap in AI literacy (PEGA, 2022). This gap underscores the urgent need for educational programs that demystify AI and equip students with the skills necessary to succeed in a digitalized and automated society in an ethically responsible way (Cantú-Ortiz et al., 2020; Kandlhofer et al., 2016; Ng et al., 2021; Southworth et al., 2023).

Given the interdisciplinary nature of AI, higher education must integrate AI-related theories (e.g., machine learning, deep learning, natural language processing, and computer vision) and applications across all disciplines, rather than treating AI as an isolated or supplementary topic (Ng et al., 2021). The University of Florida's (UF) new curriculum model is designed to ensure that students acquire the AI knowledge and skills essential for their personal and professional success. The comprehensive approach at UF fosters interdisciplinary engagement and enhances students' career readiness by encouraging them to explore new career paths and develop relevant skills. The integration of AI into higher education curricula is not only about preparing students for future careers but also about enhancing the quality of university research and its impact on teaching. Through comprehensive AI literacy programs, universities can ensure that students are well-prepared to meet the challenges and opportunities presented by this rapidly evolving technological landscape.

1.1.2 The necessity of AI education

The rapid technological innovations of the 21st century have fundamentally transformed the way our world operates, generating new expertise and workforce demands that were previously nonexistent (Buckingham Shum & Deakin Crick, 2016; Cantú-Ortiz et al., 2020; Kandlhofer et al., 2016; St Louis et al., 2021). This shift, often referred to as the Fourth Industrial Revolution, has profoundly impacted how we live and work, merging the physical, digital, and biological worlds in ways that present both opportunities and challenges (Ahmad, 2019; Ng et al., 2021; World Economic Forum, 2022). The World Economic Forum (2022) emphasized the need to incorporate ethics into the development and application of new technologies, advocating for policies that promote ethical practices in innovation. AI, as a cross-disciplinary and increasingly integral component of our modern world, is at the forefront of this revolution.

Governments and global organizations have recognized the critical role that AI expertise plays in ensuring future growth and stability. For instance, the 2021 National Security Commission on Artificial Intelligence (NSCAI) Final Report highlights the urgency of accelerating AI innovation to benefit national interests while safeguarding against its potential misuse. The limitations or biases present in data directly impact the AI solutions derived from them (Borenstein & Howard, 2021; Long & Megerko, 2020). The report identifies a significant gap in human talent in AI-enabled technology, which poses the greatest barrier to national security, and calls for a digitally literate workforce with AI readiness by 2025 (NSCAI, 2021). In addition to governmental recognition, academic institutions have also acknowledged the importance of AI education across disciplines. The National Academies of Sciences, Engineering, and Medicine (NASEM) published a report titled *Data Science for Undergraduates: Opportunities and Options*, which underscores the necessity for students of all backgrounds and professional goals to gain proficiency in data science. This proficiency includes understanding data, making informed decisions, and using data analysis tools responsibly and effectively (NASEM, 2018, p. 138). Since data science often serves as the foundation for developing AI methods, the integration of data science education into broader AI literacy initiatives is crucial.

The NASEM report (2018) further emphasizes the need to extend these educational opportunities across all disciplines, rather than confining them to a select group of students. This approach mirrors how universities have integrated essential skills like communication across campus disciplines, recognizing their broad relevance to student success and future employability (Cantú-Ortiz et al., 2020; Kandlhofer et al., 2016; St Louis et al., 2021). For the purposes of this chapter, AI is considered inclusive of related fields like data science and GeoAI (Geospatial Artificial Intelligence), which applies AI and data-driven methods to geospatial and remote sensing data. This framing acknowledges the interconnectedness of these disciplines in modern education.

As higher education in the United States continue to evolve with the primary mission of preparing students for the workforce, colleges and universities must continually adapt their curricula to align with employer's needs. One of the most pressing needs, both globally and within the United States, is for individuals entering the workforce to possess advanced knowledge and skills in AI to meet the challenges of the 21st century (Buckingham Shum & Deakin Crick, 2016; Cantú-Ortiz et al., 2020; Kandlhofer et al., 2016). Reports from the NSCAI (2021), NASEM (2018, p. 138), and the World Economic Forum (2022) consistently underscore the mandate for universities to modify their curricula to address these evolving workforce requirements.

1.1.3 AI in higher education

The pervasive integration of AI into various facets of human life necessitates not only the ability to use AI but also a deeper understanding of its applications, limitations, and biases. As AI becomes an integral part of our daily lives, there is a pressing need for comprehensive education across all sectors to prepare individuals to thrive in this rapidly evolving, AI-enabled world. While some experts argue that a fundamental understanding of AI is unnecessary (Lee, 2017), more argue that it is essential to equip everyone with a basic knowledge of AI and its expanding applications (Borenstein & Howard, 2021; Dai et al., 2020; Kandlhofer et al., 2016; Long & Megerko, 2020; Ng et al., 2021). This knowledge should extend beyond the STEM fields to include a broad range of disciplines (e.g., art, education, history, law, and music), ensuring that all students, regardless of their major, can engage with AI in meaningful ways.

K-12 education systems in various countries have begun incorporating AI into their curricula, offering a model from which higher education institutions can learn. For instance, China, Singapore, and several other countries have developed formal AI education programs, aiming to integrate AI into their national curricula (Chiu & Chai, 2020; Kandlhofer et al., 2016; Su et al., 2022; Xia et al., 2022). These initiatives highlight the advanced set of AI skills and knowledge students will have entering higher education in the near future, further emphasizing the need for higher education institutions to be ready to offer more advanced AI curriculum to remain competitive and relevant for the next generation of students.

Several universities have made strides in integrating AI into their curricula. For example, Stanford University's Human-Centered Artificial Intelligence (HAI) institute and Northwestern University's AI education initiatives focus on interdisciplinary approaches and partnerships to advance AI education (https://hai.stanford.edu/; https://ai.northwestern.edu/education/index.html). However, these programs are often limited to specific departments, such as computer science and engineering, leaving many students without access to essential AI knowledge and skills. Other institutions, including Harvard, MIT, and Carnegie Mellon, have begun to offer courses focused on the ethics and implications of AI, acknowledging the broader societal impacts of AI technologies.

The lack of comprehensive AI curricula across universities links to a number of factors, such as investment needed, lack of knowledge beyond the traditional spaces, concern regarding the use of AI in education (both by students and faculty) and as such, there has been resistance to the uptake of AI, or the uptake has been slow. This significantly underscores the critical need for models like the UF's AI Across the Curriculum initiative (Southworth et al., 2023). This initiative aims to infuse AI education across all disciplines, ensuring that students from diverse academic backgrounds graduate with the knowledge and skills necessary to navigate and succeed in the 21st-century workplace (Ahmad, 2019; Borenstein & Howard, 2021; Cantú-Ortiz et al., 2020; Dai et al., 2020; Kandlhofer et al., 2016; Long & Megerko, 2020; Ng et al., 2021; St Louis et al., 2021). While the initiative is currently being deployed, its implementation will be closely evaluated to ensure its effectiveness and to refine the approach as needed.

Since 2019, UF has made significant investments in AI development, including building a world-class faculty, assembling a technical staff of AI specialists, and developing cutting-edge infrastructure such as supercomputing resources. These efforts are designed to position UF as a leader in AI education, enabling students to engage with AI technologies in a comprehensive and interdisciplinary manner. The AI Across the Curriculum initiative represents a forward-thinking approach to higher education, preparing students to be active participants in the technological advancements of the future.

1.1.4 AI at UF: AI Across the Curriculum

Traditional educational models, such as discipline-siloed curricula, lecture-centric instruction, and content-heavy survey courses, which often focus on narrow, discipline-specific knowledge, are increasingly seen as insufficient for equipping students with the interdisciplinary skills required in the 21st century (Buckingham Shum & Deakin

Crick, 2016; Ng et al., 2021). The AI Across the Curriculum initiative at UF represents a paradigm shift in how higher education approaches the teaching of AI. Rather than confining AI education to computer science or engineering programs, UF has adopted a more inclusive approach that seeks to embed AI concepts into a wide range of disciplines, including the humanities, social sciences, and natural sciences (Cantú-Ortiz et al., 2020; Southworth et al., 2023). This approach is driven by the understanding that AI is not merely a technical skill, but a set of tools and ways of thinking that can be applied across various domains to solve complex problems.

One of the key areas where this interdisciplinary approach is particularly relevant is in geospatial research. The rise of GeoAI has opened new possibilities for analyzing spatial data, from predicting environmental changes to optimizing urban planning and disaster response (Kandlhofer et al., 2016; Luckin et al., 2022). However, the effective use of these tools requires a deep understanding of both AI techniques and the specific challenges associated with spatial data. This need has led to the development of educational programs that combine AI with geospatial analysis, equipping students with the skills necessary to harness the power of GeoAI in their future careers.

1.1.5 Objective

The objective of this chapter is to provide a comprehensive exploration of how Big Data and GeoAI can be effectively integrated into undergraduate and graduate education, with a focus on preparing students for the future of geospatial research. This chapter seeks to address several key challenges associated with teaching AI and Big Data, particularly to students who may not have a background in coding or data science. These challenges include making AI concepts accessible to a broad audience, ensuring that students develop practical skills through hands-on experience, and aligning educational outcomes with the needs of the modern workforce (Southworth et al., 2023). Specifically, the concepts and ideas presented in this chapter will help to explore some of the broader challenges of integrating AI across the curriculum. Namely:

1. **Approaches to Integrating AI and Big Data into the Curriculum:** Examining different strategies for embedding AI literacy across various disciplines, drawing on the AI Across the Curriculum initiative at UF as a case study.
2. **Making AI and Big Data Accessible to Students:** A significant challenge in teaching AI is making complex concepts accessible to students who may not have prior experience with programming or data science.
3. **Incorporating AI and Big Data into Lab Settings:** Practical experience is crucial for developing a deep understanding of AI and Big Data and educators need to create lab environments that allow students to work with real-world datasets, apply AI techniques, and gain hands-on experience with GeoAI tools.
4. **Shaping Undergraduate Student Research with AI and Big Data:** Students are often at the forefront of innovation in geospatial research. AI and Big Data can be integrated into undergraduate-level research projects, providing students with the opportunity to develop cutting-edge solutions to complex problems and transforming traditional research methodologies, enabling more sophisticated analyses and deeper insights.

5. **Implications for Curriculum Development:** Highlight the integration of AI and Big Data into the curriculum, including the need for new courses, the potential for interdisciplinary programs, and the challenges of keeping pace with rapidly evolving technologies. Also, the importance of aligning educational outcomes with industry needs, ensuring that students are equipped with the skills necessary to succeed in the workforce.
6. **Future Job Prospects for Students Trained in AI and Big Data:** As AI and Big Data become increasingly important in the workforce, students trained in these areas will have a distinct advantage in the job market.

This chapter aims to provide educators and institutions with a potential roadmap for integrating Big Data and GeoAI into the curriculum. By highlighting the experiences and lessons learned from the AI Across the Curriculum initiative at UF, this chapter offers practical guidance for preparing students to thrive in an AI-driven world. The goal is not only to equip students with the technical skills they need but also to foster a deeper understanding of the transformative potential of AI and Big Data in geospatial research and beyond.

2. AI literacies: A comprehensive framework

2.1 Introduction to AI literacies

AI literacies provide a structured framework that defines the competencies required to understand, engage with, and critically assess AI technologies and their broader implications. AI literacies are not just about technical skills; they encompass a wide array of abilities that enable individuals to interact with AI in an informed, responsible, and ethical manner. As AI becomes increasingly embedded in various aspects of daily life; from voice-activated assistants like Siri and Alexa to automated decision-making systems in healthcare, finance, and law, the need for a well-rounded AI literacy framework becomes more urgent. This framework ensures that students are not only proficient in the technical aspects of AI but are also capable of understanding its societal impacts, ethical considerations, and future.

2.2 The core literacies of AI

The AI literacy framework at UF is built on four core literacies, with an additional fifth category termed "Enabling AI," which encompasses supporting skills necessary for AI proficiency. These literacies collectively provide a holistic approach to AI education, ensuring that students acquire a comprehensive understanding of AI and its implications (Fig. 15.1). These literacies developed from work published by Ng et al. (2021).

2.2.1 Knowing and understanding AI

This literacy focuses on foundational knowledge about what AI is, how it works, and the fundamental principles that underlie AI technologies. It involves understanding key concepts such as machine learning, neural networks, natural language processing, and the role of data in AI systems.

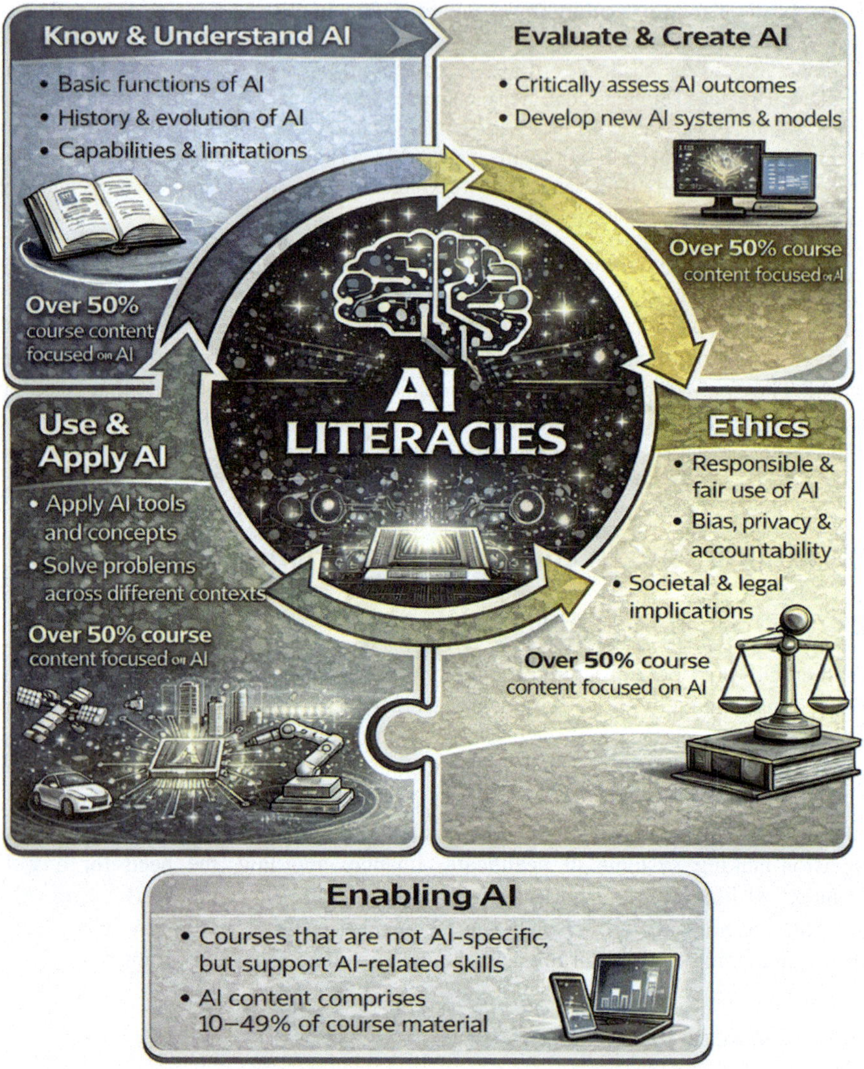

Figure 15.1 University of Florida model of AI course literacy showing the primary categories and their corresponding components.

Significance: Knowing and understanding AI is the bedrock of AI literacy. Without this fundamental knowledge, students cannot effectively engage with AI technologies, let alone apply or critique them. This literacy ensures that students can grasp the basics of AI, enabling them to understand how AI systems make decisions, learn from data, and interact with the world. For example, students learn about the differences between supervised, unsupervised, and reinforcement learning, and how these methods are applied in real-world AI applications (Ng et al., 2021).

2.2.2 Using and applying AI

This literacy involves the practical application of AI knowledge. It includes the ability to use AI tools and platforms to solve problems, automate tasks, and generate insights from data. This literacy also covers the practical skills needed to implement AI solutions in various contexts, whether through programming, using pre-built AI models, or integrating AI into existing systems.

Significance: Using and applying AI is crucial for transforming theoretical knowledge into practical skills. This literacy empowers students to leverage AI technologies in their chosen fields, whether they are developing new AI models, applying AI to business analytics, or using AI for scientific research. For instance, students might engage with platforms like TensorFlow or PyTorch to develop AI applications, or they might use AI tools for data analysis in social sciences or humanities projects (Laupichler et al., 2021).

2.2.3 Evaluating and creating AI

This literacy focuses on the critical assessment and creative development of AI systems. It involves higher-order thinking skills including analysis, evaluation, and synthesis, such as assessing the effectiveness, reliability, and fairness of AI models and algorithms, as well as designing and creating new AI tools, applications, or models that address specific challenges.

Significance: Evaluating and creating AI goes beyond using existing tools; it involves understanding the strengths and limitations of AI systems and contributing to their development. This literacy is vital for students who aspire to be innovators in the field of AI, as it equips them with the skills needed to assess AI technologies critically and develop new AI solutions. For example, students might evaluate the biases in AI algorithms or create new AI models that are more efficient or ethical (Long & Megerko, 2020; Ng et al., 2021).

2.2.4 AI ethics

This literacy covers the moral, ethical, and societal implications of AI technologies. It includes understanding the potential risks and benefits of AI, the ethical challenges posed by AI systems, and the responsibilities of AI practitioners. This literacy also involves knowledge of frameworks and guidelines for the ethical use of AI.

Significance: AI Ethics is perhaps the most critical literacy in the framework, as it addresses the broader impact of AI on society. As AI technologies become more pervasive, the ethical considerations surrounding their use, such as issues of privacy, fairness, accountability, and transparency, become increasingly important. This literacy ensures that students are not only technically proficient but also ethically aware, enabling them to make responsible decisions in their professional lives. For example, students might explore case studies of AI misuse or participate in debates about the regulation of AI technologies (Laupichler et al., 2021).

2.3 Development of the AI literacy model

The AI literacy model at UF was developed as part of a broader initiative to integrate AI across all disciplines and to make AI education accessible to every student, regardless of their major. This was driven by the mindset that AI has benefits for all disciplines which was a vision of the UF leadership and UF alumnus Chris Malachowsky. This model is grounded in the belief that AI literacy is not merely a technical skill but a critical competency that intersects with ethical, social, and professional dimensions (Ng et al., 2021; Southworth et al., 2023). The development of the AI literacy model involved a comprehensive review of existing AI education programs, both within the United States and internationally, and an analysis of the competencies required by the contemporary workforce. This review revealed a significant gap in AI education, particularly outside of traditional STEM fields, where AI is often taught in isolation from broader societal and ethical considerations (Cantú-Ortiz et al., 2020; Dai et al., 2020). In response, UF's model was designed to provide a more holistic approach, incorporating not only the technical aspects of AI but also its broader implications for society (Kandlhofer et al., 2016; Luckin et al., 2022).

2.4 Implementation of AI literacies at UF

2.4.1 Integration into the curriculum

The AI literacies framework was systematically integrated into UF's curriculum through a series of initiatives led by the Artificial Intelligence Academic Initiative Center, known as AI^2 (AI squared) Center, which was established to provide leadership and organization for AI-related academic activities across campus. The integration process involved the identification and categorization of AI-related courses into the five literacy categories, ensuring that students could select courses that matched their interests and career goals while gaining the necessary AI competencies (Ng et al., 2021). To support this integration, the AI^2 Center formed a subcommittee under the University Curriculum Committee, tasked with reviewing and evaluating AI-related courses. This subcommittee developed guidelines for identifying AI literacies within courses and worked closely with faculty to ensure that the courses met the required standards. The review process has started with an estimated list of 230 existing AI-related courses to evaluate (Southworth et al., 2023). The curriculum development initiative was further supported by seed grants provided by the AI^2 Center. This approach ensured that AI literacy topics were widely available across the university and that all AI literacy areas were represented in the curriculum.

3. AI certificate programs at the University of Florida

3.1 Introduction to AI certificate programs

As part of the AI Across the Curriculum initiative, UF has developed a suite of certificate programs (e.g., AI and Data Analytics in Tourism, Hospitality, and Event Management, AI in Public Health and Healthcare, and AI-Based SmartAg Systems) aimed at providing students with foundational and advanced knowledge in AI. These certificate programs are designed to be accessible to students across all disciplines, reflecting the interdisciplinary nature of AI and its broad applications in various fields. The certificate programs serve as a critical component of UF's efforts to prepare students for the 21st-century workforce, where AI competency is increasingly essential.

3.2 AI Fundamentals and Applications Certificate

3.2.1 Overview

The *AI Fundamentals and Applications* Certificate is the flagship offering in UF's suite of AI certificate programs. It is unique in its university-wide availability, allowing undergraduates from any major to enroll and gain essential AI skills. This certificate was specifically designed to lower the barrier to entry for AI education, making it accessible to students who may not have a background in computer science or other STEM fields. The program underscores UF's commitment to creating a digitally literate workforce capable of engaging with AI technologies in a variety of professional contexts (Southworth et al., 2023).

3.2.2 Curriculum structure

The *AI Fundamentals and Applications* Certificate is comprised of three core courses:

1. **EEL 3872: Fundamentals of Artificial Intelligence**—Offered by the College of Engineering, this course provides a comprehensive introduction to AI, covering essential topics such as machine learning algorithms, data processing, and AI applications in real-world scenarios. It lays out the foundation for students to understand the technical aspects of AI and how these can be applied across different industries.
2. **PHI 3681: Ethics, Data, and Technology**—Taught by the Department of Philosophy in the College of Liberal Arts and Sciences, this course addresses the ethical implications of AI and data-driven technologies. The course is taught by the Department of Philosophy because the expertise on ethics of AI and data science is in this department. It explores issues such as bias in AI algorithms, the ethical use of data, and the societal impact of AI innovations. This course is crucial for ensuring that students not only understand how to use AI but also recognize the responsibilities that come with its application.
3. **Elective Course**—Students can choose from a range of elective courses offered by various colleges at UF, allowing them to tailor their AI education to their specific interests. For example, a student majoring in Business might choose QMB 3302: Foundations of Business Analytics and Artificial Intelligence, while a student in Biomedical Engineering might opt for BME 4760: Biomedical Data Science. This flexibility ensures that the certificate program is relevant to students' career goals while providing them with a strong foundation in AI.

3.2.3 Impact and accessibility

The *AI Fundamentals and Applications* Certificate is designed to be broadly accessible, ensuring that students from non-technical backgrounds can engage with AI concepts and applications. This is critical in a world where AI is becoming ubiquitous across various industries, from healthcare to finance to the arts. By offering this certificate, UF is democratizing AI education, making it possible for all students to develop the skills needed to thrive in an AI-driven world.

3.3 Specialized AI certificate programs

3.3.1 GeoAI: Geographic Artificial Intelligence and Big Data Certificate

One of the specialized undergraduate certificate programs offered at UF is the *GeoAI: Geographic Artificial Intelligence and Big Data* Certificate. This program, developed by the Geography Department in the College of Liberal Arts and Sciences, is designed to address the growing demand for expertise in the application of AI to geospatial data. GeoAI is an emerging field that combines geographic information systems (GIS), remote sensing, and AI techniques to analyze and solve complex spatial problems (Hu et al., 2019; Janowicz et al., 2020). The GeoAI certificate program provides students with skills that are increasingly in demand as industries and governments seek to leverage AI for environmental monitoring, urban planning, disaster management, and other geospatial challenges. The curriculum includes courses on GIS, remote sensing, machine learning, and big data analytics, equipping students with the technical expertise needed to work with large and complex datasets (Hu et al., 2019).

3.3.2 Curriculum structure

The *GeoAI: Geographic Artificial Intelligence and Big Data* Certificate is comprised of four courses:

1. **GIS 2114: The World and Big Data or GIS 2002: The Digital Earth**—Offered by the Department of Geography, both course options provide an introduction to digital literacy and geospatial data. Students explore relevant concepts and techniques (e.g., cloud computing) of big (spatial) data as well as its applications in the real world. Students also gain practical experience in data analysis and visualization to understand the Earth's environment and human society based on the vast quantities of geographic information in our ever-changing world.
2. **One GIScience techniques course**—Students select from a range of upper division courses taught within the Department of Geography, allowing them to tailor their technical skills to their interests and intended careers. For example, a student majoring in Computer Science or Statistics might choose GIS 4115C: Spatial Surface Modeling and Geostatistics, while a student majoring in Broadcast Meteorology might complete MET 3753: Pragmatic Python for Weather. The flexibility of the GIScience technique course list ensures the certificate program is broadly relevant to student interests and career pathways.
3. **One GIScience applications course**—Students again can choose from a range of upper division application courses offered by the Department of Geography at UF, allowing them to tailor their AI education to their specific interests. For example, a premed student might choose GIS 4424C: Applications in GIS for Zoonoses and Disease Ecology, while a student in Engineering might consider GIS 4324: GIS Analysis of Hazard Vulnerability.
4. **Capstone GeoAI Course**—All students completing the GeoAI certificate are required to take GIS 4123C: GeoAI—Geographic Artificial Intelligence as a capstone experience. This course focuses on using real-world geospatial data through hands-on computing

labs. Students explore how GeoAI integrates into topics such as image classification, object detection, scene segmentation, simulation and interpolation, retrieval and question answering, on-the-fly data integration, and geo-enrichment.

3.3.3 Industry relevance and career prospects

The GeoAI certificate is particularly relevant given the projected growth in the geospatial industry, which is expected to expand by approximately 16% by 2050 (Janowicz et al., 2020). Graduates of this program will be well-positioned to take advantage of opportunities in both the public and private sectors, where there is a growing need for professionals who can apply AI to geospatial data to address pressing environmental and societal issues.

3.4 Integration with broader AI initiatives

The certificate programs at UF are not isolated initiatives but are integrated into the broader AI Across the Curriculum framework. This integration ensures that students pursuing these certificates are part of a larger educational ecosystem that includes academic pathways, research opportunities, and industry engagement. For instance, students enrolled in the *AI Fundamentals and Applications* Certificate are encouraged to participate in the AI Undergraduate Scholars and Medallion Programs, which provide additional research and experiential learning opportunities. Furthermore, the certificates are designed to be flexible, allowing students to build on their AI education as they progress through their academic careers. For example, a student who completes the *AI Fundamentals and Applications* Certificate may choose to pursue the more specialized GeoAI certificate or another advanced AI-related program offered at UF. This approach not only enhances the students' knowledge and skills but also provides them with credentials that are recognized and valued by employers. To illustrate how these initiatives collectively function within the university, Fig. 15.2 presents the structure of the University of Florida's AI Across the Curriculum ecosystem. The model highlights the interconnected nature of AI education at UF, including the AI literacy framework, interdisciplinary certificate programs, and experiential learning opportunities. These elements collectively enable the integration of AI concepts across a wide range of academic disciplines while simultaneously preparing students for emerging workforce needs. By linking foundational literacies, applied training, and interdisciplinary applications, the model demonstrates how AI education can extend beyond traditional computer science programs to create a campus-wide ecosystem for AI learning and innovation.

4. Evaluation of AI literacy implementation

The implementation of AI literacies across the curriculum at UF is an ongoing process, with continuous assessment and refinement based on feedback from students and faculty. The AI^2 Center plays a central role in this evaluation, using data collected

Figure 15.2 Conceptual model of the University of Florida AI across the curriculum ecosystem.

from course evaluations, student performance metrics, and feedback from industry partners to assess the effectiveness of the AI literacy framework. One of the key metrics used in this evaluation is the alignment of course content with the identified AI literacy categories and related Student Learning Outcomes (SLO's). The AI^2 Center ensures that courses designated as AI-related meet the required standards for their respective literacy categories and that they provide students with the competencies needed to succeed in the AI-enabled workforce. Outcomes from required SLOs are recorded and reported annually. Additionally, the center tracks the prevalence of different AI literacy topics across the curriculum to identify gaps and areas for further development.

5. Challenges and future directions

While the integration of AI literacies into the curriculum at UF has been a significant achievement, it has not been without challenges. One of the primary challenges has been ensuring that AI literacy topics are accessible to students from all disciplines, particularly those without a strong background in STEM. This challenge is especially pronounced in classrooms where students bring vastly different academic backgrounds, levels of technical preparation, and career goals. In practice, however, this type of heterogeneity is not new, particularly within disciplines such as geography, where courses routinely enroll majors and non-majors from the natural sciences, social sciences, humanities, and professional programs, and where technology-intensive methods (e.g., GIS, remote sensing, and data analysis) have long been taught to diverse student populations. Drawing on this experience, UF has emphasized interdisciplinary collaboration, scaffolded instruction, and flexible course design, allowing students to engage with AI concepts at multiple levels of depth. Introductory pathways focus on conceptual understanding and ethical reasoning, while more advanced or technically prepared students pursue deeper engagement through programming, modeling, or applied projects. Looking forward, UF plans to expand the AI literacy framework by developing new courses and programs that address emerging trends in AI and its applications. The university also aims to enhance its partnerships with industry to ensure that the AI curriculum remains relevant and aligned with the needs of the modern workforce (Cantú-Ortiz et al., 2020; Kandlhofer et al., 2016; Southworth et al., 2023).

6. Conclusion

The chapter has provided an in-depth exploration of the University of Florida's innovative approach to integrating AI across its curriculum along with the broader rationale for the need for such campus wide initiatives. Through a detailed examination of the theoretical framework, methodology, and case studies, we have outlined how UF aims to prepare its students for the challenges of the 21st century by embedding AI literacies into the educational experience. As illustrated in Fig. 15.2, the AI Across the Curriculum initiative functions as an integrated educational ecosystem that connects AI literacies, certificate programs, experiential learning opportunities, and interdisciplinary applications across the university. The two case studies: (1) AI-focused

Industry Coupled Undergraduate Research Experience (ICURE) courses and (2) the Summer Undergraduate Research Experience, illustrate the practical application of AI pedagogy and the university's commitment to experiential learning. The AI-focused ICURE classes emphasize the importance of industry partnerships and interdisciplinary collaboration, providing students with hands-on opportunities to solve real-world problems using AI. The Summer Undergraduate Research Experience offers a deep dive into AI applications within the geosciences, allowing students to engage with cutting-edge research under the mentorship of faculty. In addition, the previously discussed GeoAI Certificate Program provides a structured pathway for students to gain proficiency in GeoAI, ensuring they are well-prepared for the growing demand in this field. Each of these initiatives not only enhances students' technical skills but also fosters critical thinking, problem-solving, and ethical considerations, all of which are essential components of AI literacy.

6.1 Closing thoughts

As we look to the future, the AI Across the Curriculum initiative at UF is poised to serve as a model for other institutions seeking to integrate AI into their educational frameworks. The program's focus on transdisciplinary learning, coupled with a commitment to inclusivity and accessibility, ensures that students from all backgrounds and disciplines can engage with AI in meaningful ways. The inclusion of all undergraduate students is particularly noteworthy, as it addresses the growing concern of widening inequalities in access to emerging technologies.

The future direction of UF's AI initiative will likely involve continuous refinement and expansion, through the annual assessment and review process within the AI Across the Curriculum program that evaluates and updates AI-related courses, certificates and other related programs. As the AI landscape evolves, so will the curriculum, ensuring that students remain at the forefront of technological innovation, with expansion occurring through the regular addition of new AI-focused courses, certificates, and majors across the university. The integration of AI into non-STEM fields, as demonstrated in the AI Across the Curriculum initiative, suggests that the future of AI education will be increasingly interdisciplinary, reflecting the diverse applications of AI in society. The AI Across the Curriculum initiative represents a bold and forward-thinking approach to higher education. By embedding AI literacy into the fabric of its academic offerings, UF is not only preparing its students for the workforce but also fostering a generation of informed, ethical, and globally minded citizens. The success of this initiative will be measured not just by the number of AI-literate graduates but by the broader impact these individuals will have on society as they apply their knowledge to address the complex challenges of our time. As UF continues to lead in this space, the AI Across the Curriculum model will undoubtedly serve as a benchmark for educational institutions worldwide, setting the standard for how AI can be integrated into higher education to create a more equitable and innovative future.

Case Studies

Case study 1: AI-focused ICURE classes

Background: The University of Florida's (UF) Department of Geography is at the forefront of developing innovative educational programs that integrate Artificial Intelligence (AI) into the curriculum. As part of the broader AI Across the Curriculum initiative, the Geography Department is spearheading the creation of AI-focused Industry Coupled Undergraduate Research Experience (ICURE) courses. These courses represent a groundbreaking approach to undergraduate education, focusing on experiential learning and interdisciplinary collaboration. Designed to provide students with hands-on experience in applying AI techniques to real-world problems, the AI-focused ICURE courses will bridge the gap between academic learning and industry needs.

Implementation: The UF Geography Department is leading the implementation of AI-focused ICURE courses in partnership with the AI^2 Center and various industry collaborators. These courses will be structured to engage students in interdisciplinary research projects that address real-world challenges, specifically in the geospatial domain. Industry and governmental partners propose real-world geospatial challenges that form the basis of interdisciplinary student projects. Students collaborate across disciplines to develop AI-based solutions using machine learning, natural language processing, and geospatial analytics. Through continuous interaction with industry partners and a final project presentation, students translate research into practical solutions, preparing graduates with workforce-relevant GeoAI skills (Case Study Figure 1.1). The key components of the AI-focused ICURE courses include:

1. **Problem Identification:** Industry partners or governmental organizations, in collaboration with the Department of Geography, will propose real-world problems relevant to geospatial analysis and other fields. These problems will be carefully selected to ensure they are suitable for AI-based solutions and aligned with the undergraduate students' academic backgrounds.
2. **Interdisciplinary Teams:** Students from Geography and other academic disciplines across UF will be grouped into interdisciplinary teams to work on the projects. This approach will encourage collaboration and allow students to bring different perspectives and expertise to the problem-solving process.
3. **AI-Based Solutions:** Students will be tasked with developing AI-based solutions to the identified problems, focusing on geospatial challenges. This will involve applying machine learning, natural language processing, geospatial analysis, and other AI techniques to analyze data, identify patterns, and propose actionable solutions.
4. **Industry/Government Collaboration:** Throughout the course, students will interact regularly with industry partners, who will provide feedback and guidance on the projects. This collaboration will ensure that the solutions are practical and aligned with industry needs, particularly in the geospatial sector.
5. **Project Presentation:** At the end of the course, students will present their findings and solutions to the industry partners, Geography faculty, and their peers. This presentation will serve as both an assessment of their work and an opportunity to demonstrate the practical applications of their AI skills.

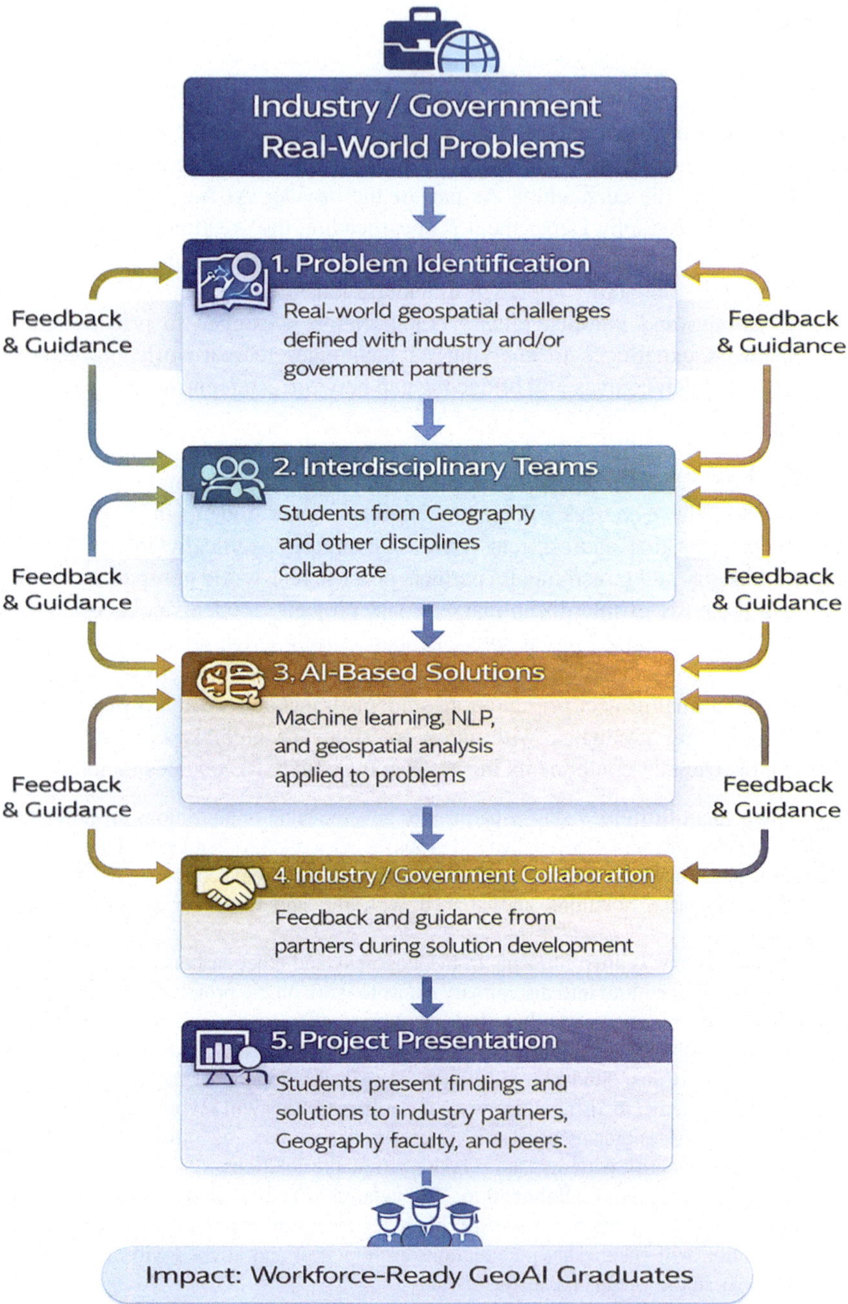

Case Study Figure 1.1 The implementation workflow for AI-Focused ICURE (or AI-CURE) courses.

Results: The AI-focused ICURE courses, led by the Geography Department, are expected to successfully engage students in complex, real-world problem-solving activities, particularly within the geospatial domain (Case Study Figure 1.1). These link back to the core requirements for the general certificate at the University in that all student presentations are expected to not only demonstrate project/research success but to also highlight connections to ethical considerations around the use of AI and related methods. Anticipated outcomes are illustrated below (Case Study Figure 1.2) and include:

1. **Enhanced AI Literacy:** Students will develop a deeper understanding of AI technologies and their applications in geography and geospatial analysis. They will gain practical experience in using AI tools and techniques, significantly enhancing their technical skills.
2. **Interdisciplinary Collaboration:** The interdisciplinary nature of the AI-focused ICURE courses will foster collaboration among students from Geography and other academic disciplines. This collaboration will enrich the learning experience and help students develop essential teamwork and communication skills.
3. **Industry/Government Engagement:** The close collaboration with industry partners will provide students with valuable insights into the practical applications of AI in various sectors, especially those related to geography. It will also offer industry partners access to innovative solutions and potential future employees with AI and geospatial expertise.
4. **Real-World Impact:** The AI-focused ICURE projects will address real-world challenges, such as optimizing logistical operations, enhancing environmental sustainability, and improving resource management within the geospatial domain. The solutions developed by the students will have the potential to be implemented by the industry partners, demonstrating the tangible impact of the AI-focused ICURE courses.

Impact: The AI-focused ICURE courses, spearheaded by the Geography Department, are expected to have a significant impact on both the students and the industry partners involved. For students, the courses will provide an opportunity to develop critical skills that are highly valued in the workforce, including technical proficiency in AI, geospatial analysis, problem-solving abilities, and the capacity to work effectively in interdisciplinary teams. The experiential learning approach will also help students build confidence in their ability to tackle complex challenges and contribute to real-world solutions. For industry partners, the AI-focused ICURE courses will offer access to fresh ideas and innovative solutions developed by talented students. The collaboration with UF's Geography Department will help these organizations build relationships with the next generation of AI and geospatial professionals, potentially leading to future partnerships and recruitment opportunities.

Overall, the AI-focused ICURE courses will serve as a successful model for integrating AI literacy into the curriculum through experiential learning, with a particular focus on geospatial challenges. They will not only enhance students' technical and analytical skills but also prepare them for the challenges of the 21st-century workforce by providing them with real-world experience and industry connections.

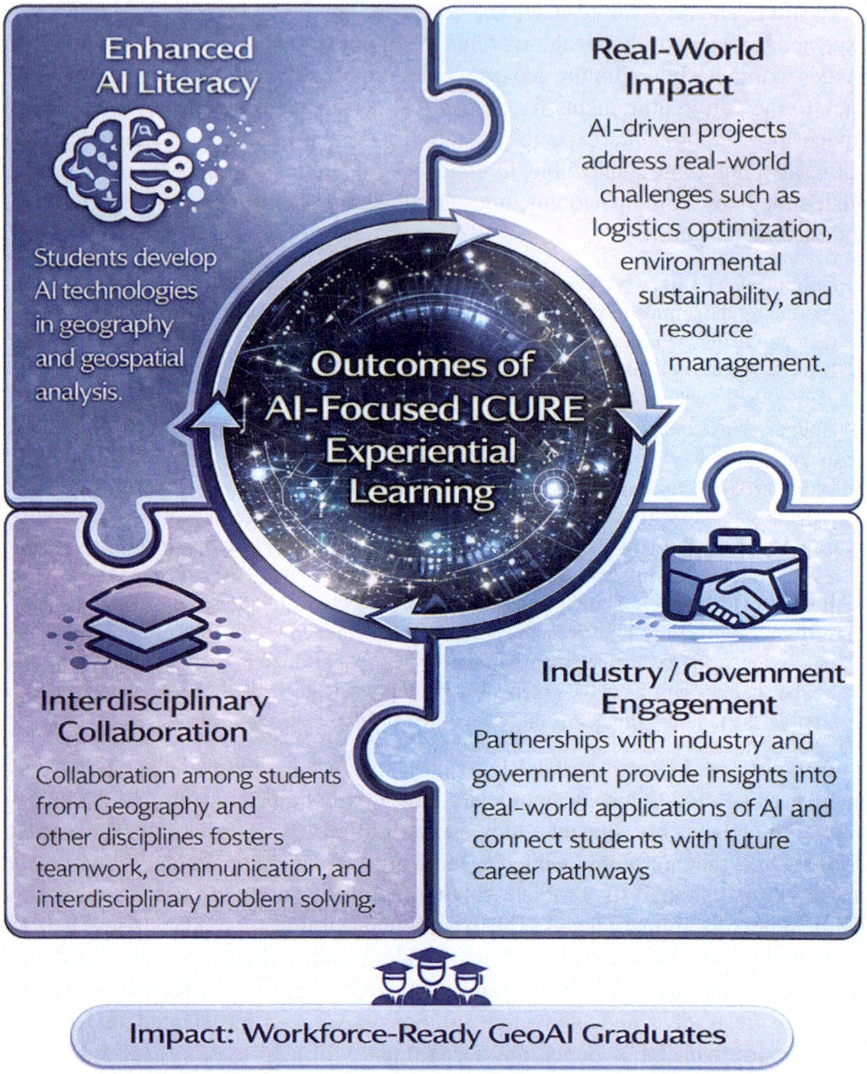

Case Study Figure 1.2 Anticipated outcomes of the AI-focused ICURE courses.

Case study 2: GeoGator Summer Undergraduate Research Experience

Background: The GeoGator Summer Undergraduate Research Experience, hosted by the University of Florida's Department of Geography, represents a forward-thinking initiative designed to immerse undergraduate students in the cutting-edge field of geospatial research. This program is a vital component of the department's commitment to integrating experiential learning into the curriculum, specifically focusing on applying AI and geospatial technologies to address pressing environmental and societal challenges, and is a program the Department is looking to fund so stipends can be paid to students. Currently students enroll as a course, but no funding is available to broaden participation. Future plans, including a new student experiential fund, have been established and initial funding received for future program cohorts. The program aims to equip students with hands-on experience in AI-driven geospatial research, enhancing their technical skills, critical thinking, and problem-solving abilities, in a more one-on-one, student-to-faculty mentor experience.

Implementation: The GeoGator Summer Research Program is a comprehensive, for-credit research experience conducted over the summer term, beginning in 2023 with great success in the first couple years. It is designed to provide students with a unique opportunity to work closely with UF Geography faculty on active research projects. These projects span various critical areas, including environmental monitoring, hurricane forecasting and analysis, climate change impact assessments, and disaster response, each of which leverages AI and geospatial technologies.

Key components of the program include:

1. **Mentorship and Research Projects:** Each student is paired with a faculty mentor from the Geography Department who guides them through the research process. These mentors are experts in applying GeoAI to geospatial challenges, ensuring that students are exposed to the latest tools and methodologies in the field. The research topics are carefully selected to align with the department's ongoing projects, allowing students to contribute meaningfully to cutting-edge research.
2. **Professional Development Seminars:** In addition to their research work, students participate in weekly professional development seminars designed to enhance their understanding of the broader context of their research. These seminars cover a range of topics, including the mentor-mentee relationship, scientific literature review, research ethics, career planning, and science communication. These sessions are crucial in helping students develop a well-rounded skill set that extends beyond technical expertise.
3. **Independent Research and Progress Reporting:** Students are expected to engage in independent research activities throughout the program. Weekly progress reports are submitted to their mentors, fostering accountability and ensuring that students stay on track with their research objectives. These reports also provide an opportunity for students to reflect on their learning and receive ongoing feedback.
4. **Culminating Presentation:** The program concludes with a symposium during the Fall semester, where students present their research findings in a poster session. This event allows students to showcase their work to faculty, peers, and industry professionals, offering them a platform to communicate their research effectively and receive valuable feedback (Case Study Figure 2.1).

The GeoGator Summer Undergraduate Research Experience is anticipated to yield several important outcomes:

1. **Enhanced GeoAI Competency:** Students will gain practical experience in applying AI to geospatial data, significantly enhancing their competencies in both AI and geographic information systems (GIS). This experience will prepare them for careers in the growing field of GeoAI, where such skills are in high demand.
2. **Research Contributions:** The students' work will contribute to the department's ongoing research projects, potentially leading to new insights and innovations in a wide variety of scientific fields. These contributions may be published in academic journals or presented at conferences, furthering the impact of their research.

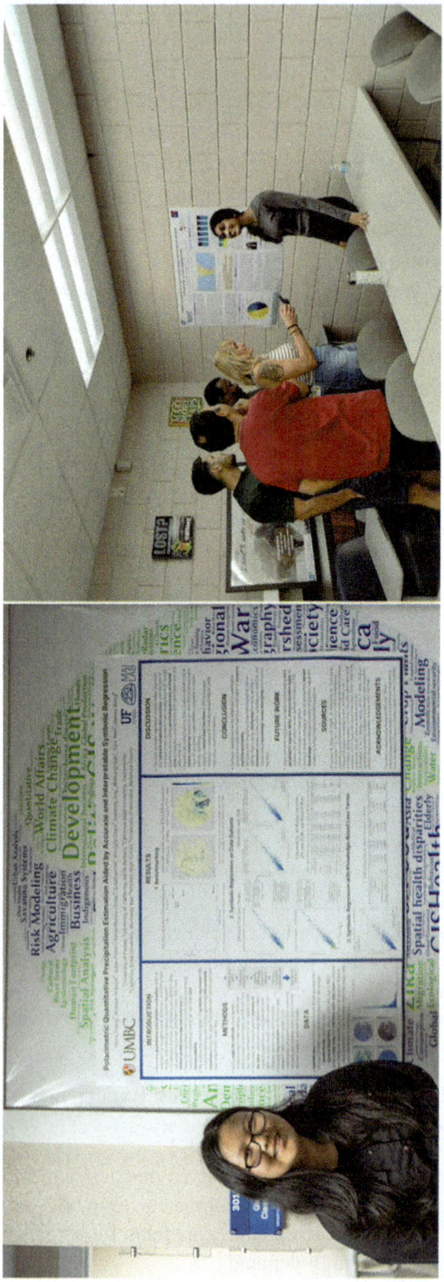

Case Study Figure 2.1 The GeoGator summer undergraduate research experience: Images from the culminating presentation at the fall 2023 symposium.

3. **Professional Growth:** The program's emphasis on professional development will ensure that students are well-prepared for the next steps in their academic and professional careers. Whether they choose to pursue graduate studies or enter the workforce, the skills and experience gained during the GeoGator program will be invaluable.
4. **Increased Engagement with GeoAI:** By focusing on GeoAI, the Geography Department will strengthen its position as a leader in this emerging field. The program will attract students who are passionate about using AI to solve real-world problems, thereby fostering a community of innovative thinkers and practitioners within the department.

Impact: The GeoGator Summer Undergraduate Research Experience is expected to have a lasting impact on both the students who participate and the broader academic community at UF. For students, the program offers a unique opportunity to engage in cutting-edge research, develop valuable skills, and build a professional network within the field of GeoAI. Once more, these impacts can also be linked back to the core requirements for the general certificate at the University in that all student presentations are expected to not only demonstrate project/research success but to also highlight connections to ethical considerations around the use of AI and related methods. For the Geography Department, the program will enhance its research output, attract top talent, and solidify its role as a leader in AI-driven geospatial research.

Overall, the GeoGator program exemplifies the department's commitment to integrating AI literacy and experiential learning into the geography curriculum, ensuring that students are not only prepared for the challenges of the 21st-century workforce but also equipped to make meaningful contributions to solving some of the world's most pressing geospatial challenges.

References

Ahmad, T. (2019). Scenario based approach to re-imagining future of higher education which prepares students for the future of work. *Higher Education, Skills and Work-based Learning, 10*(1), 217–238.

Borenstein, J., & Howard, A. (2021). Emerging challenges in AI and the need for AI ethics education. *AI Ethics, 1*, 61–65. https://doi.org/10.1007/s43681-020-00002-7

Buckingham Shum, S., & Deakin Crick, R. (2016). Learning analytics for 21st century competencies. *Journal of Learning Analytics, 3*(2), 6–21. https://doi.org/10.18608/jla.2016.32.2

Cantú-Ortiz, F. J., Galeano Śanchez, N., Garrido, L., Terashima-Marin, H., & Brena, R. F. (2020). An artificial intelligence educational strategy for the digital transformation. *International Journal on Interactive Design and Manufacturing, 14*, 1195–1209. https://doi.org/10.1007/s12008-020-00702-8

Chiu, T. K. F., & Chai, C. (2020). Sustainable curriculum planning for artificial intelligence education: A self-determination theory perspective. *Sustainability, 12*, 5568. https://doi.org/10.3390/su12145568

Dai, Y., Chai, C. S., Lin, P. Y., Jong, M. S. Y., Guo, Y., & Qin, J. (2020). Promoting students' well-being by developing their readiness for the artificial intelligence age. *Sustainability, 12*(16), 6597. https://doi.org/10.3390/su12166597

Hu, Y., Li, W., Wright, D., Aydin, O., Wilson, D., Maher, O., & Raad, M. (2019). Artificial intelligence approaches. In J. P. Wilson (Ed.), *The geographic information science and technology body of knowledge*. https://doi.org/10.22224/gistbok/2019.3.4, 3rd Quarter 2019 Edition

Janowicz, K., Gao, S., McKenzie, G., Hu, Y., & Bhaduri, B. (2020). GeoAI: Spatially explicit artificial intelligence techniques for geographic knowledge discovery and beyond. *International Journal of Geographical Information Science, 34*(4), 625–636. https://doi.org/10.1080/13658816.2019.1684500

Kandlhofer, M., Steinbauer, G., Hirschmugl-Gaisch, S., & Huber, P. (2016). Artificial intelligence and computer science in education: From kindergarten to university. *IEEE Frontiers in Education Conference (FIE)*, 1–9. https://doi.org/10.1109/FIE.2016.7757570

Laupichler, Lee, I., Ali, S., Zhang, H., DiPaola, D., & Breazeal, C. (2021). Developing middle school students' AI literacy. In *SIGCSE '21: Proceedings of the 52nd ACM technical symposium on computer science education* (pp. 191–197). https://doi.org/10.1145/3408877.3432513

Lee, K.-F. (2017). *The real threat of artificial intelligence*. https://www.nytimes.com/2017/06/24/opinion/sunday/artificial-intelligence-economic-inequality.html. (Accessed 5 May 2026).

Long, D., & Megerko, B. (2020). What is AI literacy? Competencies and design considerations. In *CHI '20: Proceedings of the 2020 CHI conference on human factors in computing systems* (pp. 1–16). https://doi.org/10.1145/3313831.3376727

Luckin, R., Cukurova, M., Kent, C., & du Boulay, B. (2022). Empowering educators to be AI-ready. *Computers and Education: Artificial Intelligence, 3*, Article 100076. https://doi.org/10.1016/j. caeai.2022.100076

Ma, Y., & Siau, K. L. (2018). Artificial intelligence impacts on higher education. *MWAIS 2018 Proceedings, 42*. https://aisel.aisnet.org/mwais2018/42.

National Academies of Sciences, Engineering, and Medicine. (2018). *Data science for undergraduates: Opportunities and options* (p. 138). Washington, DC: The National Academies Press. https://doi.org/10.17226/25104

National Security Commission on Artificial Intelligence. (2021). Final report. Retrieved from https://www.nscai.gov/2021-final-report/. (Accessed 18 October 2022).

Ng, D. T. K., Leung, J. K. L., Chu, S. K. W., & Qiao, M. S. (2021). Conceptualizing AI literacy: An exploratory review. *Computers and Education: Artificial Intelligence, 2*, Article 100041. https://doi.org/10.1016/j.caeai.2021.100041

PEGA. (2022). 101 artificial intelligence statistics. Retrieved from https://techjury. net/blog/ai-statistics/#gref. (Accessed 18 October 2022).

Russell, S. J., & Noruig, P. (1995). *Artificial intelligence—A modern approach: The intelligent agent book*. Englewood Cliffs, N.J.: Prentice Hall.

Southworth, J., Migliaccio, K. W., Glover, J., Glover, J., Reed, D., McCarty, C., Brendemuhl, J., & Thomas, A. (2023). Developing a model for AI across the curriculum: Transforming the higher education landscape via innovation in AI literacy. *Computers and Education: Artificial Intelligence, 4*, Article 100127. https://doi.org/10.1016/j.caeai.2023.100127

St Louis, A. T., Thompson, P., Sulak, T. N., Harvill, M. L., & Moore, M. E. (2021). Infusing 21st century skill development into the undergraduate curriculum: The formation of the iBEARS network. *Journal of Microbiology and Biology Education, 22*(2). https://doi.org/10.1128/jmbe.00180-21. e00180-21

Su, J., Zhong, Y., & Ng, D. T. K. (2022). A meta-review of literature on educational approaches for teaching AI at the K-12 levels in the Asia-Pacific region. *Computers and Education: Artificial Intelligence, 3*, Article 100065. https://doi.org/10.1016/j.caeai.2022.100065

World Economic Forum. (2022). Global issue: Artificial intelligence. In *Curation: Desautels faculty of management*. McGill University. Retrieved from: Strategic Intelligence we forum.org. (Accessed 18 October 2022).

Xia, Q., Chiu, T. K. F., Lee, M., Sanusi, I. T., Dai, Y., & Chai, C. S. (2022). A self-determination theory (SDT) design approach for inclusive and diverse artificial intelligence (AI) education. *Computers and Education, 189*, Article 104582. https://doi.org/10.1016/j. compedu.2022.104582

Further reading

Markauskaite, L., Marrone, R., Poquet, O., Knight, S., Martinez-Maldonado, R., Howard, S., Tondeur, J., De Laat, M., Buckingham, S., Dragan Gašević, S., & Siemens, G. (2022). Rethinking the entwinement between artificial intelligence and human learning: What capabilities do learners need for a world with AI? *Computers and Education: Artificial Intelligence, 3*, Article 100056. https://doi.org/10.1016/j.caeai.2022.100056

Zimmerman, M. (2018). *Teaching AI: Exploring new frontiers for learning*. Portland, OR: International Society for Technology in Education.

Future Trends and Emerging Technologies in Remote Sensing, Big Data, and GeoAI

16

Jane Southworth[1], *Erin Bunting*[2], *Hannah Herrero*[3] *and Cerian Gibbes*[4]
[1]Department of Geography, University of Florida, Gainesville, FL, United States; [2]Department of Geography, Environment, and Spatial Sciences, Michigan State University, East Lansing, MI, United States; [3]Department of Geography & Sustainability, University of Tennessee, Knoxville, TN, United States; [4]Department of Geography & Environmental Studies, University of Colorado Colorado Springs, Colorado Springs, CO, United States

1. From structural transformation to structural responsibility

Chapter 1 traced the emergence of GeoAI as a structural reorganization of geospatial science across four dimensions: data complexity, analytical ambition, computational infrastructure, and epistemic orientation. The chapters that followed demonstrated how these shifts consolidated across domains. This concluding chapter asks a different question: what responsibilities accompany that consolidation? GeoAI has matured from a methodological innovation into an integrated analytical ecosystem, and maturity carries consequence.

Maturity, however, introduces new questions. As GeoAI systems become more predictive, more automated, and more embedded within governance and planning frameworks, the implications of methodological choices expand. Predictive accuracy is no longer the sole metric of success. Transparency, sustainability, equity, and institutional accountability become inseparable from technical performance. To reflect on these emerging responsibilities, contributing authors were invited to participate in a short survey capturing distributed perspectives across the domains represented in this volume. Authors were asked four questions:

1. What excites you most about the future of GeoAI and remote sensing?
2. What do you see as the main limitations or gaps in the field?
3. Which fields or disciplines do you envision collaborating more closely with GeoAI and remote sensing in the future?
4. How do you view disparities in access to data and computational resources globally, and what environmental or ethical impacts do you associate with this field?

The responses reveal both optimism and caution. Contributors envision foundation models that generalize across sensors, digital twins that simulate landscape dynamics in near real time, AI-enabled hazard forecasting systems, biodiversity monitoring at

Remote Sensing, Big Data, and GeoAI. https://doi.org/10.1016/B978-0-443-26741-3.00012-X

planetary scale, and increasingly integrated socio-ecological intelligence systems. At the same time, they identify structural pressures related to data scarcity, computational asymmetry, infrastructure centralization, environmental cost, interpretability, and workforce preparation.

This chapter synthesizes those perspectives and situates them within the structural framework introduced in Chapter 1. If earlier chapters documented how GeoAI evolved, this chapter examines what that evolution now requires. The goal is to articulate the responsibilities that accompany maturation by interpreting the structural trajectory revealed across Chapters 1–15 and examining what that trajectory now demands of the field.

1.1 The prescriptive turn: From mapping to planetary intelligence

Remote sensing historically answered descriptive questions: What is there? Where is it located? How has it changed? The integration of machine learning expanded that scope toward detection and classification at increasing scales and speeds. Yet the most consequential shift emerging across this volume is not automation alone, but the movement from descriptive analytics toward prescriptive and anticipatory systems. GeoAI is increasingly positioned not only to observe the Earth system, but to inform decisions about it. Across survey responses, contributors identified this transition as the most compelling frontier of the field. Rather than emphasizing incremental improvements in classification accuracy, they described a future defined by integrated, predictive, and simulation-driven systems. Several themes emerged consistently.

1.1.1 Theme 1: From monitoring to anticipation

Chapters in this volume document the move from per-scene land cover mapping to dense time-series modeling and cross-sensor integration. What authors now anticipate is a further extension: systems capable of forecasting environmental states before impacts unfold. Contributors highlighted:

- Climate-health predictive integration that links atmospheric variability with disease vulnerability
- AI-assisted hazard forecasting that integrates physics-based models with learned surrogates
- Biodiversity monitoring systems capable of detecting resilience thresholds
- Agricultural forecasting frameworks that integrate weather, soil, crop phenology, and market signals

The field is shifting from retrospective assessment toward anticipatory environmental intelligence. As one contributor noted, "*GeoAI is no longer just about making better maps. It represents a shift toward actually understanding how the Earth works.*" That perspective reflects a broader movement toward embedding AI outputs within system analysis, adaptation strategies, and institutional decision pipelines.

1.1.2 Theme 2: Toward foundation and generalizable models

Another recurring theme in the survey responses was the aspiration for foundation-level GeoAI systems. Several authors described the need for models capable of learning transferable representations across sensors, biomes, and regions. Rather than building task-specific classifiers for each application domain, contributors envision multi-modal architectures trained on large-scale geospatial corpora that can be adapted efficiently to new contexts. While this ambition parallels trends in broader AI research, it carries distinct constraints in geospatial domains. Unlike natural language data, Earth observation data are physically grounded, sensor-dependent, and scale-sensitive. The development of generalizable GeoAI systems will require harmonization across spectral regimes, spatial resolutions, temporal frequencies, and cultural contexts of land use. Survey responses suggest cautious optimism. Contributors expressed excitement about the potential for cross-sensor transfer learning and multi-resolution embeddings, while simultaneously noting the substantial engineering and governance challenges such systems entail. The prescriptive turn therefore depends not only on larger models, but on integrative design principles that respect physical and social constraints.

1.1.3 Theme 3: Digital twins and simulation environments

Several authors referenced the growing discourse around digital twins of landscapes, cities, and ecological systems. In this framing, GeoAI becomes a component of dynamic simulation environments that integrate real-time sensing with predictive modeling. Rather than producing static outputs, such systems could continuously update environmental states, simulate intervention scenarios, and evaluate policy alternatives. The implications are profound. Urban heat mitigation strategies could be evaluated before implementation. Flood mitigation designs could be stress-tested under compound hazard scenarios. Conservation strategies could be simulated across alternative climate pathways. The integration of physics-informed models with machine learning surrogates, discussed in earlier chapters, positions GeoAI as a bridge between observation and scenario planning. However, contributors were careful to emphasize that simulation capacity does not eliminate uncertainty. Prescriptive systems amplify the importance of communicating model limitations, error propagation, and scenario assumptions. The prescriptive turn thus intensifies rather than reduces the need for epistemic transparency.

1.1.4 Theme 4: Planetary-scale environmental intelligence

Perhaps the most ambitious trajectory emerging from the survey responses is the notion of planetary-scale environmental intelligence systems. Continuous monitoring of forest loss, urban heat exposure, agricultural productivity, coastal vulnerability, and atmospheric dynamics is already underway in fragmented form. What contributors envision is greater integration across these systems. This integration would link environmental monitoring directly to sustainable development goals, biodiversity targets, and resilience frameworks. GeoAI systems could provide near real-time indicators that inform institutional dashboards, adaptation funding decisions, and international reporting mechanisms. Such ambition reflects the maturation traced earlier in this volume. Remote sensing is no longer confined to academic analysis. It is increasingly embedded within policy, finance, and governance structures. As one author reflected,

"We can see how policies, land use decisions, or climate shocks ripple through ecosystems and communities." This encapsulates the prescriptive turn: analytical outputs are becoming instruments within institutional systems.

1.1.5 Theme 5: The risks of acceleration

Despite enthusiasm, contributors also expressed caution. The prescriptive turn introduces risks of over-automation, model opacity, and premature institutional reliance on insufficiently validated systems. Several respondents warned against equating predictive capability with policy readiness. The movement toward anticipatory analytics requires rigorous validation across diverse geographies, transparent uncertainty quantification, and participatory design processes that involve domain experts and affected communities. As predictive systems gain influence, the consequences of error magnify. Thus, the prescriptive turn is not merely technological; it is ethical and institutional.

1.1.6 Theme 6: From analytical capability to institutional embedding

Across the chapters in this volume, a structural progression becomes clear. Early chapters establish the data archives and computational infrastructures that made large-scale geospatial analytics possible. Middle chapters consolidate methodological capacity through machine learning, time-series integration, and multi-sensor fusion. Later chapters move further, embedding GeoAI within operational systems for health, climate resilience, disaster management, agriculture, and governance. Survey responses confirm that contributors view this embedding as both inevitable and desirable. Yet they also emphasize that institutional integration demands new competencies. GeoAI practitioners must increasingly engage beyond technical domains, collaborating with policymakers, public health experts, conservation planners, legal scholars, and community stakeholders. The design of GeoAI systems now intersects with regulatory frameworks, ethical standards, funding mechanisms, and participatory decision processes.

This shift expands the epistemic boundaries of the field. GeoAI is not simply becoming more accurate; it is becoming more consequential. As systems move closer to operational and policy environments, analytical outputs influence allocation decisions, resilience strategies, and governance priorities. The question is no longer whether GeoAI can scale or predict, but how it should be embedded responsibly within institutional structures. The prescriptive turn therefore marks not the endpoint of methodological evolution, but the beginning of institutional consequence. GeoAI systems increasingly shape decisions as much as they describe landscapes. This transition from analytical capability to institutional embedding sets the stage for the structural constraints and responsibilities examined in the sections that follow.

1.2 The emerging data wall

If the previous decade of GeoAI research was defined by rapid expansion in data volume and computational scale, the coming decade may be defined by confronting its structural limits. Across survey responses and the chapters in this volume, a consistent theme emerges: abundance does not automatically translate into sufficiency. Petabyte-scale archives now exist for optical, radar, climate, and urban data streams, yet limitations in training data, representativeness, and generalization persist as binding constraints. As noted by one author "*One of the biggest challenges right now is simply getting good enough data to train these models … Want to train a model to spot illegal mining activity? … The data bottleneck is real.*" The field is approaching what might be described as a data wall: not a shortage of imagery, but a limit to how effectively it can be translated into reliable, transferable knowledge.

1.2.1 Archive abundance versus labeled scarcity

Open-access satellite archives and cloud-native platforms have democratized access to imagery at unprecedented scales. Dense time-series stacks are now available for much of the planet. Yet supervised learning frameworks remain heavily dependent on labeled datasets, and these remain unevenly distributed across geographies, land cover types, and socio-economic contexts.

Several contributors emphasized that labeled data creation remains labor-intensive, costly, and often regionally biased. Agricultural classification datasets are abundant in highly monitored temperate regions, while drylands, informal settlements, and biodiversity hotspots frequently lack robust annotation. Urban datasets often prioritize cities in high-income nations. Climate-health linkages are better documented where institutional data infrastructures are strong and sparse elsewhere. As one contributor noted, "*Computational resources and data availability could further strengthen global inequity.*"

The imbalance between archive scale and labeled availability produces asymmetries in model performance. Models trained on temperate agricultural systems may underperform in heterogeneous tropical landscapes. Hazard detection systems optimized for well-documented disaster events may struggle in data-poor regions. The promise of planetary-scale AI therefore encounters uneven epistemic terrain. As one respondent observed, "*We're building tools that let us observe the whole planet, but not everyone gets to be part of building them.*"

1.2.2 Generalization and transfer limits

Closely related to training scarcity is the challenge of generalization. Deep learning models often perform strongly within training domains but degrade under domain shift. Variations in sensor characteristics, atmospheric conditions, seasonal timing, ecological regimes, and socio-political context complicate transferability. While transfer learning and domain adaptation offer partial mitigation, foundational models for geospatial data remain early relative to other AI domains. Unlike natural language data, spatial phenomena are physically grounded, scale-dependent, and ecologically constrained. A forest is not merely a pixel pattern; it is an ecological system shaped by climate, soil, disturbance, and governance.

The data wall is therefore not simply quantitative: it is structural. It reflects the difficulty of learning representations that generalize across heterogeneous Earth systems without sacrificing interpretability or contextual integrity.

1.2.3 Synthetic data and simulation

One proposed response to labeled scarcity is synthetic data generation through physics-based modeling and digital twin simulation environments. Hybrid systems that combine mechanistic models with learned components offer promising avenues for augmentation and validation. Yet contributors cautioned that synthetic augmentation must be applied carefully. Simulated data encode assumptions, and those assumptions propagate through models. Synthetic realism is not equivalent to ecological or social realism. Overreliance on simulated training sets risks embedding structural bias at scale. The challenge, therefore, is not merely producing more data, but producing data that meaningfully represent diverse landscapes, governance contexts, and socio-ecological conditions.

1.2.4 The representation gap

Beyond technical generalization lies a deeper concern: whose landscapes are represented in GeoAI systems, and whose are marginalized? Several contributors emphasized that data gaps often align with geopolitical and economic inequalities. Regions with limited remote sensing infrastructure, sparse ground truth networks, or restricted data access may be underrepresented in training and training bodies. This affects not only model accuracy but fairness and global equity. As one author stated, "*This disparity can restrict participation in GeoAI research and slow locally relevant applications such as disaster monitoring, agricultural management, and biodiversity conservation.*"

GeoAI systems increasingly inform conservation funding, agricultural planning, disaster prioritization, and climate adaptation strategies. If models perform unevenly across regions, decision support may inadvertently reinforce existing disparities. The data wall therefore intersects directly with governance and justice. Addressing it requires more than algorithmic innovation; it demands collaborative data generation, open labeling initiatives, inclusive partnerships, and cross-regional benchmarking frameworks. The author from above then went on to say, "*Expanding open satellite archives, cloud-based analysis platforms, and shared computational resources can help reduce these barriers and make remote sensing research more inclusive.*"

1.2.5 Scaling without saturation

A subtler theme emerging from survey responses concerns diminishing returns. Incremental increases in model complexity do not necessarily overcome structural data constraints. As archives grow denser and architectures grow deeper, performance gains may plateau without parallel improvements in representativeness, harmonization, and evaluation rigor. This does not imply stagnation. Rather, it signals a reorientation. The next phase of GeoAI development may depend less on larger models and more on better-curated, more diverse, and more ethically constructed datasets. It may depend on metadata transparency, uncertainty communication, and rigorous cross-domain validation.

In this sense, the data wall functions as a maturation checkpoint. It challenges the field to move beyond scale-centric optimism toward structural robustness.

1.2.6 A turning point, not a barrier

Importantly, contributors did not frame the data wall as an insurmountable obstacle. Instead, it was described as a transition point. The next frontier is not more data, but better data—though, as one respondent noted succinctly, "*the data bottleneck is real.*" Recognition of these limits reflects growing disciplinary sophistication. Early enthusiasm centered on archive expansion and computational acceleration. The current moment calls for representational integrity, cross-context validation, collaborative infrastructure building, and equitable data partnerships.

The data wall therefore tempers the prescriptive turn described in Section 1.1. If GeoAI is to serve anticipatory and decision-support functions responsibly, it must confront uneven training landscapes, domain shift, and representation bias directly. In doing so, the field shifts from technological acceleration toward epistemic accountability.

1.3 Centralization, infrastructure, and dependency

The maturation of GeoAI has been inseparable from the rise of centralized computational infrastructure. Chapters 3, 4, 6, 10, and 11 demonstrate that dense time-series modeling, multi-sensor fusion, and operational hazard analytics are now fundamentally cloud-dependent. Platforms such as Google Earth Engine, Amazon Web Services, Microsoft Planetary Computer, and other large-scale cloud ecosystems have transformed what is technically possible. Planetary-scale processing, once confined to well-funded supercomputing environments, is now accessible through browser-based interfaces and scalable APIs.

This shift has been profoundly enabling. It has accelerated discovery, lowered entry barriers, and facilitated reproducible workflows. Without cloud-native architectures, much of what is now routine in GeoAI would be infeasible. Yet centralization introduces structural implications that extend beyond technical convenience.

1.3.1 Platform dependency

The transition from desktop GIS to cloud-native analysis has altered not only computational scale but institutional dependency. Where early geospatial workflows were locally executed and archived, contemporary systems often rely on externally hosted repositories and remote compute environments. Several survey respondents expressed concern that reliance on a small number of dominant platforms introduces vulnerability. Data access policies, pricing structures, API changes, and long-term archival commitments remain largely governed by corporate entities or centralized institutions. While access is often generous or subsidized, long-term stability is not guaranteed. One author noted, "*I believe open-source data and platforms, such as Google Earth Engine, are essential for broad and equitable access to Earth observation data and computational resources.*" This sentiment reflects a broader structural asymmetry: many workflows are now built upon infrastructures that researchers do not govern.

This observation is not a critique of any specific provider. Rather, it highlights concentration of authority. When planetary-scale archives reside within a limited number of computational ecosystems, research continuity becomes partially contingent on platform-level decisions. Democratized access thus coexists with infrastructural centralization.

1.3.2 *Data sovereignty and geopolitical asymmetry*

Centralized infrastructure intersects directly with geopolitical concerns. Satellite imagery and derived socio-environmental datasets frequently cross-national boundaries in storage and processing. In many contexts, data sovereignty is emerging as a policy priority, raising questions about where data are stored, who controls access, and under which legal frameworks they operate. Several contributors emphasized that lower-income nations may depend heavily on external cloud infrastructure to access and process their own environmental data. This creates asymmetrical power relationships in which technical capability resides outside the regions being analyzed. As one respondent stated, "*Global disparities in access to data and computational resources remain a significant challenge for the field.*"

As GeoAI outputs increasingly inform disaster management (Chapter 11), conservation and SDG monitoring (Chapter 10), and climate-health analytics (Chapter 12), infrastructure dependency becomes more than a technical issue, it becomes political. This does not imply that centralized platforms are inherently problematic. Rather, it signals the need for institutional awareness and strategic diversification. Hybrid infrastructures, regional cloud partnerships, and investment in local computational capacity may become essential components of equitable GeoAI development. As stated succinctly by one respondent, "*Ultimately, the promise of GeoAI is greatest when it is not only technically powerful but also accessible, responsible, and aligned with the needs of diverse communities worldwide.*"

1.3.3 *Environmental cost of computation*

A further dimension of centralization concerns environmental sustainability. Large-scale deep learning models and dense time-series pipelines require substantial energy consumption. GPU-intensive workflows associated with deep neural networks and multi-sensor fusion can carry significant carbon footprints. Although environmental cost is not always foregrounded in geospatial AI discussions, contributors emphasized that a field dedicated to environmental monitoring must remain attentive to its own computational externalities. One survey response observed, "*Large-scale AI models require significant energy and computing power, raising concerns about sustainability and carbon footprint.*" Another contributor noted, "*Leveraging a large system with a high compute capacity to train a model that can then be deployed on the edge will allow us to integrate systems and scale without having to build many new large data centers.*"

Chapters addressing time-series analytics (Chapter 4), multi-sensor fusion (Chapter 6), and ethical and legal considerations (Chapter 13) illustrate the computational scale now common in the field. As resolution increases and architectures deepen, energy demand rises nonlinearly. This does not require abandoning computational ambition. Rather, it calls for efficiency-aware design: model pruning, optimization, judicious deployment, and transparency regarding computational cost in published research.

Fig. 16.1 visualizes an emerging structural tradeoff. As model complexity, spatial and temporal resolution, and computational scale increase, environmental cost rises while interpretability often declines. The upper-right quadrant represents a high-performance but governance-sensitive zone characterized by elevated energy demand and reduced explainability. The transition from classical machine learning to deep learning and foundation models intensifies this tension. Model design choices therefore become institutional and ethical decisions, not merely technical ones. The question is not whether to innovate, but where along this curve responsible innovation should stabilize (Fig. 16.1). As one responder summarized, "*As the field advances, it must prioritize open science, capacity building, and energy-conscious computing. Innovation should be paired with responsibility, transparency, and equitable access.*"

1.3.4 Infrastructure as epistemic shaping

Infrastructure does more than enable analysis; it shapes it. Platform-provided datasets, preprocessing pipelines, and default workflows influence methodological norms. When analysis-ready data products become standardized, certain forms of inquiry are encouraged while others become more difficult. Chapters 2 and 3 document the historical shift from scarce archives to open-access planetary repositories; a transformation that democratized remote sensing research. That transformation now extends into centralized computational ecosystems. Although centralized systems are powerful, epistemic diversity depends on maintaining methodological plurality. If platform convenience narrows experimentation, the field risks gradual homogenization. Conversely, if cloud ecosystems remain open, interoperable, and adaptable, they can continue to catalyze innovation.

Centralized infrastructure should therefore be understood as both breakthrough and responsibility. It has made operational GeoAI systems possible; disaster forecasting,

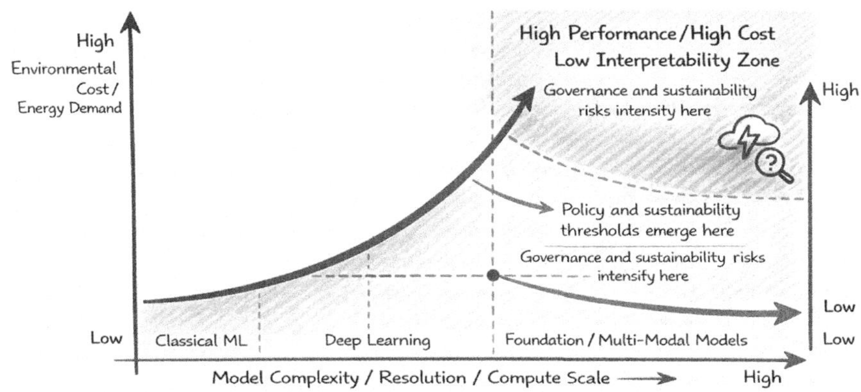

Figure 16.1 Sustainability–compute tradeoff in contemporary GeoAI systems.

conservation monitoring, and climate-health modeling at scale would be infeasible without cloud-native architectures. At the same time, reliance on concentrated computational ecosystems introduces governance, sustainability, and autonomy considerations. Maturity requires awareness of these structural dependencies. We do not advocate for withdrawal from centralized platforms. Rather, we suggest that the present era demands infrastructural alignment with collective benefit and long-term resilience. As one contributor summarized, "*It is the responsibility of those with technological resources to improve accessibility through collaboration.*"

Platform dependency, geopolitical asymmetry, environmental cost, and interpretability decline are not isolated concerns; they interact structurally. The same infrastructures that enable anticipatory analytics also shape whose knowledge is represented, whose institutions control computation, and how environmental costs are distributed. GeoAI has become infrastructure dependent. The next step is to ensure that this infrastructure remains transparent, accountable, and aligned with the public good.

1.4 Automation, interpretability, and trust

Across this volume, a clear methodological trajectory emerges: from supervised classification toward increasingly automated, predictive, and in some cases operational GeoAI systems. Chapters 4, 5, 8, 9, and 11 illustrate this evolution. Dense time-series models detect vegetation resilience; hybrid physics–AI systems approximate coastal hazards; anticipatory disaster systems ingest multi-stream data; agricultural intelligence platforms predict yield and risk. Automation now sits at the core of GeoAI workflows. Yet automation alters the epistemic contract among analyst, model, and decision-maker. As one responder states, "*remote sensing analyses may influence land management, policy decisions, or resource allocation, so transparency, data stewardship, and responsible interpretation are essential.*" As GeoAI systems move closer to institutional decision processes, interpretability and trust become structural requirements rather than methodological preferences. As stated by one contributor "*If you're a city official about to order an evacuation based on an AI prediction, you'd better have a solid explanation beyond 'the computer said so.' That transparency just isn't there yet.*"

1.4.1 From analyst-guided to representation learning

Earlier remote sensing workflows were explicitly analyst-guided. Feature selection, training data definition, and classification thresholds required visible intervention. Even as statistical machine learning entered the field, interpretability remained comparatively high; random forests, support vector machines, and ensemble approaches offered traceable feature importance and rule-based logic.

The rise of deep learning and representation learning introduced a different paradigm. Convolutional neural networks, recurrent architectures, and transformer-based models internalize feature abstraction. Performance often improves, particularly in complex, high-dimensional environments. However, interpretability may decline.

As one contributor observed, "*One major limitation is the emphasis on predictive accuracy over interpretability.*" They continued, "*If models cannot explain environmental mechanisms or quantify uncertainty clearly, their real-world impact is constrained.*" This tension is not unique to GeoAI, but its implications are amplified in spatially explicit, decision-proximate contexts. Applications described throughout this volume from urban heat exposure mapping (Chapter 8), conservation monitoring (Chapter 10), disaster anticipation (Chapter 11), and climate-health integration (Chapter 12), operate close to planning and resource allocation. In such settings, opacity carries institutional consequence.

1.4.2 Decision-support versus autonomous systems

GeoAI currently functions primarily as decision-support rather than fully autonomous control. Human oversight remains embedded in workflow design, validation, and interpretation. This "*human-in-the-loop*" model appears across methodological and application chapters. However, the boundary between support and autonomy is shifting. Automated disaster alerts may trigger mobilization. Agricultural forecasts may influence insurance pricing or planting decisions. Conservation monitoring may shape enforcement priorities. As automation intensifies, transparent validation frameworks become essential. One respondent captured this emerging challenge: "*Deep learning models can also lack transparency and uncertainty reporting, making results harder to trust, and integrating diverse environmental and socioeconomic datasets remains technically complex.*" The critical issue is not model complexity itself, but the clarity with which assumptions, uncertainties, and failure modes are communicated to downstream users.

1.4.3 Uncertainty, validation, and institutional trust

Interpretability is inseparable from uncertainty communication. Several chapters, particularly those addressing environmental monitoring and climate analytics, emphasize validation across spatial and temporal scales. Time-series smoothing, sensor harmonization, and multi-modal fusion introduce layers of uncertainty that may compound if poorly documented. Yet institutional environments often favor simplified outputs. A single vulnerability index, hazard probability, or yield forecast integrates more easily into policy frameworks than a distribution of confidence intervals. One contributor reflected, "*Bridging the gap between advanced methods and practical, scalable, and usable tools remains an important challenge for the field.*"

The maturity challenge for GeoAI lies here: resisting oversimplification for operational convenience. Trust is not built through opacity masked by confidence, but through explicit articulation of strengths and limits. As articulated by one responder, "*As models become more complex, it can be difficult to clearly explain results to decision-makers, which may reduce trust and slow adoption.*" Chapters 5 and 6 reinforce this need for governance-aware methodology, while Chapters 14 and 15 highlight institutional embedding within socio-ecological systems. In these contexts, interpretability becomes foundational rather than optional.

1.4.4 Bias, equity, and training data limitations

Automation intersects directly with the training data asymmetries identified earlier in this chapter. Petabyte-scale archives coexist with uneven labeled datasets. Training data are often concentrated in well-monitored geographies and socio-economic contexts. As one respondent noted, "*Disparities in access to geospatial data and computational resources are among the most consequential yet under-discussed challenges in GeoAI and remote sensing today.*" When models are trained disproportionately on data-rich regions, predictive reliability may decline in underrepresented areas.

Bias in GeoAI is not solely demographic; it may be ecological, climatic, infrastructural, or institutional. Ensuring representative training quantities, transparent cross-context validation, and localized calibration becomes increasingly essential as automation expands, particularly in high-stakes applications across the Global South. As stated clearly by one contributor "*Location Bias: A model trained on cities in the U.S. often fails in Iran or Nigeria …*"

1.4.5 Toward responsible automation

The chapters in this volume do not advocate retreat from automation. On the contrary, they demonstrate its transformative value. Predictive modeling, multi-sensor integration, and operational analytics now form the backbone of environmental intelligence systems. However, the prescriptive turn described in Section 1.1 extends directly into automation design. Interpretability frameworks, uncertainty quantification, documentation standards, and governance alignment are not secondary add-ons; they are structural components of mature GeoAI systems. One contributor summarized this maturation imperative clearly: "*Ethical practice should emphasize open science, reproducibility, and collaboration with local stakeholders to ensure that geospatial technologies support equitable decision-making and environmental protection rather than reinforcing existing inequalities.*"

Automation therefore represents both achievement and responsibility. As GeoAI systems become embedded within institutional processes, trust becomes infrastructural. It must be deliberately constructed, transparently communicated, and continuously evaluated. The transition from analyst-guided classification to representation-driven abstraction has expanded analytical capability. The next phase requires ensuring that this expanded capability remains transparent, equitable, and institutionally credible.

1.5 Equity, geopolitics, and global asymmetry

As GeoAI matures, questions of equity and geopolitical asymmetry shift from peripheral concerns to central realities. The convergence of satellite archives, cloud platforms, machine learning architectures, and institutional embedding has produced unprecedented analytical capacity. Yet that capacity is unevenly distributed.

A paradox emerges. Remote sensing data are increasingly open; computational infrastructures are increasingly centralized; advanced analytical methods are increasingly resource-intensive. Access has expanded in principle, but capacity remains differentiated in practice.

1.5.1 Data abundance and unequal capacity

The transition from data scarcity to archive abundance, particularly following open access policies such as Landsat continuity, has been widely celebrated. Chapters 2 and 3 demonstrate how archive expansion democratized access to decades of Earth observation data. Yet exploiting these archives increasingly requires high-performance infrastructure and advanced modeling expertise. As one contributor observed, "*GeoAI and remote sensing are moving beyond retrospective cartography toward continuous monitoring, forecasting, and automated decision-ready insights.*" Another noted, "*Regions most vulnerable to environmental change often have the least access to high-performance computing and advanced modeling tools.*"

Dense time-series modeling, multi-sensor fusion, and predictive analytics, described in Chapters 4, 6, 10, 11, and 12, require sustained computational resources and technical literacy. Institutions in high-income regions often integrate these capabilities within research centers and agencies; lower-resource contexts may face bandwidth, licensing, or training constraints. As one contributor observed, "*In water resources and agriculture, disparities in access to Earth Observation data, broadband, and GPU or cloud computing have created a significant capability gap. Well-resourced organizations can conduct near-real-time monitoring and forecasting for drought, irrigation demand, and flood risk, while many institutions in the Global South face slower workflows, higher costs, and dependence on external platforms.*" This disparity becomes particularly consequential when GeoAI informs climate adaptation, conservation prioritization, or disaster response in the regions most vulnerable to environmental change.

1.5.2 Platform centralization and dependency

The shift toward cloud-native ecosystems introduces a second dimension of asymmetry. Centralized platforms concentrate processing power, storage, and algorithmic services within a small number of institutional or corporate actors. This concentration enables planetary-scale analysis, but it also creates dependency structures. Chapter 3's discussion of cloud-native ingestion and pipeline automation highlights efficiency gains alongside structural reliance. Chapter 15 further underscores how platform dependency intersects with governance and educational frameworks. A survey respondent articulated the concern directly: "*High performance computing, cloud-based tools, such as Google Earth Engine, and new data pipelines in Python leveraging GPU processing are all required to overcome processing constraint.*" This is not a critique of specific platforms, but recognition that infrastructure design has geopolitical implications. Decisions about data continuity, pricing models, API access, or export controls shape who can analyze what, and under what conditions. The GeoAI era is therefore not only computational; it is infrastructural and political.

1.5.3 Training data bias and global representation

Equity concerns extend beyond infrastructure to training data composition. As discussed in Section 1.4, labeled datasets are often geographically uneven. Urban environments in North America and Europe may be overrepresented; rural, informal, or ecologically diverse regions may be underrepresented. This imbalance affects predictive reliability. Agricultural models trained on temperate industrial systems may not generalize to smallholder farming in tropical contexts. Urban heat models may not capture the socio-spatial complexity of rapidly expanding cities in the Global South. Ecological baselines differ dramatically across biomes. One contributor emphasized, "*This digital divide is increasingly significant as GeoAI and RS advance.*" Ensuring equitable GeoAI systems therefore requires attention to data diversity, local calibration, and participatory validation. Cross-scale integration must not override local specificity.

1.5.4 Environmental and computational sustainability

Equity is also temporal. High-resolution sensors, dense time-series stacks, and GPU-intensive models carry environmental costs. Training large architectures consumes energy; cloud infrastructures require data centers; rising computational demand contributes indirectly to carbon emissions. Several respondents raised this issue explicitly. One author stated, "*Many GeoAI models also require significant computational resources and may not generalize well across different regions or ecosystems.*" Another noted, "*There are social and environmental costs associated with large-scale data storage and energy-intensive model training that will require innovative research and solutions as the field grows.*"

The tension between resolution and sustainability, introduced earlier in this volume, becomes especially salient here. Greater spatial and temporal precision can enhance predictive capacity, but it also increases computational burden. One contributor stated "*I think the pre-trained Earth Foundation Models that already capture spatiotemporal patterns of land surface will be very exciting. They can substantially reduce computational costs while significantly improving model scalability.*" Balancing analytical ambition with environmental responsibility is emerging as a defining challenge.

1.5.5 Toward equitable GeoAI

Equity in the GeoAI era is multi-dimensional: access to data, access to infrastructure, representativeness and adequacy of training data, energy consumption, and institutional capacity. Addressing these asymmetries does not require slowing innovation. It requires embedding equity considerations within system design. Chapters 13, 14, and 15 illustrate how technical systems can be contextualized within governance, socio-ecological, and ethical frameworks. Education and workforce development, discussed below, further shape who participates in GeoAI development and deployment.

One contributor offered a forward-looking synthesis: "*In short, we are building a nervous system for the planet, with remote sensing as the eyes and GeoAI as the brain, moving us from passive observation to active, informed stewardship.*" As GeoAI becomes increasingly decision-proximate, its legitimacy will depend not only on predictive performance, but on distributive justice, infrastructural transparency, and sustainable practice. The maturation of the field therefore includes geopolitical awareness and institutional reflexivity. The transformations that enabled GeoAI must now be matched by sustained attention to equity.

1.6 Education, workforce, and the institutional future of GeoAI

The maturation of GeoAI is not only methodological or infrastructural; it is institutional. As demonstrated throughout this volume, remote sensing and artificial intelligence are now embedded within climate analytics, agricultural intelligence, disaster response, socio-ecological modeling, and governance systems. The sustainability of this integration depends not only on algorithmic advancement, but on educational reform, workforce development, and institutional adaptation. As one contributor observed, "*The introduction of Geospatial tech into the classroom is relatively young, but the technology has advanced rapidly; this has left the pedagogy of GIScience lagging. Given the commonly cited statistic that 80% of data is now spatial, we need to enhance the way we teach GeoAI to enable broader (responsible) use of the technology. Also, model transparency and scalability.*"

Several contributors emphasized that the future trajectory of GeoAI will be shaped as much by pedagogy as by performance metrics. One respondent noted, "*While computational constraints and scalability challenges remain, another major gap lies in the lack of effective educational resources that enable users to readily adopt and apply these methods in practice.*" These concerns reflect the structural shifts outlined earlier in this volume. As remote sensing evolves from single-scene classification toward integrated, predictive systems, the required skill set expands. Practitioners must understand spectral integrity, calibration, and spatial uncertainty, while also navigating machine learning architectures, cloud ecosystems, and ethical evaluation. The boundary between geospatial science and computer science can no longer function as a strict curricular divide. Chapters 3, 12, and 15 illustrate this institutional inflection. Cloud-native workflows and governance alignment require professionals who can move between technical infrastructure and policy frameworks. The field increasingly demands hybrid expertise: domain knowledge, computational fluency, systems thinking, and ethical awareness. As one respondent articulated, "*Bringing these perspectives together can help ensure that geospatial innovations are not only technically robust but also grounded in human, social, and institutional contexts.*"

1.6.1 Workforce, incentives, and equity

Workforce transformation intersects directly with equity. Access to advanced GeoAI training remains uneven across institutions and regions. High-performance computing clusters, GPU resources, and cloud credits are not universally available. Without deliberate investment in educational infrastructure and open training materials, the asymmetries discussed in Section 1.5 risk deepening. Institutional incentive structures also shape the trajectory of the field. Academic reward systems often prioritize methodological novelty over integration, publication speed over reproducibility, and performance benchmarks over interpretability. Yet the prescriptive turn described in Section 1.1 suggests that integration, governance alignment, and cross-disciplinary collaboration are increasingly central to GeoAI's societal role. As one contributor wrote, "*Strengthening connections between practitioners and academics will be essential to ensure that technological advances translate into meaningful, real-world impact.*" Reproducibility, transparent workflows, cross-sector collaboration, and community engagement must therefore be recognized as scholarly contributions. Education must embed ethics, sustainability, and interdisciplinary communication alongside model development.

1.6.2 Returning to the definition of GeoAI

In Chapter 1, we proposed a unifying definition of GeoAI as:

The integration of artificial intelligence (including machine learning) with geospatial data, models, and technologies to discover spatial–temporal patterns, generate predictive insights, and enable scalable, data-driven decision-making across scientific, operational, and policy domains.

The trajectory documented across this volume affirms that definition. Yet it also clarifies something further: integration now extends beyond technical components. It includes institutional alignment, workforce preparation, governance awareness, and equitable access. The definition remains analytically sound; what has expanded is its sphere of responsibility. GeoAI is no longer only a methodological integration. It is a socio-technical system embedded within decision environments.

1.6.3 Institutional maturation

The chapters in this volume demonstrate that this transition is already underway. Time-series modeling is coupled with resilience planning. Multi-sensor fusion informs conservation policy. Disaster analytics operate within agencies. Socio-ecological modeling integrates environmental data with governance structures. One contributor offered a clear articulation of this trajectory: "*The future of GeoAI and remote sensing will be shaped not only by advances in algorithms and sensors, but by deeper collaboration with disciplines that help translate geospatial insights into real-world impact.*"

Fig. 16.2 synthesizes this progression as a layered maturation, from algorithmic innovation to governance-integrated systems. The lower tiers reflect advances in data scale, representation learning, and computational infrastructure; the upper tiers represent decision proximity, institutional embedding, and ethical accountability. Movement upward signals not only technical sophistication but expanding societal responsibility. As one contributor summarized, "*As GeoAI evolves from a mapping tool into a societal decision infrastructure for climate adaptation, sustainability, and risk management, its success will depend less on technical capability alone and more on our willingness to bring the right people into the room—across disciplines, across cultures, and across ways of knowing.*"

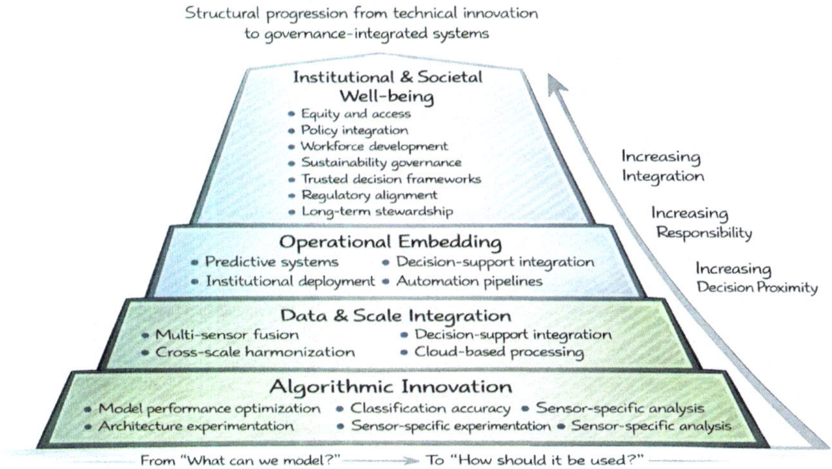

Figure 16.2 GeoAI maturity framework: From algorithmic innovation to governance-integrated systems.

1.6.4 Closing the arc

Across Sections 1.1 through 1.6, a consistent pattern emerges. GeoAI has evolved:

- From descriptive mapping to anticipatory systems
- From single-sensor archives to multi-modal cloud infrastructures
- From disciplinary experimentation to institutional embedding

With that evolution comes obligation. The same capabilities that enable climate-health analytics, disaster forecasting, conservation intelligence, and agricultural prediction also intensify tensions around infrastructure, sustainability, governance, and equity. Technical advancement alone does not constitute progress. Long-term maturity depends on alignment between computational capability and institutional accountability. Figs. 16.2 and 16.3 illustrate this structural transition. Much of the field's recent growth has occurred along the axis of technical expansion. The next phase requires upward movement toward institutional responsibility. Governance-integrated systems represent consolidation rather than constraint: a recognition that predictive power must be matched by transparency, equity, and sustainable design.

This volume has traced GeoAI from its structural origins to its institutional inflection point. Remote sensing remains foundational, but it now operates within integrated socio-technical systems that influence allocation, resilience, and policy. The future of GeoAI will not be defined by deeper networks or finer resolution alone. It will be defined by whether its intelligence is deployed transparently, equitably, and responsibly.

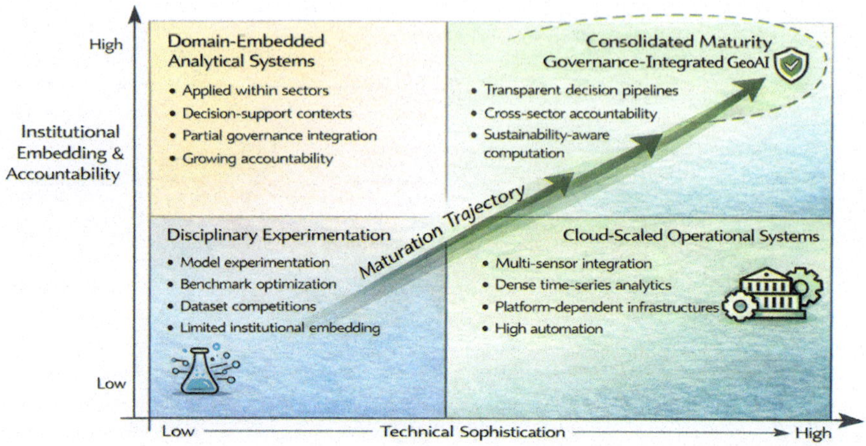

Figure 16.3 GeoAI maturity matrix: Technical sophistication and institutional embedding.

Index

'*Note*: Page numbers followed by "f" indicate figures and "t" indicate tables.'

O

P

R

Printed and bound by CPI Group (UK) Ltd, Croydon, CR0 4YY
14/09/2026
02216410-0005